HANDBOOK OF ENVIRONMENTAL PHYSIOLOGY OF FRUIT CROPS

VOLUME I: TEMPERATE CROPS

HANDBOOK OF ENVIRONMENTAL PHYSIOLOGY OF FRUIT CROPS

VOLUME I: TEMPERATE CROPS

Edited by

Bruce Schaffer, Ph.D.

University of Florida
Tropical Research and Education Center

Peter C. Andersen, Ph.D.

University of Florida
North Florida Research and Education Center

Library of Congress Cataloging-in-Publication Data

Handbook of environmental physiology of fruit crops / edited by Bruce Schaffer and Peter C. Andersen

p. cm.

Contents: v. 1. Temperate -- v. 2. Sub-tropical and tropical.

Includes bibliographical references and index.

ISBN 0-8493-0175-0 Vol. I

ISBN 0-8493-0179-3 Vol. II

1. Fruit -- Ecophysiology -- Handbooks, manuals, etc. 2. Tropical fruit -- Ecophysiology -- Handbooks, manuals, etc. I. Schaffer, Bruce. II. Andersen, Peter C.

SB357.28.H35 1994

582.13′.045222--dc20 93-45704

CIP

This book contains information obtained from authentic and highly regarded sources. Reprinted material is quoted with permission, and sources are indicated. A wide variety of references are listed. Reasonable efforts have been made to publish reliable data and information, but the author and the publisher cannot assume responsibility for the validity of all materials or for the consequences of their use.

Direct all inquiries to CRC Press, Inc., 2000 Corporate Blvd., N.W., Boca Raton, Florida 33431.

International Standard Book Number 0-8493-0175-0

Library of Congress Card Number 93-45704

Printed in the United States of America 1 2 3 4 5 6 7 8 9 0

Printed on acid-free paper

THE EDITORS

Bruce Schaffer, Ph.D., is Professor of Plant Physiology at the University of Florida, Tropical Research and Education Center in Homestead.

Dr. Schaffer received his B.Sc. degree in Entomology and Zoology in 1978 and his M.S. degree in Forest Biology in 1981, both from Colorado State University in Fort Collins. He obtained a Ph.D. degree in Horticulture in 1985 from Virginia Polytechnic Institute and State University in Blacksburg. After working as a Post-doctoral Associate in Plant Physiology at the University of Heidelberg, Germany, he was appointed Assistant Professor of Plant Physiology at the University of Florida, Tropical Research and Education Center in 1985. He became an Associate Professor in 1990, and Professor in 1994.

Dr. Schaffer is a member of the International and American Societies for Horticultural Science and the honor societies Sigma Xi, Gamma Sigma Delta, Phi Sigma, and Chi Sigma Xi. He has been an Associate Editor of *HortScience,* is currently an Associate Editor of the *Journal of the American Society for Horticultural Sciences,* and is on the Editorial Review Board of *Tree Physiology.*

Dr. Schaffer has been a recipient of research grants from several organizations, including the United States Department of Agriculture and the Charles A. Lindbergh Fund. He has published more than 60 research papers, primarily focusing on whole-plant physiology. His major research interests are environmental physiology of tropical fruit and vegetable crops and sustainable tropical horticulture.

Peter C. Andersen, Ph.D., is Professor at the University of Florida, North Florida Research and Education Center in Monticello.

Dr. Anderson obtained a B.Sc. degree (cum laude) in Biology from Florida State University in 1974, an M.S. degree in Horticultural Sciences from the University of Florida in 1978, and a Ph.D. degree in Horticultural Sciences from Oregon State University in 1982. From 1982 to 1984 he was employed as Assistant Research Professor of Plant Physiology at the Southwest Missouri State University Experiment Station in Mountain Grove. In 1984 he accepted a position as Assistant Professor of Horticulture at the University of Florida, North Florida Research and Education Center in Monticello. He was promoted to Associate Professor in 1989, and Professor in 1994.

Dr. Andersen is a member of the American Society for Horticultural Sciences and the Scandinavian Society for Plant Physiology. He has published more than 100 research articles, the majority of which deal with aspects of environmental physiology. His current research interests include topics in environmental physiology and plant-insect-disease interactions of horticultural crops.

CONTRIBUTORS

Peter C. Andersen
University of Florida
North Florida Research and Education Center
Monticello, FL 32344

Guido Bongi
National Research Council
Istituto Ricerche Olivicoltura
Perugia 06128, Italy

James G. Buwalda
The Horticulture and Food Research Institute
of New Zealand Ltd.
Ruakura Research Centre
Hamilton, New Zealand

Rebecca L. Darnell
Department of Horticultural Sciences
University of Florida
Gainesville, FL 32611

Frederick S. Davies
Department of Horticultural Sciences
University of Florida
Gainesville, FL 32611

Nick K. Dokoozlian
Department of Viticulture and Enology
University of California, Davis
and Kearney Agricultural Center
Parlier, CA 93648

James A. Flore
Department of Horticulture
Michigan State University
East Lansing, MI 48829

Alan P. George
Department of Primary Industries
Maroochy Horticultural Research Station
Nambour 4560, Queensland, Australia

Ian Goodwin
Institute of Sustainable Irrigated Agriculture
Tatura 3616, Victoria, Australia

Peter H. Jerie
Institute of Sustainable Irrigated Agriculture
Tatura 3616, Victoria, Australia

Alan N. Lakso
Department of Horticultural Sciences
Cornell University
New York State Agricultural Experiment Station
Geneva, NY 14456

Kirk D. Larson
South Coast Research and Extension Center
University of California
Irvine, CA 92718

Peter D. Mitchell
Institute of Sustainable Irrigated Agriculture
Tatura 3616, Victoria, Australia

Alistair D. Mowat
The Horticulture and Food Research Institute
of New Zealand Ltd.
Ruakura Agricultural Centre
Hamilton, New Zealand

Alberto Palliotti
National Research Council
Istituto Ricerche Olivicoltura
Perugia 06128, Italy

Bruce Schaffer
University of Florida
Tropical Research and Education Center
Homestead, FL 33031

Garth S. Smith
The Horticulture and Food Research Institute of
New Zealand Ltd.
Hamilton, New Zealand

Robert L. Wample
Department of Horticulture and
Landscape Architecture
Washington State University
Irrigated Agriculture Research and
Extension Center
Prosser, WA 99350

Larry E. Williams
Department of Viticulture and Enology
Univesity of California, Davis
and Kearney Agricultural Center
Parlier, CA 93648

CONTENTS

Chapter 1

Introduction

Bruce Schaffer and Peter C. Andersen

Environmental variables modulate virtually all aspects of plant growth and development. The sum of the interactions between plant genotype and the environment determines plant phenotype and plant productivity. An understanding of environmental physiology is crucial in selecting optimum sites for a given species or cultivar, to minimize the deleterious impacts of suboptimal environmental conditions and to modify or create microenvironments that are conducive to maximum productivity. Moreover, the establishment of a database on environmental physiology is essential in an era of agricultural sustainability where agricultural inputs are becoming increasingly scrutinized.

Previous review articles concerning physiological, and growth and developmental responses of fruit crops to the environment have typically addressed a specific or multiple plant response to a single environmental factor, or alternatively, a specific plant response to multiple environmental variables. In addition, few articles integrate environmental physiology with plant growth and development. Often a large volume of literature from many sources must be compiled to obtain a comprehensive coverage of the environmental physiology of a specific crop.

The purpose of this volume of *The Handbook of Environmental Physiology of Fruit Crops* is to provide a comprehensive reference source on the responses of temperate fruit crops to the environment. This book addresses plant responses to abiotic variables (light, temperature, water, wind, salinity, and air pollution) and when possible integrates whole plant physiology with production horticulture. The volume was designed to appeal to a wide audience of researchers, extension workers, and students interested in fruit crop physiology and is written on a level that is commensurate with a general understanding of plant physiology.

We have relied on the broad horticultural definition of fruit crops to include woody or herbaceous perennial trees, shrubs, or vines with the edible portion (fruit) an expanded and ripened ovary with attached and subtending reproductive structures.[1,2] At times we have strayed from this definition. For example, we have included strawberry, which is often grown as an annual. Most of the commercially-important temperate fruit crops are contained in this volume. However, some crops were omitted due to either the lack of available information and/or the inability to locate a qualified author. A few "warm temperate" crops, such as persimmon, are discussed in this volume although they can be cultivated in subtropical and tropical regions. Although an attempt was made to solicit chapters from highly qualified individuals, clearly there are many more individuals who would have been eminently qualified as authorities on certain crop species.

This book is generally divided according to fruit crops species or related species of fruit crops. In one chapter, "Temperate Nut Species", many species have been combined in the discussion because of the limited available database. Crop origin, genetic diversity, botany, and ecology were often included to facilitate a discussion of environmental physiology. There is a great deal of flexibility in the orientation, content, and style of each chapter. The terminology and abbreviations used to define physiological variables are consistent within each chapter. However, there are some differences in terminology among chapters, reflecting the lack of a commonly accepted terminology among plant physiologists. Although the authors attempted to comprehensively discuss plant responses to abiotic factors, to some extent chapters reflect the authors' particular research specialty. For certain crops such as grape, apple, and stone fruit there has been much research on environmental physiology focusing on leaf gas exchange, water relations, and carbon partitioning. The extensive database for those crops presented the authors with much flexibility to determine the scope and content of their material and allowed for an integration of several physiological responses. In contrast, little information was available for some of the fruit and nut species, and consequently, the discussion often relied more heavily on growth and developmental responses.

In addition to serving as a useful reference source, this volume of *The Handbook of Environmental Physiology of Fruit Crops* will underscore the limitations of our knowledge in this area and will help delineate areas in need of further research.

0-8493-175-0/94/$0.00+$.50

The physiological, growth and developmental responses in relation to the interaction among multiple environmental stresses is a topic that warrants increasing emphasis. For most fruit and nut species the database concerning the influences of environmental variables must be expanded to allow for the development of models that maximize plant productivity and resource use efficiency.

LITERATURE CITED

1. Barden, J. A. and Halfacre, R. G., *Horticulture,* McGraw Hill, New York, 1979.
2. Soule, J., *Glossary for Horticultural Crops,* John Wiley and Sons, New York, 1985.

Chapter 2

Apple

Alan N. Lakso

CONTENTS

0-8493-175-0/94/$0.00+$.50

I. INTRODUCTION

The environmental physiology of the apple (*Malus domestica* Borkh.) has been extensively researched compared to many other fruit crop species. Since a comprehensive review of this entire knowledge base is beyond the scope of a review of this size, this chapter will be interpretive rather than comprehensive, emphasizing (1) knowledge developed primarily since the last major reviews on radiation, water, temperature, etc. (mostly work in the 1980s), (2) integration of knowledge with a view toward whole-plant and crop physiology, (3) interactions of physiological responses to multiple environmental factors in the field, and (4) identification of remaining roadblocks and new opportunities to apply this knowledge to solve problems of apple production.

The reader is referred to several excellent reviews of apple physiology and morphology on which this chapter is based[1-9] as well as general reviews and books that are applicable.[10-13] The reader is referred to several useful books and reviews and references therein for methods and techniques.[14-18] This review will emphasize carbon and water relations of apple trees growing in temperate zones. Since the apple is grown for fruit production, the discussions will primarily concern the range of conditions normally occurring in a producing orchard rather than in extreme conditions.

A. LIMITING FACTORS

Before discussing individual environmental factors that affect productivity, it is worthwhile to consider the concepts of limiting factors. Monteith[19] has emphasized that each resource (light, CO_2, water, nitrogen, phosphorus, boron, etc.) and each process (light interception, photosynthesis, respiration, leaf area development, translocation, nutrient uptake, nitrogen reduction, protein synthesis, hormone transport, ion exchange processes, etc.) is *essential* for productivity. However, that does not mean that any one resource or process is necessarily *controlling the variations* in crop productivity. It is important to understand that many processes can be essential without ever being important to the variations in crop productivity (for example, water is essential, but accounts for little of the variations in productivity in irrigated orchards). The primary role of the crop physiologist is to determine which resources or processes are limiting crop performance at any given time and ideally determine a feasible way to overcome the limitation.

There is no *a priori* reason to believe that any one resource or process is always limiting to a crop. It is much more likely that there are periods of differing resource or process limitations during the developmental stages of the crop. Environmental resources (light, water, etc.) or conditions (temperature, humidity) may limit crop development at certain times depending on the crop sensitivity and how extreme the environment may be. Additionally, plant status, such as crop load, may greatly change the relative sensitivity of the plant to an environmental stress or resource limitation.

Interactions of plant development and the environment should be kept in mind when evaluating the plant's response to any particular environmental factor. A useful analogy is that of pulling at any one connection point in a net. Other nearby connection points will react similarly, while those further away will react less. The entire net, however, will adjust to the new set of forces. The plant is a network of processes that must be considered together. We must consider the whole when considering any one part. As Jules Henri Poincaré reportedly said: "Science is built up with facts, as a house is built up with stones. But a collection of facts is no more a science than a heap of stones is a house."

This review will attempt to address not only the physiological "stones" that have been studied for many years, but also the physiological "house" we call an apple tree.

B. GROWTH AND DEVELOPMENT

Before considering specifics of physiological responses of the apple to the environment, a brief overview of the inherent growth and development patterns can provide a context with which to evaluate the physiological information (see review by Forshey and Elfving[2] for review of general growth relationships in the apple).

The apple is a hardy, deciduous, temperate fruit crop species that is very adaptable to different climates, growing commercially from the tropics[20] to the high latitudes in Norway.[21] A perennial structure with carbon and nutrient reserves allows very rapid leaf area development in the spring compared to annual crops (see Oliveira and Priestley[22] for a general review of carbon reserves). Early development, combined with slow leaf aging for apple,[23] translates to a long duration of photosynthetically-active leaf area. This is the basis of high yield potential since dry matter production by crops is generally positively related to the amount of total radiation intercepted per hectare per season.[24] In general, apple is considered day neutral for most processes. The few reports of light quality effects indicate that fruit set may be stimulated by night breaks with red light,[25] and that UV light may be important in anthocyanin development in the fruit.[26] Leaf retention appears to be due in part to the lack of photoperiodism. Leaf senescence and abscission in the autumn are apparently regulated by the advent of cool temperatures, not daylength.[27,28] The growth and productivity of apple trees are largely limited by the length of the growing season at each location, perhaps one reason why apple is so widely adaptable.

The structure of the apple canopy is varied, containing shoots of differing lengths. Short rosette-type shoot complexes, referred to here as the "spurs", and longer single shoots with leaves spaced with long internodes, referred to as the "extension shoots", develop concurrently in the spring.[8] The spurs develop with a rosette of primary leaves unfolding by bloom, followed by the development of generally one lateral shoot, the "bourse" shoot, after bloom. The bourse shoots are typically also quite short and terminate within a few weeks of bloom; however, in high-vigor situations they may grow to be quite long, similar to the extension shoots. Extension shoots typically develop from terminal or lateral buds of the extension shoots of the previous year. These shoots may terminate early (sometimes called "lateral or terminal short shoots") and become the spurs of the following year or they may extend as described above. Depending on the age of the tree, spur versus standard growth habit, pruning, vigor, and growing conditions, the distribution of shoot length and the spur:shoot ratio will vary dramatically.[2,29–31] The balance of the types of growth habit has implications for the type of canopy and for the penetration of light within the canopy (see Radiation section).

The apple bears many more flowers compared to fruits at harvest. A dwarf tree with heavy bloom may have up to 800 flower clusters of an average of 6 flowers per cluster, a total of up to about 5000 flowers. The same tree will typically retain perhaps 200 fruits to harvest, only 4–10% of the potential fruits. There are many potential reasons why the abscission rate is so high, but nonetheless, the apple has a great variability in the numbers of sinks that can be retained on the tree. Under near-ideal conditions fruit yields can be extremely high, reaching record sustained yields approaching 180 tons per hectare in New Zealand.[32] Converted to dry matter of about 30 tons per hectare, apple is comparable with potato, having the highest C_3 crop productivity.[33] The high dry matter yield is also due in part to the very inexpensive energetic costs for the primarily carbohydrate apple fruit compared to other crops, such as nut crops and avocado, that accumulate energetically expensive proteins or oils.[34,35] Heavily cropping apple trees can exhibit very high harvest indices (percentage of total dry weight gain partitioned into the fruit). Palmer[36] reported a harvest index of about 65%, including root mass, with young, dwarf Mutsu/M27 trees. In mature heavy-cropping 'Jonamac'/M26 trees, the estimated harvest index exceeded 80% including roots.[37] These high harvest indices are largely due to the very low investment of carbon required for the perennial structure in established dwarf trees.

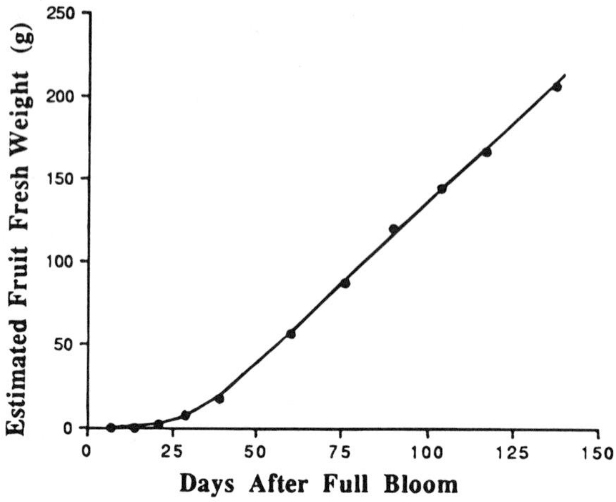

Figure 1 Seasonal growth of 'Empire' apples with early thinning to light crop in New York and fitted "expolinear" model based on the expolinear growth function (Equation 1) described by Goudriaan and Monteith.[40] Regression parameter estimates were C_m = 1.95 g day^{-1}, R_m = 0.167 g g^{-1} day^{-1}, t_b = 30 days.

The growth of the apple fruit has been described in textbooks and reviews as being sigmoid.[1,10,12,38] It is not clear, however, if the sigmoid pattern is that inherent to the fruit (i.e., the growth pattern if the fruit was not limited by any resource) or whether it reflects a variety of late-season limitations of resources or temperature. When 'Empire' apple fruits were allowed to grow on very light cropping trees with early thinning and apparently optimal water and nutrients, fruit weight increased exponentially during the cell division period, typically the first 3–6 weeks after bloom, then increased linearly until normal harvest (Figure 1). This pattern was shown by Blanpied in 1966.[39] Cold temperatures in the last weeks before harvest slowed growth, but plotting growth against growing degree days during the last portion of the season linearized the growth. This growth pattern has been described well with a growth equation developed by Goudriaan and Monteith[40] called the "expolinear" function:

$$W = (C_m/R_m)\ln\{1 + \exp[R_m(t - t_b)]\} \tag{1}$$

where W = crop (or fruit) weight, C_m = maximum absolute growth rate (in weight gain per day in the linear phase), R_m = maximum relative growth rate (in weight gain per unit weight per day), t = time in days, t_b = x axis intercept of the linear growth phase (termed the "lost time").

The implications of this growth habit suggest that: (1) after the cell division period is over, the growth rate or demand of the fruit (i.e., the slope of the weight vs. time curve) remains essentially constant until harvest; and (2) the slope of the linear portion of the growth curve depends on accumulation of cell numbers during the exponential period. This emphasizes resources needed to support good fruit growth during cell division in order to provide good potential for the rest of the season.

C. AN ECOPHYSIOLOGY VIEWPOINT

When considering the environmental physiology of the apple, it is useful to consider two factors. First, clonal propagation stops the natural selection process (except for breeders' selection for fruit color and quality factors and uniformity of bearing), thus little selection for physiological responses has occurred in commercial cultivars. Few physiological differences in environmental responses can be found between a 200-year-old cultivar such as 'McIntosh' and a newly released cultivar. This suggests that an understanding of the physiological behavior of apple may be enhanced by considering natural selection pressures during evolution.

Although not entirely clear, the center of origin of the apple is probably in southern China or Georgia and Armenia. The descriptions of Vavilov[41] in the Caucasus Mountains suggest that the apple evolved in exposed wooded canyons reasonably close to sources of water, but in competition with other grasses, shrubs, and trees in a temperate climate with occasional droughts. Under such conditions it might be expected that overall strategies of productivity may be based on the need for initial vegetative development of the shoot system in competition for sunlight and of the root system to reach a relatively large, deep water supply found below the dense rooting zone of grasses and other annuals. Since the production of fruits is detrimental to shoot growth and especially to root growth,[42,43] the extended juvenility period of apple seedlings would appear advantageous to competitiveness. Once an adequately exposed canopy and a deep, extensive root system have developed, the tree may be expected to be able to produce fruit without harming the competitive status of the tree for water, nutrients, or radiation.

Therefore, it may be reasonable that early season stresses such as water deficits, shade, or nutrient deficiencies would induce fruit abscission to allow for a vegetative response to compete for the limited resource. The general concept from ecology that the reproductive development is the first priority of the plant needs to be viewed in a multi-year perspective for perennials like apple. In this case, survival for many years and successful competition against other plants for limited resources afforded by good vegetative growth may lead to the maximization of reproduction over the life of the tree. In each individual year, however, the drop of fruit may, ironically, be the method by which the apple maximizes lifetime fruit production.

This may explain to some extent the variability of fruit set and yield that is the bane of agricultural practitioners and marketers. It may be helpful to consider that the fruits may not be programmed to set, but rather to abscise. Only under conditions of adequate resources (nutrients, water, radiation, etc.) that do not threaten the survival of the tree will the fruit be able to continue to grow and mature. Zucconi[44] has proposed that fruit retention is a function of maintenance of fruit growth rate and that limitation of growth from any source will lead to the reduction of auxin transport from the fruit to the abscission zone[45] and thus cause abscission. This view is consistent with the great range of conditions or stresses observed to cause fruit abscission and yield variation. Lang[46] has suggested similarly that the study of dormancy should be founded in an understanding of the ecophysiology of the species and the range of environmental stresses under which it evolved. Perhaps an ecophysiological view may provide some insights to apple physiology that are not as apparent when tree behavior is viewed from the agricultural perspective alone.

II. RADIATION

Of all the environmental factors, solar radiation should be considered first due to its primary role as the source of energy that drives the biological production of dry matter that ultimately limits fruit yield. First, we should consider the quantitative aspects of yield and how they are regulated by light availability and utilization. It is helpful to break yield into several components: biological yield, fruit yield, and economic yield (i.e., total dry matter, dry matter in fruit, and the economic return from the fruit, respectively).

A. BASES OF ORCHARD PRODUCTIVITY

As illustrated in Equation 2, biological yield of dry matter has as primary components: the amount of incident light, the fraction of light intercepted, the photosynthetic transduction of light energy into fixed carbon minus the carbon lost to respiration.[47]

$$\text{Bio Yield} = (\text{Light Available})(\%\ \text{Intercepted})(\text{Photosynthesis}) - \text{Respiration} \quad (2)$$

Inducing variation in specific components via management can be, and has been, used to cause desired effects on yields. The light available to an orchard is a function of the climate of a given location and year, thus is not amenable to manipulation, although the variations among locations and years on yield can be great. The amount of radiation intercepted by an orchard, however, is an important regulator of potential productivity of apple orchards, although total light interception is often sacrificed due to specific localized needs for exposure for fruit yield and fruit quality (see Section II.D).

B. ORCHARD LIGHT INTERCEPTION

Crop physiologists, led by Monteith,[19,24] have demonstrated a fundamental relationship between crop dry matter production and seasonal integrated light interception. The slope of the relationship (dry matter produced per unit of light intercepted) has been called radiation use efficiency or conversion efficiency. Palmer[6] has reported in two different studies that apple production systems conform well to

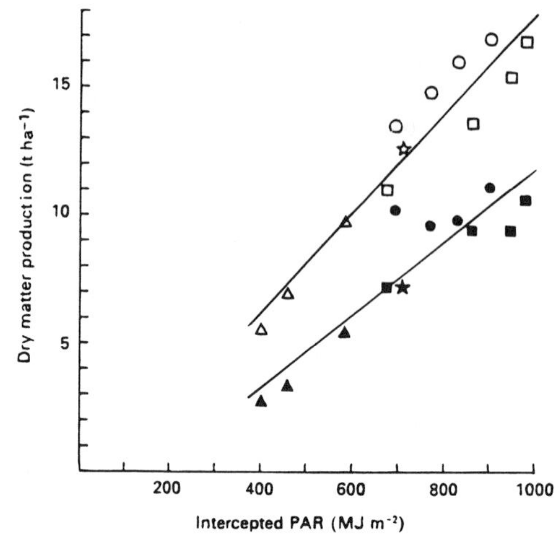

Figure 2 The relationship between seasonal intercepted PAR and dry matter accumulation in the fruit (solid symbols) and total tree (open symbols) in 'Golden Delicious'/M9 spindle and 'Crispin'/M27 bed plantings in England. (From Palmer,[6] reproduced with permission)

this relationship (Figure 2). Caution should be used, however, when interpreting the dry matter to accumulated light interception correlations. A good linear correlation may imply that different productivities are caused only by differences in light interception, reasonable considering the essential role of light as the energy source for crops. Yet, to show a stronger causal relationship, short-term (e.g., weekly) dry matter gains and light interception should be plotted against each other rather than plotting the accumulated values over a season.[48,49] Nonetheless, this concern does not diminish the importance of light interception and manipulation of light for comparisons of potential, if not actual, productivity of different apple training systems with similar management.[6,50] If the canopies are managed to be open for good light penetration, differences among systems are commonly due to differences in total light interception. From the literature, fruit yields versus mid-season percent of light interception in well-managed orchards show a generally good relationship (Figure 3). The data from the low interception/low yield portion of the plot show a linear correlation, while at interceptions over 50% the yields vary more, indicating that factors other than light may be limiting. Interestingly, although poor yields

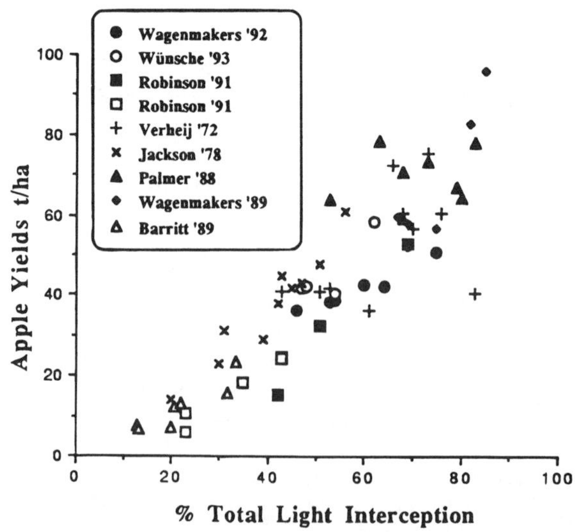

Figure 3 The relationship between apple training system yields and percent of total light interception from several reports from the literature. (References 36, 52, 57, 60, 119, 274–276)

in dense orchards with high light interception are often observed, those situations rarely appear in the literature.

1. Factors Affecting Light Interception

A great deal of progress has been made in the last 25 years in understanding the factors related to light interception by orchards. A variety of light interception models have been developed based on radiation geometry and canopy characteristics such as height, width, spacing, orientation, and leaf area density. Initially, these were "solid" models that assumed no transmission through the canopy, so that they predicted the maximum possible interception for each orchard design (see reviews by Jackson,[3] Palmer,[6] Johnson and Lakso,[51] and Wagenmakers[52]). Palmer extended the models by incorporating leaf area to estimate patterns of exposure on the canopy surfaces and within canopies.

The classic assumption for annual crops of full cover with leaf area distributed randomly in space could not be utilized for discontinuous canopies as in orchards.[53] Consequently, Jackson and Palmer[54] proposed a simplified approach (Equations 3 and 4) for orchards in which they distinguished the light that was transmitted directly to the orchard floor between the rows from the light that was transmitted through the tree canopies.

$$T = T_f + T_c \tag{3}$$

and

$$T_c = (1 - T_f)\, e^{-KL'} \tag{4}$$

where T = fractional transmission of light by the orchard; T_f = fractional transmission of light passing *between* the trees; T_c = fractional transmission of light passing *through* the canopies; K = extinction coefficient (amount of light intercepted per unit LA) and is the slope of the ln light transmission vs. leaf area index (LAI); L' = adjusted leaf area index ($LAI/1-T_f$) which expresses leaf area on the basis of mean ground area shaded by the tree.

The fraction of light that was transmitted directly to the floor was estimated with solid models, either mathematical or physical,[3,55] allowing the relationship of light transmission to leaf area to be determined for radiation incident upon the canopy. Light transmission through the canopy was then related to the light actually incident on the volume alloted to the canopy. The LAI was divided by the fraction of light incident to the canopy (i.e., essentially the mean daily shadow area) to be more representative of actual LA density. This approach has recently also been utilized for a simplified canopy photosynthesis model of the apple where the assumption of LA being distributed over the whole orchard area was not appropriate.[56]

In most species there is typically one extinction coefficient, K, given for that type of canopy. However, Jackson[3,57] has summarized the estimates of K to vary between 0.33 in more mature trees, and 0.77 for young trees with primarily extension shoots. With the variation in the balance of spurs versus shoots in different apple canopies, extinction coefficients should not be expected to be constant. In mature trees having a spur-type habit and relatively few extension shoots, K values of 0.3–0.4 are reasonable since the clumped foliage of the rosette-type spurs would be expected to intercept less light per unit leaf area than the more uniformly displayed leaves on extension shoots. Clearly, this variation in canopy K should be taken into account when modeling light interception and distribution patterns in apple trees.

2. Value of Orchard Models

Orchard light utilization models have helped guide apple crop physiology. First, the models were helpful to understand the implications of the changing canopy sizes and spacing patterns that became prevalent in the 1970s with higher density plantings that emphasized earlier bearing. From the solid models it became apparent that even though the newer plantings were of higher density and intercepted light earlier in their life; at maturity light interception actually declined compared to older larger tree orchards. This decline was primarily due to restricting tree height to 2–2.5 m, while maintaining alleyways of about 2.5 m or more for equipment. The older, taller standard trees could grow over the alleyways and intercept light, allowing equipment to pass under the canopy. The solid models have demonstrated that modern dwarf pyramid from orchards limit light interception to a potential maximum of about 60%.[54]

Table 1 **Simulation of comparable light interception (all approximately 62%) by orchards of different canopy dimensions, spacings, leaf area indices, and common indirect canopy indices for truncated triangular hedgerows listed below**

Canopy Dimensions (m)			Indirect Canopy Indices		
Height	Basal Width	Clear Alley (m)	LAI	Volume ($m^3\ m^{-2}$)	Surface ($m^2\ m^{-2}$)
3.0	1.2	2.5	4.3	0.73	1.79
3.0	1.5	1.0	2.4	1.35	2.72
1.5	0.3	0.5	2.5	0.42	3.94

Adapted from Jackson, J.E. and Palmer, J.W., *Sci. Hortic.*, 13, 1, 1990.

In actual orchards, modern plantings of Slender Spindle forms typically intercept between 40 and 50% of the incident light in mid-season because the rows are not complete hedgerows and a relatively porous canopy is required for maximum yield and quality.[58,59]

Understanding the limits to high density plantings led, in part, to attempts to increase productivity by reducing the amount of alleyway relative to tree size with double, multiple rows, and full fields of single rows with very narrow between-row spacing. Although successful in increasing light interception, and in some cases total yields, generally these approaches have generally failed due to the difficulties of management and to poor fruit quality caused by poor light exposure in dense plantings. A different approach has been to move from the pyramid tree either to taller, less sloped hedgerow forms (i.e., Italian palmette, French l'axis centrale) or to Y- or V-shaped hedgerows that grow over the alleyway above the equipment. In a comparison of three similar pyramid forms of differing sizes versus a Y-shaped trellis hedgerow, Robinson et al.[58] and Robinson and Lakso[60] found that all pyramid form systems had similar light interception values at about 45–50%, but that the Y-trellis form had 60–70% interception, and higher yields, while allowing equivalent or better equipment access.

Row orientation effects on orchard light interception and distribution have been elucidated. A north-south (N-S) orientation of a solid hedgerow distributes the light on both sides of an orchard row and latitude will not have much effect on light interception or distribution. For individual trees at higher latitudes with lower solar altitudes, more tree-to-tree shading within the row will occur at mid-day than at lower latitudes with high solar altitudes (on the summer solstices for each hemisphere the sun at noon is perpendicular to the earth at 23° from the equator). Similarly, it seems logical that east-west (E-W) rows should have less light interception than N-S rows since the solar track runs E-W. Modeling has shown, however, that this last assumption is not necessarily true,[61] since E-W rows intercept significantly less light than N-S rows only at lower latitudes, 10–25° from the equator, where the solar track is almost parallel to the E-W row. At higher latitudes the solar track is relatively flat so that in the morning and afternoon the sun shines on the north side of an E-W hedgerow.

It should be noted that the effects discussed above are primarily relevant to more vertical tree forms. The flatter the tree form, the less the effects of row orientation. The light interception of a flat horizontal surface is independent of orientation. It is important when measuring light interception to understand that measurements taken at one time during a day may be representative of the daily integral for horizontal canopies, but not for more vertical canopies. The light interception of a tall, very thin tree wall at noon will be low, but clearly not true for the rest of the day.

In a particularly useful analysis, Jackson and Palmer[54] simulated orchard light interception with families of curves useful to estimate light interception by hedgerow shapes with varying row spacing, tree height, tree thicknesses, clear alley width, and leaf area densities. A good example (Table 1) is given that demonstrates that light interception can be equal with greatly varying combinations of the above factors, pointing out that none of these factors alone (nor other indirect terms such as tree volume or surface area) can predict light interception or potential yields.

C. SHORT-TERM RESPONSES TO RADIATION

The light interception of an orchard is the maximum energy supply that determines the *potential* dry matter yield. But the energy must be efficiently converted to dry matter by the process of photosynthesis (and is partially utilized by respiration) to produce the *actual* dry matter yield.

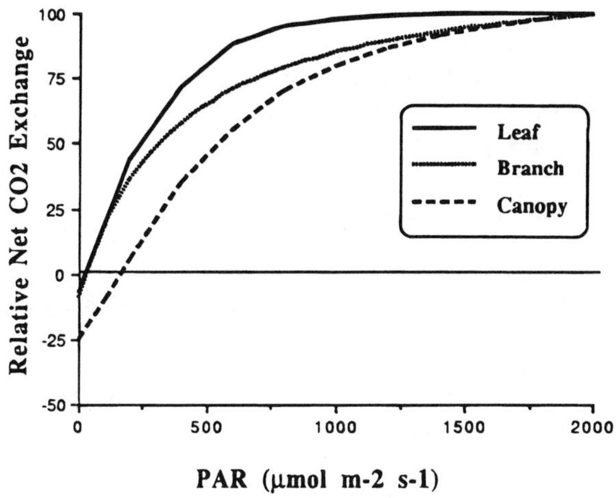

Figure 4 Light response curves of CO_2 exchange for a single apple leaf, a branch (from data of Proctor et al.[64]) and for a whole canopy (data from Corelli Grappadelli and Magnanini[62]).

1. Photosynthesis

Although incident light is not controllable, the light responses of apple leaves and whole canopies need to be known to understand the effects of different light levels since light is the single most important factor controlling tree carbon fixation of healthy trees. A large number of studies have shown that the apple leaf photosynthesis is a C_3 type with a hyperbolic light response that typically saturates at 25–50% of full sunlight (500–1000 μmol quanta m^{-2} s^{-1}), with a light compensation at about 20–60 μmol quanta m^{-2} s^{-1} photosynthetic photon flux (PPF) (Figure 4). Good rates of photosynthesis per unit leaf area for healthy, exposed leaves are approximately 15 μmol CO_2 m^{-2} leaf area s^{-1}, which is comparable to the Prunus species and grapes, but higher than citrus.[13]

The photosynthetic response of branches or the whole canopy to light intensity, however, is much more gradual than that of individual leaves, and may not saturate even at full sunlight levels[62–65] (Figure 4). The lack of clear light saturation in the whole canopy is due primarily to the variation in leaf angles and exposures such that in a full canopy only a portion of the leaves are exposed to saturating light levels at any time. Other leaves are exposed to sun, but at low angles of incidence so the light intensity is low, while the remaining leaves are shaded and dependent on diffuse light only. Canopy structure may be expected to affect the light response as well. Horizontal canopies with leaves that have adapted to the light orientation by displaying more horizontally would be expected to have a clearer mid-day light saturation than a vertical canopy. Note also that the dark respiration is much greater in whole-canopy measurements than for single leaf measurements due to the inclusion of non-photosynthetic organs in the whole canopy chamber (Figure 4).

For a whole canopy, the distribution of light between direct and diffuse components may be important as well as the total light since many leaves may be dependent only on diffuse light. Under very clear conditions with only 10–15% diffuse light, the shaded leaves may be at much lower light intensities than under bright, hazy conditions even though the total light intensities are not very different (easily seen as the darkness of shadows under different conditions). Allen et al.[66] suggested that for C_3 plants whole-canopy photosynthesis may actually be higher under bright, hazy conditions than under clear conditions due to higher light levels on the shaded leaves. Although it is expected that this may be the case also for apple trees, it has not been documented. If true, climates with lower light, but higher diffuse components, may not have less whole-tree photosynthesis compared to clear climates. Fortunately, relatively simple, inexpensive methods have been developed to measure whole-tree gas exchange under close-to-ambient conditions in clear plastic film "balloon" chambers.[62] These methods should be increasingly useful in fruit crop physiology.

Estimating whole-canopy responses of trees is not straightforward, however. Measured CO_2 exchange of whole canopies is *not* strictly photosynthesis in that respiration of non-photosynthetic organs is included in the measurements. Additionally, light response curves for a whole tree should not be estimated by plotting diurnal changes in whole-canopy gas exchange against diurnal changes in incident light. First, the amount of light intercepted by the canopy may change during the day, especially in the case of vertical canopies. Second, there may be changes with time, temperature, or humidity during the day that may modify the apparent light response if these factors are ignored. It is more appropriate to do short-term measurements with shade materials to avoid many of these problems, although shading may reduce canopy temperatures, possibly reducing respiration rates (see Section II.B.1).

2. Short-Term Carbon and Water Balance Effects

Since many aspects of apple tree growth and physiology are affected by the photosynthetic rates, there will be many indirect, thus somewhat delayed, effects related to the carbohydrate availability. One recently documented effect is that of a reduction in root respiration rate and proportion of the total root length (white in color) following a reduction in shoot irradiance.[67] Similarly, growth rates of young apple fruitlets during the fruit drop period have been found to decline with shading of the canopy, but increase with short-term CO_2 enrichment, implicating carbon supply limitations.[5,68] Reductions in fruit growth rates have been found in our studies to be an indicator of impending fruit abscission in young fruitlets as reported earlier.[44,69] Short-term shading has been shown to reduce cell division rates in young fruits[70] as well as to cause significant abscission if imposed during the cell division period prior to the final fruit drop.[68,71–75]

The exposure of leaves to high radiation loads also induces higher rates of transpiration (E) and water use. In a tree such as the apple that has a relatively high resistance to water flow, short-term high E rates induce low leaf water potentials (Ψ_l) in exposed leaves.[4,11,76] Although this may be expected to lower the leaf turgor, osmotic adjustment has been found to occur in relation to exposure so that mean daily turgor was essentially unchanged with exposure.[5] Water relations will be covered in detail later.

D. LONG-TERM RESPONSES OF EXPOSURE TO RADIATION

1. Photosynthesis

Long-term exposure of leaves to sunlight has significant effects on both the development and the "aging" of mature leaves. As with most plants, apple leaves that develop in the sun develop into "sun" leaves that are thicker, denser, contain more nitrogen, and exhibit higher photosynthetic rates than those developing in the shade.[13,77,78] Later in the season the decline of light-saturated photosynthesis rates appears to be remarkably slow compared to many fruit crops, with exposed leaves maintaining relatively constant rates until harvest in cropping field trees[23] (also see Flore and Lakso[13] for a discussion of methods of determining photosynthetic "aging"). Leaves that are shaded, however, show more rapid declines in light-saturated photosynthesis rates, apparently related to the time-shade integral, with little recovery if re-exposed by summer pruning.[23,79] It is thought that the very slow aging of apple leaves compared to other temperate deciduous species may be an important mechanism that allows the apple to adapt to long growing seasons with high productivity. This would be especially advantageous since most new leaf production usually stops in mid-summer, and productivity depends on the existing leaves.

a. Fruiting

In addition to the short-term instantaneous responses to light, many important processes in the production of apple fruit are strongly modulated by the amount of light exposure over long periods.[3,6,59,80,81] The initiation and differentiation of floral buds is strongly stimulated by good exposure of the spurs to light. Little or no bloom develops on spurs with less than about 15% exposure to light, while flowering is maximal at exposures above about 60%. Similarly, the leaf areas and productive potential of spurs require good light exposure, and may be the actual basis of the apparent correlation between spur age and declining spur vigor or fruit quality.[82] Yet, this author has observed vigorous spurs with large leaf areas and large fruits on spur complexes greater than 20 years old if the spurs were very well exposed throughout their life. In the fruiting season, fruit set, size, dry matter, soluble solids, and color are positively related to exposure, although the effects are probably a combination of carry-over and short-term processes (e.g., good exposure the previous year may give high leaf area/spur which can provide the short-term supply of carbohydrates needed for fruit growth).

b. Potential Mechanisms of Light Exposure Effects

The mechanisms of physiological responses to light exposure have not been elucidated. Several possibilities are worthy of discussion.

First, the natural shade/exposure effects may be due to changes in light quality with filtration of the light through the leaves of the canopy. The light spectrum changes within the canopy, especially the red:far red ratio[83] that is of importance to phytochrome-mediated morphogenesis. It does not appear, however, that important exposure effects (such as flowering, leaf photosynthetic characteristics, fruit growth rates, fruit size, and fruit color) are phytochrome-mediated since studies with artificial, neutral density shade cloths are able to reproduce the exposure effects with only changes in light intensity.[3]

Second is the direct effect of light striking the organ of interest. Outside of the direct effect of light on the development of red color in fruit, there is little evidence of morphogenetic mechanisms requiring high light intensities. However, the effects of light on temperature, photosynthesis, and carbon balance may be important. The proximity or the specific type of the photosynthesizing organ may be important to flowering or fruit development. In the case of fruit growth and fruit set, it appears that much of the early fruit growth after bloom is dependent on photosynthates from spur leaves. Leaves on growing shoots primarily export to the shoot tip until the shoots develop at least 10–12 leaves, and many more leaves are required if the shoot is in the shade.[5,68,84–87] Increases of available photosynthates due to light exposure of spur leaves at that time could explain the exposure effects on fruit growth and set.

Since apple fruits contain chlorophyll, direct exposure of the fruits allows some photosynthetic activity that should help the carbon balance of the fruit. A series of studies at the Institut für Obstbau und Gemüsebau, University of Bonn, and by others have shown that exposure to light can significantly reduce the loss of carbon via respiration by the fruit, and that the fruit may be able to respond to light throughout the season.[88–93] The relative importance of the direct effects of carbon balance as affected by light exposure of the fruit needs to be better documented, especially the potential role for fruit set.

A third alternative mechanism of exposure effects is not related to visible or photosynthetically active light, but to radiation-induced differences in foliage temperature, E, and the resultant allocation patterns of root-supplied hormones and/or nutrients.[94,95] Within the canopy, the exposed foliage receives much higher radiant energy loads that induce higher transpirative rates as the energy is dissipated (see Nobel[96] to review energy balances). The higher E rates are expected to partition a greater portion of the xylem sap and its contents into the more exposed sites in the canopy. Although there is evidence that root-derived regulators or nutrients may be important to flowering, the potential direct importance for fruit growth and set is not as clear except possibly for the general role of cytokinins as stimulators of cell division. In apple as well as other fruit crops, nitrogen contents of leaves have been found to increase with light exposure, and leaf photosynthetic rates were related to the nitrogen contents as well.[97–102] Partitioning of xylem-derived solutes may provide a better nutrient supply in general to exposed foliage and fruits. Additionally, the direct exposure of the fruit has been shown to increase fruit temperatures[103] and would be expected to increase fruit transpirational flux as well. Higher fruit temperatures may stimulate the translocation of nutrients and hormones to the fruit sink, thus increasing sink strength.

c. *Flower Bud Development*

Flower bud development is a difficult process to understand, despite considerable research previously reviewed.[104–106] The regulation of flowering in general has been accepted to be a hormonal mechanism, specifically diffusible gibberellins from apple fruits inhibiting flower bud development on that spur. Unfortunately, the research on hormonal regulation of flowering has not thoroughly addressed the problem of the light exposure requirement for flowering.[3,6] Recent reports suggest that root-supplied xylar cytokinins or nitrogenous regulators such as arginine or putrescine can induce flowering, but the results have not been consistent.[107–111] Although carbohydrate availability has been disregarded as a primary regulator of flowering, Hansen and Grausland[112] reported that CO_2 enrichment of young trees increased bloom, and in preliminary studies in our lab we have found increases in return bloom in heavily cropping 'Empire' apple trees injected with sorbitol solutions four times during the month after bloom.[113]

As discussed in the introduction, there is no reason to assume any one resource or regulator is always limiting. For example, light exposure-induced partitioning of xylar cytokinin or other regulators may induce flower bud development, while increased availability of photosynthates produced by the leaves of the spur and the nitrogen from reserves or current uptake may provide quantitative support for flower bud differentiation. Thus, any of these regulators may limit the process under a particular set of conditions. Regardless of the merit of this example, progress is more likely to be made if the various physiological mechanisms are integrated rather than isolated. It is clear that there is much to do before we will understand this complex, but critical process.

d. *Fruit Growth and Set*

Fruit growth and fruit set are more amenable to study than the microscopic processes of flower bud development, thus we have a somewhat better information base. As with flowering, there is a debate between hormone vs. nutrient (i.e., carbon or nitrogen) mechanisms of regulation. The hormone hypotheses have been based on correlations of endogenous hormones or exogenous growth regulators to the observed processes. Due to complexity of measuring endogenous hormones and/or the interests of the investigators, these studies have not typically included treatments of environmental factors such as light, temperature, CO_2, etc. So it is difficult to make clear conclusions as to hormonal interactions with the environment. The reviews of Dennis[1] and Browning[114] are suggested reading for understanding the hormonal approach. Although there are many methodological difficulties and inconsistencies in the literature, hormones are surely involved in environmental responses and cannot be either ignored or accepted as autonomous controllers of fruit growth and set.

e. *An Example of Utilization of Physiology*

The accumulation of research on carbon partitioning in relation to fruit growth, set, and yield provides a useful example of how such physiological knowledge may be integrated to address orchard management questions. An example is offered addressing the physiological bases of variation in productivity per hectare of different apple planting designs, training systems, and pruning regimes.

Clearly, total radiation is an ultimate limitation to yield, but at maturity, many orchards have declining productivity due to excessively dense canopies[3,6] (Figure 3). As discussed earlier, apple fruit development during the cell division period appears to be supported with carbon initially from spur leaves. The extension shoot leaves support the growth of the shoot tip until there are at least 10–12 unfolded leaves in full sun, later if in low light conditions. It is hypothesized that a basis of the low yields in orchards with dense canopies (assuming otherwise healthy trees with appropriate flowering, fertility, etc.) is a combination of two potential factors: (1) the total light interception may be low, limiting potential productivity, and/or (2) limitation of carbon availability to the fruit when final fruit numbers and yield potential are being set, normally about 4 weeks after bloom. This limitation of carbon to the fruit is viewed as a supply/demand balance that depends on the numbers of fruits versus actively growing shoots, the amount of spur leaf area, and the relative exposure of the spur canopy that supports early fruit development. Combining all these considerations, it is proposed that the yield of different orchard systems depends on the total radiation intercepted by the spur canopy per hectare at about 3–4 weeks after bloom.

A modification of the classic point quadrat method for canopy analysis in which metal needles, simulating sunbeams, are pushed into a canopy[115–117] was used to test this hypothesis. The laser modification of Vanderbilt et al.[118] and a new solar tracking device to move a laser over apple trees was used to estimate which types of leaves or organs intercepted light.[119] Combined with measurements of total light interception, estimates of light interception/hectare by spur canopy and shoot canopy were made in several orchard systems with different productivities. Dry matter yield of fruit was positively correlated with the spur light interception/hectare and negatively correlated with shoot light interception at one month after bloom (Figure 5), supporting the hypothesis.[119] The relationships at full canopy were not as strong, especially if pruning that season led to differences in spur versus shoot light interception at full canopy that were not developed at one month after bloom. In mature canopies that do not change dramatically from year to year, spur canopy light interception/hectare may be a useful indicator of not only current carbon balances for the fruit development, but also the longer-term exposure effects on the reproductive behavior.

III. TEMPERATURE

A. GENERAL EFFECTS

The effect of temperature on growth and productivity is the most complex topic in environmental physiology, since it integrates all processes. Since comprehensive coverage of this topic is beyond the scope of this chapter, only selected examples will be examined (see Jones[17] for a general review of temperature effects). Before addressing high and low temperature extremes, some general effects in the intermediate range of temperatures will be discussed.

The growth of apple fruits and other organs such as shoots are positively related to temperatures below about 25–30°C, but at higher temperatures growth may be limited by other factors. In the lower

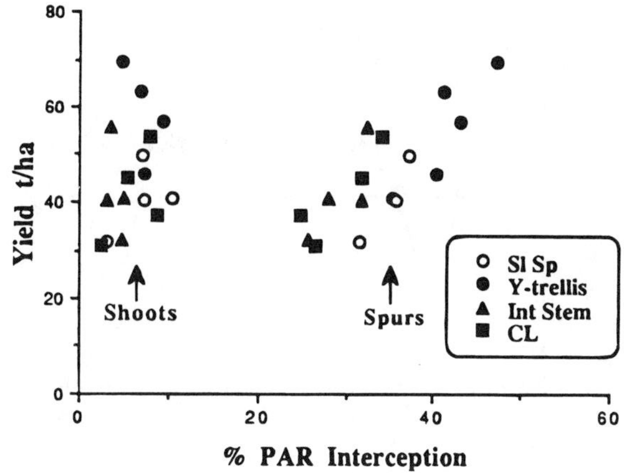

Figure 5 Relationships between yield per hectare of representative 'Empire' apple trees of different training systems and the light interception by the shoot canopy and by the spur canopy. (From Wünsche et al.[119] with permission)

range of temperatures, active growth of organs during periods of variable temperatures can be related or normalized by expressing growth rates against degree-days (normally the accumulation of daily temperature means minus a base temperature) instead of time. This is particularly useful as an index of development in the spring when growth in a perennial plant, with carbon and nutrient reserves, is probably more limited by temperatures than by resource availability. Mid-summer development tends to be less tightly regulated by temperature, suggesting that resource availability may be more limiting.

1. Effects of Yearly Patterns of Temperature

Long-term temperature effects on apple may be integrated over many months or years. The length of the growing season has clear effects on apple productivity as longer growing seasons typically produce larger crops (as long as there is adequate chilling in the winter). For example, the long, mild growing season in New Zealand can produce apple fruit yields almost twice that achievable in the shorter growing season in New York State. Although climatic differences in temperature have great effects on apple productivity, unfortunately, little is known concerning the physiological bases of such effects. Short-term temperature effects may cause long-term manifestations (e.g., the long-term effects of a single crop-destroying freeze are obvious). Subtle temperature effects generally require many years of observation of natural systems in a range of climates, either year to year or geographically, or by statistical correlation studies (called "fishing for correlations" by some).

A good example of "statistical fishing" has been that of correlations of apple yields to temperature periods. As early as 1927, Mattice,[120] a statistician for the U.S. National Weather Service, had begun to note correlations of apple yields to spring temperatures. This work was followed in more detail by scientists in England who found that in addition to the expected value of warm temperatures before and during pollination, variation in yearly yields of Cox's 'Orange Pippin' in England was negatively related to temperatures in February–April.[121–123] Similar results were found for regional yields of different cultivars of apples in New York State.[124] Though the relationships are purely correlations, these statistical approaches have been useful in pointing out effects not obvious otherwise. The effects can be reproduced in potted trees and the results also apply to fruit set data from apple breeders.[123] This regression approach has been extended in New York to relate regional yield variation to the temperatures during three periods with differing signs of correlation with temperature: previous year harvest to leaf fall period (positive), mid-winter to budbreak (negative), and budbreak to bloom (positive) (Figure 6). Heim et al.[125] compared tree growth with a consistent group of young trees grown in England versus in southern France. Seasonal dry matter accumulation differed little up to harvest, but post-harvest growth accounted for about 5% of the seasonal total in England, but more than 20% in France. Observations made while observing apple production in various years and climates lead this author to conclude that the post-harvest period may be important in the ability of the tree to recover from a crop that season, and to allow vegetative buds to develop into reproductive buds. Researchers working on biennial bearing have noted that following an "on" year, the subsequent year "off" yield depends on the warmth and radiation

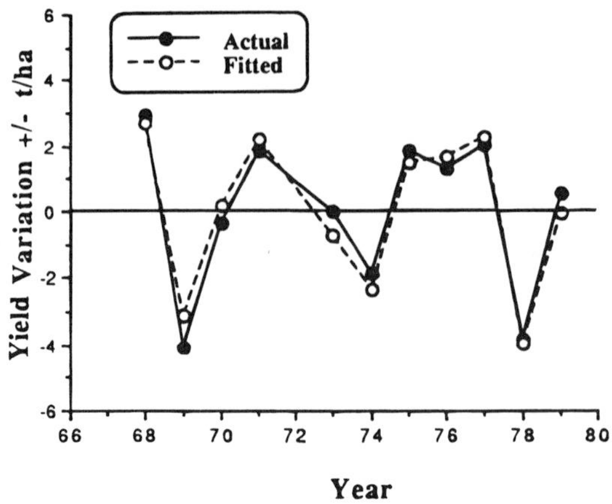

Figure 6 Yield variation over years of Western New York regional apple yields and fitted regression values with regression using temperatures during 3 periods: previous year harvest–leaf fall, January 1–budbreak, and budbreak–bloom. (From Lakso[113])

during the "on" year postharvest period.[126,127] In the spring Tromp and Penders[128] have found that rapid spring development to bloom led to higher final fruit set compared to slow development in cool temperatures.

It is most striking that a large amount of the variation in regional yields appears to be related to environmental conditions during times in which there are no leaves or fruit on the tree!

2. Growing Season Temperature Effects

After bloom there is much growth occurring, both vegetative and reproductive, that is driven by respiration as well as the supportive processes of photosynthesis, E, and nutrient uptake. All of these processes are affected by temperature, but not with the same sensitivity.

Respiration is unavoidable since it is required as the energy conversion process for growth and for the maintenance of tissues.[129] Consequently, when there are high growth rates there will be high respiration rates. For example, DeJong and Goudriaan[130] found a very high correlation between the relative growth rate of peach fruits and the respiration rates. Similar work needs to be done with apple to not only confirm the relationship, but also to allow broader modeling of respiratory costs from growth rate measurements that are much easier to take than respiration measurements.[130–132]

In general, respiration responds to temperature logarithmically over the normal range of temperatures encountered during the season, 0–42°C. The general form of the response is

$$R = a\, e^{kT} \tag{5}$$

where R = respiration rate; a = R at T = 0°C (the intercept of lnR versus T); k = temperature coefficient of R (the slope of lnR versus T); T = temperature in °C.

The temperature coefficients of respiration varies with the type of tissue and with the time of the year.[64,78,90,133,134] Generally, respiration rates for leaves are based on surface areas (g m^{-2} leaf area s^{-1}), while fruit respiration is typically based on fruit fresh weight (g g^{-1} fresh weight s^{-1} or day^{-1}). It has been found that respiration rates of different ages of the structural wood of the tree vary if expressed on a weight or volume basis, but they are essentially the same if expressed on a wood surface area basis.[133] We have confirmed this with winter respiration rate measurements. The primary activity, i.e., active living cells, of the stems of the tree are in the cambium and phloem which represent the surface wrapped around the xylem. Although helpful, there is little information on the surface area of apple tree structures,[30] though correlations of surface area to weight of branch and trunk diameters should be derivable.

Leaf dark respiration rates (g m^{-2} leaf area s^{-1}) tend to have a relatively constant temperature coefficient **k** of about 0.09 while the intercept, **a,** varies seasonally from 0.025 to 0.080.[64,78,134] With many crops the leaf dark respiration rate has been found to be increased in proportion to the photosynthetic rate of the leaf during the previous light period.[53,129]

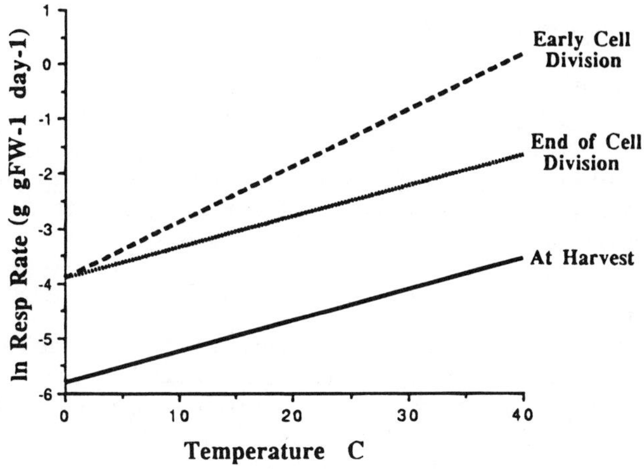

Figure 7 Apple fruit respiration responses to temperature at different stages of fruit development. (Data adapted from Jones[90] and Lakso and Johnson[143])

Fruit respiration rates have been of great interest to postharvest physiologists for many years, but fruit respiration while on the tree has received less attention. Respiration temperature coefficients **a** and **k** (the intercept and slope of lnR versus T) both vary seasonally. During the cell division period, **k** begins at 0.100 then falls to 0.055 during the cell expansion phase in mid and late summer[90] (Figure 7). The intercept **a** began at 0.020 and fell to 0.003 by late season. This suggests that the respiration rate response to temperature changes significantly during the seasonal development of the fruit. Although the respiration rate declines over the season, the increase in fruit weight leads to a relatively stable total respiration rate per fruit (Figure 8). A significant effect of light on decreasing apparent respiration via re-fixation of CO_2 in the light has been found[90,92,93,135,136] and may be very significant to the carbon balance of fruits in the first half of the season. Pruning to provide good exposure for fruit might be expected to improve the carbon balance, but increased fruit temperatures may counterbalance this effect by increasing respiration compared to shaded fruits.

Wood respiration (g m^{-2} surface area s^{-1} basis) for apple was reported by Butler and Landsberg[133] to have a constant **k** slope coefficient at 0.085 throughout the year. In preliminary experiments the author has found similar results during the dormant period. The **a** intercept coefficient varied from 0.004 at harvest to 0.010 near bloom, then declined during the remainder of the season.

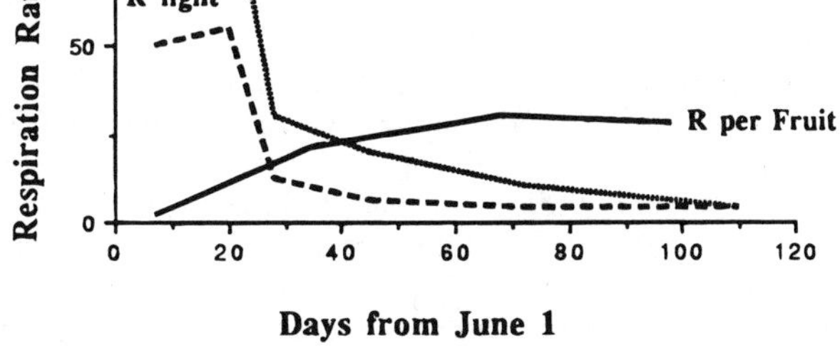

Figure 8 Apple fruit respiration rates in the dark, in the light and per fruit (ng s^{-1} per fruit) over a season. (Data adapted from Jones[90])

A significant gap in the knowledge of respiration of the apple tree is the respiration of the root system. Recent initial studies by Ebert et al.[137–139] have shown that root respiration rates tend to increase linearly with temperature, the slope of the response to temperature was higher in the summer than in the winter, respiration was inhibited by low oxygen concentrations, and phosphorus or potassium deficiencies reduced respiration, but mostly via reduced root growth. A transfer of nitrogen source from nitrate to ammonium gave a rapid increase in root respiration that lasted for many days. A particularly interesting finding was that the diurnal pattern of root respiration lagged behind that of leaf photosynthesis by about 4–6 hours, peaking at 10–12 and 14–16 hours, respectively.[138]

3. Seasonal Respiratory Costs

Even though respiration rates are reported in the literature, there have been few attempts to quantify the total respiratory costs of the tree or any particular organ. Jones[90] measured fruit respiration over the season in England and estimated that about 15% of the carbon imported seasonally by an exposed fruit was used for fruit respiration while the other 85% was accumulated as dry matter in the fruit. This value was reasonably similar to Walton and DeJong's estimate of about 23% of imported carbon used for respiration for the development of the kiwifruit in the much warmer Central Valley of California.[140] Blanke and Lenz[93] estimated a much lower value of about 6%. It is difficult to compare the estimates since the basis of calculations and temperatures used were not explained.

Estimates of total seasonal or yearly respiratory costs for whole trees are few. Hansen[141] estimated about 40–45% of fixed ^{14}C-carbon used for respiration, but it appears that this refers to the growing season when labeling experiments were done. Estimates for a 3-year-old tree for about one month in mid-summer by Proctor et al.[64] in England were only 6% of the fixed carbon, but that appears to be very low compared to most estimates for perennial plants. Likely this low value was due to cool temperatures, little vegetative growth, and very long days in that experiment, all conditions that would lower the respiratory costs. A very detailed study of apricot by Evenari et al.[142] may be relevant as they similarly estimated that about 45–50% of the total seasonal photosynthetic fixation was respired in irrigated or non-irrigated mature trees in the Negev desert.

Simulations of carbon balance of apple trees were run from budbreak to harvest with the simplified dry matter production model of Lakso and Johnson.[143] Assuming a semi-dwarf apple tree with a commercial crop (about 40 tons/ha) and average weather data for New York State and for Hawke's Bay in New Zealand, carbon fixation and respiration were estimated. The largest differences in carbon balance between locations were due to the cooler night temperature effects on lowering dark respiration in New Zealand. The accumulated total respiration estimates amounted to about 40–45% of fixed carbon for New Zealand to 45–50% for New York. Of the different plant organs, the dark respiration of leaves at night was estimated to account for about 60%, fruits about 30%, and structure about 10% of the total seasonal respiration for the semi-dwarf trees. Evenari et al.[142] estimated about 55% of seasonal respiration was by leaves in the apricot study.

Clearly, the respiratory activity of the apple tree appears to be both qualitatively and quantitatively important to carbon balance, yet we have inadequate information on whole-plant respiration in the apple. This is an obvious area of research that needs much attention. Establishment of sound relationships between respiration and growth rates of different organs is an approach that should be explored due to its potential to utilize the relatively abundant data on growth rates already available. Also, modeling will be another useful method to estimate respiratory sums as shown above.

4. Photosynthesis

Photosynthesis of apple leaves does not appear to have a strong response to temperature over a fairly wide range of temperatures from 15–35°C, with the temperature optimum generally near 30°C.[13,134,144] The decline at temperatures higher than 32–35°C may be partly due to temperature-induced increases in vapor pressure deficits (VPD) that can affect stomata. Excluding possible stomatal effects, declines in net photosynthesis of C_3 plants at high temperatures are due to the increases in photorespiration with temperature.[17] Lange et al.[145] found the seasonal adjustment of the temperature optimum of apricot leaf photosynthesis was almost 10°C higher in mid summer in the Negev desert. There is a need to determine if such temperature adaptations occur in apple. Measurements taken by the author in the field under many different temperatures suggest that apple leaf photosynthesis rates may be maintained up to about 37°C, but seasonal adjustments of temperature optima have not been documented yet for apple.

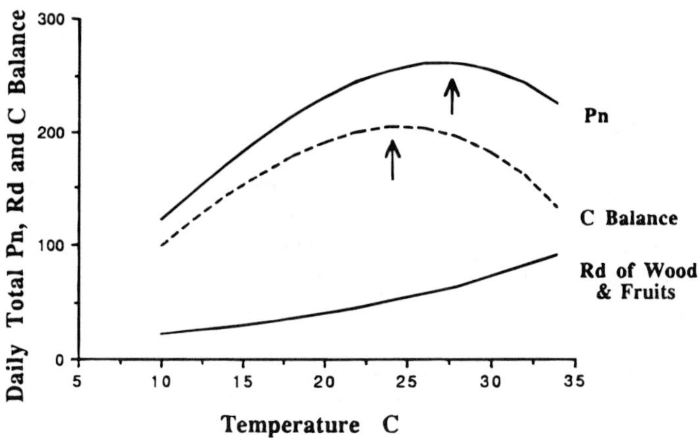

Figure 9 Simulated apple tree net photosynthesis (includes leaf respiration), respiration of the wood and fruits, and the net balance ($P_n - R_d$) responses to temperature in the light. (Simulations with model of Lakso and Johnson[143])

Since the classic work of Heinicke and Childers,[63,146] the net CO_2 exchange of whole apple canopies has been measured occasionally.[65,147] With the development of an inexpensive "balloon"-type chamber,[62] whole-canopy measurements will likely increase in the future. The photosynthetic temperature response peaks at about 28–30°C, but the respiratory response is a continuous logarithmic increase. The net effect is that a whole-canopy temperature response would be expected to be optimal at a lower temperature than that of the photosynthetic component (Figure 9). Results from the early studies of whole-tree gas exchange by Heinicke and Childers[63,146] indicated significant decreases in net CO_2 uptake during the day if temperatures exceeded about 25–26°C. This may help explain why apples produce higher yields than might be expected in the cool climates of northern Europe or why productivity is particularly high in the long cool season in New Zealand.

B. HIGH TEMPERATURE EFFECTS

High temperature or heat stress should be viewed in relative terms since the physiological responses we are generally concerned about are not as much related to any absolute temperature, as much as related to "supra-optimal" temperatures. A 35°C day in the early spring may induce physiological problems for a tree with tender foliage while in mid-summer these effects may be minimal. Consequently, the emphasis will be on response to supra-optimal temperatures, although for simplicity the term "heat stress" will be used.

1. Fruit Set

Temperatures above the optimum for photosynthesis lead to a strong decline in net gas exchange for the whole tree due to the simultaneous decline in the carbon fixation (supply) and increase in respiration (demand). The response of the tree to such a potential limitation should also depend on how critical the carbon balance is at the time of the heat stress. During the period before final fruit drop about a month after bloom, when it is proposed that the carbon balance may be critical,[68] high temperatures, especially at night, have been found to cause fruit abscission[120,148–150] or poor fruit growth.[151,152] Tukey[151,152] found greater fruit growth and development at 24°C versus 9 or 31°C, with the high temperature being particularly detrimental. Higher temperatures appeared to stimulate early fruit development for the first few weeks, but were detrimental at about 1 month after bloom. Similarly, Tamura et al.[153] found that raising night temperatures to 23°C was stimulatory for the first 3 weeks after bloom, then was detrimental, so that by 30–35 days the fruit were the same size. The fruit growth stimulation was due to increased early cell division that ended 7–10 days earlier than the controls, but gave the same final fruit size and cell numbers.

The mechanism responsible for the variable effects of temperature on fruit development is not clear. Changes in availability of carbohydrates at 3–4 weeks after bloom, but not earlier, has been suggested.[68] Fruit ethylene production is stimulated by high night temperatures has been suggested to be the mechanism by Kondo and Takahashi,[154] but Fukui et al.[149] presented evidence that the ethylene may be a response to rather than a cause of abscission. The same high temperatures in mid-season, however, have essentially no significant effect on fruit ethylene production, indicating a differential sensitivity with stage of development.

Root temperatures may also be excessive in apple. Gur et al.[155] reported that temperatures above 30°C caused marked reductions in both root and shoot growth, with severe effects on leaves at root temperatures of 35°C and above. Respiration of the roots peaked at 35°C, apparently with damage at higher temperatures. Less resistant rootstocks accumulated larger quantities of ethanol in the roots, along with acetaldehyde. Root and leaf cytokinins were also reduced at the higher temperatures. Although these root temperatures would be unusual in most apple-growing regions, they are conceivable in potted plants used for experiments and perhaps in young trees in the field.

2. Water Relations with High Temperatures

Many of the effects of heat on plants are confounded by the effect of higher temperatures on the VPD. Assuming that the absolute water vapor pressure of the air remains constant (e.g., equal to 50% RH at 25°C) and leaf temperature increases with air temperature, the VPDs at 25, 30, 35, and 40°C would be about 1.6, 2.1, 2.8, and 3.7 kPa (1 kPa = 10 mbars), respectively. This increase in VPD can have dramatic effects on the evaporative demand, Ψ_l if E increases, stomatal conductance (g_s), and photosynthesis. This author has found that reductions in apple g_s at higher temperatures began at 33–34°C if the RH was kept constant at 40% so that the VPD increased with increasing temperature; the decline began at about 37°C if the RH was also raised to maintain a constant 2-kPa VPD.[113]

With high temperatures, high VPD, and high E, stomates normally close partially, which may conserve water, but reduces evaporative cooling. Leaf temperatures will increase unless there is enough wind to cool the leaf via sensible heat transfer (see Jones[17] or Nobel[96] to review energy balance). Since the leaf temperature is very dependent on radiation load and boundary layer, particular care should be paid to measurement of leaf temperature. The use of infrared remote thermometry is the preferred method of measuring leaf temperature, since the boundary layers and heat capacities of instruments such as porometers or photosynthesis chambers may influence leaf temperatures markedly. Ideally, temperature response curves should be produced, then the natural leaf temperatures from IR thermometry be used to estimate leaf status at those temperatures. Due to the dependence on the evaporative demand, radiation load, and wind, the importance of water stresses in the response to heat stress will be variable and unique with each situation. Therefore, it is important to be aware of the interactions of temperature and water relations on leaf temperature and physiology.

Whole-plant responses due to high temperature-induced increases in E are also expected. Decreases in stem and shoot tip Ψ_l would also be expected to reduce vegetative growth rates as well, although little documentation is available. Specific heat-induced lesions such as sunburn on fruit or branches or sunscald on branches in the spring can have important effects on tree productivity. Sunburn of the fruit or branch surface occurs when the fruit is exposed to high direct sunlight and is usually more severe in more arid climates[156] or on trees with poor water status. This suggests that sunburn effects are due to excess water stress of the fruit or branch surface. Sunscald of branches or trunks in the winter or spring is actually more related to radiation-induced warming of the bark, followed by a precipitous drop in temperature in the evening, leading to cold damage of the tissue.

C. LOW-TEMPERATURE EFFECTS

Extreme cold temperatures are an important limitation to long-term commercial success of apple growing in many marginal temperate regions. Spring frosts and severe winter cold have obvious detrimental effects on apple survival and production, but the emphasis in this section will be on less severe cold effects.

1. Cold During the Growing Season

The effects of cool or cold temperatures during the growing season depend heavily on the stage of development of the tree. In general, temperatures dropping below 15 or 20°C, but not below about 5°C, merely slow all aspects of metabolism and growth. During most of the growing season this range of temperatures does not tend to change any basic patterns of plant development, just slows the rate of development. However, late in the growing season as the temperatures decline, cool temperatures begin to induce basic changes in physiology. The complex processes of dormancy and leaf abscission are induced by decreasing temperatures. These allow the tree to develop the cold hardiness required for the winter.

2. Dormancy

The apple in general requires a chilling period during part of the year to allow the proper annual growth cycle to occur. Without adequate chilling budbreak is poor and/or very erratic, bloom is not synchronous, and yields are poor. The dormancy required to obtain the necessary chilling is a particularly complex physiological process that is difficult to study due to the lack of visible, macro-scale growth or gas exchange that can be easily measured. Many of the critical processes are at the cell or membrane level in complex, microscopic tissues and are influenced over long periods of fluctuating weather that make each situation unique. Therefore, much of the research has been directed toward controlled environment studies and the development of models to relate temperature patterns to observed phenomena.

The reader is referred to several reviews on the physiology and terminology of dormancy by Samish,[157] Lang,[46] Cannell,[158] and several papers in a dormancy symposium.[159–162] Apple appears to behave similarly to other fruit crops, thus research in the general mechanisms is likely applicable across species.

In the broadest sense, dormancy is the absence of active growth in an organ with a meristem. The terminology proposed by Lang[159] has gained general acceptance. During the normal yearly cycle of dormancy, there are three general periods of dormancy: (1) the initiation of dormancy in the late summer and autumn, "paradormancy"; (2) winter dormancy that is internally controlled in that growth will not occur even with higher temperatures, "endodormancy"; and (3) late-winter dormancy that is primarily controlled by an unfavorable environment for growth, "ecodormancy". Prediction of rest completion is of importance to define the chilling requirement for a species or cultivar, an important limiting factor for cultivar choice in warmer climates that may have insufficient winter chilling temperatures.

Most commercial cultivars of apple require about 1200 ± 300 chilling units (CU), defined as hours at optimum chilling temperature (considered to be between 3–8°C), although some cultivars have been selected for low chilling requirements. Apple requirements are similar to pear, but tend to be longer than most of the stone fruits.[10,12] Hauagge and Cummins[163] found that the range of chilling requirements of different cultivars and species over several years ranged from about 200 CU for the cultivar 'Anna' to about 1500 CU for 'Wright #1' under New York conditions. In one year 'Anna' only required 88 CU. It has also been reported that in Zimbabwe under tropical conditions 'Anna' may require no chilling and can produce two crops per year without dormancy-breaking sprays.[164] Therefore, great range exists in apple germplasm for the chilling requirements, and genetic studies on heritability patterns suggest that the low chilling requirement of Anna is due to a shallow dormancy.[163,165–168] This germplasm may provide excellent material for further physiological studies.

The cellular physiological transduction of the environmental signals has been studied in many species from many mechanistic approaches (hormonal, carbohydrate, protein, enzymes, lipid phase change, gene expression) and with exogenous treatments or manipulations of photoperiod, temperature regimes, dormancy-breaking chemicals, or pruning/defoliation. The reader is referred to the reviews of Lang[46] and Cannell[158] for a more complete review of this research. A particularly interesting new series of studies implicate changes in the free versus bound water in apple buds in relation to the stages of dormancy and perhaps cold hardiness.[169] Lipase activity has been found to be related to the water state changes and that membrane lipid composition may be important.[170] The availability of new technologies, such as magnetic resonance imaging, promise new opportunities for studying such subtle physiological changes.

Although most dormancy studies primarily have treated the buds, an interesting series of recent papers by Young and colleagues[171–177] have explored the root-shoot interactions in dormancy and spring development and have shown that chilling of the rootstock can have different effects on behavior of the top compared to scion chilling.

3. Cold Hardiness

The reviews by Burke and Gusta,[178] Ashworth and Wisniewski,[179] Ashworth,[180] and Flore on Stone Fruit in this volume are recommended for general review on woody plants and fruit crops. The resistance to extreme cold temperatures occurs during the dormant period. Although chilling and dormancy appear to be qualitatively required for adequate cold hardiness, they do not appear to be quantitatively related to cold hardiness. In general, death due to freezing is thought to be due to cellular disruption from ice crystal formation, while supercooled tissues remain alive. The migration of water from the supercooled cells to the intercellular spaces where ice crystals may form avoids the crystal formation within the cells. But if too much water migrates from the cells, dehydration effects may cause damage.

Apple is one of the hardiest fruit crops with maximum hardiness of close to −40°C.[181–183] Studies of apple clones and heritability of cold hardiness indicate that there is genetic variation among rootstock and scion clones, but year-to-year variability is high.[184–186] This level of hardiness is greater than many other fruit crops and allows apples to be grown over a wide range of higher latitudes. Since economic pressures have forced the elimination of orchards in marginal sites or regions, the most obvious problems due to mid-winter cold have decreased. However, hardiness due to cold temperatures after early spring deacclimation (i.e., during ecodormancy) is still a problem in more continental regions that have periods of warm weather followed by sharp drops in temperature. Apple rootstocks have been found to differ greatly with Robusta 5, having great fall and winter hardiness, but it dehardens rapidly in the spring.[186] MallingMerton 106 also has good mid-winter hardiness and dehardens in the spring, but this rootstock has a much greater ability to reharden after a warm period.

Cold hardiness develops with increasingly cooler temperatures in the autumn. Cold acclimation in the autumn is concurrent with the cessation of growth, and deepens to maximum hardiness (during endodormancy) in mid-winter.[183,187,188] Hardiness is also dependent on other factors that affect the health and physiological status of the tree. Maximum hardiness develops when the trees are healthy, but not actively growing. Water deficits, flooding, poor nutrition, heavy cropping, and low light, especially in combination, are all detrimental to winter hardiness.[189] A well-known observation in the orchard is that trees pruned in early winter before the minimum temperatures have more winter damage than unpruned trees at the same location. The reasons for this are not clear, but it has guided the timing of pruning of commercial orchards.

Not all tissues of the apple tree attain equal hardiness as might be expected from the differences in tissue structure and location. Roots are typically less hardy than the tops of the trees, but the roots normally do not experience the severe cold temperatures due to the heat capacity of the soil. It appears that soil moisture content may play a buffering or insulation role, as root winter damage has been found to be related to sprinkler irrigation patterns.[190] Within the stem tissues it appears that bark and the xylem respond differently in the freezing process. The bark tissue appears to freeze extracellularly, while the cells in the xylem parenchyma and pith may supercool to almost −40°C.[191]

The hardiness levels in apple shoots have been found to be correlated with sugar levels in the wood,[192] although hardiness does not apparently depend on the concentration of sugars in the tissues, since the sugar and hardiness levels can vary independently.[193] Recent work with cryopreservation of apple germplasm has indicated that during acclimation apple buds increase the amount of bound, unfreezable water; but once hardy, they need resistance to dehydration from the freezing that occurs in nearby free water or less hardy tissues.[194–197] Thus, intracellular water relations seem to be an important factor in cold acclimation of apple. Lipid metabolism was suspected to be involved in hardiness, but glyco- and phospholipid compositions were not found to be related to apple bark hardiness.[198,199] There do appear to be correlations with the peroxide-scavenging enzyme systems and some ultrastructural changes in the cell walls and membranes.[200–202]

While dormant, apple does have the ability, within limits, to reharden after a warm dehardening period.[187,188] The amount of rehardening ability declines as spring approaches. Once growth occurs in the spring, the tissue hardiness is lost rapidly, with developing spur leaves and flowers varying in temperature for 10% kill from about −9 to −12°C at budbreak to about −2 to 3°C at bloom.[203,204]

IV. WATER DEFICITS

The normal functioning of the apple tree depends on appropriate water relations. Nonetheless, in the orchard water may not necessarily be a limiting factor for growth and productivity. The study of physiological responses to water deficits is probably more important in humid climates than in arid climates where irrigation is used routinely. In humid climates where irrigation is not normally used and economic feasibility is still questionable, apple trees in unirrigated orchards will experience more water stress during periodic droughts than irrigated trees in drier climates. Conversely, however, the opportunity to use water relations to control tree growth by irrigation management is greater in arid climates where water availability can be controlled.

A. TERMINOLOGY AND FUNDAMENTALS

This discussion will emphasize the water relations of apple, especially relative to other crops used to establish the textbook concepts of water relations. The basics of water relations [i.e., water potential components, relative water content (RWC), hydraulic conductivity] can be reviewed in a variety of textbooks and reviews,[205-209] although the author recommends the excellent books by Nobel[96] for a biophysical approach and Jones[17] for a whole-plant/crop approach. Reviews of general fruit tree water relations by Jones et al.[11] and specifically for apple by Landsberg and Jones[4] are recommended. After a brief discussion of the fundamentals, the responses of apple trees to water deficits will be reviewed in a format to place the various responses in context.

A few terms and concepts used here need definition. *Drought* is a meteorological term that refers to the lack of significant rainfall for a period; it is not necessarily detrimental if there is no need for the water (e.g., in the winter or in irrigated orchards). *Stress* will be used in the common way to describe an imbalance between the supply of and the demand for water. In fact, the greatest concern is not drought or stress, but the response to stress. This describes the changes in physiology in relation to changes in the water status of the plant. It is quite common for plants to experience similar levels of stress, but the responses may differ greatly (e.g., apple functions normally at mid-day water potentials of −1.5 MPa while many annual crops would be near death at the same potentials). Also, a change in soil moisture stress does not necessarily cause the same change in stress in the plant. Thus, it is important to recognize that there are environmental stresses that can lead to physiological stress responses; but it will be seen that the relationships among these are not always the same.

1. Total Water Potential (Ψw)

There is little evidence that Ψw itself has a direct effect on metabolism or enzyme activity within the range normally found in plants; normally turgor, osmotic, or other factors such as RWC or cell volume appear to be sensed more directly by plant cells.[17,210] The importance of total potential is primarily for determining the direction and gradient of the movement of water through the tree.[205,207]

2. Osmotic Potential ($\Psi\pi$)

Osmotic potential is a colligative property, depending on the concentration of solutes in the cell and not on the mass of the solutes. This has some important implications. If a plant needs to accumulate solutes in the cells, 1 mol of a large molecule to reduce $\Psi\pi$ requires much more energy and carbon than 1 mol of a much smaller compound. Accumulations of ions may be more energy efficient than larger molecules such as sugars, but large concentrations of ions may be detrimental to metabolism. It appears that $\Psi\pi$ may also be adjusted by breaking large compounds into smaller compounds in response to water stress. For example, an invertase enzyme in a cell will convert 1 mol of sucrose into 1 mol of glucose and fructose; thus the $\Psi\pi$ decreases. In some systems invertase is stimulated by abscisic acid, which itself is stimulated by water stress.[211] This may be very important to the Ψw of apple fruits or other solute storage organs that accumulate small molecules, giving low $\Psi\pi$ to develop and maintain turgor for expansion growth.

3. Turgor Potential (Ψp)

Turgor is likely the Ψp component most easily sensed by the plant cell. Thus, the regulation of turgor is very important. Hence, it is important to estimate osmotic and turgor components, and not just the total potential to understand the water status and the total plant response to stress.

Typical expected values for water potential components and RWC for exposed leaves on field apple trees under sunny conditions, moderate evaporative conditions, and adequate soil moisture are typically: $\Psi w = -1.5$ to -2.0 MPa, $\Psi s = -2.5$ to -3.0 MPa, and $\Psi p = 0.8$ to 1.2 MPa, with RWC = 85–90%. Of course, with soil moisture stress or differences in evaporative demand, radiation, or VPD the values will adjust. The discussion later will address these adjustments.

4. Significance of Plant Resistance to Control of Leaf Water Potential (Ψ_l)

Leaf water potential depends on three factors: (1) soil water potential (Ψ_{soil}), which provides the upper limit for Ψ_l; (2) the resistance of the plant-soil system to the flow of water; and (3) the rate of E from the leaves. In most grasses and annual plants, there is generally a low resistance to water movement through the plant. These plants have many roots in very good contact with the soil, the plant is small, and is efficient at moving water through its system. Apple trees, however, have a high total soil-leaf resistance due to (a) the size of the plant and distance the water has to flow, (b) the root systems are not dense,[212] do not explore the soil intensively to find moisture, hence more soil resistance; and (c)

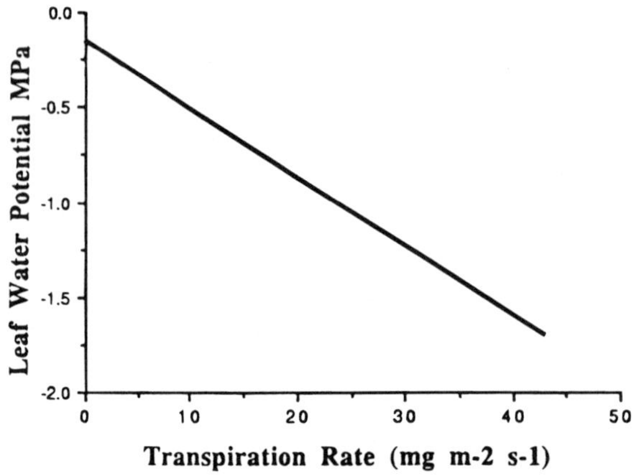

Figure 10 The reduction in leaf water potential of apple leaves in the field related to transpiration rate. (Adapted from Landsberg et al.[213])

the root system itself has a high resistance.[4,11,213] A change in the rate of E induces lower Ψ_l in apple trees (Figure 10) while many annual plants show relatively little change in Ψ_l with changes in E.

A knowledge of the soil/plant resistance is important to understand which factors control leaf water potential. Unlike many annuals where control of leaf water potential is primarily a function of soil water potential, in apple there is a strong control of leaf water potential by transpiration. Consequently, any factor that affects transpiration (vapor pressure deficit, radiation, wind, stomatal opening) strongly affect leaf or stem water potential in the top of the apple tree. This places much greater relative emphasis on evaporative conditions and factors controlling stomatal opening than primarily on soil water status as in annuals. The experimental implications of this are that for apple trees:

1. soil moisture readings are not adequate to determine the stress level within the tree
2. atmospheric conditions must be monitored to understand the stress levels
3. water potentials can be highly variable since they depend heavily on the variable environmental conditions
4. atmospheric conditions must be much more constant to reduce variation when making comparative readings of water potentials between treatments.

B. ORCHARD MICROCLIMATE AND MANAGEMENT CONSIDERATIONS

In the past decade much more appreciation has developed for the importance of structural characteristics of whole crops to water use and water relations.

1. Whole-Crop Water Use

The Penman-Monteith equation for crop E (see Jones[17] and Jarvis[214] for review) shows that E in the field depends primarily on three factors: net radiation, the humidity gradient from the crop to the bulk air, and the g_s. Whole-crop E for short, dense field crops or grasses is dependent on stomatal aperture when g_s is low. But as the stomata begin to open, whole-canopy E soon becomes relatively insensitive to changes in g_s. In orchards or forests, on the other hand, the E responds to the g_s. The reason for this is that low crops have a high crop boundary layer resistance so that the transfer of water vapor between the crop and the air above is slowed. If the stomates open further, the net result is primarily an increase in the humidity in the air within and above the crop, reducing the vapor pressure gradient and the boundary layer conductance (g_b), thus countering the effect of increased stomatal opening. A dense, short crop essentially has its own microclimate. In that case crop E becomes primarily controlled by net radiation, not g_s. It is clear that crop water use results and methods from annual crops may not be directly applicable to apple orchards, since fruit or apple orchards, conversely, have small boundary layers due to their stature and roughness of the canopy. The water transpired from the leaves and the soil or cover crops can mix readily with the bulk air above so the humidity does not build up near the leaves. Therefore, whole-crop E is a strong function of the canopy conductance and vapor pressure gradient as well as of radiation.

Table 2 **Summary of responses to water deficits in apple trees following a format modified from Turner**[218]

Mechanism	Response in Apple Tree
I. Drought avoidance/escape	Early development of spur canopy before drought development; perennial habit allows very rapid leaf development
II. Drought tolerance by maintaining high water potentials	
(A) Reduction of water loss	
(1) Stomatal/cuticle resistance	Coupling of stomatal opening to photosynthesis
(2) Reduction of radiation absorbed	Folding of leaves
(3) Reduction of leaf area	Reduction of leaf area development initially, then leaf abscission
(B) Maintenance of water uptake	
(1) Increased root density and depth	(Responses not clear)
(2) Increased root conductivity	Increase of root conductance
III. Drought tolerance at low water potentials	
(A) Maintenance of turgor	
(1) Osmotic adjustment	Osmotic adjustment of mature leaves, but no adjustment of shoot tips
(2) Increase in elasticity	Small if any
(3) Decrease in cell size	(Responses not clear)

Wind can alter crop water use by affecting the boundary layer. Thus, a field crop in wind will respond as an orchard does on a calm day. Also, as a crop canopy closes with development, there will be decreases in crop g_b and thus changes in crop E responses to crop conductance. Similarly, a densely planted dwarf orchard almost completely free of wind (as with surrounding windbreaks) may act more like a field crop. Jarvis[214] has pointed out that studies of fruit tree water use in closed glasshouses may not give results applicable to the orchard or forest. These effects need to be considered when planning and interpreting physiological studies of water use.

Apple tree water use is also related to total leaf area on the tree.[215] Mature orchards typically have leaf area indices on a whole-orchard basis (hectares of per hectare of land) of 1.5 to 4.5.[3,6,50] The relationship of leaf area to water use is not expected to be linear, as increases in leaf area per tree generally cause greater percentages of shaded leaves. Although interior shaded leaves do not transpire as much as exposed leaves, they do respond to the VPD so that the E in the shade may be 30–50% of that of exposed leaves, but at the lower light levels photosynthesis may be only 10% of maximum.[216] Consequently, the water use efficiency (WUE) of shaded leaves is poor. Overall WUE of a whole canopy may be better in more open canopies with fewer shaded leaves, but that has not been documented.

The relative influence of temperature, radiation, VPD, and wind speed will vary with the configuration and prevailing conditions in each orchard. Butler[217] evaluated the relative importance of radiation versus humidity components on water use in English apple hedgerows and concluded that the humidity component was the most important determinant of E, but that radiation was important as well. This would be expected for a crop well-coupled to the atmosphere as discussed above. For example, a comparison among mature training systems in New York has recently found that compared to a Slender Spindle, a Y-trellis system intercepted about 20–25% more light per hectare, but also developed twice the leaf area index[58,60,119] both would be expected to increase water use for the Y-trellis. In conclusion, there are many interactions of environmental factors and plant responses that must be considered concurrently when studying plant water relations in general, and especially under orchard conditions.

C. OVERVIEW OF APPLE WATER RELATIONS

The study of the individual aspects of water relations does not provide the picture of the integrated responses of a species to water deficits. A useful method to place these components into context has been proposed by Turner[218] and is similar to that used by Jones.[17] This format breaks plant responses into broad categories that first deal with avoidance of drought stress, then avoiding stresses within the plant, then finally tolerating unavoidable stresses (Table 2). The discussion will follow this format.

1. Drought Avoidance/Escape

Drought avoidance/escape refers to mechanisms by which the plant avoids growing in drought conditions. The best examples are desert plants which lie dormant until rainfall comes, then complete their life cycle and produce seed before drought can develop. Apples have very rapid leaf area development during the spring due to the carbohydrate and nitrogen reserves that allow pre-formed leaves and shoots to develop very quickly and intercept radiation early in the spring. Flowering and initial fruit development also occur early in the season before possible summer drought can develop. Thus, the rapid development can be viewed as a drought-avoidance mechanism.

2. Drought Tolerance by Maintaining High Water Potentials

Under mild to moderate stress there are mechanisms that keep the water potentials in the top of the tree relatively high by reducing water loss and/or maintaining uptake from the soil. This involves developmental plasticity (i.e., the ability of the plant to change its development in response to its environment).

a. Reduction of Water Loss

To reduce water loss through non-stomatal diffusion out of the plant, waxy cuticles develop. In general, apple leaves have quite thick waxy cuticles that have very low water vapor conductances, thus the vast majority of leaf E is via the stomata, which are found only on the abaxial surface on apple leaves.

i. Stomatal and Cuticle Resistances

Stomatal regulation, thus, has been a major concern of water relations research in apples and other fruit crops.[4,11] Several factors affect stomatal behavior. Stomata respond directly to light, but the author has found that there is light saturation at about one tenth of full sunlight. Consequently, the light requirement for stomatal opening is lower than the light effect on photosynthesis even in shaded positions within the canopy. Thus, the direct light limit on stomata is normally not considered an important regulator of stomata. For many years it was thought that Ψ_l was the primary regulator of stomata, based on the obvious closure of stomata with increasing water deficits. The documentation of active osmotic adjustment in apple[29,219,220] led to the interpretation that Ψp, instead of Ψ_l, was regulating stomatal closure at variable Ψ_l.[221] Under non-extreme conditions in the orchard the typical diurnal pattern of g_s shows a peak in mid to late morning followed by a gradual decline for the rest of the day while the Ψw typically shows a symmetric pattern with minimum values in mid-day (mirroring the radiation pattern). This difference of patterns suggests that the water potential components were not controlling stomata under normal conditions.

Examination of the diurnal pattern of air VPD typically shows a mirror pattern to that of g_s, thus a high correlation is found between the g_s and VPD.[134,222–224] It appeared that humidity may be the controlling factor. However, based on the experience of the author there are times that contradict this conclusion.[113,225] The typical diurnal pattern of g_s can still be seen with trees placed in growth chambers with constant radiation and VPD. Also, trees with strongly growing sinks (vegetative, fruit, or both) may show only a mild afternoon decline in conductance, while under the same conditions non-cropping plants may show a strong decline. Over many experiments and observations, the stomatal behavior of apple leaves appears to be better correlated with the photosynthetic rate than with humidity or Ψ_l.

Response of stomata to changes in the photosynthesis has been found in many species.[208] Apple photosynthesis and g_s appear to be well coupled, especially under orchard conditions. An example is that the relationship of photosynthesis and g_s appears to be the same in data from a drought study as in a crop load study with high water status (Figure 11). Consequently, at any given time it is not at all clear whether photosynthesis is controlling stomatal behavior, or vice versa. An indirect method to estimate the relative stomatal versus non-stomatal changes with treatment was to calculate the internal CO_2 concentration (C_i) from the CO_2 and water vapor fluxes[14,17] to determine if the effect was primarily stomatal closure (C_i decreasing) or non-stomatal reduction (C_i increasing). In most cases with apple in the field, the C_i remains essentially constant regardless of the type of stress. However, this type of analysis has been questioned recently with the finding that in some situations, especially during stress that induces high abscisic acid, stomata close in patches rather than uniformly.[226,227] Stomatal patchiness has not been documented yet in apple. This may give errors in the calculations of C_i from gas exchange data, so that calculated C_i values may be overestimated in stressed plants, and the stress effects may be essentially only stomatal. The simultaneous use of gas exchange and chlorophyll fluorescence

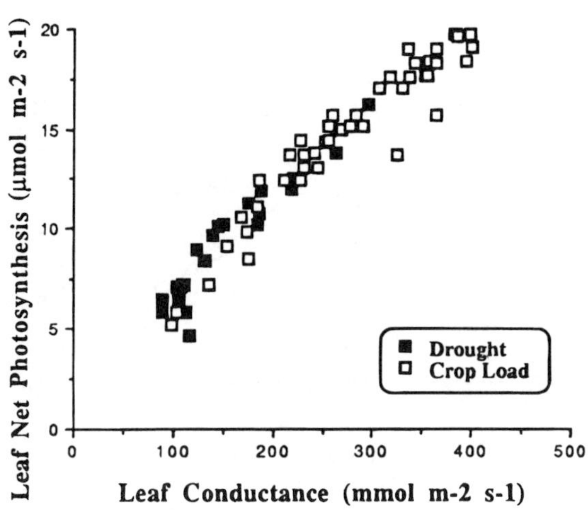

Figure 11 The relationship between photosynthesis and leaf conductance in separate studies of drought stress and of varying crops loads with 'Empire'/M9 apple trees in the field. (Data from Lakso[37])

measurements are needed as recently reported by Massaci and Jones[228] on apple. Their results show that with rapid water stress development only the gas exchange was reduced (i.e., stomatal effect only); but with long-term stress both stomatal and the non-photochemical quenching photosynthetic component declined. Thus the relative roles of stomatal versus non-stomatal limitations in apple appear to depend on how rapidly the stress develops.

An important current area of research relates to regulation of g_s and photosynthesis by root-produced signals, generally thought to be abscisic acid. A variety of split root studies and other approaches have shown that leaf gas exchange tends to be more closely related to the soil water availability of the driest portions of the root system rather than to the Ψ_l components as previously thought.[229] This concept has been documented in apple recently by Gowing et al.,[230] who showed in a split root study that apple leaf E and leaf growth rates declined with the decline in soil water availability in the dry half while top water relations were maintained at normal levels with water from the wet half. Additionally, they showed that the apparent signal was from the dry half of the root system, since control growth and E rates were restored when the dry half was re-watered or excised. This mechanism of regulation appears to be widespread, as similar studies in grapes[231,232] have been reported.

There are several ramifications if stomata are regulated by photosynthesis. First, it has been shown that low crop load may lead to decreases in tree water use, due in part to less leaf area, but in greater part to lower E rates.[233,234] Second, since g_s is well coupled to photosynthesis, stomatal opening will only allow the E based on the photosynthetic rates, maintaining high WUE (photosynthesis per unit of E). Third, reductions in g_s can be caused by factors other than drought stress, so the crop water stress index (CWSI),[235] based on leaf-to-air temperature differentials caused by stomatal closure, can give erroneous readings of stress as recently demonstrated by Andrews et al.[236] For example, a non-cropping plant will transpire less, use less water, have higher Ψ_l, and maintain higher soil water reservoirs; yet the CWSI method would indicate that irrigation is necessary.

ii. Reduction of Radiation Absorbed

Apple leaves respond to water stresses by a folding of mature leaves along the midrib and by the reduction in final leaf size in expanding leaves. The degree of folding of the mature leaves is positively related to the light exposure. For the whole canopy the radiation not intercepted by the folded exterior leaves may be intercepted by interior leaves so that the total light interception and water use may not change. The better distribution of light into the canopy should increase whole-canopy photosynthesis, thereby increasing whole-tree WUE, although this has not been experimentally verified in apple. Differences in orchard light interception can lead to different potentials for water use, but the light interception is only a component of water use, with humidity also playing an important role. Canopy form-induced differences in orchard light interception would tend to have its greatest effects on water use under clear skies and with low VPD values.

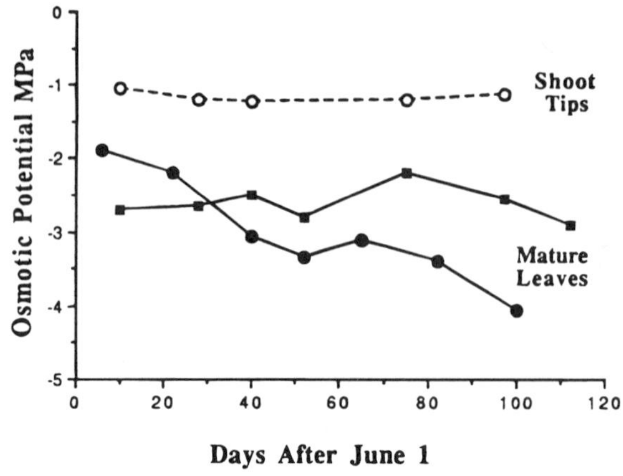

Figure 12 Seasonal patterns of osmotic potentials of expanding shoot tips and mature leaves in 'McIntosh'/MM111 apple trees in the field in New York. The different patterns for mature leaves are two different seasons with different rainfall patterns. (Data from Lakso et al.[29])

iii. Reduction of Leaf Area

In apple trees, developing extension shoots grow more or less according to the amount of summer rainfall. Growing shoot tips do not osmotically adjust to increased water stress to the extent of mature leaves.[29] The result is that when stress develops, new leaf production slows or ceases while the existing canopy adapts to tolerate the stress. As stress reaches severe levels, leaf abscission occurs, which can reduce leaf area dramatically.

b. Maintenance of Water Uptake

With declining soil water availability, water stress in the top of the tree can be avoided by increased root exploration for water or by reducing the hydraulic resistance of the root system. Unfortunately, there has been little research on responses of apple root systems to water deficits.[4] It has been found, however, that the hydraulic conductivity of apple root systems can change in relation to the apparent evaporative demand on the top. Two-year-old container-grown apple trees growing outdoors for the first nine weeks of the second season had about three times the root conductivity of similar trees growing in a humid glasshouse for the same period.[237,238] This difference led to higher Ψ_l for the outdoor plants when both treatments were placed together in a high evaporative demand. Perhaps this adaptation may contribute to the wide adaptability of the apple to varying climates.

3. Drought Tolerance at Low Water Potentials

a. Maintenance of Turgor

Maintaining cell turgor is the primary way of tolerating unavoidably low Ψw. Osmotic adjustment or changes in the physical properties of cells, such as elasticity or cell size, are the most common mechanisms to maintain turgor (see Jones[17] for general review).

Osmotic adjustment refers to the changes, in either direction, of $\Psi\pi$ within a cell or tissue. Passive osmotic adjustment refers to the changes in $\Psi\pi$ due solely to dehydration of cells leading to a concentration of existing solutes. Active osmotic adjustment refers to the changes that occur by active accumulations of produced or imported solutes in cells. Usually, active adjustment is measured at 100% RWC to eliminate the passive dehydration effect. Osmotic adjustment of mature leaves is a striking characteristic of the apple compared to many crops. The amount of active osmotic adjustment has been found to be up to 2.5 MPa over a season[29,219,220] (Figure 12). The adjustment is apparently reversible, since the seasonal pattern of $\Psi\pi$ varies from year to year with different rainfall patterns (Figure 12). Osmotic adjustment maintains turgor to maintain leaf function such as g_s and photosynthesis with changes in water stress levels.[29,221] The advantage of this adaptation is that the apple is able to tolerate mild to moderate droughts while maintaining canopy photosynthesis. The disadvantage of osmotic adjustment is that if the drought is very prolonged, the apple may continue to deplete its soil water resources over too long a period. Species that are more conservative and sacrifice productivity by stomatal closure to conserve water earlier would survive prolonged droughts better. Since under orchard conditions prolonged severe droughts are not allowed to develop in commercial orchards, osmotic adjustment may allow the apple to tolerate the occasional droughts that do occur in non-irrigated orchards.

Carbohydrates appear to be the primary osmotic solutes that change with increased water stress.[220,239] In an excellent detailed study of the biochemistry of osmotic adjustment in apple, Wang and Stutte[239] found that water stress induced the accumulation of sorbitol in apple leaves while starch levels declined. Sorbitol accounted for most of the osmotic adjustment, although glucose and fructose accumulated to some extent. It appears that water stress induces the preferential role of sorbitol as the osmoticum. Since sorbitol has been used as the osmoticum of choice for enzyme and organelle extractions for many years, it is suggested that it plays a role as a compatible solute (a solute that has minimal effect on metabolism as it accumulates). A fascinating report from research on desert plants indicates that sorbitol accumulation may also confer high-temperature resistance.[240] Clearly, more research is needed in this area to determine if there are interactions among multiple stresses.

4. Cell Elasticity and Size

Changes in cellular physical properties, such as elasticity, have not been examined in great detail in apple. The studies that have been done indicate that there may be changes in bulk leaf elasticity, but that the changes are not enough to make a significant effect on the water relations of apple leaves.[11,220,241,242] The role of cell size in apple water relations has not apparently been determined.

V. FLOODING

The detrimental effects of flooding are not due to excess water, but due to lack of oxygen (anoxia) and possibly the accumulation of soil CO_2. The success of hydroponics with aeration demonstrates this point. In general, there has been relatively little work on flooding in apple, not because of the lack of effect on the tree, but because of improvements in site selection and affordable tile drainage. Also, economic pressures over the years have provided selection pressure on growers that undoubtedly has eliminated many orchards in poor sites with poor production due to frosts, flooding, or drought. Consequently, severe problems of flooding are uncommon in most apple industries, but short periods of flooding with less dramatic effects may still occur.

There have been several thorough reviews of flooding in plants, both for woody plants in general[243] and for fruit crops.[9,244] The reader is referred to these excellent reviews. This discussion will briefly summarize the general findings and what is known about flooding effects on apple tree physiology.

A. GROWTH AND PRODUCTIVITY

Apple has generally been characterized as having moderate tolerance to flooding compared to other fruit crops.[9,244] Childers and White[245] found that within several days of flooding E, photosynthesis, leaf growth, and root growth were reduced and leaf respiration was increased.

The timing and duration of flooding are important. Flooding for short periods of less than a month during the winter or early spring when the trees are dormant is common, yet appears to have little or no obvious effects on apple trees.[246] Once growth begins, the sensitivity to flooding increases dramatically, with one week of flooding reducing young leaf expansion and increasing wilting.[246] Olien[248] and Olien and Lakso[247] flooded cropping trees for six weeks in the spring, summer, and fall and found that spring and summer flooding were most detrimental to vegetative growth and yield, with spring flooding reducing yield most dramatically. The tree water relations reflected apparent reductions in the hydraulic conductance of the root system as the g_s and E decreased, but the Ψ_l remained the same or became more negative. Leaf N, P, K, and Cu levels on a dry weight basis in mid-season were reduced. Effects tended to accumulate over three years of flooding. The effects of flooding, low O_2, or high CO_2 on nutrients are rather inconsistent, and are likely confounded by different amounts of growth reduction and possibly different dry matter accumulations in the leaves.[247,248] Overall, reduced growth and gas exchange, and changes in water relations appropriate to lower root hydraulic conductances, reductions in nutrient uptake, and poorer set and yield reflect reductions in root system growth and function. The sensitivity of the tree to flooding appears to correlate to the periods of most active growth in the spring and summer, as might be expected. In these responses, apple appears consistent with other fruit crops.[9,244] Even after the soil drains, there are carry-over effects of the damage to the root system on growth and physiology.

The detailed physiological effects of flooding on the regulation of growth and gas exchange processes have not apparently been examined in apple, but have been studied in many plants.[9,243,244]

B. DIFFERENCES IN ROOTSTOCK RESPONSE

Due to the long use of clonal rootstocks in apple, there have been many observations and/or tests of rootstock differences to flooding. As with any complex response, the rankings show a great range of resistance, but have not been consistent. The rankings are summarized in the review by Schaffer et al.,[244] and more detailed discussion of rootstock responses can be found in Cummins and Aldwinckle.[185] The physiological bases of these variations by rootstock are not known, although the high-temperature effects shown by Gur et al.[155] earlier may provide some insights.

C. SENSITIVITY AND RESPONSE TO PATHOGENS

A major concern with flooding, in addition to direct effects on the tree, is the increase in root pathogens, especially *Phytophthora* species, especially *Phytophthora cactorum.*[185,244] It is possible that some of the detrimental effects of short-term spring flooding may be due more to providing an infection period for pathogens than to the direct effects on the tree. In normal orchard conditions, it may be difficult to separate the effects of flooding and pathogens. Clearly, site selection and drainage that avoid flooding may help eliminate two related problems. Very little is known about the interactions of pathogen, environment, and host plant response in this case.

VI. OTHER ENVIRONMENTAL FACTORS

A. SALINITY

Apple trees are generally considered to be quite sensitive to salinity compared to other fruit crops. Yet, there is relatively little work on salinity effects on apple. This is likely due to the fact that apples have historically been grown in somewhat cooler, humid, or sub-humid areas or in arid regions that have used irrigation waters from mountainous areas. Salinity may become more of a problem in the future if apples are planted in more arid climates with limited quantities of high-quality irrigation water.

Long-term studies in Germany found that: (1) Na and Cl contents of apple leaves increased with treatment levels while K decreased,[249] (2) yields were reduced via early season effects of Na treatment on fruit size,[250] (3) leaf photosynthesis and E were reduced by Na treatment,[251] (4) in general, more damage was caused by early season treatment, and (5) the effects were more severe on trees grafted on M9 compared to M26 rootstock. Simultaneous treatments with K did not affect Na accumulation in the leaves or the water consumption by the tree.[252,253] The M9 and M26 rootstocks differed in the relative distribution of Na. The lack of many reports of long-term effects in the field leave the question of salinity effects open.

B. POLLUTANTS

1. Air Pollutants

A significant effort has gone into examining air pollutant effects on horticultural crops, but mostly on herbaceous crops.[254] Apple is responsive to both ozone and sulfur dioxide, with high doses of each leading to severe foliar injury, leaf abscission, and reductions in g_s and growth in greenhouse-grown trees.[255] The injury when both pollutants were present was less than additive, since apparently each pollutant caused some stomatal closure that reduced the sensitivity to the other pollutant. Using lower ozone levels closer to ambient in outdoor chambers, Retzlaff et al.[256] found that apple leaf photosynthesis and g_s declined linearly with increasing ozone up to 0.12 ppm, but that no visual symptoms appeared and only slight reductions in trunk growth rates were found. Since no visual symptoms were apparent, yet photosynthesis was reduced, it is not safe to assume that apple does not suffer from ozone injury in the field. Compared to the other fruit crops, apple was moderate in response.

2. Acid Deposition

Since acid deposition has become a concern in many areas of the world where apples are grown, a few studies have examined its effects on apple. Fortunately, apple seedlings or trees in the field are relatively resistant to the effect of acid rain unless there are extended treatments with pHs of 3 or less.[257–260] When the lower pHs did cause lesions, the reductions in dry weight growth and photosynthetic rates were approximately proportional to the percent leaf area damaged. Yield or fruit quality in the field do not appear to suffer due to ambient acid rain episodes.[257]

C. ATMOSPHERIC CO_2

Predictions of global climate change are currently controversial because of the complexity of such climate models and the presence of natural global temperature cycles. The most agreed-upon change is the increase in atmospheric CO_2 that is much more under the control (or lack of control!) of humans. Unless there is an increase in political determination, the CO_2 concentration in the atmosphere is likely to continue to rise. At the current increase rates per year,[261] the atmospheric CO_2 would reach 425–475 ppm within about 40–50 years. Despite the extensive research being conducted on CO_2 effects on other crops, the author is aware of little being done on apples at this time. The reasons are several, but the major problems with experiments in the field are adequate replication due to tree size, cost, and the long-term enrichments needed to determine long-term effects.

An early study with seedlings found significant increases in shoot growth with elevated CO_2.[262] A study of CO_2 enrichment (CDE) of cropping versus non-cropping young apple trees in chambers in a greenhouse in Denmark showed little effect of elevated CO_2 on non-cropping trees.[263] The effects on cropping trees were more striking, especially when supplemental lighting was provided. However, the CDE did not begin until after fruit numbers and potential sink demand, i.e., fruit cell numbers, were already established, thus the potential response was limited to increases in cell size only. Hansen[264] enriched container-grown trees in growth cabinets for several weeks at the start of the season and found that the tree began dependence on current photosynthates by about 3 weeks after budbreak. The effects were primarily on extension shoot growth, since the spurs apparently developed earlier with reserves. Tree gas exchange was not measured in these early studies. Corelli Grappadelli and Magnanini[62] show a light response curve of apple canopy gas exchange in ambient and 900 ppm CO_2. The brief elevated CO_2 about doubled the net CO_2 uptake at full light, but the effect decreased at lower light levels even though enhancement still occurred at very low light and respiration was inhibited.

Will apple tree productivity respond very much to increased CO_2 levels? Not all crops respond equivalently to CO_2 enrichment.[265,266] The knowledge gained over the past decade indicates that some plants, especially in restrictive containers, demonstrate a compensatory decrease in gas exchange so that the net response to elevated CO_2 is small. It appears that much of the responsiveness to CO_2 enrichment is related to the source/sink balances and the ability of the plant to respond to the increased source with increased sink numbers or activity, as Kramer reviewed in 1981.[267] The few other longer term studies with fruit crops have evaluated citrus and have shown marked increases in growth and yields with little evidence of reductions in photosynthetic responses to CO_2[268–272] (see Chapter 4, Volume II). This suggests that fruit trees that have a supply of sinks may be very responsive to CO_2 enrichment. Clearly, such studies must be done with apple trees.

The potential of response of apple to elevated CO_2 may depend on how much additional vegetative growth may occur and how many additional fruits may be produced. Studies of the carbon balance of apple trees and potential limitations of carbon supply to fruit[5,56] suggest that apple fruit production may be quite responsive to increased CO_2. If additional fruit are set due to increased carbon supply, then the increased sink demand may maintain the photosynthetic sensitivity to CO_2. There may be many interactions that may not be obvious. Increased growth may lead to denser canopies, causing changes in canopy microclimate, fruit quality, or need for pruning. It will be extremely interesting to observe the progress in this area.

VII. CONCLUSIONS

In conclusion, the apple is an adaptable, productive species that poses a major challenge to understand, especially if the goal is to improve agricultural productivity. Concerning environmental physiology, the apple is essentially a wild-type plant that has been changed little by breeding. It appears to draw its inherent productivity from long and adaptable leaf area duration, moderate photosynthetic rates, and stress tolerance mechanisms, such as good WUE, plasticity of canopy development, osmotic adjustment, and slow physiological aging. When considering the evolutionary pressures of occasional stress periods, it is difficult to imagine a better set of adaptations. Fortunately for humans, these adaptations have led to many mechanisms of good sustained productivity as long as we can understand how to keep the tree from switching from the fruit-producing mode to the fruit-abscising survival mode.

A. NEEDS FOR FUTURE RESEARCH

To make progress in improving the productivity and efficiency of the apple in the orchard, it will be increasingly necessary to develop a sound understanding of the physiology of the plant. Progress was made in the past by overcoming major limitations of nutrients by fertilization, water by irrigation, pest damage by chemical pest management, erratic production by thinning, and low yields by improved pruning, training, and orchard design. Progess in the future will require a much more subtle set of changes or improvements giving smaller increments of productivity. Physiology will need to change from the explanatory science that primarily follows practice to a science that understands principles and leads practice. This mandate within the areas covered in this review will require a crop physiology approach that evaluates the problem, determines the likely principles involved and the level of the problem (i.e., orchard, tree, organ, tissue, cell, or molecular), develops good testable hypotheses, and uses whatever techniques are necessary to conduct the research. Clearly, a broad view of the entire physiology and the environment is needed. Modeling will become more important as a means to integrate information more efficiently and to estimate some potential outcomes to evaluate.

Many more studies that evaluate interactions of multiple stresses and that integrate the different physiological process (especially in carbon, water, or nutrient balances) are needed. Good examples are the effects of increasing CO_2 and of foliar biotic stresses on productivity and fruit quality. In the latter case, effective chemical pest management for a generation has made biotic stress physiology almost irrelevant. But with the loss of available pesticides, modern pest management practices are founded on the concept that the tree can tolerate biotic stresses to some level called the "threshold" before reductions in yield or quality occur. This requires a detailed understanding of stress physiology to be able to utilize as few pesticides as possible while maintaining the risk for the grower at a reasonable level. Recent studies in our laboratory suggest that the effects of some foliar pests may be mediated via the carbon balance of the tree; thus, many other factors that affect the carbon balance (radiation, crop load, temperature, CO_2, and other stresses) are expected to interact with pest thresholds.

The challenges outlined are not easy, but the high value of the apple and other fruit crops allows unique opportunities to put physiological understanding into practice to improve fruit production and quality for everyone.

ACKNOWLEDGMENTS

The author would like to acknowledge the many and varied contributions to this chapter of many generous colleagues, especially L. Corelli Grappadelli, J. A. Flore, T. L. Robinson, R. S. Johnson, J. Wünsche, M. Al-hazmi, A. H. D. Francisconi, B. Schaffer, P. Andersen, and E. Lakso.

REFERENCES

1. Dennis, F. G., Apple, in *Handbook of Fruit Set and Development,* S. P. Monselise, Ed., CRC Press, Boca Raton, FL, 1986, 1.
2. Forshey, C. G. and Elfving, D. C., The relationship between vegetative growth and fruiting in apple trees, *Hortic. Rev.,* 11, 229, 1989.
3. Jackson, J. E., Light interception and utilization by orchard systems, *Hortic Rev.,* 2, 208, 1980.
4. Landsberg, J. J. and Jones, H. G., Apple Orchards, in *Water Deficits and Plant Growth,* T. T. Kozlowski, Ed., Academic Press, New York, 1981, 419.
5. Lakso, A. N., Robinson, T. L., and Pool, R. M., Canopy microclimate effects on patterns of fruiting and fruit development in apples and grapes, in *Manipulation of Fruiting,* C. J. Wright, Ed., Butterworths, London, 1989, 263.
6. Palmer, J. W., Canopy manipulation for optimum utilisation of light., in *Manipulation of Fruiting,* C. J. Wright, Ed., Butterworths, London, 1989, 245.
7. Pratt, C. S., Apple flower and fruit: morphology and anatomy, *Hortic. Rev.,* 10, 273, 1988.
8. Pratt, C. S., Apple trees: morphology and anatomy, *Hortic. Rev.,* 12, 265, 1990.
9. Rowe, R. N. and Beardsell, D. V., Waterlogging of fruit trees, *Hortic. Abstr.,* 43, 533, 1973.
10. Westwood, M., *Temperate Zone Pomology,* W. H. Freeman, San Francisco, 1978.
11. Jones, H. G., Lakso, A. N., and Syvertsen, J. P., Physiological control of water status in temperate and subtropical fruit trees, *Hortic. Rev.,* 7, 301, 1985.
12. Faust, M., *Physiology of Temperate Zone Fruit Trees,* Wiley-Interscience, New York, 1989.
13. Flore, J. A. and Lakso, A. N., Environmental and physiological regulation of photosynthesis in fruit crops, *Hortic. Rev.,* 11, 111, 1989.

14. Pearcy, R. W., Ehleringer, J., Mooney, H. A., and Rundel, P. W., *Plant Physiological Ecology, Field Methods and Instrumentation,* Chapman and Hall, London, 1989.
15. Campbell, G. S., Soil water potential measurement: an overview, *Irrig. Sci.,* 9, 265, 1988.
16. Coombs, J., Hall, D. O., Long, S. P., and Scurlock, J. M. O., *Techniques in Bioproductivity and Photosynthesis,* Pergamon Press, Oxford, 1985.
17. Jones, H. G., *Plants and Microclimate,* Cambridge University Press, Cambridge, 1992.
18. Turner, N. C., Measurement of plant water status by the pressure chamber technique, *Irrig. Sci.,* 9, 289, 1988.
19. Monteith, J. L., Does light limit crop productivity?, in *Physiological Processes Limiting Plant Productivity,* C. B. Johnson, Ed., Butterworths, London, 1981, 23.
20. Janick, J., The apple in Java, *HortScience,* 9, 13, 1974.
21. Meland, M., personal communication, 1988.
22. Oliveira, C. M. and Priestley, C. A., Carbohydrate reserves in deciduous fruit trees, *Hortic. Rev.,* 10, 403, 1988.
23. Porpiglia, P. J. and Barden, J. A., Seasonal trends in net photosynthetic potential, dark respiration, and specific leaf weight of apple leaves as affected by canopy position, *J. Am. Soc. Hortic. Sci.,* 105, 920, 1980.
24. Monteith, J. L., Climate and the efficiency of crop production in Britain, *Phil. Trans. R. Soc. London B,* 281, 277, 1977.
25. Greene, D. W., Craker, L. E., Brooks, C. K., Kadkade, P., and Bottecelli, C., Inhibition of fruit abscission in apple with night-break red light, *HortScience,* 21, 247, 1986.
26. Arakawa, O., Characteristics of color development in some apple cultivars changes in anthocyanin synthesis during maturation as affected by bagging and light quality, *J. Jpn. Soc. Hortic. Sci.,* 57, 373, 1988.
27. Jonkers, H., Autumnal leaf abscission in apple and pear, *Fruit Sci. Rep.,* 7, 25, 1980.
28. Lakso, A. N. and Lenz, F., Regulation of apple tree photosynthesis in the autumn by temperature, in *Regulation of Photosynthesis in Fruit Trees,* A. Lakso and F. Lenz, Eds., N.Y. State Agric. Exp. Sta. Spec. Publ., 1986, 34.
29. Lakso, A. N., Geyer, A. S., and Carpenter, S. G., Seasonal osmotic relations in apple leaves of different ages, *J. Am. Soc. Hortic. Sci.,* 109, 544, 1984.
30. Forshey, C. G., Weires, R. W., Stanley, B. H., and Seem, R. C., Dry weight partitioning of 'McIntosh' apple, *J. Am. Soc. Hortic. Sci.,* 108, 149, 1983.
31. Forshey, C. G., Weires, R. W., and Vankirk, J. R., Seasonal development of the leaf canopy of 'MacSpur Mcintosh' apple trees, *HortScience,* 22, 881, 1987.
32. McKenzie, D. W. and Rae, A. N., Economics of high density apple production in New Zealand, *Acta Hortic.,* 65, 41, 1978.
33. Loomis, R. S. and Gerakis, P. A., Productivity of agricultural ecosystems, in *Photosynthesis and Productivity in Different Environments,* J. P. Cooper, Ed., Cambridge University Press, Cambridge, 1975, 145.
34. Penning de Vries, F. W. T., The cost of maintenance processes in plant cells, *Ann. Bot.,* 39, 77, 1975.
35. Penning de Vries, F. W. T., Use of assimilates in higher plants, in *Integration of Activity in the Higher Plant,* D. H. Jennings, Ed., Symp. Soc. Exp. Biol., 1977, 459.
36. Palmer, J. W., Annual dry matter production and partitioning over the first 5 years of a bed system of Crispin-M.27 apple trees at four spacings, *J. Appl. Ecol.,* 25, 569, 1988.
37. Lakso, A. N., unpublished data, 1989.
38. Bollard, E. G., The physiology and nutrition of developing fruits, in *The Biochemistry of Fruits and Their Products,* A. C. Hulme, Ed., Academic Press, London, 1970, 387.
39. Blanpied, G. D., Changes in the weight, volume and specific gravity of developing apple fruits, *Proc. Am. Soc. Hortic. Sci.,* 88, 33, 1966.
40. Goudriaan, J. and Monteith, J. L., A mathematical function for crop growth based on light interception and leaf area expansion, *Ann. Bot.,* 66, 695, 1990.
41. Vavilov, N. I., Wild progenitors of the fruit trees of Turkistan and the Caucasus and the problem of origin of the fruit trees, *Proc. 9th Intl. Hortic. Congress, Group B,* 271, 1931.
42. Hansen, P., The effect of cropping on the distribution of growth in apple trees, *Tidsskr. Planteavl,* 75, 119, 1971.

43. Maggs, D. H., The reduction in growth of apple trees brought about by fruiting, *J. Hortic. Sci.,* 38, 119, 1963.
44. Zucconi, F., Regulation of abscission in growing fruit, *Acta Hortic.,* 120, 89, 1981.
45. Luckwill, L. C., Hormones and the productivity of fruit trees, *Scientific Hortic.,* 31, 60, 1980.
46. Lang, G. A., Dormancy—models and manipulations of environmental/physiological regulation, in *Manipulation of Fruiting,* C. J. Wright, Ed., Butterworths, London, 1989, 79.
47. Scurlock, J. M. O., Long, S. P., Hall, D. O., and Coombs, J., Introduction, in *Techniques in Bioproductivity and Photosynthesis,* J. Coombs, D. O. Hall, S. P. Long and J. M. O. Scurlock, Eds., Pergamon Press, Oxford, 1985, xxi–xxiv.
48. Russell, G., Jarvis, P. G., and Monteith, J. L., Absorption of radiation by canopies and stand growth, in *Plant Canopies: Their Growth, Form and Function,* G. Russell, B. Marshall and P. G. Jarvis, Eds., Cambridge University Press, Cambridge, 1989, 21.
49. Demetriades-Shah, T. H., Fuchs, M., Kanemasu, E. T., and Flitcroft, I., A note of caution concerning the relationship between cumulated intercepted solar radiation and crop growth, *Agric. For. Meteorol.,* 58, 193, 1992.
50. Palmer, J. W., Avery, D. J., and Wertheim, S. J., Effect of apple tree spacing and summer pruning on leaf area distribution and light interception, *Sci. Hortic.,* 52, 303, 1992.
51. Johnson, R. S. and Lakso, A. N., Approaches to modeling light interception in orchards, *HortScience,* 26, 1002, 1991.
52. Wagenmakers, P. S., Planting systems for fruit trees in temperate climates, *Crit. Rev. Plant Sci.,* 10, 369, 1991.
53. Charles-Edwards, D. A., *Physiological Determinants of Crop Growth,* Academic Press, Sydney, 1982.
54. Jackson, J. E. and Palmer, J. W., A computer model study of light interception by orchards in relation to mechanised harvesting and management, *Scientia Hortic.,* 13, 1, 1980.
55. Middleton, S. G. and Jackson, J. E., Solid model estimation of light interception by apple orchard systems, *Acta Hortic.,* 240, 83, 1989.
56. Lakso, A. N., The simplified dry matter production model for apple: estimation of canopy photosynthesis in discontinuous canopies, *Acta Hortic.,* 313, 45, 1992.
57. Jackson, J. E., Utilization of light resources by high density planting systems, *Acta Hortic.,* 65, 61, 1978.
58. Robinson, T. L., Lakso, A. N., and Carpenter, S. G., Canopy development yield and fruit quality of 'Empire' and 'Delicious' apple trees grown in four orchard production systems for ten years, *J. Am. Soc. Hortic. Sci.,* 116, 179, 1991.
59. Robinson, T. L., Lakso, A. N., and Ren, Z., Modifying apple tree canopies for improved production efficiency, *HortScience,* 26, 1005, 1991.
60. Robinson, T. L. and Lakso, A. N., Bases of yield and production efficiency in apple orchard systems, *J. Am. Soc. Hortic. Sci.,* 116, 188, 1991.
61. Palmer, J. W., The effects of row orientation tree height time of year and latitude on light interception and distribution in model apple hedgerow canopies, *J. Hortic. Sci.,* 64, 137, 1989.
62. Corelli Grappadelli, L. and Magnanini, E., A whole tree system for gas exchange studies, *HortScience,* 28, 41, 1993.
63. Heinicke, A. J. and Childers, N. F., The daily rate of photosynthesis, during the growing season of 1935, of a young apple tree of bearing age, *Cornell University Agricultural Experiment Station Memoir,* 201, 1937.
64. Proctor, J. T. A., Watson, R. L., and Landsberg, J. J., The carbon balance of a young apple tree, *J. Am. Soc. Hortic. Sci.,* 101, 579, 1976.
65. Sirois, D. J. and Cooper, G. R., The influence of light intensity, temperature and atmospheric carbon dioxide concentration of the rate of apparent photosynthesis of a mature apple tree, *Maine Agric. Experiment Station Bulletin,* 626, 1964.
66. Allen, L. H. J., Stewart, D. W., and Lemon, E. R., Photosynthesis in plant canopies: effect of light response curves and radiation source geometry, *Photosynthetica,* 8, 184, 1974.
67. Buwalda, J. G., Fossen, M., and Lenz, F., Carbon dioxide efflux from roots of calamondin and apple, *Tree Physiol.,* 10, 391, 1992.
68. Lakso, A. N. and Corelli Grappadelli, L., Implications of pruning and training practices to carbon partitioning and fruit development in apple, *Acta Hortic.,* 322, 231, 1992.
69. Magein, H., Growth and abscission dynamics of Cox's Orange Pippin and Golden Delicious apple fruits, *J. Hortic. Sci.,* 64, 265, 1989.

70. Blanpied, G. D. and Wilde, M. H., A study of the cells in the outer flesh of developing McIntosh apple fruits, *Bot. Gaz.,* 129, 173, 1968.
71. Byers, R. E., Lyons, C. G., Jr., Yoder, K. S., Barden, J. A., and Young, R. W., Peach and apple thinning by shading and photosynthetic inhibition, *J. Hortic. Sci.,* 60, 465, 1985.
72. Byers, R. E., Barden, J. A., and Carbaugh, D. H., Thinning of spur 'Delicious' apples by shade, terbacil, carbaryl and ethephon, *J. Am. Soc. Hortic. Sci.,* 115, 9, 1990.
73. Byers, R. E., Carbaugh, D. H., Presley, C. N., and Wolf, T. K., The influence of low light on apple fruit abscission, *J. Hortic. Sci.,* 66, 7, 1991.
74. Schneider, G. H., Abscission mechanism studies with apple fruitlets, *J. Am. Soc. Hortic. Sci.,* 103, 455, 1978.
75. Heinicke, A. J., Factors influencing the abscission of flowers and partially developed fruits of the apple (*Pyrus malus* L.), *Cornell University Agricultural Experiment Station Bulletin,* 393, 1917.
76. Jones, H. G., Physiological aspects of the control of water status in horticultural crops, *HortScience,* 25, 19, 1990.
77. Barden, J. A., Apple leaves, their morphology and photosynthetic potential, *HortScience,* 13, 644, 1978.
78. Watson, R. L. and Landsberg, J. J., The photosynthetic characteristics of apple leaves (cv. Golden Delicious) during their early growth, in *Photosynthesis and Plant Development,* R. Marcelle, H. Clijsters and M. V. Poucke, Eds., Dr. W. Junk, The Hague, 1979, 39.
79. Lakso, A. N., Robinson, T. L., and Carpenter, S. G., The palmette leader, a tree design for improved light distribution, *HortScience,* 24, 271, 1989.
80. Rom, C. R., Light thresholds for apple tree canopy growth and development, *HortScience,* 26, 989, 1991.
81. Barritt, B. H., Rom, C. R., Konishi, B. J., and Dilley, M. A., Light level influences spur quality and canopy development and light interception influence fruit production in apple, *HortScience,* 26, 993, 1991.
82. Robinson, T. L., Seeley, E. J., and Barritt, B. H., Effect of light environment and spur age on 'Delicious' apple fruit size and quality, *J. Am. Soc. Hortic. Sci.,* 108, 855, 1983.
83. Lakso, A. N., Correlations of fisheye photography to canopy structure, light climate and biological responses to light in apple trees, *J. Am. Soc. Hortic. Sci.,* 105, 43, 1980.
84. Johnson, R. S. and Lakso, A. N., Carbon balance model of a growing apple shoot. II. Simulated effects of light and temperature on long and short shoots, *J. Am. Soc. Hortic. Sci.,* 111, 164, 1986.
85. Hansen, P., ^{14}C-Studies on apple trees. VII. The early seasonal growth in leaves, flowers and shoots as dependent upon current photosynthates and existing reserves, *Physiol. Plant.,* 25, 469, 1971.
86. Quinlan, J. D., The pattern of distribution of ^{14}carbon in a potted apple rootstock following assimilation of ^{14}carbon dioxide by a single leaf, *Rep. E. Malling Res. Sta. for 1964,* 117, 1965.
87. Tustin, S., Corelli, G. L., and Ravaglia, G., Effect of previous-season and current light environments on early-season spur development and assimilate translocation in Golden Delicious apple, *J. Hortic. Sci.,* 67, 351, 1992.
88. Noga, G. and Lenz, F., Einfluβ von verschiedenen Klimafaktoren auf den CO_2-Gaswechsel von Äpfeln während der Light- und Dunkelperiode, *Gartenbauwissenschaft,* 47, 193, 1982.
89. Noga, G., and Lenz, F., Transpiration von Äpfeln während der Light- und Dunkelperiode in Abhängigkeit von verschiedenen Klimafaktoren, *Gartenbauwissenschaft,* 47, 274, 1982.
90. Jones, H. G., Carbon dioxide exchange of developing apple (*Malus pumila* Mill.) fruits, *J. Exp. Bot.,* 32, 1203, 1981.
91. Clijsters, H., On the photosynthetic activity of developing apple fruits, *Qual. Plant. Mater. Veg.,* 19, 129, 1969.
92. Cordes, D. and Lenz, F., Carbon dioxide gas exchange of apple fruits after defoliation, *Gartenbauwissenschaft,* 53, 133, 1988.
93. Blanke, M. M. and Lenz, F., Fruit photosynthesis, *Plant Cell Environ.,* 12, 31, 1989.
94. Neumann, P. M. and Stein, Z., Xylem transport and the regulation of leaf metabolism, *What's New Plant Physiol.,* 14, 33, 1983.
95. Neumann, P. M. and Stein, Z., Relative rates of delivery of xylem solute to shoot tissues: Possible relationship to sequential leaf senescence, *Physiol. Plant.,* 62, 390, 1984.
96. Nobel, P. S., *Physicochemical and Environmental Plant Physiology,* Academic Press, San Diego, 1991.
97. Ferree, D. C. and Forshey, C. G., Influence of pruning and urea sprays on growth and fruiting of spur-bound delicious apple trees, *J. Am. Soc. Hortic. Sci.,* 113, 699, 1988.

98. DeJong, T. M. and Doyle, J. F., Seasonal relationships between leaf nitrogen content photosynthetic capacity and leaf canopy light exposure in peach (*Prunus persica*), *Plant Cell Environ.*, 8, 701, 1985.
99. DeJong, T. M. and Day, K. R., Relationships between shoot productivity and leaf characteristics in peach canopies, *HortScience,* 26, 1271, 1991.
100. Klein, I., Weinbaum, S. A., DeJong, T. M., and Muraoka, T. T., Relationship between fruiting specific leaf weight and subsequent spur productivity in walnut, *J. Am. Soc. Hortic. Sci.,* 116, 426, 1991.
101. Klein, I., DeJong, T. M., Weinbaum, S. A., and Muraoka, T. T., Specific leaf weight and nitrogen allocation responses to light exposure within walnut trees, *HortScience,* 26, 183, 1991.
102. Mooney, H. A. and Gulmon, S. L., Environmental and evolutionary constraints on the photosynthetic characteristics of higher plants, in *Topics in Plant Population Biology,* D. T. Solbrig, S. Jain, G. B. Johnson and P. H. Raven, Eds., Columbia University Press, New York, 1979, 316.
103. Thorpe, M. R., Radiant heating of apples, *J. Appl. Ecol.,* 11, 755, 1974.
104. Buban, T. and Faust, M., Flower bud induction in apple trees: internal control and differentiation, *Hortic. Rev.,* 4, 174, 1982.
105. Miller, S. S., Plant bioregulators in apple and pear culture, *Hortic. Rev.,* 10, 309, 1988.
106. Sedgley, M., Flowering of deciduous perennial fruit crops, *Hortic. Rev.,* 12, 223, 1990.
107. Rohozinski, J., Edwards, G. R., and Hoskyns, P., Effects of brief exposure to nitrogenous compounds on floral initiation in apple trees, *Physiol. Veg.,* 24, 673, 1986.
108. Skogerbo, G., Effects of root pruning and trunk girdling on xylem cytokinin content of apple, *Norw. J. Agric. Sci.,* 6, 499, 1992.
109. Skogerbo, G., Effects of xylem cytokinin application on flower bud development in apple, *Norw. J. Agric. Sci.,* 6, 473, 1992.
110. Tromp, J. and Ovaa, J. C., Seasonal changes in the cytokinin composition of xylem sap of apple, *J. Plant Physiol.,* 136, 606, 1990.
111. Gao, Y. P., Motosugi, H., and Sugiura, A., Rootstock effects on growth and flowering in young apple trees grown with ammonium and nitrate nitrogen, *J. Am. Soc. Hortic. Sci.,* 117, 446, 1992.
112. Hansen, P. and Grauslund, J., Flower bud formation on apple trees. Some effects of cropping, growth and climate, *Tidsskr. Planteavl,* 84, 215, 1980.
113. Lakso, A. N., unpublished data, 1984
114. Browning, G., The physiology of fruit set, in *Manipulation of Fruiting,* C. J. Wright, Ed., Butterworths, London, 1989, 195.
115. Levy, E. B. and Madden, E. A., The point method of pasture analysis, *N.Z. J. Agric.,* 46, 267, 1933.
116. Warren Wilson, J., Inclined point quadrats, *New Phytol.,* 59, 1, 1960.
117. Warren Wilson, J., Stand structure and light penetration. I. Analysis by point quadrats, *J. Appl. Ecol.,* 2, 383, 1965.
118. Vanderbilt, V. C., Bauer, M. E., and Silva, L. F., Prediction of solar irradiance distribution in a wheat canopy using laser technique, *Agric. Meteorol.,* 20, 147, 1979.
119. Wünsche, J. N., Lakso, A. N., Robinson, T. L., and Lenz, F., unpublished data, 1993.
120. Mattice, W. A., The relation of spring temperatures to apple yield, *Mon. Weather Rev.,* 56, 456, 1927.
121. Beattie, B. B. and Folley, R. R. W., Production variability in apple crops. II The long-term behavior of the English crop, *Scientia Hortic.,* 8, 325, 1978.
122. Jackson, J. E. and Hamer, P. J. C., The causes of year-to-year variation in the average yield of Cox's Orange Pippin apple in England, *J. Hortic. Sci.,* 55, 149, 1980.
123. Jackson, J. E., Hamer, P. J. C., and Wickenden, M. F., Effects of early spring temperatures on the set of fruits of Cox's Orange Pippin apple and year-to-year variation in its yields, *Acta Hortic.,* 139, 75, 1983.
124. Lakso, A. N., unpublished data, 1987.
125. Heim, G., Landsberg, J. J., Watson, R. L., and Brain, P., Eco-physiology of apple trees: dry matter production and partitioning by young 'Golden Delicious' apple trees in France and England, *J. Appl. Ecol.,* 16, 179, 1979.
126. Jonkers, H., personal communication.
127. Zatyko, I., [Autumn flower formation in apples and possibilities of influencing it.] (Hu), *Gyumolcstermesztes,* 1, 5, 1974.
128. Tromp, J. and Penders, L. H. M. M., Leaf diffusion resistance as affected by defruiting and ringing in apple, *Gartenbauwissenschaft,* 51, 11, 1986.
129. Amthor, J. S., *Respiration and Crop Productivity,* Springer Verlag, New York, 1989.

130. DeJong, T. M. and Goudriaan, J., Modeling peach fruit growth and carbohydrate requirements reevaluation of the double-sigmoid growth pattern, *J. Am. Soc. Hortic. Sci.,* 114, 800, 1989.
131. Walton, E. F. and DeJong, T. M., Estimation of the bioenergetic cost to grow a kiwifruit berry, *Acta Hortic.,* 276, 231, 1990.
132. DeJong, T. M. and Walton, E. F., Carbohydrate requirements of peach fruit growth and respiration, *Tree Physiol.,* 5, 329, 1989.
133. Butler, D. R. and Landsberg, J. J., Respiration rates of apple trees, estimated by CO_2-efflux measurements, *Plant Cell Environ.,* 4, 153, 1981.
134. Watson, R. L., Landsberg, J. J., and Thorpe, M. R., Photosynthetic characteristics of the leaves of 'Golden Delicious' apple trees, *Plant Cell Environ.,* 1, 51, 1978.
135. Lankes, C., Light and temperature effects on carbon dioxide gas exchange of apple trees during winter season, *Gartenbauwissenschaft,* 52, 9, 1987.
136. Lenz, F., Carbon dioxide-gas exchange of attached fruit in apples, *Acta Hortic.,* 240, 217, 1989.
137. Ebert, G., Blanke, M., and Lenz, F., Kontinuierliche Messung des Gasaustausches von Wurzeln, *Gartenbauwissenschaft,* 51, 212, 1986.
138. Ebert, G., and Lenz, F., Jahresverlauf der Wurzelatmung von Apfelbäumen und ihr Beitrag zur CO_2-Bilanz, *Gartenbauwissenschaft,* 56, 130, 1991.
139. Ebert, G., Die Wurzelatmung von Apfelbäumen. I. Einfluβ von Umweltfaktoren, *Erwerbsobstbau,* 33, 227, 1991.
140. Walton, E. F. and DeJong, T. M., Estimating the bioenergetic cost of a developing kiwifruit berry and its growth and maintenance respiration components, *Ann. Bot.,* 66, 417, 1990.
141. Hansen, P., Carbohydrate allocation, in *Environmental Effects on Crop Physiology,* J. J. Landsberg and C. V. Cutting, Eds., Academic Press, London, 1977, 247.
142. Evenari, M., Lange, O. L., Schulze, E.-D., Kappen, L., and Buschbom, U., Net photosynthesis, dry matter production, and phenological development of apricot trees (*Prunus armeniaca* L.) cultivated in the Negev highlands (Israel), *Flora,* 166, 383, 1977.
143. Lakso, A. N. and Johnson, R. S., A simplified dry matter production model for apple using automatic programming simulation software, *Acta Hortic.,* 276, 141, 1990.
144. Lakso, A. N. and Seeley, E. J., Environmentally induced responses of apple tree photosynthesis, *HortScience,* 13, 646, 1978.
145. Lange, O. L., Schulze, E.-D., Evenari, M., Kappen, L., and Buschbom, U., The temperature-related photosynthetic capacity of plants under desert conditions. I. Seasonal changes of the photosynthetic response to temperature, *Oecologia,* 17, 97, 1974.
146. Heinicke, A. J. and Childers, N. F., Influence of respiration on the daily rate of photosynthesis of entire apple trees, *Proc. Am. Soc. Hortic. Sci.,* 34, 142, 1936.
147. Wibbe, M. and Lenz, F., Wievel CO_2 nimmt ein Apfelbaum an heissen Tagen im Sommer auf?, *Erwerbsobstbau,* 31, 88, 1989.
148. Dennis, F. G., Factors affecting yield in apple, with emphasis on 'Delicious', *Hortic. Rev.,* 1, 395, 1979.
149. Fukui, H., Imakawa, S., and Tamura, T., Relation between early drop of apple fruit, ethylene evolution and formation of abscission layer, *J. Jpn. Soc. Hortic. Sci.,* 53, 303, 1984.
150. Kondo, S. and Takahashi, Y., Effects of high temperature in the nighttime and shading in the daytime on the early drop of apple fruit 'Starking Delicious', *J. Jpn. Soc. Hortic. Sci.,* 56, 142, 1987.
151. Tukey, L. D., Some effects of night temperatures on the growth of McIntosh apples, I, *Proc. Am. Soc. Hortic. Sci.,* 68, 32, 1956.
152. Tukey, L. D., Some effects of night temperatures on the growth of McIntosh apples, II, *Proc. Am. Soc. Hortic. Sci.,* 75, 39, 1960.
153. Tamura, T., Fukui, H., Imakawa, S., and Mino, Y., Effect of temperature at early stage of fruit development on the development of fruit and seed in apple, *J. Jpn. Soc. Hortic. Sci.,* 50, 287, 1981.
154. Kondo, S. and Takahashi, Y., Relation between early drop of apple fruit and ethylene evolution under high night temperature conditions, *J. Jpn. Soc. Hortic. Sci.,* 58, 1, 1989.
155. Gur, A., Bravdo, B., and Mizrahi, Y., Physiological responses of apple trees to supraoptimal root temperatures, *Physiol. Plant.,* 27, 130, 1972.
156. Flore, J. A. and Dennis, F. G., Jr., Disorders caused by environmental factors, in *Compendium of Apple and Pear Diseases,* A. L. Jones and H. S. Aldwinckle, Eds., American Phytopath. Society, St. Paul, MN, 1990, 84.
157. Samish, R. M., Dormancy in woody plants, *Annu. Rev. Plant Physiol.,* 5, 183, 1954.

158. Cannell, M. G. R., Chilling, thermal time and the date of flowering of trees, in *Manipulation of Fruiting*, C. J. Wright, Ed., Butterworths, London, 1989, 99.
159. Lang, G. A., Dormancy: a new universal terminology, *HortScience*, 22, 817, 1987.
160. Dennis, F. G., Jr., Two methods of studying rest: temperature alternation and genetic analysis, *HortScience*, 22, 820, 1987.
161. Fuchigami, L. H. and Nee, C.-C., Degree growth stage model and rest-breaking mechanisms in temperate woody perennials, *HortScience*, 22, 836, 1987.
162. Powell, L. E., Hormonal aspects of bud and seed dormancy in temperate-zone woody plants, *HortScience*, 22, 845, 1987.
163. Hauagge, R. and Cummins, J. N., Phenotypic variation of length of bud dormancy in apple cultivars and related *Malus* species, *J. Am. Soc. Hortic. Sci.*, 116, 100, 1991.
164. Jackson, J. E., Cropping and multiple cropping of apples in medium altitude tropical Zimbabwe, in *Abstracts of the 23rd International Horticulture Congress, Int. Soc. Hortic. Sci.*, Firenze, 1990, Abst # 4325.
165. Hauagge, R. and Cummins, J. N., Genetics of length of dormancy period in *Malus* vegetative buds, *J. Am. Soc. Hortic. Sci.*, 116, 121, 1991.
166. Hauagge, R. and Cummins, J. N., Age growing temperatures and growth retardants influence induction and length of dormancy in *Malus*, *J. Am. Soc. Hortic. Sci.*, 116, 116, 1991.
167. Hauagge, R. and Cummins, J. N., Seasonal variation in intensity of bud dormancy in apple cultivars and related *Malus* species, *J. Am. Soc. Hortic. Sci.*, 116, 107, 1991.
168. Hauagge, R. and Cummins, J. N., Relationships among indices for the end of bud dormancy in apple cultivars and related *Malus* species under cold winter conditions, *J. Am. Soc. Hortic. Sci.*, 116, 95, 1991.
169. Faust, M., Liu, D., Millard, M. M., and Stutte, G. W., Bound versus free water in dormant apple buds a theory for endodormancy, *HortScience*, 26, 887, 1991.
170. Liu, D., Norman, H. A., Stutte, G. W., and Faust, M., Lipase activity during endodormancy in leaf buds of apple, *J. Am. Soc. Hortic. Sci.*, 116, 689, 1991.
171. Young, E., Effects of rootstock and scion chilling during rest on resumption of growth in apple and peach, *J. Am. Soc. Hortic. Sci.*, 109, 548, 1984.
172. Young, E. and Werner, D. J., Effects of shoot root and shank chilling during rest in apple and peach on growth resumption and carbohydrates, *J. Am. Soc. Hortic. Sci.*, 110, 769, 1985.
173. Young, E. and Werner, D. J., 6 BA applied after shoot and-or root chilling and its effect on growth resumption in apple and peach, *HortScience*, 21, 280, 1986.
174. Young, E., Motomura, Y., and Unrath, C. R., Influence of root temperature during dormancy on respiration carbohydrates and growth resumption in apple and peach, *J. Am. Soc. Hortic. Sci.*, 112, 514, 1987.
175. Young, E., Winter root temperature effects of vegetative budbreak spur development and flowering of 'Delicious' apple, *J. Am. Soc. Hortic. Sci.*, 113, 301, 1988.
176. Young, E., Timing of high temperature influences chilling negation in dormant apple trees, *J. Am. Soc. Hortic. Sci.*, 117, 271, 1992.
177. Belding, R. D. and Young, E., Shoot and root temperature effects of xylary cytokinin levels during budbreak in young apple trees, *HortScience*, 24, 115, 1989.
178. Burke, M. J. and Gusta, L. V., Freezing and injury in plants, *Annu. Rev. Plant Physiol.*, 27, 507, 1976.
179. Ashworth, E. N. and Wisniewski, M. E., Response of fruit tree tissues to freezing temperatures, *HortScience*, 26, 501, 1991.
180. Ashworth, E. N., Deep supercooling in woody plant tissues, in *Advances in Plant Cold Hardiness*, P. H. Li and L. Christersson, Eds., CRC Press, Boca Raton, FL, 1993, 203.
181. Quamme, H. A., Stushnoff, C., and Weiser, C. J., The relationship of exotherms to cold injury in apple stem tissue, *J. Am. Soc. Hortic. Sci.*, 97, 608, 1972.
182. Quamme, H. A., Weiser, C. J., and Stushnoff, C., The mechanism of freezing injury in xylem of apple twigs, *Plant Physiol.*, 51, 273, 1973.
183. Ketchie, D. O. and Kammereck, R., Seasonal variation of cold resistance in *Malus* woody tissue as determined by differential thermal analysis and viability tests, *Can. J. Bot.*, 65, 2640, 1987.
184. Holubowicz, T., Cummins, J. N., and Forsline, P. L., Responses of *Malus* clones to programmed low-temperature stresses in late winter, *J. Am. Soc. Hortic. Sci.*, 107, 492, 1982.
185. Cummins, J. N. and Aldwinckle, H. S., Breeding apple rootstocks, *Plant Breed. Rev.*, 294, 1983.

186. Forsline, P. L., Winter hardiness of common New York apple varieties and rootstocks as determined by artificial freezing, *Proc. N.Y. State Hortic. Soc.*, 128, 20, 1983.
187. Ketchie, D. O. and Beeman, C. H., Cold acclimation in 'Red Delicious' apple trees under natural conditions during four winters, *J. Am. Soc. Hortic. Sci.*, 98, 257, 1973.
188. Howell, G. S. and Weiser, C. J., Fluctuations in the cold resistance of apple twigs during spring dehardening, *J. Am. Soc. Hortic. Sci.*, 95, 190, 1970.
189. Proebsting, E. L., Adapting cold hardiness concepts to deciduous fruit culture, in *Plant Cold Hardiness and Freezing Stress. Mechanisms and Crop Implications,* P. H. Li and A. Sakai, Eds., Academic Press, New York, 1978, 267.
190. Quamme, H. A. and Brownlee, R. T., Observation of winter injury to the roots of apple trees associated with sprinkler irrigation pattern, *Can. J. Plant Sci.*, 69, 617, 1989.
191. Ashworth, E. N., Echlin, P., Pearce, R. S., and Hayes, T. L., Ice formation and tissue response in apple twigs, *Plant Cell Environ.*, 11, 703, 1988.
192. Raese, J. T., Williams, M. W., and Billingsley, H. D., Cold hardiness, sorbitol, and sugar levels of apple shoots as influenced by controlled temperature and season, *J. Am. Soc. Hortic. Sci.*, 103, 796, 1978.
193. Coleman, W. K., Estabrooks, E. N., O'Hara, M., Embleton, J., and King, R. R., Seasonal changes in cold hardiness sucrose and sorbitol in apple trees treated with plant growth regulators, *J. Hortic. Sci.*, 67, 429, 1992.
194. Tyler, N. and Stushnoff, C., Dehydration of dormant apple buds at different stages of cold acclimation to induce cryopreservability in different cultivars, *Can. J. Plant Sci.*, 68, 1169, 1988.
195. Tyler, N. J. and Stushnoff, C., The effects of prefreezing and controlled dehydration on cryopreservation of dormant vegetative apple buds, *Can. J. Plant Sci.*, 68, 1163, 1988.
196. Tyler, N., Stushnoff, C., and Gusta, L. V., Freezing of water in dormant vegetative apple buds in relation to cryopreservation, *Plant Physiol.*, 87, 201, 1988.
197. Vertucci, C. W. and Stushnoff, C., The state of water in acclimating vegetative buds from malus and amelanchier and its relationship to winter hardiness, *Physiol. Plant.*, 86, 503, 1992.
198. Bervaes, J. C. A. M., Ketchie, D. O., and Kuiper, P. J. C., Lipid composition of pine needle chloroplasts and apple bark tissue as affected by growth temperature and daylength changes. 2. Glycolipids neutral lipids and chlorophyll, *Physiol. Plant.*, 71, 425, 1987.
199. Ketchie, D. O., Bervaes, J. C. A. M., and Kuiper, P. J. C., Lipid composition of pine needle chloroplasts and apple bark tissue as affected by growth temperature and daylength changes. 1. Phospholipids, *Physiol. Plant.*, 71, 419, 1987.
200. Kuroda, H. and Sagisaka, S., Ultrastructural changes in cortical cells of apple *Malus-pumila* Mill. associated with cold hardiness, *Plant Cell Physiol.*, 34, 357, 1993.
201. Kuroda, H., Sagisaka, S., Asada, M., and Chiba, K., Peroxide-scavenging systems during cold acclimation of apple callus in culture, *Plant Cell Physiol.*, 32, 635, 1991.
202. Kuroda, H., Sagisaka, S., and Chiba, K., Seasonal changes in peroxide-scavenging systems of apple trees in relation to cold hardiness, *J. Jpn. Soc. Hortic. Sci.*, 59, 399, 1990.
203. Proebsting, E. L. J., Low temperature resistance of developing flower buds of six deciduous fruit species, *J. Am. Soc. Hortic. Sci.*, 103, 192, 1978.
204. Ballard, J. K., Proebsting, E. L., and Tukey, R. B., *Critical Temperatures for Blossom Buds. Apples,* Washington State University Extension Bulletin, 913, 1992.
205. Boyer, J. S., Water transport, *Annu. Rev. Plant Physiol.*, 36, 473, 1985.
206. Kozlowski, T. T., Kramer, P. J., and Pallardy, S. G., *The Physiological Ecology of Woody Plants,* Academic Press, San Diego, 1991.
207. Kramer, P. J., *Water Relations of Plants,* Academic Press, New York, 1983.
208. Schulze, E.-D., Carbon dioxide and water vapor exchange in response to drought in the atmosphere and in the soil, *Annu. Rev. Plant Physiol.*, 37, 247, 1986.
209. Turner, N. C., Adaptation to water deficits: a changing perspective, *Aust. J. Plant Physiol.*, 13, 175, 1986.
210. Kaiser, W. M., Correlation of changes in photosynthetic activity and changes in total protoplast volumes in leaf tissue from hygro-, meso- and xerophytes under osmotic stress, *Planta,* 154, 538, 1982.
211. Claussen, W., Loveys, B. R., and Hawker, J. S., Influence of sucrose and hormones on the activity of sucrose synthase and invertase in detached leaves and leaf sections of eggplants *Solanum melongena*, *J. Plant Physiol.*, 124, 345, 1986.

212. Atkinson, D., The distribution and effectiveness of the roots of tree crops, *Hortic. Rev.,* 2, 424, 1980.
213. Landsberg, J. J., Beadle, C. L., Biscoe, P. V., Butler, D. R., Davidson, B., Incoll, L. D., James, G. B., Jarvis, P. G., Martin, P. J., Neilson, R. E., Powell, D. B. B., Slack, E. M., Thorpe, M. R., Turner, N. C., Warrit, B., and Watts, W. R., Diurnal energy, water and CO_2 exchanges in an apple (*Malus pumila*) orchard, *J. Appl. Ecol.,* 12, 659, 1975.
214. Jarvis, P. G., Coupling of transpiration to the atmosphere in horticultural crops: the omega factor, *Acta Hortic.,* 171, 187, 1985.
215. Angelocci, L. R. and Valancogne, C., Leaf area and water flux in apple trees, *J. Hortic. Sci.,* 68, 299, 1993.
216. Andersen, P. C., Leaf gas exchange of 11 species of fruit crops with reference to sun-tracking-non-sun-tracking responses, *Can. J. Plant Sci.,* 71, 1183, 1991.
217. Butler, D. R., Estimation of the transpiration rate in an apple orchard from net radiation and vapor pressure deficit measurements, *Agric. Meteorol.,* 16, 277, 1976.
218. Turner, N. C., Drought resistance and adaptation to water deficits in crop plants, in *Stress Physiology in Crop Plants,* H. M. a. R. C. Staples, Ed., Wiley-Interscience, New York, 1979, 343.
219. Goode, J. E. and Higgs, K. H., Water, osmotic and pressure potential relationships in apple leaves, *J. Hortic. Sci.,* 48, 203, 1973.
220. Fanjul, L. and Rosher, P. H., Effects of water stress on internal water relations of apple leaves, *Physiol. Plant.,* 62, 321, 1984.
221. Lakso, A. N., Seasonal changes in stomatal response to leaf water potential in apple, *J. Am. Soc. Hortic. Sci.,* 104, 58, 1979.
222. Thorpe, M. R., Warrit, B., and Landsberg, J. J., Responses of apple leaf stomata: a model for single leaves and a whole tree, *Plant Cell Environ.,* 3, 23, 1980.
223. Warrit, B., Landsberg, J. J., and Thorpe, M. R., Responses of apple leaf stomata to environmental factors, *Plant Cell Environ.,* 3, 13, 1980.
224. Fanjul, L. and Jones, H. G., Rapid stomatal responses to humidity, *Planta,* 154, 135, 1982.
225. Lakso, A. N., Morphological and physiological adaptations for maintaining photosynthesis under water stress in apple trees, in *Stress Effects on Photosynthesis,* R. Marcelle, H. Clijsters and M. V. Poucke, Eds., Dr. W. Junk, The Hague, 1983, 85.
226. Downton, W. J. S., Loveys, B. R., and Grant, W. J. R., Non-uniform stomatal closure induced by water stress causes putative non-stomatal inhibition of photosynthesis, *New Phytol.,* 110, 503, 1988.
227. Downton, W. J. S., Loveys, B. R., and Grant, W. J. R., Stomatal closure fully accounts for the inhibition of photosynthesis by abscisic acid, *New Phytol.,* 108, 263, 1988.
228. Massacci, A. and Jones, H. G., Use of simultaneous analysis of gas-exchange and chlorophyll fluorescence quenching for analyzing the effects of water stress on photosynthesis in apple leaves, *Trees,* 4, 1, 1990.
229. Davies, W. J. and Zhang, J., Root signals and the regulation of growth and development of plants in drying soil, *Annu. Rev. Plant Physiol. Plant Mol. Biol.,* 42, 55, 1991.
230. Gowing, D. J. G., Davies, W. J., and Jones, H. G., A positive root-source signal as an indicator of soil drying in apple, *J. Exp. Bot.,* 41, 1535, 1990.
231. Loveys, B. R., Abscisic acid transport and metabolism in grapevine (*Vitis vinifera* L.), *New Phytol.,* 98, 575, 1984.
232. Loveys, B. R., Diurnal changes in water relations and abscisic acid in field-grown *Vitis vinifera* cultivars. III. The influence of xylem-derived abscisic acid on leaf gas exchange, *New Phytol.,* 98, 563, 1984.
233. Sritharan, R. and Lenz, F., The influence of long-term water stress and fruiting on photosynthesis and transpiration in apple, *Gartenbauwissenschaft,* 54, 150, 1989.
234. Buwalda, J. G. and Lenz, F., Effects of cropping nutrition and water supply on accumulation and distribution of biomass and nutrients for apple trees on M9 root systems, *Physiol. Plant.,* 84, 21, 1992.
235. Hatfield, J. L., Measuring plant stress with an infrared thermometer, *HortScience,* 25, 1535, 1990.
236. Andrews, P. K., Chalmers, D. J., and Moremong, M., Canopy-air temperature differences and soil water as predictors of water stress of apple trees grown in a humid temperate climate, *J. Am. Soc. Hortic. Sci.,* 117, 453, 1992.
237. Davies, F. S. and Lakso, A. N., Water stress responses of apple trees. I. Effects of light and soil preconditioning treatments on tree physiology, *J. Am. Soc. Hortic. Sci.,* 104, 392, 1979.

238. Davies, F. S. and Lakso, A. N., Water stress responses of apple trees. II. Resistance and capacitance as affected by greenhouse and field conditions, *J. Am. Soc. Hortic. Sci.,* 104, 395, 1979.
239. Wang, Z. and Stutte, G. W., The role of carbohydrates in active osmotic adjustment in apple under water stress, *J. Am. Soc. Hortic. Sci.,* 117, 816, 1992.
240. Seemann, J. R., Downton, W. J. S., and Berry, J. A., Temperature and leaf osmotic potential as factors in the acclimation of photosynthesis to high temperature in desert plants, *Plant Physiol.,* 80, 926, 1986.
241. Davies, F. S. and Lakso, A. N., Diurnal and seasonal changes in leaf water potential components and elastic properties in response to water stress in apple trees, *Physiol. Plant.,* 46, 109, 1979.
242. Ranney, T. G., Whitlow, T. H., and Bassuk, N. L., Response of five temperate deciduous tree species to water stress, *Tree Physiol.,* 6, 439, 1990.
243. Kozlowski, T. T., Responses of woody plants to flooding, in *Flooding and Plant Growth,* T. T. Kozlowski, Ed., Academic Press, Orlando, FL, 1984, 129.
244. Schaffer, B., Andersen, P. C., and Ploetz, R. C., Responses of fruit crops to flooding, *Hortic. Rev.,* 13, 257, 1992.
245. Childers, N. F. and White, D. G., Influence of submersion of the roots on transpiration, apparent photosynthesis and respiration of young apple trees, *Plant Physiol.,* 17, 603, 1942.
246. Heinicke, A. J., The effect of submerging the roots of apple trees at different times of the year, *Proc. Am. Soc. Hortic. Sci.,* 29, 205, 1932.
247. Olien, W. C. and Lakso, A. N., Effect of rootstock on apple (*Malus domestica*) tree water relations, *Physiol. Plant.,* 67, 421, 1986.
248. Olien, W. C., Effect of seasonal soil waterlogging on vegetative growth and fruiting of apple trees, *J. Am. Soc. Hortic. Sci.,* 112, 209, 1987.
249. Dinkelberg, W. and Lüdders, P., Influence of seasonally different sodium treatment on mineral content in leaves of apple trees, *Angew. Bot.,* 64, 237, 1990.
250. Dinkelberg, W. and Lüdders, P., Influence of seasonally different sodium stress on vegetative growth and yield of the apple cultivar Golden Delicious, *Gartenbauwissenschaft,* 56, 275, 1991.
251. Dinkelberg, W. and Lüdders, P., Influence of seasonally different sodium treatment on photosynthesis and transpiration of apple trees, *Gartenbauwissenschaft,* 55, 130, 1990.
252. Schreiner, M. and Lüdders, P., Mechanisms of salt resistance in apple, *Gartenbauwissenschaft,* 57, 253, 1992.
253. Schreiner, M. and Lüdders, P., The effect of sodium salts on water consumption and mineral uptake of golden delicious apple trees with various levels of potassium nutrition, *Gartenbauwissenschaft,* 57, 93, 1992.
254. Ormrod, D. P., Gaseous air pollution and horticultural crop production, *Hortic. Rev.,* 8, 1, 1986.
255. Shertz, R. D., Kender, W. J., and Musselman, R. C., Foliar response and growth of apple trees following exposure to ozone and sulfur dioxide, *J. Am. Soc. Hortic. Sci.,* 105, 594, 1980.
256. Retzlaff, W. A., Williams, L. E., and DeJong, T. M., The effect of different atmospheric ozone partial pressures on photosynthesis and growth of nine fruit and nut tree species, *Tree Physiol.,* 8, 93, 1991.
257. Forsline, P. L., Musselman, R. C., Kender, W. J., and Dee, R. J., Effects of acid rain on apple tree productivity and fruit quality, *J. Am. Soc. Hortic. Sci.,* 108, 70, 1983.
258. Forsline, P. L., Dee, R. J., and Melious, R. E., Growth changes of apple seedlings in response to simulated acid rain, *J. Am. Soc. Hortic. Sci.,* 108, 202, 1983.
259. Rinallo, C., Effects of acidity of simulated rain on the fruiting of Summerred apple trees, *J. Environ. Qual.,* 21, 61, 1992.
260. Rinallo, C., Modi, G., Ena, A., and Calamassi, R., Effects of simulated rain acidity on the chemical composition of apple fruit, *J. Hortic. Sci.,* 68, 275, 1993.
261. Mitchell, J. F. B., The "greenhouse" effect and climate change, *Rev. Geophys.,* 27, 115, 1989.
262. Krizek, D. T., Zimmerman, R. H., Kleuter, H., and Bailey, W. A., Growth of crabapple seedlings in controlled environments: effect of CO_2 level and time and duration of CO_2 treatment, *J. Am. Soc. Hortic. Sci.,* 96, 285, 1971.
263. Hansen, P., Fruit trees and climate. I. Temperature, carbon dioxide concentration and growth in apple trees, *Tidsskr. Planteavlr,* 79, 303, 1975.
264. Hansen, P., The effect of carbon dioxide concentration on the early growth of apple trees, *Tidsskr. Planteavlr,* 79, 227, 1975.
265. Kimball, B. A., Carbon dioxide and agricultural yield: an assemblage and analysis of 430 prior observations, *Agron. J.,* 75, 779, 1983.

266. Wolfe, D. and Erickson, J., CO_2 fertilization: Will it benefit agriculture, in *Agricultural Dimensions of Global Climate Change,* H. M. Kaiser and T. E. Drennen, Eds., St. Lucie Press, Delray Beach, FL, in press.
267. Kramer, P. J., Carbon dioxide concentration, photosynthesis, and dry matter production, *BioScience,* 31, 29, 1981.
268. Downton, W. J. S., Grant, W. J. R., and Loveys, B. R., Carbon dioxide enrichment increases yield of Valencia orange, *Aust. J. Plant. Physiol.,* 14, 493, 1987.
269. Idso, S. B., Kimball, B. A., and Allen, S. G., CO_2 enrichment of sour orange trees: 2.5 years into a long-term experiment, *Plant Cell Environ.,* 14, 351, 1991.
270. Idso, S. B., Kimball, B. A., and Allen, S. G., Net photosynthesis of sour orange trees maintained in atmospheres of ambient and elevated CO_2 concentration, *Agric. For. Meteorol.,* 54, 95, 1991.
271. Idso, S. B. and Kimball, B. A., Seasonal fine-root biomass development of sour orange trees grown in atmospheres of ambient and elevated CO_2 concentration, *Plant Cell Environ.,* 15, 337, 1992.
272. Idso, S. B. and Kimball, B. A., Downward regulation of photosynthesis and growth at high CO_2 levels. No evidence for either phenomenon in three-year study of sour orange trees, *Plant Physiol.,* 96, 990, 1991.
273. Corelli Grappadelli, L., Lakso, A. N., and Flore, J. A., Early season patterns of carbohydrate partitioning in exposed and shaded apple branches, *J. Am. Soc. Hortic. Sci.,* 119, in press.
274. Barritt, B. H., Influence of orchard system on canopy development, light interception and production of third-year Granny Smith apple trees, *Acta Hortic.,* 243, 121, 1989.
275. Verheij, E. W. M. and Verwer, F. L. J. A. W., Light studies in a spacing trial with apple on a dwarfing and semi-dwarfing rootstock, *Scientia Hortic.,* 1, 25, 1973.
276. Wagenmakers, P. S. and Callesen, O., Influence of light interception on apple yield and fruit quality related to arrangement and tree height, *Acta Hortic.,* 243, 149, 1989.

Chapter 3

Blueberries, Cranberries, and Red Raspberries

Frederick S. Davies and Rebecca L. Darnell

CONTENTS

0-8493-175-0/94/$0.00+$.50

I. INTRODUCTION

Blueberries, cranberries, and raspberries are small fruits encompassing a range of species that are distributed worldwide. Blueberries and cranberries are in the family Ericaceae and the genus *Vaccinium* which contains about 400 species of shrubs, woody vines and small trees.[1] Two-thirds of the species are found in Malaysia, with the remaining ones distributed in southeast Asia, Japan, Africa, Europe, and North and South America. *Vaccinium* has been further subdivided into 10 sections.[2] The blueberries are in *Cyanococcus* which includes lowbush (*V. angustifolium* Ait.), highbush (*V. corymbosum* L.), and rabbiteye species (*V. ashei* Reade), although Vander Kloet[1] combines rabbiteye blueberries into *V. corymbosum*. Highbush blueberries are grown commercially in more than 30 states and provinces in North America; lowbush blueberries are grown primarily in Maine and five eastern Canadian provinces; and rabbiteye blueberries are produced primarily in Georgia, Florida, Mississippi, Texas, South Carolina, Louisiana, and Alabama. There are also several wild *Vaccinium* species that are not of commercial importance but have potential for use in breeding to improve existing cultivars, including *V. darrowi, V. tenellum, V. myrsinites*, and *V. elliotti* among others.[3] The cranberry (*V. macrocarpon* Ait.) belongs to the section *Oxycoccus*. Cranberries are grown primarily in Massachusetts, Wisconsin, New Jersey, British Columbia, Quebec, and Nova Scotia.

Raspberries are in the family Rosaceae and the genus *Rubus* which consists of shrubs, many of which are biennial bearing and woody.[4] *Rubus* is further subdivided into 12 sections including *Idaeobatus* (red raspberry), *Eubatus* (blackberry), *Cylactis* (related polar and alpine species), and *Anoplobatus* (related flowering species).[5] Red raspberry is of greatest worldwide importance with over 200 species.[6] Raspberry fruits readily separate from the receptable, whereas blackberry fruit adhere to the receptacle.

Red raspberries are of two major genetic origins: *Rubus idaeus* subsp. *vulgatus* Arrhen. is native to northern Europe and Asia and *R. idaeus* subsp. *strigosus* Michx. to North America. The black raspberry, *R. occidentalis* L., is also native to North America.[5,7] The commercially important red raspberries are natural hybrids of the European and American types and thus have acquired advantageous traits of each group. The former have naturally superior fruit quality and the latter have superior drought tolerance and cold hardiness. Red raspberries are produced primarily in the Unified Countries, New Zealand, Chile, eastern Europe, Scotland, in the western United States (Oregon and Washington), and British Columbia.

These two families of small fruits differ from most other fruit species discussed in this book because they represent multi-stemmed bushes and prostrate-growing shrubs rather than large trees. Most species remain relatively small at maturity compared to most tree crops, although some *V. ashei* plants may exceed 8 m in height if left unpruned.[8] In addition, some raspberry species are distinctly biennial bearing. Furthermore, *Vaccinium* species differ from other commercial fruit crops in their requirement for low pH soils. Consequently, the environmental physiology of these small fruits may differ in some aspects from that of large trees.

In this chapter we will discuss the effects of light, temperature, and water on the growth and development of commercially important *Vaccinium* and *Rubus* species. Some references will also be cited from related species from the wild, particularly for *Vaccinium*, where several heath and heather species have been studied in detail.

II. IRRADIANCE

A. *VACCINIUM*

1. Irradiance Level

In nature, *Vaccinium* species are found in a wide range of light environments, ranging from full sun to partial or dense shade.[9] Many *Vaccinium* species grow in the wild as understory plants while others are found along river banks in full sunlight.[8] Optimum commercial growth and production, however, are greatest under full sun conditions.[10]

Leaf net CO_2 assimilation (A) and light saturation curves have been reported for leaves of lowbush, rabbiteye, and highbush blueberries. In general, *A* of lowbush and highbush blueberry averages 9–12 μmol m^{-2} s^{-1},[11-13] substantially higher than rates in rabbiteye blueberry, which average 5–8 μmol m^{-2} s^{-1}.[14-16] Leaf *A* of cranberry averages 9–15 μmol m^{-2} s^{-1},[17] although rates as high as 25 μmol m^{-2} s^{-1} have been reported.[11] This range in cranberry leaf *A* rates may reflect temperature, seasonal, and/or cultivar differences. Leaf *A* increases with increasing photosynthetic photon flux (PPF),

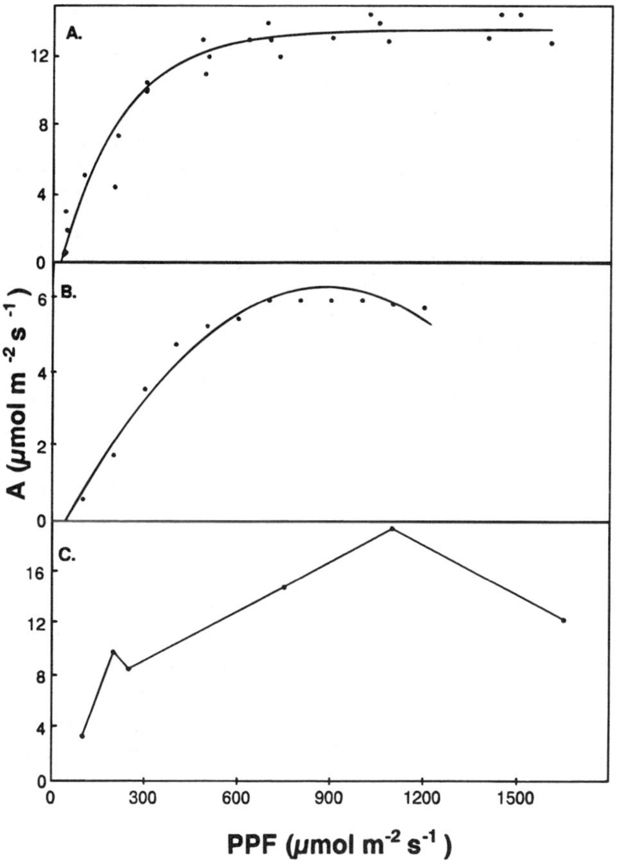

Figure 1 Effects of photosynthetic photon flux (PPF) on leaf net CO_2 assimilation (A) in 'Jersey' highbush blueberry (A), 'Woodard' rabbiteye blueberry (B), and 'Searle' cranberry (C). From Moon et al., *J. Amer. Soc. Hort. Sci.*, 112,134,1987 (A); Teramura et al., *HortScience*, 14, 723, 1979 (B); Stang et al., *Acta Hort.*, 165, 325, 1985 (C). With permission.

reaching light saturation at 600–800 μmol m^{-2} s^{-1} for both blueberry[13,14] and cranberry[17] (Figure 1). Photosynthetic light saturation in lowbush blueberry occurs between 1000 and 1500 ft-c,[11] which although not readily convertible to SI units, is in the same range as light saturation for other *Vaccinium* species.

Photosynthetic photon flux of 250 to 300 μmol m^{-2} s^{-1}, sufficient for 50–60% of maximum photosynthesis, has been measured in the shaded canopy interior of 5-year-old rabbiteye blueberry bushes.[14] This, coupled with the low light compensation point of 25–35 μmol m^{-2} s^{-1},[18] suggests that irradiance level effects on photosynthesis may not be a major yield limitation in blueberry. The relatively small size and multiple stemmed structure permits greater light penetration into the canopy than might occur for large, densely-foliated trees such as citrus or mango.

Stang et al.[17] reported that *A* of cranberry shoots grown for 30 or 60 days under 100, 80, 60, or 15% full sun were similar when measured under saturating PPF. The authors suggested that cranberry, unlike other species that exhibit reduced rates of photosynthesis in shade-developed leaves, has a "unique and rapid ecological adaptation to changing light levels" that is well-suited for the light competition that occurs in a dense cranberry planting. However, the proportion of the total photosynthetic area of the shoots that developed under the shaded conditions during the 30 or 60 day experiment was not indicated. If this proportion were small, then differences in leaf *A* due to the light environment during development may have been masked. Further work is needed to clarify this concept.

Recent research suggests that fruit photosynthesis may contribute a significant portion of the total fruit carbon requirement in blueberries, especially during early fruit development. Net CO_2 assimilation occurred in fruit from petal fall through color break, declining from about 19 μmol g^{-1} FW h^{-1} to 0.2 μmol g^{-1} FW h^{-1} during development.[19] During the first 10 days after anthesis, fruit photosynthesis was estimated to contribute 50% of the total carbon required for dry weight gain and respiratory loss. From anthesis to fruit ripening, fruit photosynthesis contributed 15% of the total fruit carbon requirement. As with leaves, *A* of fruit is light dependent. However, light saturation for fruit photosynthesis in highbush blueberry occurred at a much lower PPF, approximately 300–350 μmol m^{-2} s^{-1}, with a light compensation point at 100 μmol m^{-2} s^{-1}.[20] This low light saturation level suggests that PPF may be saturating for fruit photosynthesis throughout much of development, even on the shaded side of the fruit.

Although PPF effects on photosynthesis may not be a major limitation to productivity under most conditions, evidence suggests that light is still a major factor regulating growth and development of *Vaccinium* species. Stang et al.[17] reported that increasing levels of shade decreased cranberry leaf thickness, due to a decrease in the number and size of mesophyll and palisade cells. Similarly, reproductive development is markedly influenced by light. Hall[21] found that flower bud number in lowbush blueberry decreased linearly with decreasing irradiance level from 100 to 11% full sun. In a subsequent study, Hall and Ludwig[22] reported that flower bud initiation of lowbush blueberry was significantly decreased when irradiance level was decreased below 60% full sun (a PPF of about 1200 μmol · m^{-2} · s^{-1}). These data indicate that PPF well above that required for light-saturated photosynthesis is required for optimum flower bud initiation in lowbush blueberry.

Fruit development period and fruit maturity are also influenced by light. Aalders et al.[23] found that the fruit development period was lengthened and fruit sugar content decreased in lowbush blueberry by decreasing irradiance level. Partial defoliation under high irradiance level also lengthened the fruit development period, but had no effect on fruit sugar content. Light also appears to be the primary factor in fruit development of rabbiteye blueberry.[24] Fruit from the top of the canopy, receiving full sun, ripened faster than fruit from other locations. The correlation between amount of light striking the canopy and fruit ripening rate in that area of the canopy was linear, as was the correlation between amount of light and fruit soluble solids. Overall, light exposure accounted for 64% of the variability in fruit maturity (as measured by fruit color) and 30% of the variability in soluble solids (Figure 2). Three hours of photosynthetic light saturation (25–30% full sun) were considered sufficient to ensure that light was not the overriding limitation in fruit development. However, the importance of light distribution over the course of the day was not addressed.

Ripening of highbush blueberry fruit (as measured by color development) was delayed when fruit clusters were bagged to exclude light.[25] This is evidence for light regulation of anthocyanin biosynthesis, and further indicates a distinct role of light in fruit development separate from its role in photosynthesis.

2. Photoperiod and Light Quality

A substantial amount of data indicates that growth and development in *Vaccinium* is dependent not only on irradiance level, but also on photoperiod. Flower bud initiation in lowbush blueberry is promoted under short days.[26,27] In general, at least 6 weeks of photoperiods less than 12 hours is needed for normal flower bud formation. The photoperiod response of highbush blueberries is similar to that of lowbush. Eight weeks of 8-, 10-, or 12-hour photoperiods resulted in flower bud initiation in highbush cultivars, whereas fewer flower buds were formed under 14- and 16-hour photoperiods.[27] Reproductive development in rabbiteye blueberry is also affected by photoperiod.[28] Six weeks of 8-hour photoperiods in the fall promoted flower bud initiation and decreased the length of the bloom period the following spring compared to 11–12 hour daylengths (Table 1). This suggests that short days not only increase the number of flower buds initiated, but may also result in more synchronized flower bud differentiation.

Concomitant with an increase in flower bud initiation under short days is a decrease in total shoot growth.[22,27] Conversely, long days promote an increase in shoot growth, shoot number, and leaf area in blueberry[29,30] and cranberry.[31]

Little information is available on light quality effects on growth and development of *Vaccinium* species. Since photoperiod effects appear to be phytochrome mediated,[32] it is possible that light quality would have a profound effect on development. Stushnoff and Hough[33] found that blueberry seeds exposed to red light had a much higher germination percentage than those exposed to white light. There is no information available on red:far red effects on other aspects of *Vaccinium* development.

High levels of UV-B radiation had no consistent effect on fruit number or berry weight in rabbiteye blueberries.[34] However, the waxy bloom on fruit decreased as UV-B radiation increased. Additionally, fruit exposed to high levels of UV-B developed a corky layer on the surface, having symptoms similar to sunscald.

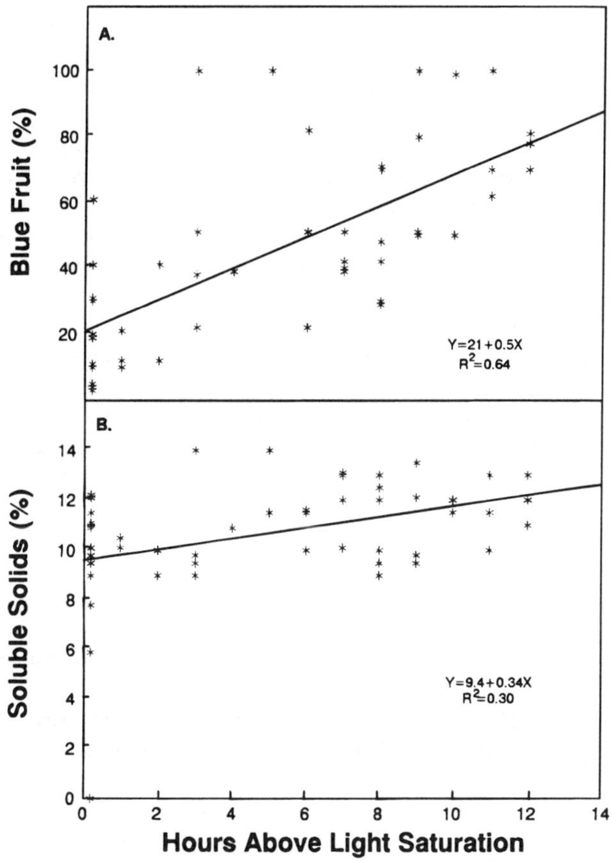

Figure 2 Relationship between the number of hours above light saturation and the development of blue fruit (A) and fruit soluble solids (B) for 'Tifblue' rabbiteye blueberry. From Patten and Neuendorff, *Proc. Texas Blueberry Growers Assoc.*, Beaumont, 1989, 109.

Table 1 **Daylength effect on flower bud initiation, bloom period, flower budbreak, and total flower number of 'Beckyblue' rabbiteye blueberry**

	Flower buds/ plant		Bloom period (days)		Flower buds broken (% flower buds/plant)	Total florets (no.)	
Daylength[a]	**1989**	**1990**	**1989**	**1990**	**1989[b]**	**1989**	**1990**
Natural	21.9	183.7	12.7	37.5	43.8	56	1016
SD	48.0	214.9	9.0	32.9	61.7	177	1194
Significance	**	*	**	**	**	**	**

[a] Daylength during the 5 weeks just before defoliation and chilling. Natural = decreasing daylength from 12 to 11 hr; SD = decreasing daylength from 10 to 8 hr. [b] No data for 1990. *p = 0.10. **p = 0.05.
From Darnell, R. L., *J. Amer. Soc. Hort. Sci.*, 116, 856, 1991. With permission.

B. *RUBUS* (RED RASPBERRY)

1. Irradiance Level

Information on irradiance level effects on *A* in raspberry is limited. Maximum *A* of 6–10 μmol m^{-2} s^{-1} at 25°C under saturating light conditions (24 klux) have been reported for mature, detached leaves of field-grown red raspberry plants.[35] Goulart and Braun[36] measured PPF effects on A in five cultivars of greenhouse-grown, primocane fruiting red raspberry. They found little variation in light response curves among the cultivars. Leaf A measured at an average leaf temperature of 25.6°C increased with PPF up to 900 to 1000 $\mu mol\ m^{-2}\ s^{-1}$, with maximum values averaging 5–7 $\mu mol\ m^{-2}\ s^{-1}$. However, average values of 9 to 12 $\mu mol\ m^{-2}\ s^{-1}$ have been found for raspberries in the field in New York.[37]

Light microclimate may have significant effects on vegetative and reproductive growth within the raspberry canopy. Individual raspberry canes are biennial; primocanes arising from adventitious buds in roots or basal buds on previous canes grow vegetatively throughout the season until flower bud initiation in the fall. The following year, the axillary buds on these canes produce lateral fruiting branches. Thus, a raspberry canopy is comprised of both vegetative primocanes and reproductive floricanes. Palmer et al.[38] compared light interception and intensity in vertical vs. horizontal raspberry training systems. Horizontal systems, planted at spacings of 5 m between rows, intercepted about 60% of the incident light in a pattern that was evenly distributed throughout the day. Vertical systems at the same spacing, however, intercepted about 40% of the incident light, with a mid-day decrease to about 20%. Although yield data were not taken, the authors concluded that horizontal training systems would increase productivity compared to vertical systems planted at the same spacing by promoting more even light distribution in the canopy throughout the season. In contrast, fruit became sunburned when light was maximized under Missouri growing conditions.[39] Braun et al.[40] estimated the light microclimate within the canopy of red raspberry and correlated irradiance level and distribution with growth and yield. Total floricane canopy volume was similar throughout the season, but the extent of light exposure shifted during the season as primocanes grew. During the season, 60 to 70% of the total floricane leaf area and 55 to 60% of the flowers and fruit were in the area of the canopy exposed to 0 to 25% of the PPF. Thus, on a whole plant basis, the area of the canopy exposed to 0 to 25% PPF was the most productive. However, on a leaf area basis, productivity increased with increasing light. Fruit set in canopy areas exposed to greater than 25% PPF averaged 90%, while fruit set in areas exposed to 25% or less averaged 35 to 50%.

Annual cropping systems, where both primocanes and floricanes are present each year, have been compared to biennial cropping systems, where primocane growth alternates yearly with floricane growth. This biennial system is accomplished by removal of either floricanes or primocanes prior to the start of the growing season. Total light interception by the canopy was greater in the annual system than the biennial; however, light distribution was poorer.[41] There was a decrease in the number of primocanes produced, and an increase in primocane internode length in the annual system, consistent with the far-red enrichment that would occur as the fruiting canes filtered out the red light.[42] As primocane growth developed, the lower fruiting laterals in the annual system became heavily shaded, resulting in excess leaf loss and a decrease in fruit set and yield.[41,43] On the other hand, the biennial system resulted in increased primocane production and increased nodes per cane in the vegetative year, which translated into higher yields the following year.[44] However, even though productivity was decreased on the annual cropping system vs. the fruiting year of the biennial cropping system, yields averaged over a 2-year cycle were greater for the annual system.

Red raspberries, exposed to clear days and high temperatures, may develop symptoms of sunscald, where red coloration fails to develop on the sun-exposed portion of the fruit. Renquist et al.[45] found that 30 or 60% shade applied for the last 3 weeks or the last few days of fruit development prevented sunscald symptoms in the field. Use of fans to decrease fruit temperature did not significantly affect the development of sunscald. Therefore, they concluded that high light exposure just prior to fruit harvest was the major factor in the development of symptoms. In a subsequent study, Renquist et al.[46] observed that sunscald symptoms did not develop when fruit were exposed to high temperatures (37 to 44°C) in the absence of UV radiation. Significant injury occurred only when 4 to 7 hours of UV radiation was combined with high temperatures, indicating that methods of alleviating sunscald must include attenuation of UV radiation interception by the fruit.

In general, there is a paucity of information on light effects on growth and development of raspberry compared to other fruit crops. Although correlations between light interception and yield have been drawn, there are limited data available on irradiance level, quality, or photoperiod effects. Further work on light effects on reproductive and vegetative growth of raspberry is warranted.

III. TEMPERATURE

A. *VACCINIUM*

1. Growing Season Temperatures

In general, *A* of lowbush, rabbiteye, and highbush blueberries increases as temperatures increase from 10 to 25–30°C, depending on species.[12,13,47] In cranberry, leaf photosynthetic rates, as measured by O_2 evolution, increased linearly as temperature increased from 3.5 to 25.0°C.[48] Temperature optima for *A* in both highbush (Figure 3A) and rabbiteye (Figure 3B) blueberry are similar, ranging from 20–25°C.[12,13] As temperature increased to 30°C, *A* decreased significantly in both species, a somewhat unexpected result since rabbiteye blueberry plants are native to the southeastern United States and therefore might be expected to be better adapted to high temperatures than highbush blueberry plants. Crosses between *Vaccinium corymbosum* ('Bluecrop') and *V. darrowi* (Florida 4B), a wild diploid blueberry selection with a temperature optimum for *A* about 8–10°C higher than *V. corymbosum*,[48] resulted in an F_1 hybrid and a progeny of the F_1 hybrid with photosynthetic heat tolerance similar to that of Florida 4B.[49] In a later study, Hancock et al.[50] identified several benotypes from F_2 and backcross populations derived from Florida 4B × 'Bluecrop' that also had increased heat tolerance. This suggests that the photosynthetic heat tolerance of *V. darrowi* can be transferred to commercial blueberry cultivars, which may be an important factor in potential yield increases, since high temperatures in the field may limit *A*. However, correlations between *A* and yield are often poor because other factors such as number of flower buds formed and amount of cross pollination also affect yields. Thus, field testing of progeny with increased photosynthetic heat tolerance is required.

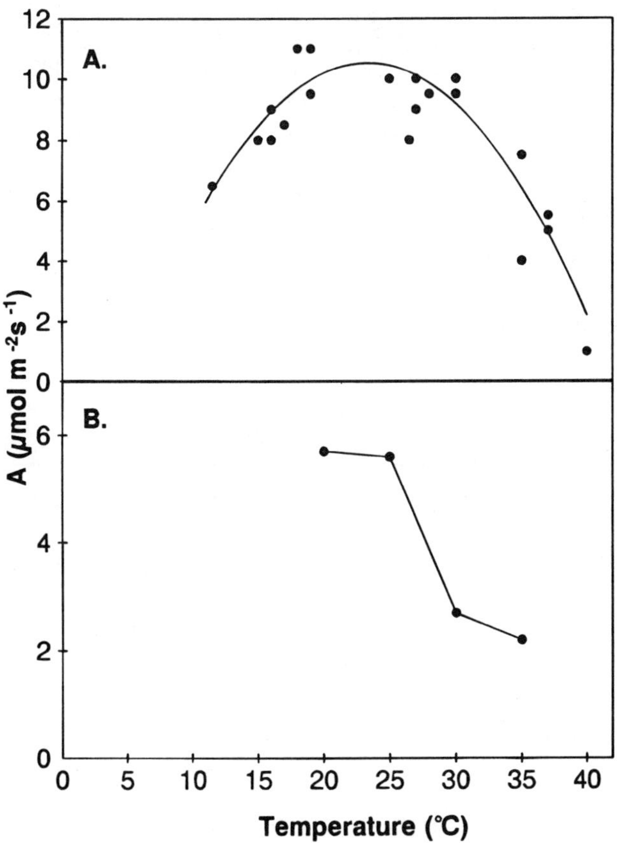

Figure 3 Effects of temperature on leaf net CO_2 assimilation (A) in 'Jersey' highbush blueberry (A) and 'Woodard' rabbiteye blueberry (B) From Moon et al., *J. Amer. Soc. Hort. Sci.*, 112, 134, 1987 (A); Davies and Flore, *J. Amer. Soc. Hort. Sci.*, 111, 565, 1986 (B). With permission.

There are limited data on temperature ranges for vegetative and reproductive growth of *Vaccinium*. Hall and Ludwig[22] found that shoot growth of lowbush blueberry increased as temperature increased from 10 to 21°C. Root growth of highbush blueberry was limited at soil temperatures below 8°C or above 18°C, with optimum growth occurring between 14 and 18°C.[51] However, total shoot growth of highbush blueberry increased as soil temperature increased from 13 to 32°C.[52] Several studies indicate that increases in blueberry growth due to mulching are due, in part, to the resultant decrease in soil and root temperatures.[53-55] Optimum temperatures for cranberry growth range from 16 to 27°C, and high temperature injury may occur at temperatures near 27°C if drying winds are present.[56] Hawker and Stang[57] in Wisconsin characterized vegetative and reproductive growth in cranberry by summation of growing degree days (GDD) within the range of 9 to 32°C. They found that accumulation of GDD in three locations within geographical and climatically distinct regions in Wisconsin were similar in their sigmoidal pattern (Figure 4A). Shoot length increased rapidly until the end of June (700 GDD),

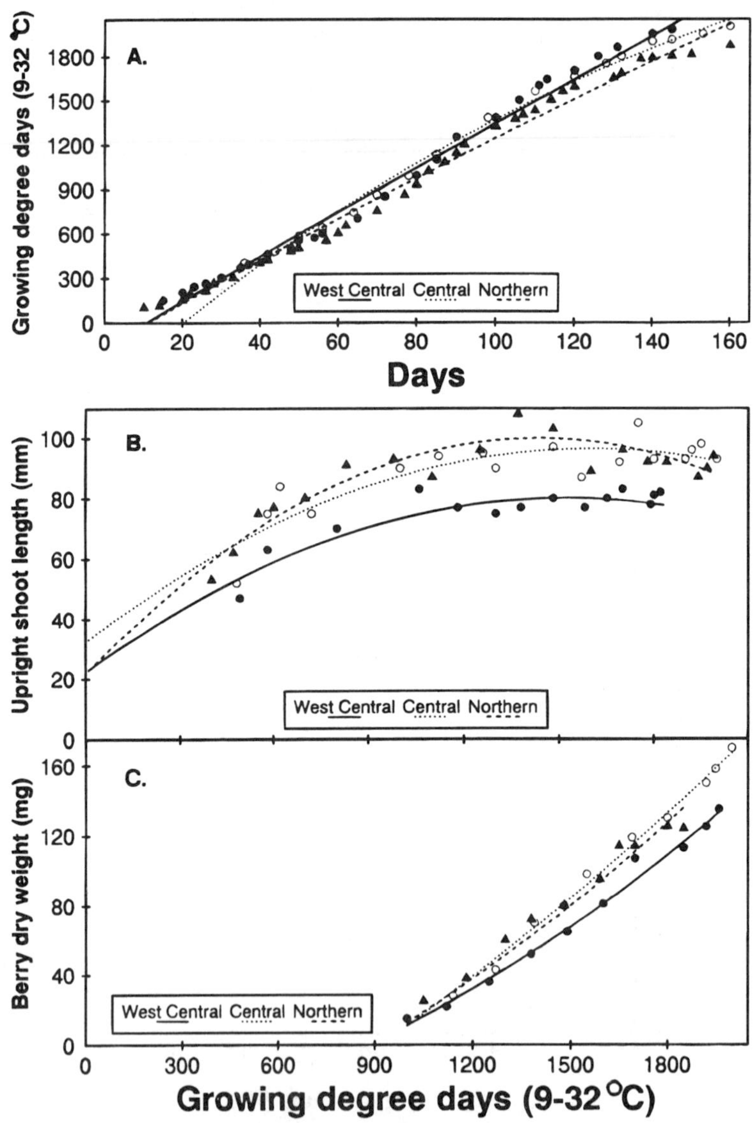

Figure 4 Seasonal growing degree days (GDD) accumulation at three Wisconsin cranberry producing locations (A), relationship between GDD and shoot growth of 'Searles' cranberry (B), and relationship between GDD and fruit dry weight for 'Searles' cranberry (C). From Hawker and Stang, *Acta Hort.*, 165, 311, 1985. With permission.

and terminal growth ceased by the end of July (1300 GDD) (Figure 4B). Shoot dry weight accumulation mirrored shoot length. Shoot length of plants growing in the most northerly location averaged 15 cm shorter than the other two locations; however, differences in shoot growth were not related to lower GDD accumulation since the rate of accumulation did not differ significantly among the three locations. Therefore, factors in addition to GDD, such as water relations or photoperiod, affect growth of cranberry plants. Cranberry floral buds bloomed about mid-June (510 GDD), and a rapid linear increase in berry dry weight occurred from mid-July (1000 GDD), when shoot growth had almost ceased, until the end of the growing season (2100 GDD) (Figure 4C). Berry anthocyanin accumulation began by the end of August (1600 GDD), with the fastest rate occurring at the more northerly location.

Reproductive development in *Vaccinium* may be limited by temperature. In cranberry, increased upright production, flower bud number, and fruit yield were correlated with minimum/maximum temperatures above 10 and 18°C, respectively, during the summer.[58] However, excessively high temperatures in spring, when coupled with moisture deficits, decreased fruit set.[56]

Within the temperature ranges studied (10 to 25°C), the rate of blueberry flower bud development increased as temperature increased, but there was little effect of temperature on flower bud initiation.[22,26,59] No differences in fruit set or berry weight were found when blueberry plants were grown under controlled, diurnal temperature conditions of 21/10 or 16/10°C,[60] although the fruit development period increased under the cooler conditions. However, Knight and Scott[61] reported that fruit set and berry size increased, and fruit development period decreased under warm (16–27°C) greenhouse conditions compared to cool (8–24°C) conditions. Increased fruit set was attributed to increased rate of pollen tube growth under the warmer conditions.

Empirical evidence suggests that temperature effects on the length of the fruit development period of blueberry occur primarily during growth stages I and III. Plants with late blooms followed by warm temperatures may exhibit shorter stage I periods than when early blooms are followed by cool temperatures.[62] Observation suggests a similar temperature effect on stage III, while the length of a stage II is thought to be genetically linked, as well as strongly influenced by the presence of viable seed. Carlson and Hancock[63] suggested that fruit development of highbush blueberry may be inhibited to some extent by temperatures ranging from 21 to 32°C, depending on cultivar. This may reflect the adverse effects of high temperatures on *A*.

Information on low temperature effects on *Vaccinium* fruit set is sparse. Rigby and Dana[64] reported that the time between pollination and fertilization of cranberry flowers under greenhouse temperature conditions of 18°C night and 24–32°C day was 24 to 72 hours. They suggested that the period for pollen tube growth and ovule fertilization in the field would be longer than this due to cooler night temperatures during bloom. In fact, night temperatures below freezing during cranberry bloom are not uncommon.[65] Low night temperatures may contribute to the low fruit set percentages observed in cranberry if pollen tube growth is delayed and the ovules deteriorate prior to fertilization.

Low temperatures after pollination also decrease fruit set and fruit development in blueberry. Hall et al.[66] found that 4 hours at −2°C or 2 hours at −3°C decreased fruit set in lowbush blueberry by 42 and 77%, respectively. Additionally, plants exposed to −1°C for 6 hours, or to −2°C for 4 hours produced small and late ripening berries. Injury was comparable on plants subjected to freezing temperatures immediately after pollination and plants subjected to freezing temperatures 6 days after pollination.

Fruit set in *Vaccinium* is also indirectly influenced by temperature effects on bee activity. Temperatures below 10°C inhibit bee flight.[67] In general, low temperatures, high winds, and/or rain during bloom will decrease bee activity and, therefore, pollination and fruit set.

Development of anthocyanins in *Vaccinium* leaves and fruit is highly temperature dependent. Hall and Stark[68] showed that anthocyanin content increased as temperature decreased from day/night temperatures of 23.9/18.3 to 12.8/7.2°C in detached cranberry fruit and intact cranberry plants. Anthocyanin content of lowbush blueberry leaves increased as temperature decreased from diurnal day/night temperatures of 25.6/15.6 to 15.6/4.4°C.[59] However, anthocyanin accumulation did not occur in harvested highbush blueberry fruit at temperatures below 10°C.[25] Between 10 and 15°C, only red pigmentation developed. At 16°C and above, blue pigmentation developed, with the rate dependent on temperature. The response difference in anthocyanin development between harvested cranberry and blueberry fruit may be due to a light interaction; the harvested blueberry fruit were held in the dark, while the cranberry fruit were exposed to a 12-hour photoperiod during their respective temperature treatments.

2. Cold Hardiness

The ability of plant tissue to withstand low winter temperatures is dependent on the degree of acclimation attained by that tissue prior to exposure to low temperatures. The timing and extent of acclimation depends on environmental and genetic factors. In the northeastern United States, highbush blueberry shoot tissue began acclimating in October and continued to acclimate through January, when maximum hardiness was attained.[69] Dehardening began soon after, and was gradual over the next 4 months. At maximum hardiness, shoot tissue tolerated −25°C with little injury, but severe injury occurred in tissue subjected to −40°C. Bittenbender and Howell[70] found that maximum hardiness of highbush blueberry cultivars occurred at the end of November in Michigan. Dehardening began between mid-January and early March. Under Wisconsin growing conditions, cranberry leaf and shoot hardiness increased very gradually, from 0°C early in the growing season, to −7°C in mid-July, attaining maximum hardiness of −24°C by mid-October.[71] Dehardening began in late April, and proceeded rapidly, reaching −7°C by early May. The increase in leaf hardiness was associated with an increase in sugar concentration of the leaves. Similarly, Siekmann and Boe[72] found a 95% increase in total sugar concentration in cranberry leaves after plants were exposed to 7°C for 6 days. The increase in sugars during cold acclimation may be correlative rather than causal, however.

Although early research indicated that highbush blueberries in Minnesota were able to withstand winter temperatures only down to −20°C,[73] the short growing season in northern climates may prevent full acclimation, resulting in decreased hardiness and the reliance on snow cover as an insulator.[74] Additionally, variability in environmental factors during the season may affect the rate and extent of acclimation and the development of cold hardiness.[75,76] The degree of stem acclimation in highbush blueberry under Minnesota conditions varied from a maximum of −25 to −40°C in different years.[75] The decrease in the degree of hardening was attributed to drought conditions during the fall, which induced premature leaf senescence and prevented complete acclimation of the tissue. Bittenbender and Howell[77] used multiple regression analysis to determine which environmental factors affect cold hardiness of highbush blueberry. In the fall, bud hardiness was positively correlated with water content, supporting results from Quamme et al.[75] Both an increase in air temperature and photoperiod were correlated with decreased hardiness. Since air temperature would be expected to vary during fall, this suggests that tissue cold hardiness would also vary. In the winter and spring, hardiness was again negatively correlated with air temperature and duration of the photoperiod.[77] In contrast, water content was negatively correlated with bud hardiness,[70,77] and bud hardiness was found to increase linearly as bud water content decreased.[78]

Cold hardiness of blueberry plants is species dependent. Lowbush blueberry is inherently more cold hardy than highbush blueberry,[75,79] which is inherently more cold hardy than rabbiteye blueberry.[80,81] Recently, half-high blueberry genotypes (*V. corymbosum* L × *V. angustifolium* Ait. hybrids) have been developed that are similar in growth and fruiting characteristics to highbush blueberry, yet exhibit greater cold hardiness.[82,84]

Within a species, flower bud hardiness is dependent on cultivar, bud developmental stage, and location on bush. In general, cranberry flower buds survive temperatures of −12 to −18°C without injury if fully acclimated.[56,71] Initial injury in dormant cranberry flower buds occurs at the abscission zone between the bud and the stem,[56,71] followed by injury to the ovary and filaments.[71] Open cranberry flowers cannot survive temperatures less than 0°C.[71] Flowers that were injured but not killed by low temperatures developed more slowly and had fewer primordia than uninjured flowers,[85] and may produce smaller fruit with fewer seed.[86] Cranberry flower buds that developed first on an inflorescence were more susceptible to cold than those developed later,[56] presumably reflecting differences in bud developmental stage. Abdallah and Palta[71] reported that young cranberry fruit were injured at temperatures of 0°C; however, hardiness increased to between −3 and −8°C as fruit changed from green to red stage.

In rabbiteye blueberry, the temperature required to damage flower buds was inversely related to the stage of development.[87] Closed floral buds survived temperatures of −10 to −15°C, depending on cultivar, while open buds were killed at −1°C. A similar inverse relationship was evident for southern highbush (*V. corymbosum* interspecific hybrids) flower buds and young fruit.[81] In general, flower buds of southern highbush and highbush cultivars exhibit less cold damage than buds of rabbiteye cultivars at similar stages of development[81] (Table 2).

Terminal flower buds of blueberry are less hardy than median or basal buds,[78,88,89] and within a floral bud, apical florets are less hardy than median or basal florets.[78] This probably reflects differences in bud development due to bud location on the stem and floret location within the bud.

Table 2 **Cold injury to blueberry flowers and young fruit as a function of germplasm and developmental stage**

Type	Cultivars	Frost damage (% dead ovaries)[a] by flower developmental stage[b]			
		4	5	6	7
Rabbiteye	Brightwell	0	38	100	—[c]
	Tifblue	10	74	—	89
	Climax	30	73	100	99
Southern highbush	Blueridge	0	26	63	100
	Cape Fear	0	22	79	95
	Georgiagem	0	13	53	94
Highbush	Croatan	—	—	0	76

[a] Damage assessed after freezes of −2 to −4°C. [b] Stages of flower bud development: 4 = bud scales abscised, florets visible; 5 = individual florets separated, corollas unexpanded and closed; 6 = corollas expanded and open, stigma receptive, and anthers dehiscent; 7 = corollas abscised. (Data are from Spiers, *J. Amer. Soc. Hort. Sci.*, 103, 452, 1978. With permission.) [c] No data taken.

From Patten et al., *HortScience*, 26, 18, 1991. With permission.

The data suggest there is little difference in the temperature responses of *Vaccinium* compared to other temperate fruit crops. Within the genus, species and cultivars express some diversity in terms of temperature effects on *A* and the ability to develop cold hardiness; however, temperature ranges for overall growth and development are similar.

3. Chilling

Most fruit crops require a period of low temperature or "chilling" following the onset of dormancy in order for normal growth and development to occur. Insufficient chilling results in delayed and erratic budbreak and a reduction in the number of buds that break.[90–97] Field observations have indicated that insufficient chilling also decreases fruit set in blueberry[98]; however, more recent data suggest the reduction in set is probably due to a decrease in the number of floral buds that break, rather than to a direct effect of chilling on the physiology of fruit set.[97] Chilling accumulation may be influenced by environmental and/or cultural factors both before and during the chilling process.

a. Factors Prior to Chilling Accumulation

There is no information on the effects of environmental or cultural factors prior to dormancy onset on chilling accumulation in *Vaccinium*. In general, young, vigorously growing blueberries are delayed in the onset of dormancy and grow into late fall and early winter. There is no evidence to suggest, however, that this directly affects the chilling requirement. Spiers and Draper[92] found that 'Tifblue' rabbiteye blueberries hand-defoliated prior to chilling temperatures broke vegetative buds faster the following spring than did non-defoliated plants. The presence of leaves on 'Bonita' rabbiteye blueberry plants during chilling inhibited both the rate and amount of vegetative and floral budbreak.[99] Interestingly, budbreak was also inhibited on defoliated canes when there were foliated canes present on the same plant. This suggests that the presence of foliated canes inhibits either chilling accumulation of buds on the defoliated canes or some process after chilling accumulation that leads to budbreak. Phloem girdling experiments on these plants indicated that the "inhibitor" was phloem mobile, and was apparently translocated from leaves to other parts of the plant. These findings are of particular relevance in areas such as the southeastern United States, where insufficient winter chilling of blueberry may occur.

Flower buds formed on spring growth flushes of rabbiteye blueberry usually bloom before flower buds formed on fall growth flushes.[24] In a study of several rabbiteye cultivars, Davies[100] reported that lateral flowers, which initiated later in the season than terminal flowers, also opened later in the spring. This suggests that flower buds formed later in the fall may begin chilling accumulation at a later date than do flower buds initiated earlier. Alternatively, buds initiated later in the season may be less developed upon entering dormancy and therefore require further development after the chilling requirement is met. There have been no studies in which precise bud developmental events have been correlated to chilling accumulation.

b. Factors During Chilling Accumulation

A substantial amount of research has focused on the effects of different chilling regimes on vegetative and/or floral budbreak in *Vaccinium* species. Interpretation of temperature effects on chilling accumulation is complicated by the use of different models to estimate chill unit accumulation, as well as the use of constant vs. fluctuating temperatures, and growth chamber vs. outdoor experiments. In general, the rate and amount of floral and vegetative budbreak in blueberry and cranberry increases as constant chilling in the range of 0.6 to 15°C increases[90–97,101,102] (Figure 5). Blueberry floral buds generally break before or concomitant with vegetative buds (depending on chilling regime), implying that floral buds have a lower chilling requirement than vegetative buds.[92,97] Species vary with respect to temperature ranges and optima, as well as the number of chill units required. Chandler and Demoranville[102] reported that cranberries require 2500 hours below 7.2°C for completion of rest and normal budbreak. Although less chilling did result in some budbreak, the subsequent growth of those buds was abnormal. Eady and Eaton[90] found that the amount of budbreak in cranberry increased as chilling time increased from 650 to 2500 hours. Growth was normal at all chilling regimes, in contrast to findings by Chandler and Demoranville.[102] Rigby and Dana[103] concluded that 600 to 700 hours below 7.2°C was sufficient to

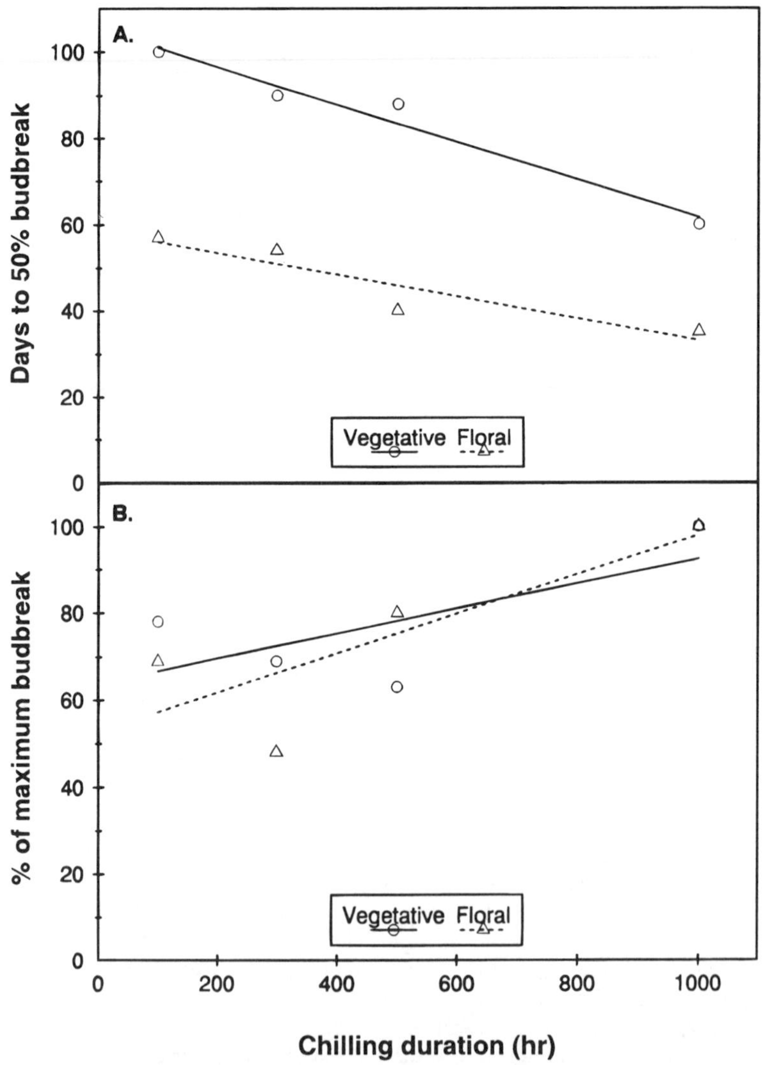

Figure 5 Days to floral and vegetative budbreak (A) and percent floral and vegetative budbreak (B) in 'Woodard' rabbiteye blueberry after various chilling times at 7°C. From Darnell and Davies, *HortScience*, 25, 635, 1990. With permission.

Table 3 **Days to vegetative budbreak of rooted cuttings of 'Woodard' rabbiteye blueberry under continuous and interrupted chilling conditions**

Chilling temp (°C)	Days to budbreak									
	Hours continuous chilling					Hours chilling + 14 days at 30°C				
	250	350	450	550	650	250	350	450	550	650
Constant										
0.6	65.5ab[a]	46.0	54.5	48.0a	44.5a	39.0ab	42.5a	30.3ab	25.8	21.0ab
3.3	62.0ab	56.3	50.0	42.0ab	42.0a	52.0a	30.3ab	24.0ab	23.5	17.8b
7.0	61.3ab	48.3	50.8	33.0c	31.0b	36.0	20.0b	25.5ab	21.8	14.0b
10.0	63.0ab	60.7	50.8	37.8bc	38.6ab	33.3ab	14.8b	17.3b	21.0	18.8b
15.0	64.7ab	53.0	50.3	44.3ab	40.3a	41.7ab	22.5b	28.0ab	22.5	19.5b
Diurnal										
0/7	69.5a	56.0	52.8	39.3bc	40.7a	48.7ab	39.8a	34.7ab	20.5	19.0b
7/15	52.3b	58.3	46.3	42.7ab	41.5a	30.5b	23.0b	24.5ab	21.3	19.5b
7/30	65.5ab	58.0	54.4	34.5bc	34.5bc	41.7ab	28.0ab	35.0a	23.7	28.8a

[a] Mean separation within columns by Duncan's multiple range test, $p = 0.05$ level.
From Gilreath and Buchanan, *J. Amer. Soc. Hort. Sci.*, 106, 625, 1981. With permission.

satisfy rest in cranberry when plants were given long days (i.e., 16 hours) subsequent to chilling. Increased chilling time was required if plants were subjected to short days (i.e., 8 hours) after chilling. They suggested that the failure of Chandler and Demoranville to demonstrate budbreak in cranberry at chilling times less than 2500 hours may have been due to inadequate day length duration subsequent to chilling.

Northern highbush blueberries have a greater chilling requirement (800–1000 hours below 7.2°C)[96,101] than do rabbiteye blueberries (300–600 hours below 7.2°C).[92–94,97] In general, the optimum chilling temperatures for rabbiteye blueberry floral and vegetative buds are higher than those for highbush buds.[104] Within cultivars, there is also variability in chilling requirement. Darrow[101] found that 'Rubel' highbush blueberry had the lowest chilling requirement (800 hours below 7.2°C) and 'Jersey' the highest chilling requirement (>1060 hours below 7.2°C) of those cultivars examined. Shine and Buchanan[95] reported that 'Woodard' and 'Aliceblue' rabbiteye blueberries had higher chilling temperature optima (7.2 and 11.0°C, respectively) and wider effective temperature ranges (−2.5 to 15.0°C and −2.5 to 13.8°C, respectively) than did 'Tifblue' (6.7°C optimum, −1.2 to 12.9°C range). Furthermore, 'Tifblue' had a longer chilling requirement than many other cultivars, even at optimum temperatures, requiring in excess of 500 hours below 7°C for any floral budbreak to occur.[92,97]

Temperature fluctuations during chilling influence the chilling response in fruit crops.[105] Eady and Eaton[90,91] found that alternating diurnal temperatures of 2/10°C were more effective than constant temperatures in promoting budbreak in cranberry. Norvell and Moore[96] chilled highbush blueberry cultivars at constant temperatures of 1, 6, or 12°C for 750 or 1000 hours, and found that only the 6°C/1000 hour chilling regime resulted in budbreak. However, budbreak was similar under chilling regimes of 1000 hours at constant 6°C or alternating temperatures of 6/1°C for 1000 hours, even though the latter treatment consisted of only 500 hours at the previously determined optimum chilling temperature of 6°C. Spiers[93] reported that 'Tifblue' rabbiteye blueberries exposed to day/night temperatures of 7/18°C or 7/23°C responded similarly in terms of vegetative budbreak to plants that received constant 7°C chilling treatments. Floral budbreak was delayed by the 7/23°C regime compared to the 7/18°C or the constant 7°C regime. This suggests that floral buds of blueberry may be more sensitive than vegetative buds to negation of chilling by high temperatures. Gilreath and Buchanan[94] found that budbreak in 'Woodard' rabbiteye blueberry under diurnal regimes of 0/7°C, 7/15°C, and 7/30°C was similar to that obtained under constant conditions (Table 3). Surprisingly, interruption of the chilling regimes midway through by 14 days at 30°C increased the rate of both floral and vegetative budbreak compared to non-interrupted constant or diurnal chilling. In this study, high temperatures (30°C) during chilling had either no effect or a promoting effect on floral budbreak, in contrast to Spiers' findings, and may reflect cultivar and/or timing differences in response to warm temperatures.

c. *Factors After Chilling Accumulation*

After chilling is completed, accumulation of heat units is required before floral and vegetative budbreak occurs in temperate fruit crops. The heat unit requirements for budbreak in *Vaccinium* have not been determined. Normal flower development in cranberry appears to depend on the accumulation of sufficient heat units after the completion of chilling. For 'Stevens', maximal normal flower development occurred when plants were exposed to 1000 to 1100 hours above 7.2°C after receiving between 1000 and 1500 chilling hours below 7.2°C.[103] Carlson and Hancock[63] reported low temperature thresholds for heat unit accumulation in several highbush blueberry cultivars ranging from −7 to 7°C, but the number of heat units required for budbreak was not determined. Mainland[104] observed that inadequate vegetative budbreak occurred in highbush and rabbiteye blueberry even when winter chilling was adequate. This was attributed to faster flower bud development compared to leaf bud development in response to warm temperatures after chilling. The more rapid flower bud development then inhibited vegetative budbreak. This suggests that flower buds may have a lower temperature threshold and/or a wider temperature range than vegetative buds for the accumulation of heat units needed for budbreak.

Species within *Vaccinium* encompass a wide array of chilling requirements, from the low chilling rabbiteye and southern highbush blueberry to the high chilling cranberry and northern highbush blueberry. Although much research has focused on the chilling responses of *Vaccinium*, the processes involved in chilling accumulation and budbreak remain unknown.

B. *RUBUS* (RED RASPBERRY)

1. Growing Season Temperatures

Temperature effects on shoot and root growth of red raspberry are similar to those of other temperate crops. Root growth has been correlated with soil temperature, with optimum growth occurring between 14 and 22°C.[106] Little root growth occurred at soil temperatures below 14°C, while no soil temperatures above 22°C were recorded during the experiment. Shoot elongation increased as air temperatures increased from 10 to 21°C.[107]

Temperature appears to have significant effects on flowering of both summer-bearing and fall-bearing (primocane-fruiting) red raspberries. In summer-bearing red raspberries, flower buds are initiated and differentiated starting in the fall, with development continuing until bloom the subsequent spring. In fall-bearing red raspberries, terminal flower buds are initiated and differentiated during the summer, giving rise to a crop in the fall. Axillary flower buds are initiated and differentiated similar to flower buds in summer-bearing cultivars. Flower bud initiation in 'Malling Promise', a summer-bearing cultivar, increased as temperatures decreased.[107] Initiation was inhibited by temperatures of 15.5°C under 9- or 16-hour photoperiods, and 12.8°C under 16-hour photoperiods. Under 9-hour photoperiods and 12.8°C, flower bud initiation occurred within 6 weeks. At 10°C, initiation occurred within 3 to 5 weeks under either photoperiod. Flower bud initiation occurred in the summer-bearing cultivar 'Latham' when grown at temperatures of 22 to 24°C, but development beyond initiation did not occur.[108] Similarly, initiation in axillary nodes of the fall-bearing cultivar 'Heritage' occurred at 22 to 24°C, but further differentiation required a cold treatment.[108] Thus, although flower bud initiation was temperature independent, differentiation was arrested at warm temperatures.

Low temperatures do not appear to be a requirement for initiation and differentiation of terminal flower buds in fall-bearing raspberries. 'Heritage' primocanes flowered terminally at 80 nodes of growth in plants grown at temperatures of 22°C or greater.[109] Exposure of developing primocanes to 25 days of 7°C at the 10–12 or 14–16 node stage shortened the vegetative growth stage and promoted earlier flowering, such that flowering occurred at 32 and 28 nodes, respectively. Thus, low temperatures were not a requirement for flower bud initiation and differentiation, but rather influenced the time or growth stage at which flowering occurred. Temperature had no influence on the number of nodes that formed flower buds. Lockshin and Elfving[110] exposed 'Heritage' plants to day/night temperatures of 29/24°C or 25.5/20°C under 16-hour photoperiods. Flowering occurred when plants reached the 24–25 node stage, regardless of temperature. The rate of cane elongation decreased under the lower temperatures, resulting in a 2-week delay in flowering compared to the higher temperature. Plants exposed to the lower temperatures also produced fewer flowering nodes per cane.

2. Cold Hardiness and Chilling

Dormancy induction in red raspberries, as in many other temperate woody plants, requires decreasing photoperiods and temperatures. More than 10 weeks of 10°C and 9-hour photoperiods were necessary for the development of full dormancy in the red raspberry, 'Malling Promise'.[107] The dormant condition was reversible if plants were returned to higher temperatures and longer photoperiods prior to 10 weeks. Induction also occurred at 10°C under 14-hour photoperiods, but full dormancy was much slower to develop.

With the induction of dormancy comes the ability to withstand low winter temperatures. As with blueberry and cranberry, the timing and extent of acclimation development in red raspberry varies from year-to-year. Under Minnesota conditions, the extent of cold acclimation in 'Latham' red raspberry was similar between two consecutive years, averaging −23°C.[111] However, in the first year, this degree of cold acclimation was reached by mid-November, while plants subjected to that temperature in the field the following year were severely injured, and did not attain full cold hardiness until mid-December. In both years, hardiness was attained rapidly, increasing from −10 to −23°C within 2 to 3 weeks. McCartney[112] found that canes of 'Latham' collected from a Minnesota field in January showed little damage when exposed to 8 hours at −29°C, and, in fact, withstood temperatures of −45°C, exhibiting some injury and bud death, but overall cane survival.

Cold acclimation in red raspberry may be rapidly lost. Brierley et al.[113] reported that if acclimated canes were exposed to 4°C for 4 hours on two consecutive days, severe injury resulted when canes were exposed to subsequent low temperatures. They concluded that daily exposure to temperatures less than 4°C was necessary to avoid cold injury in the field. In a subsequent study on dehardening, Brierley and Landon[114] found that dehardened red raspberry canes can reharden if the buds do not "become active". Canes that had been exposed to a low of −20°C in the field were collected, and subjected to different dehardening/rehardening regimes. Canes exposed to sub-freezing temperatures for 16 hours immediately after collection showed moderate injury at −34°C and severe injury at −40°C. Canes that had been dehardened for 5 days at 3°C showed severe injury when exposed to 1°C for 24 hours and were killed when exposed to −18°C for 24 hours. Canes rehardened by exposing them to 2°C for 8 hours, followed by −9°C for 3 to 10 days showed only slight injury when exposed to −18°C for 24 hours and severe injury when exposed to −23°C for 24 hours. Thus, canes were able to recover some cold acclimation after a short period of dehardening.

Dormancy release in red raspberry required a minimum of 670 hours at 3°C, and budbreak occurred more rapidly as chilling time was increased to 1300 hours.[107] In general, 800–1600 hours below 7°C is necessary to satisfy the rest requirement in raspberry.[115] There is no information on the heat unit requirement for budbreak in red raspberry.

In general, temperature responses in red raspberry appear to be better documented than light responses, particularly temperature effects on flower bud initiation and differentiation. However, there are few data available on dormancy and chilling requirements of raspberry, and no data available on heat unit requirements for growth and development.

IV. WATER STRESS AND IRRIGATION

A. *VACCINIUM*

Water relations of *Vaccinium* species vary widely due to the wide range of ecological diversity found in the genus.[8] The cultivated species, *V. angustifolium, V. corymbosum, V. ashei*, and *V. macrocarpon* also differ significantly in their water relations, in particular their responses to drought and flooding stresses. In general, however, *Vaccinium* species are moderately drought[8] and flooding tolerant[116] based on physiological and anatomical factors. Diurnal and seasonal variations in leaf water potential (Ψ), g_s, water use efficiency (WUE), relative water content (RWC), transpiration (E), and *A* will be discussed as they relate to drought and flooding tolerance.

1. Diurnal and Seasonal Water Relations

Diurnal changes in Ψ, g_s, and E are closely related and thus it is difficult to isolate one factor from another. Nevertheless, diurnal changes in these factors follow a pattern fairly typical of many other plant species. Byers et al.[117] measured changes in Ψ (Figure 6) and g_s (Figure 7) in 2-year-old 'Bluecrop' highbush blueberries in the field. They found a typical late morning decrease in leaf Ψ with a minimum at 1000–1030 hour and a minimum value of −2.79 MPa. The Ψ remained low until about 1600 hours and then became less negative, returning to predawn values of <−0.05 MPa by 2000 hours. The pattern varied for different times of the year due to differing soil moisture and environmental conditions, viz., variations in vapor pressure deficit (VPD). The minimum Ψ attained was lower than that observed for mature rabbiteye blueberries in the field (−1.5 MPa[118]) and young rabbiteye plants in the greenhouse (−2.30 MPa[119]). The minimum is lower than that which causes damage to most herbaceous plants, but is less negative than found in many tree crops.

Diurnal patterns for g_s are associated with leaf Ψ, but as expected, are much more variable because factors other than Ψ also regulate g_s. Davies and Albrigo[8] observed a gradual decrease in g_s by mid-day for 'Bluegem' rabbiteye blueberry leaves, with a gradual increase by 1600 hours. Stomatal conductances ranged from about 32–66 mmol $m^{-2}s^{-1}$, which are quite low; young, fully expanded leaves generally had lower g_s than old, mature ones. Diurnal patterns of g_s for rabbiteye leaves were similar to those observed by Byers et al.[117] for highbush leaves at some times of the year, but not at others (Figure 7). Moreover, g_s patterns were not necessarily correlated with Ψ[117] (compare Figures 6 and 7). During times of minimum Ψ, g_s was often at a maximum. However, partial stomatal closure usually occurred in the afternoon after minimum Ψ had been reached, probably serving as a mechanism to reduce water loss and maintain cell turgor. This late afternoon reduction in g_s may also result from

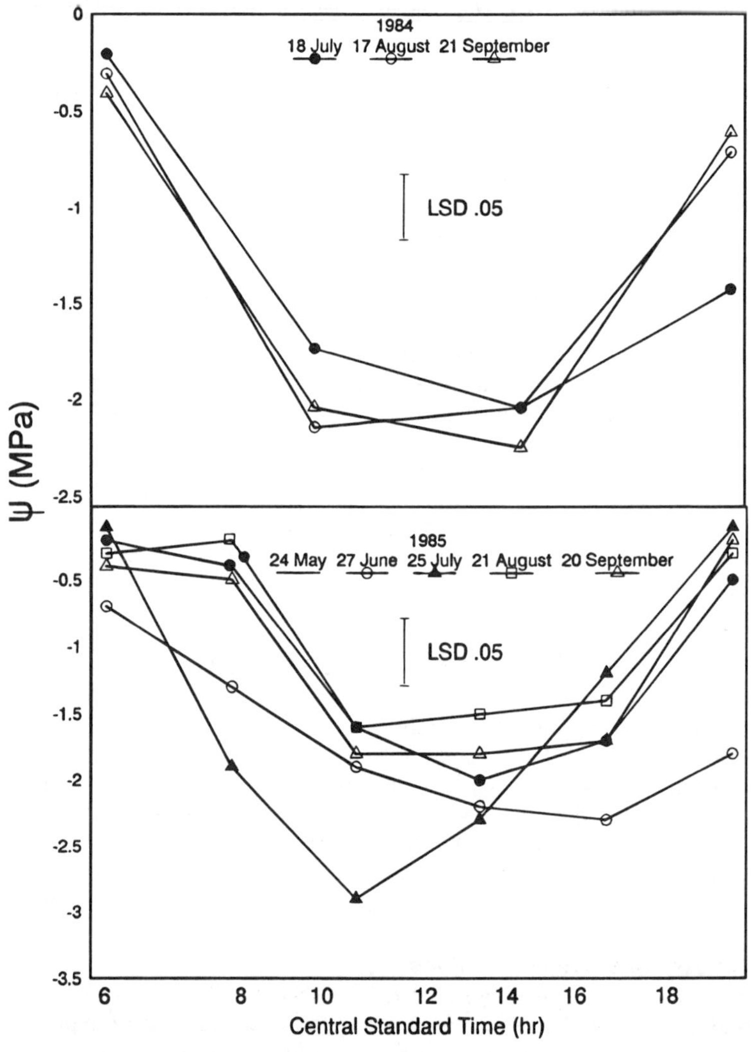

Figure 6 Diurnal patterns in young highbush blueberry leaf water potential (ψ) for various dates in 1984 and 1985. From Byers et al., *HortScience*, 23, 870, 1988. With permission.

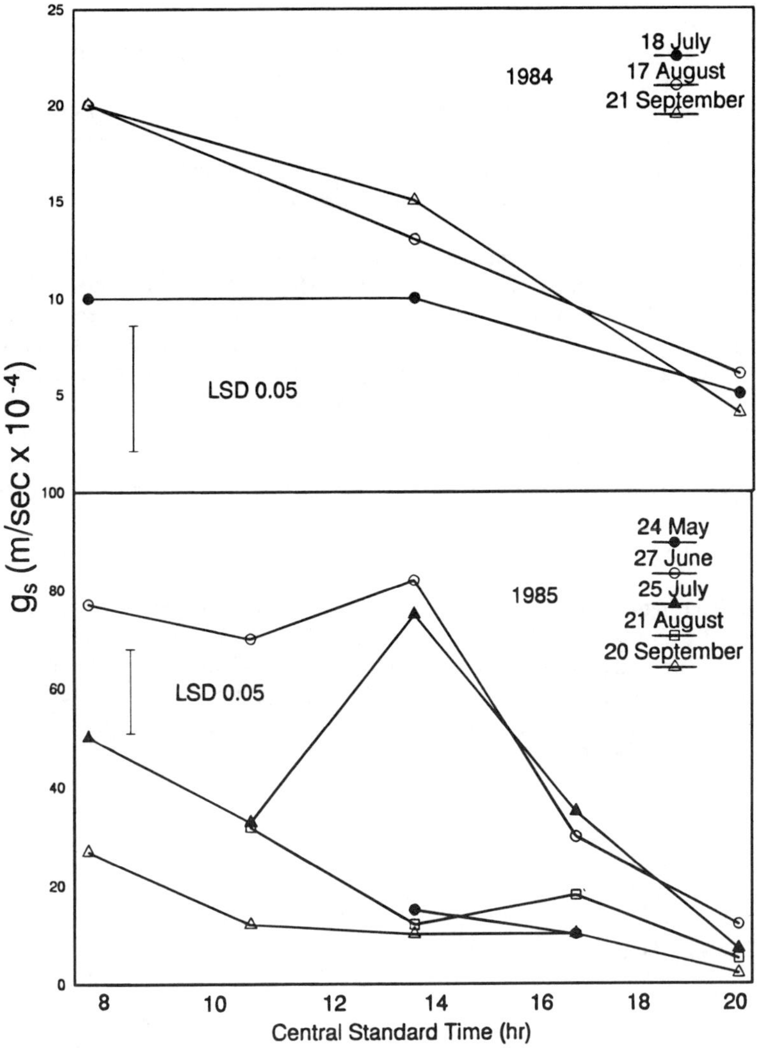

Figure 7 Diurnal patterns in young highbush blueberry leaf stomatal conductance (g_s) for various dates in 1984 and 1985. From Byers et al., *HortScience*, 23, 870, 1988. With permission.

increases in internal CO_2 as well as decreases in cell turgor. On some days (18 July), g_s remained consistently low with no characteristic diurnal pattern for the entire day, possibly due to high VPD. Similarly, Davies and Johnson[119] observed only a moderate correlation ($r^2 = 0.42$) between Ψ and g_s for rabbiteye blueberry growing in the greenhouse, although they did determine that the critical Ψ for stomatal closure occurred at about -2.22 MPa. In contrast, Byers et al.[117] did not observe complete stomatal closure even at -2.79 MPa for 'Bluecrop' plants in the field. This observation probably reflects differences in growing conditions prior to imposition of water stress. The 'Bluecrop' plants had been growing in the field while the rabbiteye plants were grown in less stressful, lower VPD greenhouse conditions.

Andersen et al.[118] measured seasonal changes in Ψ for mature, 5-year-old rabbiteye blueberry bushes in the field. Mid-day stem Ψ varied from about -0.5 to -1.5 MPa. The smaller range of values observed here compared to previous studies probably resulted from the measurement of stem rather than leaf Ψ, which due to their greater mass would show less fluctuations than in leaves. Nevertheless, a distinct seasonal variation in Ψ occurred, again related to differences in VPD and soil moisture content.

Andersen et al.[118] also measured large seasonal variations in g_s (although they actually measured leaf resistance) (Figure 8). Estimated values for g_s, calculated from midday leaf resistance, ranged from 400 mmol $m^{-2}s^{-1}$ in expanded, young leaves in April (irrigated) to as low as 20 mmol $m^{-2}s^{-1}$ in nonirrigated mature leaves later in the season (refer to next section on irrigation). There were also differences among cultivars, with 'Tifblue' generally having lower g_s than 'Woodard' and 'Bluegem'. In this study, g_s of young, expanded leaves was considerably greater than that of mature leaves, which differs from the previous findings of Davies and Albrigo.[8] However, Davies and Albrigo[8] compared g_s of fully expanded leaves during midseason, while Andersen et al.[118] compared young, expanded leaves which developed during the spring to mature leaves—a factor which may account for the discrepancies between the two studies.

Leaf E also varied seasonally as a function of g_s and most importantly VPD.[118] Midday E was greatest for rabbiteye blueberry leaves early in the season (April) (Figure 9) primarily due to high g_s in young leaves, even though VPD was lower in April than in June. Transpiration then remained relatively constant through the remainder of the season. This likely resulted from rodlet wax accumulation over the stomatal pore which stabilized water fluxes from the leaves. Transpiration was generally lower for all three cultivars for nonirrigated compared with irrigated bushes. Differences between the two irrigation treatments, however, did not occur until early May for 'Bluegem' and 'Woodard', but were apparent in April for 'Tifblue'. This difference in E under the same environmental conditions is related to differences in g_s among cultivars and reflects the greater water stress in 'Tifblue' compared to the other two cultivars.

In a controlled laboratory study, Moon et al.[13] compared E of 'Jersey' and 'Bluecrop' highbush blueberry to that of *V. darrowi*. *V. darrowi* was chosen because field observations suggest it is more drought tolerant than highbush cultivars. Transpiration of 'Jersey' ranged from 1.95–3.45, 'Bluecrop' from 2.25–3.75 and *V. darrowi* from 1.38–2.68 mmol $m^{-2}s^{-1}$ at varying controlled VPD levels, suggesting that *V. darrowi* has inherently lower E than the other cultivars. These values are similar to the range of E values for rabbiteye and highbush blueberries (0.6–4.2 mmol $m^{-2}s^{-1}$) under greenhouse conditions.[12]

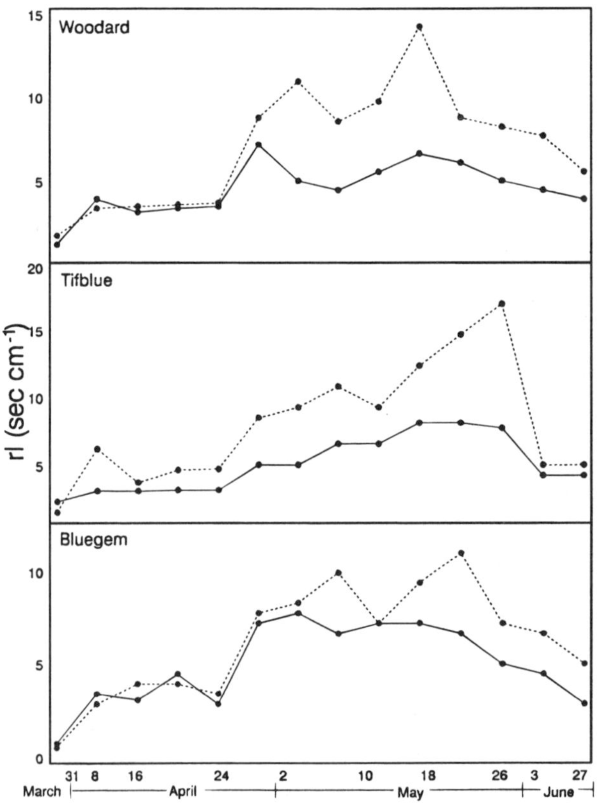

Figure 8 Midday leaf diffusive resistance (rl) for irrigated and nonirrigated plants of 3 rabbiteye blueberry cultivars, Gainesville, Florida, 1977. Dashed line = irrigated, solid line = not irrigated. Vertical bars = SD. From Andersen et al., *J. Amer. Soc. Hort. Sci.*, 104, 731, 1979. With permission.

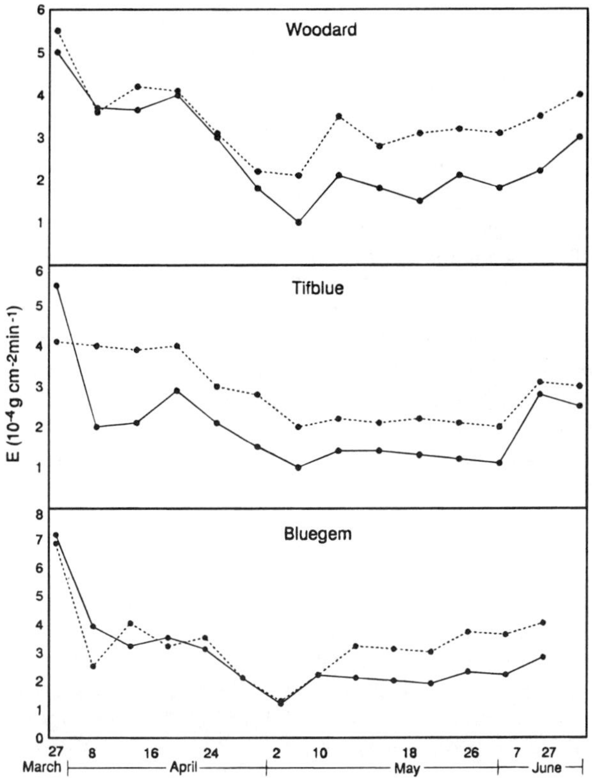

Figure 9 Midday maximum calculated leaf transpiration (E) for irrigated (---) and nonirrigated (—) plants of 3 rabbiteye blueberry cultivars, Gainesville, Florida, 1977. From Andersen et al., *J. Amer. Soc. Hort. Sci.*, 104, 731, 1979. With permission.

Andersen,[120] however, measured E as high as 6.2 mmol $m^{-2}s^{-1}$ for rabbiteye bushes in the field in Florida during the summer. Leaf temperature was 36.9°C and VPD reached as high as 5.5 kPa, which probably accounted for the high E observed in this study compared with laboratory or greenhouse studies.

Water use efficiency (WUE) is the ratio of mmol CO_2 fixed:mmol of water transpired. It is the recipricol of E ratio (g water transpired:g CO_2 fixed). Blueberry plants have moderate to high WUE depending on species or cultivar and environmental conditions during and before measurements. *V. darrowi* had a greater WUE and thus was more drought tolerant than the highbush cultivars, 'Jersey' and 'Bluecrop', averaging 6.5, 4.7, and 4.8, respectively, at 1 kPa VPD.[13] As VPD was increased to 3 kPa, however, WUE decreased by about 45% for 'Bluecrop' and 'Jersey' and 57% for *V. darrowi*. Teramura et al.[14] measured very low transpiration ratio (108) and high WUE (9.2) for detached cuttings of rabbiteye blueberry. Because detached shoots were used, however, these values are probably unrealistically high. Davies and Johnson[119] estimated whole plant WUE of 4.5 for 'Bluegem' rabbiteye blueberry in the greenhouse. This value is more consistent with WUE reported by Moon et al.[13] of 4.7 at a VPD of 1 kPa; WUE decreased significantly, however, to 2.6 at a VPD of 3 kPa. A decrease in WUE is typical under water stress conditions, as the photosynthetic capacity of the leaf decreases. Consequently, field calculated WUE was as low as 0.50 for 'Bluegem' bushes in Florida.[120] This very low value was due to the extremely high VPD (5.5 kPa) experienced in Florida during the summer. Conversely, Cameron et al.[121] found that WUE was relatively constant, ranging from 2.7–4.1 for water stressed and nonstressed 'Bluecrop' and 'Jersey' highbush blueberries growing in containers. These plants, however, were exposed to two short-term water stress cycles rather than long-term steady water stress, which could account for such differences.

2. Factors Affecting Stomatal Conductance

Stomatal conductance is influenced by anatomical characteristics of stomata and possibly stomatal density. Limited information is available on stomatal density of *Vaccinium* species. 'Bluecrop' sun leaves averaged 553 stomata/mm^2. In contrast, stomatal density of the whortleberry (*V. myrtillus*) averaged 990 stomata/mm^2 for shade leaves and 177/mm^2 for sun leaves,[122] clearly illustrating the effect of light on stomatal density. Janke[122] proposed that the increased stomatal density of sun leaves provided additional transpirational cooling compared to shade leaves. Stomata of rabbiteye blueberry have a large antechamber and are partially obscured by wax rodlets within and over the antechamber.[118,123] This anatomical adaptation decreases g_s and reduces E and may be an adaptive advantage during times of high evaporative demand for reducing plant water stress.

Environmental factors also affect g_s. Although light is necessary for stomatal opening, it is usually not a limiting factor since stomata open at relatively low light levels of about 50 μmol $m^{-2}s^{-1}$.[15] Most important to stomatal response is VPD. Davies and Flore[15] exposed 'Bluecrop' highbush blueberry plants to VPD of 1 to 3 kPa under controlled laboratory conditions with light and temperature controlled at optimum levels. They observed a significant linear decrease in g_s with increasing VPD, but found *A* to be less affected by VPD than g_s. Similarly, Moon et al.[13] measured the effect of varying VPD on leaf conductance (g_l) and *A* for 'Jersey' and 'Bluecrop' highbush blueberries and *V. darrowi*. They also observed a linear decrease in g_l with increasing VPD (Figure 10), again finding less of an effect of VPD on *A* (Figure 11). When VPD was increased from 1 to 3 kPa, g_l of highbush and *V. darrowi* decreased by about 50%, while *A* decreased only 10–20%. By comparison, E, which is strongly influenced by VPD, increased by 80–94% (1.5 to 3.4 mmol $m^{-2}s^{-1}$ for 'Jersey') over the same VPD range. Studies on *V. myrtillus* from the wild also suggest a strong dependency of g_s on VPD.[123]

A wide range of g_s values, probably reflecting both species and environmental differences, have been reported for *Vaccinium* species. Under laboratory conditions, maximum g_s were 323 for 'Bluecrop', 236 for 'Jersey', and 168 mmol $m^{-2}s^{-1}$ for *V. darrowi*.[13] Davies and Flore[12] reported maximum values of 291 for 'Bluecrop' highbush blueberry but only 219 mmol $m^{-2}s^{-1}$ for 'Woodard' rabbiteye blueberry, and in general, g_s for 'Woodard' was less than that of 'Bluecrop' under similar greenhouse conditions. However, Andersen et al.[118] measured g_s as low as 20–132 mmol $m^{-2}s^{-1}$ for rabbiteye cultivars in the field. In a subsequent study Andersen[120] measured g_s of 130 mmol $m^{-2}s^{-1}$ for 'Bluegem' leaves in the

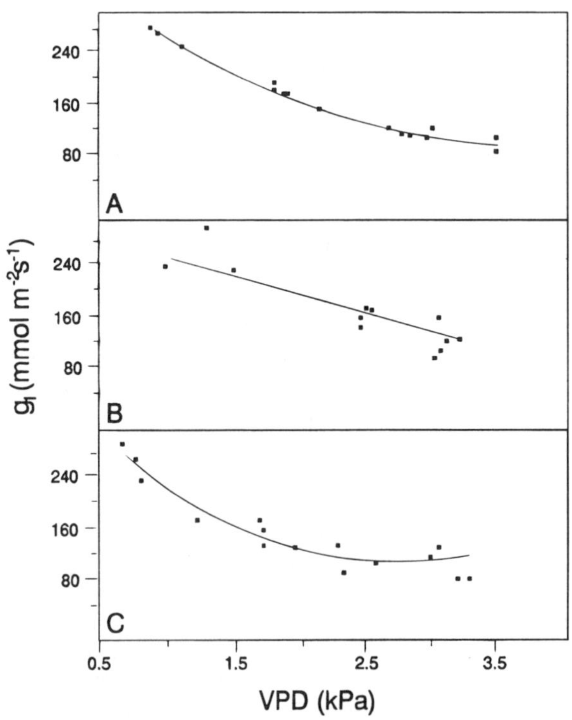

Figure 10 Effects of VPD on leaf conductance (g_l) in 'Bluecrop' (A), 'Jersey' (B), and *Vaccinium darrowi* (C). Measurements were made at saturating PPF (1000 μmol·$s^{-1}m^{-2}$) and at 26°C for 'Bluecrop' and 'Jersey' and at 30° for *V. darrowi*. Each value is the mean of 20 determinations. Different symbols represent different plants. From Moon et al., *J. Amer. Soc. Hort. Sci.*, 112, 134, 1987. With permission.

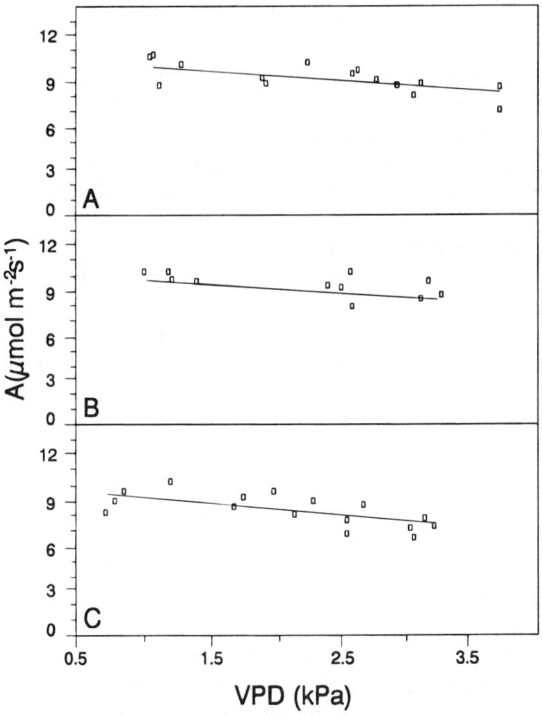

Figure 11 Effects of vapor pressure deficit (VPD) on *A* in 'Bluecrop' (A), 'Jersey' (B), and *Vaccinium darrowi* (C). Measurements were made at saturating PPF (1000 μmol·s^{-1}m^{-2}) and at 26°C for 'Bluecrop' and 'Jersey' and at 30° for *V. darrowi*. Each value is the mean of 20 determinations. Different symbols represent different plants. From Moon et al., *J. Amer. Soc. Hort. Sci.*, 112, 134, 1987. With permission.

field. The lower g_s values for rabbiteye vs. highbush cultivars again probably reflects the wax rodlet accumulation in and over the stomata and also the greater VPD usually encountered for rabbiteye plants growing in the southern rather than the northeastern United States.

3. Drought Tolerance

a. Physiological Factors

Physiological evidence strongly suggests that while *Vaccinium* species are clearly mesophytes, they are moderately drought tolerant, although some variability exists among species.[8] Water use efficiency[125] and g_s[126] values for *Vaccinium* are within normal ranges for mesophytes. The change in leaf RWC per change in Ψ is an indicator of relative drought tolerance.[127] For example, drought-sensitive plants such as tomato show a Δ RWC/Δ ψ of 25%, while xerophytic plants such as *Acacia harpophylla* show a change of only 2.7%. In comparison, *V. ashei* had a Δ RWC/Δ ψ of about 6.4%[119] and *V. myrtillus* about 10.5%,[8,128] again suggesting an intermediate position. Furthermore, critical leaf Ψ for stomatal closure of −2.2 MPa is also intermediate to that observed for herbaceous and xerophytic species.[119]

b. Morphological and Anatomical Factors

Drought tolerance in *Vaccinium* species includes morphological and anatomical adaptations.[8] Morphological adaptations are found in the prostrate growth habit of *V. darrowi* and extensive root systems of *V. darrowi* and *V. arboreum* in the wild.[8] In contrast, highbush[129] and rabbiteye[130] blueberry plants are typically shallow-rooted with most feeder roots located in the upper 30 cm of the soil profile. However, maximum rooting depth of highbush plants reached 81 cm and lowbush blueberries produced taproots to 90-cm depths.[131] Some species such as *V. darrowi* also have leaves with a small surface area which reduces whole-plant E.[8] Development of a waxy cuticle on the leaf, which reduces cuticular E, is also an important anatomical adaptation to drought. Freeman et al.[123] studied seasonal changes in leaf epicuticular waxes for 'Bluegem' rabbiteye blueberries. They observed a rapid increase in β-diketones during early leaf expansion through mid-May, when levels decreased. Moderate levels of triterpenoids and primary alcohols and low levels of fatty acids and paraffin were also present. By October, β-diketones had decreased with concomitant increases in triterpenoids and primary alcohols. Rodlet wax, which covers and occludes stomata, varied seasonally and the concentration was inversely correlated with leaf E. In contrast, *V. elliotti*, which is also moderately drought tolerant, does not have extensive rodlet wax covering of leaves, but instead produces trichomes. These may serve to decrease leaf boundary layer conductance and improve drought tolerance.[132] Certainly, drought tolerant *Vaccinium* species, such as *V. darrowi*, have several anatomical and physiological drought tolerance mechanisms.

c. *Species Variations in Drought Tolerance*

Several studies have been conducted on wild *Vaccinium* species involving species-related differences in drought tolerance. Hygen[133] measured E of cowberry (*V. vitis-idaea*), a xerophyte, whortleberry (*V. myrtillus* L.), a mesophyte, and bog bilberry (*V. uliginosum* L.), a hydrophyte. *V. vitis-idaea* L. had lower E with open stomata than the other species and also more rapid stomatal closure. It also had much higher stomatal than cuticular E. These data imply that development of a less permeable cuticle and improved stomatal control of E are drought tolerance mechanisms found in some *Vaccinium* species. In a similar study, Bannister[134] measured diurnal fluctuations in E and RWC of heather (*Calluna vulgaris* [L.] Hull), bog heather (*Erica tetralix* L.), and Scotch heather (*E. cinerea* L.). The *Erica* species had lower E and more rapid stomatal closure than *Calluna*, factors which improve drought tolerance. However, the *Erica* species were also less ecologically competitive in dry climates because they accumulated less assimilates than *Calluna* due to rapid stomatal closure. Therefore, rapid stomatal closure, while improving drought tolerance within a region, may not be adaptive because it limits the ecological range of a species.

Most of the above-mentioned wild species have not been used to incorporate drought tolerance into cultivated *Vaccinium* species, however, several related species have been and are currently being used, in breeding programs, including *V. darrowi, V. tenellum, V. myrsinites*, and *V. elliotti*.[3] Erb et al.[135] screened several *Vaccinium* species for potential drought tolerance by evaluating shoot growth and biomass accumulation of seedlings under drought conditions. They found a wide range of inherent drought tolerance with *V. darrowi, V. elliotti*, and *V. ashei* being superior to *V. corymbosum* and *V. vacillans (pallidum)*, all of which were superior to *V. myrtilloides* (Table 4). The percentage of biomass partitioned into roots did not appear to influence drought tolerance, as measured by shoot growth. The species having the greatest drought tolerance were native to the southern United States and evolved under high temperature and VPD conditions. The highbush blueberry, *V. corymbosum*, is native to the northern United States and southern Canada and *V. myrtilloides* is a lowbush species native to southern Canada.[1] Therefore, these species evolved under less extreme temperature and VPD conditions. The same researchers tested several *Vaccinium* hybrids also and found a range of potential drought tolerance with future possibilities as commercial cultivars.

4. Irrigation

Irrigation of blueberries and cranberries has not been a widely used practice in the past. Often growers believed there was sufficient rainfall to produce an adequate crop or the topography was too uneven

Table 4 **Mean shoot damage ratings and fraction of total biomass partitioned as roots for blueberry populations from open-pollinated species screened for drought resistance**

Species	n	Shoot rating mean[a,b]	Mean root biomass fraction[a]
Vaccinium darrowi	54	7.3a	0.13d
V. elliotti	54	7.2a	0.22c
V. ashei	54	6.8a	0.30b
V. corymbosum	54	4.3b	0.39a
V. vacillans (pallidum)	54	4.0b	0.29b
V. myrtilloides	44	2.7c	0.22c

[a] Mean separation by Duncan's multiple range test, 5% level. [b] Shoot damage ratings ranged from 1 = dead stems to 9 = a healthy, undamaged shoot.

From Erb et al., *J. Am. Soc. Hort. Sci.*, 113, 599, 1988. With permission.

to establish reliable irrigation systems, as is the case in some lowbush blueberry areas. Research studies suggest, however, that in many instances irrigation improves yields of blueberries and many new plantings are irrigated. Water table regulation has always been important in cranberry culture for irrigation, pest control, frost protection, and harvesting.

a. Blueberries

A limited amount of research data is available on irrigation of lowbush blueberries. Struchtemeyer[136] compared yields of irrigated and nonirrigated lowbush blueberries in Maine. Irrigation at 50–70% of available soil moisture produced a 2.6-fold yield increase by improving fruit size and number. Irrigation also produced more consistent cropping. Trevett[137] found that irrigation increased stem height after burning of lowbush blueberries; however, it had no influence on fruit bud number. Yield data were not presented. An elaborate greenhouse and field study was conducted by Benoit et al.[138] in Maine. Soil moisture tension was controlled at 2, 6, 10, or 50 kPa using tensiometers with 2 kPa being field capacity. More flowers/plant were produced in the greenhouse at 2 and 6 kPa than at the other irrigation levels. Similarly, irrigation at 2 kPa increased numbers of flowers and fruit even in the first year of the burn cycle. The improved yields resulted from increased fruit number, not weight. Moreover, irrigation also improved flower bud number in the following year.

Irrigation of highbush blueberry bushes has been used sporadically in most long-established plantings but is becoming more common in new plantings. Fry and Savage[139] conducted a study primarily to evaluate the effects of sawdust mulch on young blueberry plant growth. They also used irrigation as one of the treatments, finding no effect of irrigation on plant growth. No details were given on irrigation levels, however. Byers and Moore[140] conducted a 1-year study in Arkansas using 3-year-old 'Bluecrop' plants growing in barrels. They scheduled irrigation based on pan evaporation using crop coefficients of 1.0, 0.75, and 0.50 and also varying irrigation with developmental stage of the bush. They achieved a considerable reduction in water use with this method relative to recommended levels of 3.8 l/day; nevertheless, there were no differences in yields, fruit weight, quality, or vegetative growth related to irrigation levels. This lack of effect possibly resulted from the large plant-to-plant variability among bushes.

Similar irrigation studies were conducted in Missouri over a 3-year period using 'Blueray' bushes.[141] Irrigation was scheduled using tensiometers placed at 15- or 30-cm depths and included soil tensions of 10, 35, or 65 kPa vs. no irrigation. Results varied with location and season, sometimes with no effect of irrigation being observed due to high rainfall and the high water holding capacity of heavy, poorly drained soils. However, in general, irrigation at soil tensions less than 65 kPa increased bush height and cane diameter compared to nonirrigated plants. Leaf area, Ψ, and number of new canes, however, were not affected by irrigation levels. Dale et al.[142] working in Ontario, Canada varied irrigation levels from 18 to 4.5 l/day for 6 years after planting. They found that the highest irrigation level gave the greatest cumulative yield for both 'Blueray' and 'Collins' plants; bush height was also greatest for 'Collins' but irrigation levels did not affect growth of 'Blueray'. Unfortunately, no adjustments in irrigation rates were made for bush size and soil moisture levels were not reported. It is apparent from these two studies that irrigation effectiveness is influenced by environmental and edaphic conditions and cultivar. Therefore, although irrigation is generally effective, local conditions also must be considered in deciding whether irrigation is economically warranted.

With the advent of low volume (trickle) irrigation there has been some concern that only a relatively small area of the blueberry root system would receive water. This prompted Abbott and Gough[143] to conduct a split-root experiment using 2-year-old 'Bluecrop' plants. One side of the root system received no water, while the other received 300 ml/day. There was little if any lateral movement of water within the plant as evidenced by significantly less dry weight production and no flower bud production of the dry side. These findings support the need for adequate coverage of the root zone during irrigation, although the percentage of the root area to be covered has not been established.

Irrigation is more widely used for rabbiteye compared with other blueberry species probably because many new plantings have been established recently in the south where daytime VPDs are extreme, sometimes reaching 5.5 kPa.[144] Therefore, the need for irrigation has been more apparent than in cooler, northern regions. Buchanan et al.[145] compared yields of 5-year-old 'Tifblue', 'Bluegem', and 'Woodard' rabbiteye blueberry bushes in Florida using 1, 2, or 3 drippers (emitters, 4 l/hour) per bush. Irrigation increased soil moisture levels from 12% in nonirrigated plants to 23% for irrigated treatments. All irrigation levels generally increased yields by increasing berry weight, not berry number/bush. Yield increases of 20–25% occurred in seasons having a dry spring but no differences occurred during seasons with a wet spring, again illustrating the importance of environmental factors in success of irrigation. Similarly, Andersen et al.,[146] working in the same planting, found drip irrigation increased yields and berry weight, but decreased berry soluble solids by 24%. 'Tifblue' and 'Bluegem' yields were increased more than those of 'Woodard', possibly because 'Woodard' bushes were under less water stress than the other cultivars (see Figures 8 and 9).

Several studies have been conducted on irrigation of young (1 to 3-year-old) rabbiteye blueberry bushes. Spiers[147] found that irrigation increased survival and doubled dry weight for 'Tifblue' plants for the first 2 years after planting in Mississippi. He also observed that incorporating peat moss in the planting hole, which improved soil water holding capacity, increased bush survival and growth. In a subsequent study, Spiers[148] found that peat moss incorporation, plus irrigation and mulch increased root dry weight over nontreated bushes. Patten et al.[149] compared trickle with microsprinkler irrigation (40 or 360° patterns) on 3-year-old 'Climax' bushes in Texas, finding that the microsprinklers, particularly when combined with sawdust mulch, increased radial spread of roots from the main canes. Subsequently, Patten et al.[150] compared various trickle, microsprinkler, sod and mulch combinations on development of 'Tifblue' and 'Delite' rabbiteye blueberries for 3 years in the field (Table 5). In general, the 360° microsprinklers provided the greatest growth (canopy volume) and yields, with the 40° pattern intermediate but superior to the two drip irrigation treatments. Berry weight, however, was inversely correlated with yield, with the 360° treatment having the lowest weight, indicating that irrigation increased yield by increasing berry number. Fruit acidity was not affected by irrigation treatment. These data support previous studies by Abbott and Gough[143] on the importance of sufficient root coverage when irrigating blueberries, a factor that is even more important as bushes increase in size. Smajstrla et al.[151] and Haman et al.[152] studied effects of drip (trickle) irrigation on establishment of 2-year-old 'Climax', 'Beckyblue', and 'Sharpblue' bushes in Florida. They found that maintaining soil moisture tension at 10 kPa at 8- and 22-cm depths produced the greatest growth during the first year after planting.

In most blueberry growing regions, water quality, particularly total dissolved solids levels, is not a limiting factor in production. Occasionally, moderately high pH water (pH 7.5) is used to irrigate blueberries in Florida.[153] The water is quite high in calcium carbonates and plant vigor has been reduced, particularly on sandy soils, where this water has been used for irrigation. Water from some deep wells in Texas has a pH of 8.5 with electrical conductivities of 0.695 dsm^{-1}. Under these conditions, Haby et al.[154] found that growth of 'Tifblue' bushes was considerably reduced when irrigated with high salinity water primarily due to sodium toxicity. In subsequent studies, Patten et al.[155] compared drip and microsprinkler application of low and high saltwater to 3-year-old 'Tifblue' bushes. Although bushes receiving the sodic water appeared chlorotic at the end of the season, growth was unaffected. However, leaves accumulated high levels of sodium. Mulching and use of microsprinklers vs. drippers decreased salinity levels in the soil. Consequently, Patten et al.[156] tested methods for reducing salinity damage to rabbiteye blueberry plants. Again, mulching and sprinkler irrigation were effective in reducing bush damage and soil amendments of gypsum also proved successful. Cultivars varied in sensitivity to high salinity levels with 'Delite' and 'Brightwell' being more tolerant than 'Tifblue', 'Premier', or 'Climax'.

Therefore, irrigation of blueberries is generally effective in improving yields of mature bushes and growth of young bushes. The trend appears to be to provide as much root zone coverage as possible and to use irrigation in conjunction with peat moss and mulch to obtain maximum growth.

b. Cranberries

Irrigation of cranberry plants has traditionally been done by adjusting the water table depth due to the unusual growth habit and plant configuration of a cranberry bog.[56] Plants are irrigated by adjusting water levels in adjacent canals and by allowing vertical water movement via capillary action. Most studies suggest that the water table should be maintained from 23 to 37 cm below the soil surface. Beckwith[157] found yields were numerically highest at water table depths of 23–30 cm, but yields were similar at 15–23 or 30–37 cm depths. Yields decreased only at the 7.5–15 cm depth. Similarly, Eck[158] studied effects of water table depth on 'Early Black' cranberry growth and yields over a 5-year period. Yields were greatest in most years at the 30–38 cm depths, although yields were also acceptable at the 38–46 cm depth. Length of uprights and runner growth were not affected by water table depth. Hall[159] compared three water levels using cuttings of 'Beckwith' and 'Pilgrim' cranberries growing in containers. Plants initially grew fastest at the highest water level (5.5 cm) with growth slowing later in the season. However, plants growing at the lowest (34.5 cm) water table depth grew most during the final 5 weeks

Table 5 Influence of cultivar, sod, mulch, and irrigation treatments on the growth, yield, frost damage, and bud development of rabbiteye blueberry plants

	Canopy volume (m^3)			Yield (kg/ plant)	Frost damage[a]	Vegetative bud development[b]	Fruit wt (g)	Fruit from first harvest		Titratable acidity
Cultivar/treatment	1984	1985	1986	1985 and 1986	1986	1986	1986	1985	1986	(%)1985
Tifblue	0.13	0.31	0.75	3.02	1.04	2.24	1.52	74	58	1.45
Delite	0.07	0.21	0.55	2.22	2.20	2.20	2.15	81	68	1.52
	**	*	*	*	***	NS	**	*	*	*
Sod	0.10	0.27	0.69	2.54	1.63	2.13	1.85	75	62	1.53
No sod	0.09	0.26	0.61	2.68	1.61	2.30	1.82	79	65	1.45
	NS	NS	NS	NS	NS	**	NS	NS	NS	**
One drip emitter	0.08c	0.23c	0.59b	2.51c	1.49b	2.29b	1.96a	82a	68a	1.47b
Two drip emitters	0.07c	0.21c	0.59b	1.90d	1.66ab	2.48a	1.90a	83a	68a	1.59b
360° low volume spray emitter (LVSE)	0.15a	0.37a	0.87a	3.29a	1.83a	2.05c	1.75b	76b	64a	1.44b
40° LVSE	0.10b	0.26b	0.53c	2.76b	1.50b	2.05c	1.73b	69c	54b	1.45b
Interaction significance										
Cultivar × mulch	NS	NS	NS	NS	NS	NS	*	**	**	***
Cultivar × irrigation	**	***	*	***	NS	NS	NS	NS	**	**
Mulch × irrigation	**	***	**	***	NS	***	NS	***	NS	NS
Cultivar × mulch × irrigation	**	NS	***	***	NS	NS	NS	NS	NS	NS

Note: * = Significant at $p = 0.05$; ** = significant at $p = 0.01$; *** = significant at $p = 0.001$; NS = not significant.

[a] Rating of frost damage on 11 March 1986 (1 = no dead flowers, 5 = 100% dead flowers). [b] Rating of vegetative bud development on 11 March 1986 (1 = no vegetative bud break, 5 = majority of buds leafed out).

From Patten et al., *J. Amer. Soc. Hort. Sci.*, 114, 728, 1989. With permission.

of the season. Overall, the intermediate depth (19.5 cm) produced the greatest linear growth. Root distribution was also affected, with the high level producing a dense matted system, the intermediate level a deeper, spreading system, and the low level a root system with few, but long laterals. The superior root system at the intermediate depth was probably responsible for the improved vegetative growth and water relations.

Some growers have used sprinkling or perforated pipes for irrigation.[56,160] Chandler[161] found that sprinklers increased root growth near the surface and that root growth was improved compared to water table methods in dry soils.

The key to successful irrigation (water management) of cranberry plants is to maintain a well-distributed root zone that is not concentrated too near the soil surface. This practice provides a greater soil volume for water extraction during droughts and decreases water stress which may reduce growth and yields.

B. *RUBUS* (RED RASPBERRY)

1. Irrigation

Red raspberry plants, like blueberries, are usually rather shallow-rooted, and thus are susceptible to drought stress. Christensen[162] found 68–76% of the roots by weight in the top 25 cm of soil for 5-year-old plants. Therefore, irrigation would be expected to improve growth and yields of raspberry bushes. Goode and Hyrycz[163] varied irrigation levels using 20 or 50 kPa soil moisture tension at a 30-cm depth, or no irrigation, for 'Malling Exploit' and 'Malling Jewel' red raspberries. The irrigation treatments increased yields by 27% and produced a greater number of fruiting points/lateral than the nonirrigated treatment. Berry weight was increased by 20% and cane growth rate, height, and number also increased with irrigation. Goode[164] observed that irrigation reduced xylem tension and water stress in red raspberries, which is likely related to increased growth and yields. In contrast, Crandall and Chamberlain[165] withheld irrigation and increased the number of flower bud primordia formed in 'Puyallup' red raspberries. Late summer water stress, however, decreased carbohydrate reserves in the canes;[166] irrigation increased carbohydrate levels, but had no subsequent effect on cane diameter or fruit set in the following season. During a 10-year study in Scotland, MacKerran[167] compared growth and yields of 'Malling Jewel' red raspberry plants under 3 irrigation regimes. Irrigating at 40 kPa soil moisture tension improved growth and yields over nonirrigated plants in most regions, while irrigating at pink fruit stage alone did not increase yields. Irrigation improved berry size in only 4 of 10 years. Therefore, irrigation also increased berry number by increasing the number of fruiting canes/plant. DeBoer et al.[168] compared yields of 'Boyne' red raspberry using trickle compared with sprinkler irrigation over a 3-year period. Soil water content was monitored using tensiometers at various locations in the row, but irrigation was scheduled based on readings at a 20-cm depth. Soil water content was maintained between 30 and 35% from June to August for both irrigation systems. Net water requirements averaged 66.8 cm for the sprinkler and 32.8 cm for the trickle system. Therefore, trickle irrigation used considerably less water than the sprinkler, yet yields were comparable over the 3-year period.

V. FLOODING STRESS

A. *VACCINIUM*

1. Blueberries

Soil flooding, which may cause waterlogging, occurs periodically in many blueberry growing regions. In general, flooding is short-term, although some plantations have been inundated for 30 days or more under unusual conditions and have suffered considerable plant damage and yield reductions. During flooding, water displaces oxygen from the soil pore space. In addition, oxygen is respired by soil microbes and plant roots and oxygen levels decrease from 20% to less than 5% within 1–2 days of flooding.[169] Soils initially become hypoxic (low oxygen) and eventually anoxic (no oxygen), although this situation rarely occurs in the field. Redox potential of blueberry soils in Florida decreased from 300 mv (aerobic) (Figure 12a) to −100 mv within 1–2 days of flooding[169] (Figures 12b, c, d, e). As flood water subsided, redox potential returned to aerobic conditions in 5–10 days at a 15-cm depth (Figure 12b, c, d). In contrast, redox potentials remained at 0–200 mv during long-term flooding (Figure 12e). The time course and magnitude of changes in redox potentials, however, vary with soil type and temperature.

a. Plant Survival

Survival time varies considerably for blueberries under flooded conditions due to environmental, edaphic and plant factors.[116] Survival durations range from 5 days for 'Tifblue' rabbiteye blueberries in Florida[116] to 30 months for 'Bluecrop' highbush blueberries growing in Rhode Island[170] (Table 6), with a range of durations between these extremes. Media or soil type and soil and air temperature are major factors influencing survival time. Survival duration is generally longer for plants growing in artificial media such as peat:perlite:sand (1:1:1 v/v) than in native soil because soil oxygen is depleted and redox potential changes more slowly in media. For example, redox potential of a peat:perlite:sand media remained >75 mv for 30 months,[170] while it decreased to less than 0 mv in 1–2 days for Kanapaha sand as described previously.[169] Crane[171] flooded 'Woodard' rabbiteye blueberry plants under controlled conditions in the laboratory in either native soil (Kanapaha fine sand) or peat:perlite (1:1 v/v). The LD_{50} for blueberry plants growing in native soil at 30°C ranged from 16–56 days but was consistently 61 days for plants growing in artificial media. However, redox potential for both media/soil types remained greater than 100 mv for the entire 70-day duration of the study. In contrast, Davies and

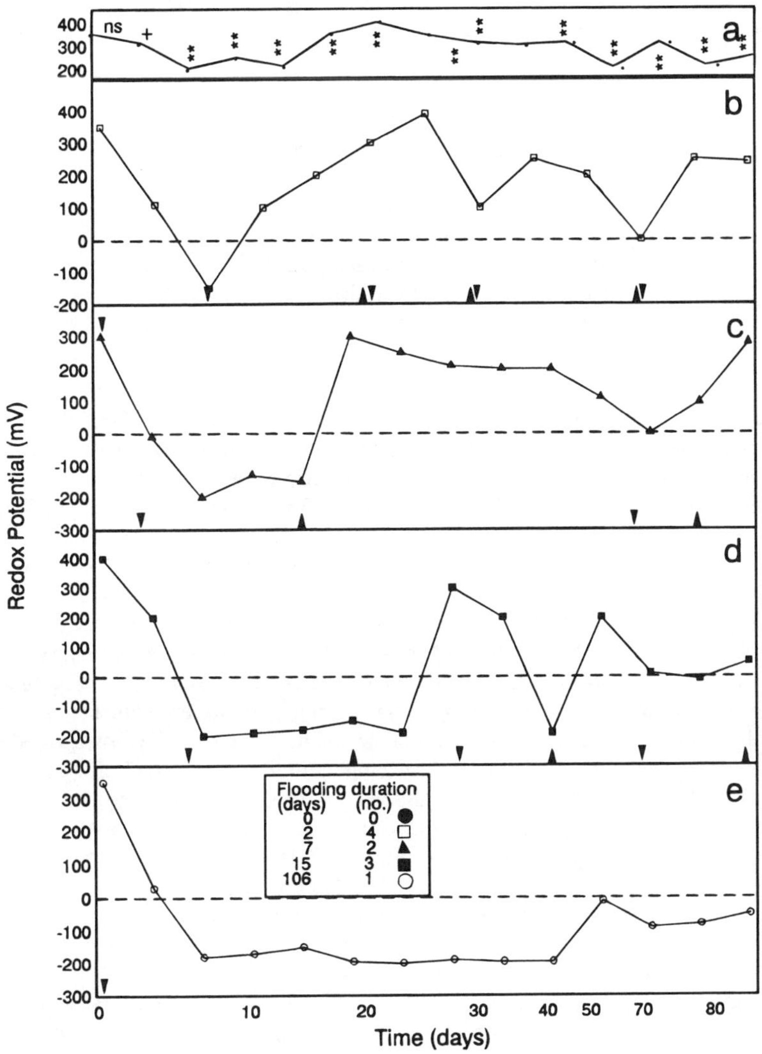

Figure 12 Soil redox potentials for flooded and nonflooded plots of 'Tifblue' rabbiteye blueberry plants at the 15-cm depth during Spring 1985. Each point is the mean of three plots (10 replications per plot). Arrowheads indicate onset (▲) and release (▼) of flooding. From Crane and Davies, *J. Amer. Soc. Hort. Sci.*, 113, 488, 1988. With permission.

Table 6 **Survival percentage of *Vaccinium* species with soil flooding under various edaphic and environmental conditions**

Species (cultivar)	Age (yrs)	Season	Soil temp. (°C)	Media/soil	Survival time (days)	Ref.
V. corymbosum (Bluecrop)	2	Spring	16–32	peat:perlite:sand 1:1:1	540	Abbott and Gough, 1987
V. corymbosum (Bluecrop)	2	Winter	10–16	peat:perlite:sand 1:1:1	900	Abbott and Gough, 1987
V. ashei (Woodard)	2	Spring	20–23	peat:pinebark 1:1	>58	Davies and Wilcox, 1984
V. ashei (Woodard)	1.5	Lab	20–30	Kanapaha sand	LD_{50} 16–117	Crane, 1987
				peat:perlite 1:1	LD_{50} 60–68	
V. ashei (Tifblue)	2–3	Spring	—	Kanapaha sand	106–117	Crane and Davies, 1988 (periodical)
		Summer	—	Kanapaha sand	78–90	
V. ashei (Woodard)	3	Spring	22	Kanapaha sand	>35	Crane and Davies, 1988
		Summer	33	Kanapaha sand	LD_{50} <5	
V. darrowi	2–3	Spring	—	Kanapaha sand	<55	Crane and Davies, 1989
		Summer	—	Kanapaha sand	<55	
V. stamineum	2–3	Spring	—	Kanapaha sand	<55	
		Summer	—	Kanapaha sand	<55	
V. ashei (Delite)	2–3	Spring	—	Kanapaha sand	<55	
		Summer	—	Kanapaha sand	<55	
(*V.* × *V.*)		Spring	—	Kanapaha sand	<55	
Sharpblue	2–3	Summer	—	Kanapaha sand	<55	
Aliceblue	2–3	Spring	—	Kanapaha sand	<55	
		Summer	—	Kanapaha sand	LD_{50} 55	

Wilcox[172] observed no differences in survival duration for 'Woodard' and 'Tifblue' plants growing in native soil or peat:pine bark (1:1 v/v). However, redox potentials were not measured and the duration of the study (56 days) may not have been sufficient for differences in plant survival to occur.

Soil temperature during flooding also has a pronounced effect on plant survival in the field. 'Bluecrop' highbush plants survived 18 months when flooding was begun in April and 30 months when flooding was begun in December.[170] Rabbiteye blueberry plants growing in Florida survived more than 35 days of flooding in spring (March) at soil temperatures of 22°C, with no plants surviving 35 days during late summer (September) flooding at soil temperatures of 33°C.[173] In a related study, 83% of 'Tifblue' rabbiteye plants survived 106–117 days of flooding during spring, while 0 to 33% survived 78–90 days of summer flooding.[169] Both studies were done under field conditions in native soil, again illustrating the wide range of survival rates commonly observed among flooding experiments. This variability in survival time also occurred under controlled laboratory conditions. Crane[171] exposed flooded 'Woodard' rabbiteye blueberry roots to 20, 25, or 30°C soil temperatures and assessed LD_{50} for survival. In all experiments LD_{50} values were greatest at 20°C and least at 30°C, which is in agreement with generally observed trends in the field. However, in the first study, LD_{50} ranged from 16–46 days, increasing to 33 to >64 days in the second experiment and to 56 to >117 days for the third study. These data suggest that plant age, size, or stage of maturity may also be factors in survival. Other studies, in contrast, suggested that neither stage of plant development nor soil temperature in the range from 20 to 31°C affected survival duration.[172] As stated previously, no soil redox potentials were measured; therefore it is difficult to assess the degree of anaerobiosis in this study.

All of the studies listed in Table 6 evaluated plant survival from a horticultural viewpoint for plants that were not infected with *Phytophthora* root rot. Roots of highbush and some rabbiteye cultivars are very susceptible to *Phytophthora parasitica* damage. For example, 'Tifblue' plants died in only 3 to 7 days in the summer if *Phytophthora* was present in the soil compared with 25–35 days in the absence of *Phytophthora.*[116] Generally, flooded blueberry plants that are infected with *Phytophthora* exhibit rapid leaf desiccation and necrosis, almost as if they were burned. These symptoms develop basipetally, and are followed by leaf abscission. In contrast, when plants are flooded without *Phytophthora* present, symptoms develop gradually. Leaves become chlorotic and redden in an acropetal manner, and older, senescing leaves are usually affected before newer growth.[116]

Despite the many flooding studies conducted on blueberries, the reason for such variability in flooding survival is unknown. Even under carefully controlled conditions using clonal material, survival durations vary from plant-to-plant. One possible causal factor that has not been studied is the actual oxygen and physiological status of the roots themselves and not of the media/soil surrounding them. Physiological changes during flooding occur very rapidly in the shoot, usually within 24 hours of flooding (see Physiological Adaptations to Flooding, Section V.A.2.C). However, there is a paucity of information on changes in root physiology during initial flooding. Small variations in root responses among plants may cause rapid and cascading responses in the rest of the plant, ultimately producing differences in survival duration. This hypothesis is supported by the fact that blueberry plant survival duration is generally less at high vs. low soil temperatures. Root respiration is temperature dependent and it is logical to expect more rapid depletion of root oxygen at high vs. low temperatures, although there are no data correlating root oxygen levels with temperature making this hypothesis rather conjectural.

b. Vegetative Growth

Although short-term flooding may result in plant death, it is more likely that the plant will survive, but vegetative growth will decrease. Flooding decreased shoot growth and number and leaf expansion of highbush[170–173] and rabbiteye blueberry bushes. Crane and Davies[173] measured leaf area and shoot growth for flooded and nonflooded rabbiteye blueberry plants in the field (Figure 13). Nonflooded shoots displayed a typical linear increase in length with growth rate slowing after 35–40 days. Leaf area increased rapidly, reached a plateau, and increased again in a double sigmoid pattern. In contrast, stems of continuously flooded plants grew very little and those of plants flooded for as little as 5 days grew at a much slower rate than shoots from nonflooded plants. Consequently, leaf area of flooded plants attained only 50% of that of nonflooded plants even after flooding was discontinued. Total and individual leaf areas decreased significantly after 35 days of continuous flooding. Similarly, Abbott and Gough[170]

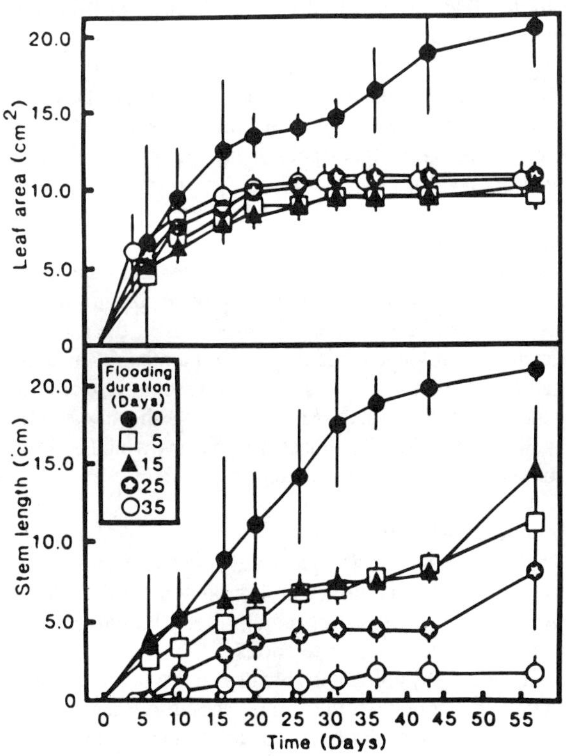

Figure 13 Effect of flooding duration on stem elongation and leaf expansion of 'Woodard' rabbiteye blueberry plants during Spring 1985. Each point is the mean of 3 leaves or stems of 6 plants per treatment ± SD. From Crane and Davies, *J. Amer. Soc. Hort. Sci.*, 113, 180, 1988. With permission.

observed that average leaf area decreased from 4.4 to 2.2 cm^2 and shoot length decreased from 5.5 to 2.3 cm in nonflooded and flooded 'Bluecrop' highbush blueberry plants, respectively. Leaf area also decreased during periodic flooding, a situation that is more likely to occur under field conditions.[169] Flooding durations of 2 days reduced leaf area of 'Woodard' and 'Tifblue' rabbiteye blueberry plants if the flooding recurred on a regular basis. Flooding of two 7-day periods or three 15-day periods also reduced leaf area compared to controls. Herath and Eaton[174] also found leaf size of 'Bluecrop' highbush plants was reduced by periodic flooding.

It appears that overall flooding duration, whether accumulated in cycles or continuously, has a detrimental effect on growth. In general, the longer the total days of inundation, the greater the growth reduction. This fact is significant from a practical viewpoint, since most rabbiteye blueberry plantations are more likely to be exposed to several short-term flooding periods rather than to continuous long-term flooding.

c. *Reproductive Growth*

Reproductive growth and development appears more sensitive to flooding than vegetative growth based on studies with 2- to 3-year-old bushes. Additionally, field observations on mature bushes also support this concept. Two-year-old 'Bluecrop' highbush plants were flooded for 4 or 30 months.[175] Flooded bushes had 60 or 74% fewer flower buds and 55 or 66% fewer flowers than nonflooded bushes after 4 or 30 months, respectively. In addition, anthesis of flooded bushes was delayed by 6 days. Similarly, Crane and Davies[173] observed even more dramatic short-term effects of flooding on reproductive growth of rabbiteye blueberry plants. Summer flooding of 2-year-old plants reduced flower bud formation in the fall by 31 to 67% after only 5 days of flooding. In a second study, summer flooding of 25 or 35 days totally inhibited flower bud formation in the fall. Percent fruit set was also decreased but only after 25 days of continuous flooding. Therefore, flower bud formation appears much more sensitive to flooding than fruit set, and most sensitive of all horticultural responses to flooding.

Since flooding reduced flower bud number and fruit set, yields should also be reduced. Yield of young rabbiteye plants in the field decreased after 25 or 35 days of flooding.[173] Abbott and Gough[175] also observed yield decreases for container-grown highbush plants. However, yield studies are not very meaningful on such young plants. Nevertheless, field observations also suggest that yields are reduced for mature bushes growing in poorly-drained areas. For example, yields were significantly reduced by 30 days of flooding for mature rabbiteye blueberry bushes in Georgia.[176] It has also been observed for many years that blueberry bushes growing in poorly drained areas are less vigorous and productive than those growing in well-drained locations.

d. *Anatomical Adaptations to Flooding*

Flood-tolerant plants have evolved anatomical and morphological adaptations such as formation of aerenchyma cells, lenticels, or adventitious rooting to provide oxygen to flooded roots.[177] Similarly, highbush blueberry plants produced aerenchyma-like structures in the root epidermal layer and aerenchyma in the shoot cortex during prolonged flooding (4–30 months).[178] In contrast, Crane and Davies[116] did not observe the formation of similar structures in flooded rabbiteye plants. Most studies on rabbiteye plants, however, were conducted for less than 4 months, while the studies on highbush flooding were conducted for as long as 30 months. Possibly these adaptations would also occur for rabbiteye plants under such long-term flooding, although it is unlikely that rabbiteye bushes could survive this long.

e. *Physiological Adaptations to Flooding*

Mesophytic plants have evolved several physiological and biochemical adaptations to flooding, including control and detoxification of glycolytic end products[179] and production of ethylene via 1-aminocyclopropane-1-carboxylicacid (ACC), which causes stomatal closure.[180] Flooding also decreases hydraulic conductivity of the root and may alter plant water relations.[181,182]

Blueberry plants also undergo physiological changes in gas exchange during flooding. Stomatal conductance decreased within 1 day of flooding in the laboratory[15] and within 1–5 days in the greenhouse.[12] Recovery of g_s to preflood levels did not occur for 18 days after plants were unflooded, despite the more rapid reoxygenation of the soil. With the decrease in g_s, A also decreased within 1 day of flooding in the laboratory[15] and within 9 days in the greenhouse[12] (Table 7). By 11 days of flooding, 'Bluecrop' bushes had negative A, particularly at leaf temperatures above 28°C. Net CO_2 assimilation returned to control levels about 11 days after plants were unflooded. 'Bluecrop' A seemed to be more

Table 7 **Effect of flooding duration on net gas exchange (A) of 'Bluecrop' and 'Woodard' blueberries (28 Aug. to 24 Sept. 1984)**

Duration (days)	Leaf temp. (°C)	PPF ($\mu mol \cdot m^{-2}s^{-1}$)	'Bluecrop'[a] Non-flooded	'Bluecrop'[a] Flooded	'Woodard'[b] A ($\mu mol \cdot m^{-2}s^{-1}$) Non-flooded	'Woodard'[b] Flooded	cv	Significance: Treatment	Significance: cv × treatment
				During flooding					
2	29	272	1.37	1.10	3.83	2.61	*	NS	NS
9	33	904	3.23	0.72	2.59	0.80	NS	*	NS
11	36	920	4.41	−1.72	5.60	1.84	*	**	NS
19	29	388	4.83	−2.80	1.38	−1.14	NS	**	**
24	34	701	5.15	−2.43	4.71	−0.56	NS	**	NS
				Flooding released					
2	28	391	2.80	−0.49	3.28	−0.82	NS	**	NS
4	26	248	3.30	−1.32	3.60	−0.20	NS	**	NS
11	24	204	3.30	1.10	2.70	2.90	NS	NS	NS

Note: * = Significant at $p = 0.05$; ** = significant at $p = 0.01$; NS = not significant.
[a]Mean of 10 plants per cultivar per duration. [b]From Davies and Flore, *J. Amer. Soc. Hort. Sci.*, 111, 565, 1986. With permission.

strongly reduced by flooding than that of 'Woodard' rabbiteye blueberry, which suggests a physiological difference between highbush and rabbiteye blueberries in their flooding, as observed in the field. The short-term effect of flooding on A appears to be stomatal, rather than nonstomatal, based on changes in the ratio of the external (Ca) to internal (Ci) CO_2 concentration.[12] The ratio of Ci:Ca averaged 0.77 to 0.80 for nonflooded plants compared to 0.65 to 0.70 for flooded plants. However, the validity of the Ci:Ca ratio depends on the homogeneity of stomatal response,[183] which was not tested in this study.

Flooding not only decreased g_s but also slowed the rate of stomatal opening.[15] Furthermore, stomata were less responsive to changes in VPD, O_2, and CO_2 levels than those of nonflooded plants. Davies and Flore[18] exposed 'Bluecrop' highbush blueberry plants to varying levels of external CO_2 and O_2 in the laboratory and measured changes in A to determine if flooding affected gas exchange characteristics of blueberry plants (Figure 14). They found that nonflooded plants showed a typical C3 plant increase

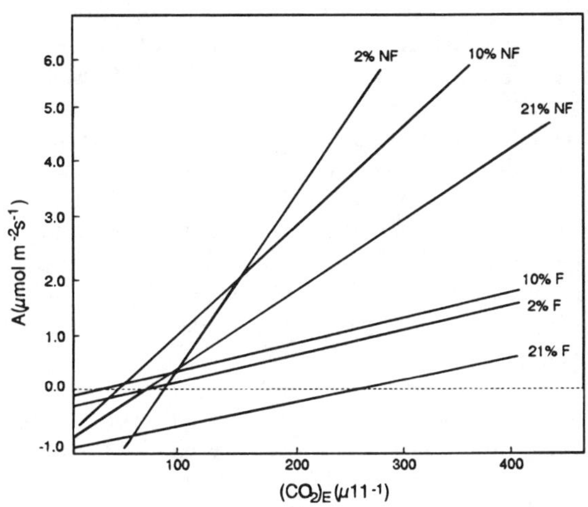

Figure 14 Effects of CO_2 concentration and O_2 percentage on net CO_2 assimilation (A) of nonflooded (NF) and flooded (F) 'Bluecrop' blueberry plants. O_2 percentages were 2, 10 or 21. From Davies and Flore, *Physiol. Plant*, 67, 545, 1986. With permission.

in A as O_2 level was reduced from 21 to 2% and as external CO_2 was increased from 100 to 400 μl l^{-1}. Flooded plants, however, showed much less responsiveness both to a reduction in O_2 and an increase in external CO_2. In addition, flooding decreased photosynthetic efficiency and apparent quantum yield of 'Woodard' rabbiteye blueberry leaves. Therefore, nonstomatal factors such as residual conductance also decreased with short-term flooding, but probably only after g_s decreased.[116] Nevertheless, with protracted flooding, g_s values approach those of residual conductance, and the latter has a significant controlling effect on A.

The biochemical changes in blueberry plants during flooding have not been widely studied, with the exception of ethylene's putative role. Crane[171] studied ethylene levels for flooded and nonflooded rabbiteye blueberry plants and soils and found only trace levels of ethylene in both treatments. Blueberry leaves showed no signs of epinasty, as found in tomato plants, further supporting the lack of involvement of ethylene in flooding responses of blueberry plants.

Water relations of flooded blueberry plants also changed, but more slowly than changes in g_s or A. Generally, leaf Ψ does not decrease for 3 to 5 weeks after flooding initiation.[12,172] This is probably a direct result of rapid stomatal closure that prevents the damaging effects of low leaf Ψ during flooding. With time, however, long-term flooding decreased Ψ.[116] In contrast, root hydraulic conductivity decreased within 4 to 6 days of flooding probably due to damage to root membranes as suggested by increases in electrolyte leakage at 6 to 10 days post-flooding.[184] Root hydraulic conductivity continued to decrease by as much as 81% of control values 2 to 3 weeks after flooding. Both hydraulic conductivity and g_s decreased within nearly equal timeframes in greenhouse and field studies and thus appear linked, possibly via substances synthesized in the roots of flooded plants and translocated to the shoots, causing stomatal closure.[185] This hypothesis, nevertheless, is speculative as no substances have been identified that cause stomatal closure during flooding of blueberry plants.

f. Time Course of Flooding Responses

Several studies conducted with rabbiteye and highbush blueberry suggest a loosely structured time course for flooding responses, consisting of initial, intermediate, and late phases (Table 8).[116] This sequence occurs in the absence of *Phytophthora* or other root pathogens. During the initial phase, g_s to CO_2 and water, A and quantum yield decrease within 1 to 2 days of flooding, followed by a decrease in root hydraulic conductivity. Because root water uptake is limited, stomatal closure protects the canopy from extensive and deleterious transpirational losses. A significant reduction in flower bud formation occurs within 5 days of flooding, and may be due to the decrease in A as well as a decrease in hormone translocation from roots. The plants recover rapidly when unflooded during the initial phases, with no lasting effects of flooding being apparent.

The second phase is characterized by a reduction or cessation of leaf and shoot expansion. Root hydraulic conductivity continues to decline, electrolyte leakage from the root occurs, and residual (nonstomatal) conductance decreases. At this stage, A becomes negative at temperatures greater than 28°C, causing the blueberry plant to use its stored carbohydrate reserves via temperature-linked increases in respiration. The plant recovers very slowly with unflooding and lasting reductions in growth and yields usually occur.

The blueberry plant continues further decline during the final phase. Leaves redden and abscission occurs acropetally; new growth flushes become chlorotic often showing iron deficiency symptoms. Fruit set is decreased and fruit may shrivel on the plant. Root hydraulic conductivity becomes extremely low and leaf Ψ becomes progressively more negative despite stomata closure. The plant is very slow to recover during unflooding and is usually permanently damaged. Plant death eventually occurs due to extreme desiccation, carbon starvation, loss of membrane integrity, and a reduction in ATP production. Highbush blueberry plants may produce aerenchyma or aerenchyma-like tissues during the final phase which extends the life of the plant.

High soil temperatures or VPD accelerate the time for each phase to occur, thus accounting for the observed differences in survival duration in summer compared with fall flooding. High temperatures cause rapid water losses (primarily cuticular) and increased respiration rates, thus using stored metabolites at a faster rate. High temperatures also cause rapid use of oxygen by microbes and roots, thus accelerating each stage of the process. The type of media (soil) affects the amount of oxygen available in pore spaces, and the types of microbes available to metabolize the oxygen. Therefore, the sequence of events in flooding, particularly survival duration, is generally longer in artificial media than in sands or clays

Table 8 **Time course of physiological and growth responses of rabbiteye blueberry plants to flooding duration**

Time (days of flooding)	Plant response
	Phase 1
0	
1	
2	$g_s(\downarrow)$, $g_s'(\downarrow)$, A($\downarrow$), $\Phi(\downarrow)$
3	RLp ($\downarrow$), Ψ_w (no change)
4	
5	Flower bud number ($\downarrow$) after summer flooding
	Phase 2
6	Leaf expansion and shoot elongation cease
7	
8	RLp ($\downarrow\downarrow$)
9	SLp ($\downarrow$), EL ($\uparrow$), g_r' ($\downarrow$), $\Phi(\downarrow)$
10	A($\downarrow$) may become zero or negative, g_s less responsive to environment
11	New shoots and/or leaves may wilt
12	
13	Leaf senescence and abscission
14	$g_s(\downarrow\downarrow)$, A($\downarrow\downarrow$), Ψ_w (no change)
	Phase 3
15	
16	RLp ($\downarrow\downarrow$), SLp ($\downarrow\downarrow$), EL ($\uparrow\uparrow$)
17	Fruit shriveling may begin
18	
19	
20–45	Continued decline in plant vigor and a reduction in yields
46–120	$\Psi(\downarrow\downarrow)$, eventual plant death

Note: *Phytophthora* root rot may rapidly accelerate flood-stress symptoms and decrease plant vigor, leading to more rapid plant death. g_s = G_s to water; g_s' = G_s to carbon dioxide; A = net CO_2 assimilation; Φ = quantum yield; RLp = root hydraulic conductivity; Ψ_w = leaf xylem Ψ; SLp = stem hydraulic conductivity; EL = electrolyte leakage; g_r' = residual conductance to carbon dioxide; $\downarrow$ decrease; $\downarrow\downarrow$ greater decrease; $\uparrow$ increase; $\uparrow\uparrow$ greater increase. From Crane and Davies, *HortScience*, 24, 203, 1989. With permission.

especially those with high organic matter levels which provide carbon substrates for microbial respiration, thus rapidly reducing O_2 levels in the soil. The above-mentioned sequence involves physiological and morphological responses to flooding which are not related to presence of *Phytophthora* root rot. In contrast, *Phytophthora* causes sudden and severe root damage and rapid plant death.

2. Cranberries

Flooding is an important part of cranberry production and effects of flooding on growth and yields have been widely studied. Short-term flooding has been used to control insects in the spring and summer and to prevent winter kill and frost damage in the spring.[56] It is also used for irrigation. Flooding of cranberry vines differs from that of blueberry bushes in that the entire plant is inundated, presenting a very different situation from blueberry root flooding. Flooding durations of more than 1 day may cause damage to vines. The critical floodwater oxygen level for cranberry plants is about 5.7 $\mu l\ l^{-1}$,[186] but varies with physiological status of vines. Alternative sources of oxygen may come from degradation of stored carbohydrates. Therefore, as with blueberry, plants with high carbohydrate levels tend to survive protracted flooding better than those with low levels.[56]

Table 9 **Effects of oxygen deficiency on production of three cranberry cultivars**

	Flowers		Fruit	
Cultivar and degree of O_2 deficiency	**Per upright (%)**	**Dead buds (%)**	**Fruit set (%)**	**Yield (mt/ha)**
Early Black				
Severe	3.1	12.6	21.6	3.0
Moderate	3.5	12.9	28.0	6.4
Slight	4.0	5.2	28.3	8.6
Howes				
Severe	3.1	16.8	33.3	5.8
Moderate	3.5	17.6	34.9	8.3
Slight	3.7	4.8	49.4	11.5
McFarlin				
Severe	3.9	16.1	20.9	4.5
Moderate	4.0	12.9	23.5	
Slight	4.0	10.0	30.0	9.3

Note: Degree of O_2 deficiency was obtained by artificially excluding light from the flooded plants. Severe deficiency resulted when O_2 content of water dropped below 2 μl/l; moderate deficiency at 2 to 4 μl/l O_2; and slight deficiency above 4 μl/l O_2.

From Franklin et al., *Mass. Agr. Expt. Sta. Bul.*, no. 402, 1943. With permission.

a. Plant Survival

Plant survival during flooding is generally not a factor with cranberry vines as it is with blueberry plants because flooding durations are controlled. Consequently, there is little information available on plant survival. However, plant damage thresholds have been extensively studied. The extent of injury is positively correlated with temperature during flooding, with less damage at low (near 0°C) than moderate temperatures.[56] Temperature affects the solubility of oxygen in water; oxygen is less soluble at high vs. low temperatures. Moreover, temperature is positively correlated with tissue respiration rate, with a Q_{10} of about two.

b. Vegetative and Reproductive Growth and Development

Symptoms of oxygen deficiency in cranberry are manifested as terminal bud dieback, stem death, and leaf abscission.[56] Flowers, small fruits, and rapidly growing buds are most sensitive to oxygen deprivation probably due to their relatively greater respiration rates and thus oxygen usage rates.[187] Mature leaves are the least sensitive to low oxygen levels. Franklin et al.[189] compared the effect of three cultivars and three oxygen depletion levels on percent flowers/upright, fruit set, yields, fruit size, and bud damage (Table 9). For all cultivars, plants grown under slight O_2 deficiency (>4 μl l^{-1}) had the highest yields and percent fruit set and the lowest bud mortality due to flooding. As O_2 levels decreased to moderate (2–4 μl l^{-1}) or severe (<2 μl l^{-1}) depletion there was a progressive drop in yields and fruit set. 'Early Black' flowers were more susceptible to low O_2 levels than flowers of 'Howes' and 'McFarlin'.

c. Physiological Adaptations to Flooding

There are some important differences in physiological flooding responses between cranberries and blueberries. Since a large portion or all of the cranberry vine is inundated, the plant's sensitivity and response to flooding is more rapid than for blueberry plants where only the roots are flooded. Therefore, flooding of more than 48 hours may cause damage. When the soil only is flooded, some O_2 is trapped in soil pores and roots, whereas entire leaf and shoot inundation allows only limited O_2 storage in intercellular spaces. At high temperatures, the O_2 is rapidly respired. Differences in flooding tolerance of cranberries among tissue types and cultivars may be a function of the level of respiration.

VI. CONCLUSIONS

Many aspects of the environmental physiology of blueberries, cranberries, and raspberries are similar to those of tree fruit crops. Light saturation (700–900 μmol $m^{-2}s^{-1}$), light compensation points (25–35 μmol $m^{-2}s^{-1}$), and temperature optima for photosynthesis are typical of other C3 temperate fruit crops. Cardinal temperatures for seasonal growth and development are also characteristic of other temperate crops. *Vaccinium* species are moderately drought tolerant, as indicated by intermediate leaf Ψ for stomatal closure and a characteristic increase in A as external CO_2 levels increase (100–400 μl l^{-1}) and O_2 levels decrease (21 to 2%). Flood-tolerance mechanisms are also similar to those of tree crops and include stomatal closure and reduced water uptake. The initial reduction in A however, appears due to stomatal, rather than nonstomatal, factors.

Nevertheless, there are some unusual features of *Vaccinium* species compared to other fruit crops. Blueberry bushes are the only fruit crop that exhibit net fruit photosynthesis throughout development. In fact, fruit photosynthesis in blueberry supplies a significant portion of the carbon requirement of young fruit, contributing as much as 50% of the carbon required for dry weight gain and respiratory loss during the first 10 days after anthesis. Blueberries are also unusual compared to most other fruit crops in that flower bud initiation is photoperiodically sensitive. Increased flower bud initiation under short photoperiods (less than 12 hours) suggests that blueberries are quantitative short-day plants.

Although light and temperature optima for photosynthesis are similar between *Vaccinium* and other fruit crops, leaf A in rabbiteye and highbush blueberry and red raspberry plants is low, ranging from 5 to 12 μmol $m^{-2}s^{-1}$, compared with much higher values for many other fruit crops. For some *Vaccinium* species, this is due in part to the development of extensive rodlet wax on leaf surfaces that cover and occlude the stomata, thus reducing g_s. Although rodlet wax formation serves as a drought tolerance mechanism, it also decreases A and may contribute to the relatively low growth rates and yield of blueberries.

REFERENCES

1. Vander Kloet, S. P., *The Genus Vaccinium in North America*, Canadian Govt. Pub. Centre, Ottawa, 1988, 201.
2. Camp, W. H., On the structure of populations in the genus *Vaccinium, Brittonia*, 4, 189, 1942.
3. Galletta, G. J., Blueberries and cranberries, in *Advances in Fruit Breeding*, Janick, J. and Moore, J. N., Eds., Purdue Univ. Press, West Lafayette, 1975, 154.
4. Bailey, L. H., *Hortus Third*, McMillan Co., New York, 1976, 984.
5. Jennings, D. L., *Raspberries and Blackberries: The Breeding, Diseases and Growth*, Academic Press, New York, 1988.
6. Pritts, M., *Compendium of Raspberry and Blackberry Diseases and Insects*, Ellis, M.A., Converse, R.H., William, R.N. and Williamson, B., Eds., Amer. Phytopath. Soc., St. Paul, Minnesota, 1991, 1.
7. Jennings, D. L., Raspberries and blackberries, in *Evolution of Crop Plants*, Simmonds, N. W., Ed., Longman, London, 1976, 251.
8. Davies, F. S. and Albrigo, L. G., Water relations of small fruits, in *Water Deficits and Plant Growth*, Kozlowski, T. T., Ed., Academic Press, New York, 1983, 89.
9. Camp, W. H., The North American blueberries with notes on other groups of Vacciniaceae, Brittonia, 5, 203, 1945.
10. Eck, P., *Blueberry Science*, Rutgers Univ. Press, New Brunswick, 1988, chap. 2.
11. Bonn, B., Forsyth, F. R., and Hall, I. V., A comparison of the rates of apparent photosynthesis of the cranberry and the common lowbush blueberry, *Nat. Can.*, 96, 799, 1969.
12. Davies, F. S. and Flore, J. A., Gas exchange and flooding stress of highbush and rabbiteye blueberries, *J. Amer. Soc. Hort. Sci.*, 111, 565, 1986a.
13. Moon, J. W., Flore, J. A., and Hancock, J. F., Jr., A comparison of carbon and water vapor gas exchange characteristics between a diploid and highbush blueberry, *J. Amer. Soc. Hort. Sci.*, 112, 134, 1987a.
14. Teramura, A. H., Davies, F. S., and Buchanan, D. W., Comparative photosynthesis and transpiration in excised shoots of rabbiteye blueberry, *HortScience*, 14, 723, 1979.
15. Davies, F. S. and Flore, J. A., Short-term flooding effects on gas exchange and quantum yield of rabbiteye blueberry (*Vaccinium ashei* Reade), *Plant Physiol.*, 81, 289, 1986b.
16. Andersen, P.C., Impact of ten spray adjuvants on leaf gas exchange of pecan, blueberry, photinia, and azalea, *HortScience*, 23, 343, 1988.
17. Stang, E. J., Struckmeyer, B. E., and Ferree, D. C., Effect of four light levels on net photosynthesis and leaf anatomy of cranberry (*Vaccinium macrocarpon* Ait.), *Acta Hort.*, 165, 325, 1985.

18. Davies, F. S. and Flore, J. A., Flooding, gas exchange and hydraulic root conductivity of highbush blueberry, *Physiol. Plant.*, 67, 545, 1986c.
19. Birkhold, K. T., Koch, K. E., and Darnell, R. L., Carbon and nitrogen economy of developing rabbiteye blueberry fruit, *J. Amer. Soc. Hort. Sci.*, 117, 139, 1992.
20. Flore, J. A. and Hancock, J. F., Blueberry fruit photosynthesis, *Carbon Economy of Fruits*, Blanke, M. M., Ed., Bonn, 1990, 7 (Abstr.).
21. Hall, I. V., Some effects of light on native lowbush blueberries, *Proc. Amer. Soc. Hort. Sci.*, 72, 216, 1958.
22. Hall, I. V. and Ludwig, R. A., The effects of photoperiod, temperature, and light intensity on the growth of the lowbush blueberry (*V. angustifolium* Ait.), *Can. J. Bot.*, 39, 1733, 1961.
23. Aalders, L. E., Hall, I. V., and Forsyth, F. R., Effects of partial defoliation and low light intensity on fruit set and berry development in the lowbush blueberry, *Hort. Res.*, 9, 124, 1969.
24. Patten, K. and Neuendorff, E., Influence of light and other parameters on the development and quality of rabbiteye blueberry fruit, *Proc. Texas Blueberry Growers Assoc.*, Beaumont, 109, 1989.
25. Shutak, V. G., Hindle, R., and Christopher, E. P., Factors associated with ripening of highbush blueberry fruits, *Proc. Amer. Soc. Hort. Sci.*, 68, 178, 1956.
26. Aalders, L. E. and Hall, I. V., A comparison of flower bud development in the lowbush blueberry, *V. angustifolium* Ait. under greenhouse and field conditions, *Proc. Amer. Soc. Hort. Sci.*, 85, 281, 1964.
27. Hall, I. V., Craig, D. L., and Aalders, L. E., The effect of photoperiod on the growth and flowering of the highbush blueberry (*Vaccinium corymbosum* L.), *Proc. Amer. Soc. Hort. Sci.*, 82, 260, 1963.
28. Darnell, R. L., Photoperiod, carbon partitioning, and reproductive development in rabbiteye blueberry, *J. Amer. Soc. Hort. Sci.*, 116, 856, 1991.
29. Perlmutter, F. and Darrow, G. M., Effect of soil media, photoperiod and nitrogenous fertilizer on the growth of blueberry seedlings, *Proc. Amer. Soc. Hort. Sci.*, 40, 341, 1942.
30. Kender, W. J., Rhizome development in the lowbush blueberry as influenced by temperature and photoperiod, *Proc. Amer. Soc. Hort Sci.*, 90, 144, 1967.
31. Eaton, G. W. and Ormrod, D. P., Photoperiod effect on plant growth in cranberry, *Can. J. Plant Sci.*, 48, 447, 1968.
32. Vince-Prue, D., Roles of phytochrome in photoperiodic floral induction, in *Phytochrome and Photoregulation in Plants*, Furuya, M., Ed., Acad. Press, Orlando, 1987, chap. IV.7.
33. Stushnoff, C. and Hough, L. F., Response of blueberry seed germination to temperature, light, potassium nitrate and coumarin, *Proc. Amer. Soc. Hort. Sci.*, 93, 260, 1968.
34. Kossuth, S. V. and Biggs, R. H., Sunburned blueberries, *Proc. Florida State Hort. Soc.*, 91, 173, 1978.
35. Donnelly, D. J. and Vidaver, W. E., Pigment content and gas exchange of red raspberry *in vitro* and *ex vitro*, *J. Amer. Soc. Hort. Sci.*, 109, 177, 1984.
36. Goulart, B. and Braun, H., unpublished, 1992.
37. Fernandez, G., unpublished, 1992.
38. Palmer, J. W., Jackson, J. E., and Ferree, D. C., Light interception and distribution in horizontal and vertical canopies of red raspberries, *J. Hort. Sci.*, 62, 493, 1987.
39. Andersen, P.C., unpublished, 1992.
40. Braun, J. W., Garth, J. K., and Brun, C. A., Distribution of foliage and fruit in association with light microclimate in the red raspberry canopy, *J. Hort. Sci.*, 64, 565, 1989.
41. Wright, C. J. and Waister, P. D., Light interception and fruiting cane architecture in the red raspberry grown under annual and biennial management systems, *J. Hort. Sci.*, 59, 395, 1984.
42. Wright, C. J. and Waister, P. D., Canopy structure and light interception in the red raspberry, *Acta Hort.*, 183, 273, 1986.
43. Wright, C. J. and Waister, P. D., With-in plant competition in the red raspberry. II. Fruiting cane growth, *J. Hort. Sci.*, 57, 443, 1982.
44. Waister, P. D., Cormack, M. R., and Sheets, W. A., Competition between fruiting and vegetative phases in the red raspberry, *J. Hort. Sci.*, 52, 75, 1977.
45. Renquist, A. R., Hughes, H. G., and Rogoyski, M. K., Solar injury of raspberry fruit, *HortScience*, 22, 396, 1987.
46. Renquist, A. R., Hughes, H. G., and Rogoyski, M. K., Combined high temperature and ultraviolet radiation injury of red raspberry fruit, *HortScience*, 24, 597, 1989.

47. Forsyth, F. R. and Hall, I. V., Effect of leaf maturity, temperature, carbon dioxide concentration, and light intensity on rate of photosynthesis in clonal lines of the lowbush blueberry, *Vaccinium angustifolium* Ait., under laboratory conditions, *Can. J. Bot.*, 43, 893, 1965.
48. Forsyth, F. R. and Hall, I. V., Rates of photosynthesis and respiration in leaves of the cranberry with emphasis on rates at low temperatures, *Can. J. Plant Sci.*, 47, 19, 1967.
49. Moon, J. W., Hancock, J. F., Jr., Draper, A. D., and Flore, J. A., Genotypic differences in the effect of temperature on CO_2 assimilation and water use efficiency in blueberry, *J. Amer. Soc. Hort. Sci.*, 112, 170, 1987.
50. Hancock, J. F., Haghighi, K., Krebs, S. L., Flore, J. A., and Draper, A. D., Photosynthetic heat stability in highbush blueberries and the possibility of genetic improvement, *HortScience*, 27, 1111, 1992.
51. Abbott, J. D. and Gough, R. E., Seasonal development of highbush blueberry roots under sawdust mulch, *J. Amer. Soc. Hort. Sci.*, 112, 60, 1987.
52. Bailey, J. S. and Jones, L. H., The effect of soil temperature on the growth of cultivated blueberry bushes, *Proc. Amer. Soc. Hort. Sci.*, 38, 462, 1941.
53. Chandler, R. B. and Mason, I. C., The effect of mulch on soil moisture, soil temperature, and growth of blueberry plants, *Proc. Amer. Soc. Hort. Sci.*, 40, 335, 1942.
54. Fry, B. O. and Savage, E. F., Evaluation of sawdust as a mulch for young blueberry plants, *Proc. Amer. Soc. Hort. Sci.*, 93, 273, 1968.
55. Patten, K. D., Neuendorff, E. W., and Peters, S. C., Root distribution of 'Climax' rabbiteye blueberry as affected by mulch and irrigation geometry, *J. Amer. Soc. Hort. Sci.*, 113, 657, 1988.
56. Eck, P., *The American Cranberry*, Rutgers Univ. Press, New Brunswick, 1990, chap. 5 and 6.
57. Hawker, G. M. and Stang, E. J., Characterizing vegetative growth and fruit development in cranberry (*Vaccinium macrocarpon* Ait.) by thermal summation, *Acta Hort.*, 165, 311, 1985.
58. Degaetano, A. T. and Shulman, M. D., A statistical evaluation of the relationship between cranberry yield in New Jersey and meteorological factors, *Agric. Forest Meteorology*, 40, 323, 1987.
59. Hall, I. V., Forsyth, F. R., and Newbery, R. J., Effect of temperature on flower bud and leaf anthocyanin formation in the lowbush blueberry, *HortScience*, 5, 272, 1970.
60. Hall, I. V. and Aalders, L. E., Fruit set and berry development of lowbush blueberry as affected by temperature, *Can. J. Plant Sci.*, 48, 321, 1968.
61. Knight, R. J., Jr. and Scott, D. H., Effects of temperatures on self- and cross-pollination and fruiting of four highbush blueberry varieties, *Proc. Amer. Soc. Hort. Sci.*, 85, 302, 1964.
62. Edwards, T. W., Jr., Sherman, W. B., and Sharpe, R. H., Fruit development in short and long cycle blueberries, *HortScience*, 5, 274, 1970.
63. Carlson, J. D. and Hancock, J. F., Jr., A methodology for determining suitable heat-unit requirements for harvest of highbush blueberry, *J. Amer. Soc. Hort. Sci.*, 116, 774, 1991.
64. Rigby, B. and Dana, M. N., Flower opening, pollen shedding, stigma receptivity, and pollen tube growth in cranberry, *HortScience*, 7, 84, 1972.
65. Birrenkott, B. A. and Stang, E. J., Pollination and pollen tube growth in relation to cranberry fruit development, *J. Amer. Soc. Hort. Sci.*, 114, 733, 1989.
66. Hall, I. B., Aalders, L. E., and Newbery, R. J., Frost injury to flowers and developing fruits of the lowbush blueberry as measured by impairment of fruit set, *Nat. Can.*, 98, 1053, 1971.
67. Westwood, M. N., *Temperate Zone Pomology*, W. H. Freeman, San Francisco, 1978, 192.
68. Hall, I. V. and Stark, R., Anthocyanin production in cranberry leaves and fruit related to cool temperatures at a low light intensity, *Hort. Res.*, 12, 183, 1972.
69. Costante, J. F. and Boyce, B. R., Low temperature injury of highbush blueberry shoots at various times of the year, *Proc. Amer. Soc. Hort. Sci.*, 93, 267, 1968.
70. Bittenbender, B. C. and Howell, G. S., Interactions of temperature and moisture content on spring deacclimation of flower buds of highbush blueberry, *Can. J. Plant Sci.*, 55, 447, 1975.
71. Abdallah, A. Y. and Palta, J. P., Changes in the freezing stress resistance of the cranberry leaf, flower bud, and fruit during growth and development, *Acta Hort.*, 241, 273, 1989.
72. Siekmann, S. and Boe, A. A., Low temperature increases reducing and total sugar concentrations in leaves of boxwood (*Buxus sempervirens* L.) and cranberry (*Vaccinium macrocarpon* Ait.), *HortScience*, 13, 439, 1978.
73. Brierley, W. G. and Hildreth, A. C., Some studies on the hardiness of certain species of *Vaccinium*, *Plant Physiol.*, 3, 303, 1928.

74. Kender, W. J. and Brightwell, W. T., Environmental relationships, in *Blueberry Culture*, Eck, P. and Childers, N., Eds., Rutgers Univ. Press, New Brunswick, 1966, chap. 4.
75. Quamme, H. A., Stushnoff, C., and Weiser, C.J., Winter hardiness of several blueberry species and cultivars in Minnesota, *HortScience*, 7, 500, 1972.
76. Eaton, G. W. and Mahrt, B. J., Cold hardiness testing of cranberry flower buds, *Can. J. Plant Sci.*, 57, 461, 1977.
77. Bittenbender, B. C. and Howell, G. S., Predictive environmental and phenological components of flower bud hardiness in highbush blueberry, *HortScience*, 10, 409, 1975.
78. Biermann, J., Stushnoff, C., and Burke, M. J., Differential thermal analysis and freezing injury in cold hardy blueberry flower buds, *J. Amer. Soc. Hort. Sci.*, 104, 444, 1979.
79. Hiirsalmi, H. M. and Hietaranta, T. P., Winter injuries to highbush and lowbush blueberry in Finland, *Acta Hort.*, 241, 221, 1989.
80. Clark, J. R., Moore, J. N., and Baker, E. C., Cold damage to flower buds of rabbiteye blueberry cultivars, *Ark. Farm Res.*, 35, 3, 1986.
81. Patten, K., Neuendorff, E., and Nimr, G., Cold injury of southern blueberries as a function of germplasm and season of flower bud development, *HortScience*, 26, 18, 1991.
82. Luby, J. J., Wildung, D. K., Stushnoff, C., Munson, S. T., Read, P. E., and Hoover, E. E., 'Northblue', 'Northsky', and 'Northcountry' blueberries, *HortScience*, 21, 1240, 1986.
83. Fear, C. D. and Lawson, V. F., Cold injury to flower buds and shoots of blueberry cultivars following extreme low fall temperatures, *Fruit Var. J.*, 41, 148, 1987.
84. Wildung, D. K. and Sargent, K., The effect of snow depth on winter survival and productivity of Minnesota blueberries, *Acta Hort.*, 241, 232, 1989.
85. Hall, I. V. and Newbery, R. J., Floral development in normal and frost-injured cranberries, *HortScience*, 7, 269, 1972.
86. Eaton, G. W., The effect of frost upon seed number and berry size in the cranberry, *Can. J. Plant Sci.*, 46, 87, 1965.
87. Spiers, J. M., Effect of stage of bud development on cold injury in rabbiteye blueberry, *J. Amer. Soc. Hort. Sci.*, 103, 452, 1978.
88. Bittenbender, B. C. and Howell, G. S., Cold hardiness of flower buds from selected highbush blueberry cultivars (*Vaccinium australe* Small), *J. Amer. Soc. Hort. Sci.*, 101, 135, 1976.
89. Hancock, J. F., Nelson, J. W., Bittenbender, H. C., Callow, P. W., Cameron, J. S., Krebs, S.L., Pritts, M. P., and Schumann, C. M., Variation among highbush blueberry cultivars in susceptibility to spring frost, *J. Amer. Soc. Hort. Sci.*, 112, 702, 1987.
90. Eady, F. and Eaton, G. W., Reduced chilling requirement of 'McFarlin' cranberry buds, *Can. J. Plant Sci.*, 49, 637, 1969.
91. Eady, R. and Eaton, G. W., Effects of chilling during dormancy on development of the terminal bud of the cranberry, *Can. J. Plant Sci.*, 52, 273, 1972.
92. Spiers, J. M. and Draper, A. D., Effect of chilling on bud break in rabbiteye blueberry, *J. Amer. Soc. Hort. Sci.*, 99, 398, 1974.
93. Spiers, J. M., Chilling regimes affect budbreak in 'Tifblue' rabbiteye blueberry, *J. Amer. Soc. Hort. Sci.*, 101, 84, 1976.
94. Gilreath, P. R. and Buchanan, D. W., Temperature and cultivar influences on the chilling period of rabbiteye blueberry, *J. Amer. Soc. Hort. Sci.*, 106, 625, 1981.
95. Shine, J., Jr. and Buchanan, D. W., Chilling requirements of 3 Florida blueberry cultivars, *Proc. Fla. State Hort. Soc.*, 95, 85, 1982.
96. Norvell, D. J. and Moore, J. N., An evaluation of chilling models for estimating rest requirements of highbush blueberries (*Vaccinium corymbosum* L.), *J. Amer. Soc. Hort. Sci.*, 107, 54, 1982.
97. Darnell, R. L. and Davies, F. S., Chilling accumulation, budbreak, and fruit set of young rabbiteye blueberry plants, *HortScience*, 25, 635, 1990.
98. Lyrene, P. M. and Crocker, T. W., Poor fruit set on rabbiteye bluberries after mild winters: Possible causes and remedies, *Proc. Fla. State Hort. Soc.*, 96, 195, 1983.
99. Darnell, R. L., unpublished data, 1992.
100. Davies, F. S., Flower position, growth regulators, and fruit set of rabbiteye blueberries, *J. Amer. Soc. Hort. Sci.*, 111, 338, 1986.
101. Darrow, G. M., Rest period requirements for blueberry, *Proc. Amer. Soc. Hort. Sci.*, 41, 189, 1942.

102. Chandler, F. B. and Demoranville, I. E., Rest period for cranberries, *Proc. Amer. Soc. Hort. Sci.*, 85, 307, 1964.
103. Rigby, B. and Dana, M. N., Rest period and flower development in cranberry, *J. Amer. Soc. Hort. Sci.*, 97, 145, 1972.
104. Mainland, C. M., Some problems with blueberry leafing, flowering and fruiting in a warm climate, *Acta Hort.*, 165, 29, 1985.
105. Erez, A. and Couvillon, G. A., Characterization of the influence of moderate temperatures on rest completion in peach, *J. Amer. Soc. Hort. Sci.*, 112, 677, 1987.
106. Atkinson, D., Seasonal changes in the length of white unsuberized roots on raspberry plants grown under irrigated conditions, *J. Hort. Sci.*, 48, 413, 1973.
107. Williams, I. H., Effects of environment on *Rubus idaeus* L. III. Growth and dormancy of young shoots, *J. Hort. Sci.*, 34, 210, 1959.
108. Vasilakakis, M. D., Struckmeyer, B. E., and Dana, M. N., Temperature and development of red raspberry flower buds, *J. Amer. Soc. Hort. Sci.*, 104, 61, 1979.
109. Vasilakakis, M. D., McCown, B. H., and Dana, M. N., Low temperature and flowering of primocane-fruiting red raspberries, *HortScience*, 15, 750, 1980.
110. Lockshin, L. S. and Elfving, D. C., Flowering response of 'Heritage' red raspberry to temperature and nitrogen, *HortScience*, 16, 527, 1981.
111. Brierley, W. G. and Landon, R. H., Some relationships between rest period, rate of hardening, loss of cold resistance and winter injury in the 'Latham' raspberry, *Proc. Amer. Soc. Hort. Sci.*, 47, 224, 1946.
112. McCartney, J. S., A study of the effects of α-naphthaleneacetic acid on prolongation of rest in the 'Latham' raspberry, *Proc. Amer. Soc. Hort. Sci.*, 52, 271, 1948.
113. Brierley, W. G., Landon, R. H., and Stadtherr, The effect of daily alterations between 27 and 39 degrees F on retention or loss of cold resistance in the 'Latham' raspberry, *Proc. Amer. Soc. Hort. Sci.*, 59, 173, 1952.
114. Brierley, W. G. and Landon, R. H., Effects of dehardening and rehardening treatments upon cold resistance and injury of 'Latham' raspberry canes, *Proc. Amer. Soc. Hort. Sci.*, 63, 173, 1954.
115. Westwood, M. N., *Temperate Zone Pomology*, W.H. Freeman, San Francisco, California, 1978, 302.
116. Crane, J. H. and Davies, F. S., Flooding responses of *Vaccinium* species, *HortScience*, 24, 203, 1989.
117. Byers, P. L., Moore, J. N., and Scott, H. D., Plant-water relations of young highbush blueberry plants, *HortScience*, 23, 870, 1988.
118. Andersen, P. C., Buchanan, D. W., and Albrigo, L. G., Water relations and yields of three rabbiteye blueberry cultivars with and without drip irrigation, *J. Amer. Soc. Hort. Sci.*, 104, 731, 1979.
119. Davies, F. S. and Johnson, C. R., Water stress, growth, and critical water potentials of rabbiteye blueberry, *J. Amer. Soc. Hort. Sci.*, 107, 6, 1982.
120. Andersen, P. C., Leaf gas exchange of 11 species of fruit crops with reference to sun-tracking/non-tracking responses, *Can. J. Plant Sci.*, 71, 1183.
121. Cameron, J. S., Brun, C. A., and Hartley, C. A., The influence of soil moisture stress on the growth and gas exchange characteristics of young highbush blueberry plants (*Vaccinium corymbosum* L.), *Acta Hort.*, 241, 254, 1988.
122. Janke, R. A., E resistance in *Vaccinium myrtillus, Amer. J. Bot.*, 57, 1051, 1970.
123. Freeman, B., Albrigo, L. G., and Biggs, R. H., Cuticular waxes of developing leaves and fruit of blueberry *Vaccinium ashei* Reade cv. Bluegem, *J. Amer. Soc. Hort. Sci.*, 104, 398, 1979.
124. Woodward, F. I., Ecophysical studies on the shrub *Vaccinium myrtillus* L. taken from a wide altitudinal range, *Oecologia*, 70, 580, 1986.
125. Nobel, P. S., *Biophysical Plant Physiology*, W. H. Freeman San Francisco, 1974.
126. Korner, C., Scheel, J. A., and Bauer, H., Maximum leaf diffusive conductance in vascular plants, *Photosynthetica*, 13, 45, 1979.
127. Maxwell, J. O. and Redmann, R. E., Leaf water potential, component potentials, and relative water content in a xeric grass, *Agropyran dasystachyum* (Hook.) Scribn, *Oecologia*, 35, 277, 1978.
128. Bannister, P., The water relations of heath plants from open and shaded habitats, *J. Ecol.*, 59, 51, 1971.
129. Gough, R. E., Root distribution of 'Coville' and 'Lateblue' highbush blueberry under sawdust mulch, *J. Amer. Soc. Hort. Sci.*, 105, 576, 1980.
130. Davies, F. S., Teramura, A. H., and Buchanan, D. W., Yield, stomatal resistance, xylem pressure potential, and feeder root density in three rabbiteye blueberry cultivars, *HortScience*, 14, 725, 1979.
131. Hall, I. V., The taproot in lowbush blueberry, *Can. J. Bot.*, 35, 933, 1957.

132. Albrigo, L. G., Lyrene, P. M., and Freeman, B., Waxes and other surface characteristics of fruit and leaves of native *Vaccinium elliotti* Chapm., *J. Amer. Soc. Hort. Sci.*, 105, 230, 1980.
133. Hygen, G., Studies in plant transpiration II, *Physiol. Plant.*, 6, 106, 1953.
134. Bannister, P., The water relations of certain heath plants with reference to their ecological amplitude. II. Field studies, *J. Ecol.*, 52, 481, 1964.
135. Erb, W. A., Draper, A. D., and Swartz, H. J., Screening interspecific blueberry seedling populations for drought resistance, *J. Amer. Soc. Hort. Sci.*, 113, 599, 1988.
136. Struchtemeyer, R. A., For larger yields irrigate lowbush blueberries, *Maine Farm Research*, July, 17, 1956.
137. Trevett, M. F., Irrigating lowbush blueberries the burn year, *Maine Farm Research*, 15, 1, 1967.
138. Benoit, G. R., Grant, W. J., Ismail, A. A., and Yarborough, D. E., Effect of soil moisture and fertilizer on the potential and actual yield of lowbush blueberries, *Can. J. Plant Sci.*, 64, 683, 1984.
139. Fry, B. O. and Savage, E. F., Evaluation of sawdust as a mulch for young blueberry plants, *Proc. Amer. Soc. Hort. Sci.*, 93, 273, 1968.
140. Byers, P. L. and Moore, J. N., Irrigation scheduling for young highbush blueberry plants in Arkansas, *HortScience*, 22, 52, 1987.
141. Fugua, B., Irrigating highbush blueberries in Missouri ozark soils, in *Proc. Fifth N. Amer. Blueberry Res. Workers Conf.*, Crocker, T. E. and Lyrene, P. M., Eds., Univ. of Florida Press, Gainesville, 168, 1984.
142. Dale, A., Cline, R. A., and Ricketson, C. L., Soil management and irrigation studies with highbush blueberries, *Acta Hort.*, 241, 120, 1989.
143. Abbott, J. D. and Gough, R. E., Split-root water application to highbush blueberry plants, *HortScience*, 21, 997, 1986.
144. Andersen, P. C., Leaf gas exchange characteristics of 11 species of fruit crops in north Florida, *Proc. Fla. State Hort. Soc.*, 102, 229, 1989.
145. Buchanan, D. W., Davies, F. S., and Teramura, A. H., Yield responses of three rabbiteye blueberry cultivars to drip irrigation and vapor guard, *Proc. Fla. State Hort. Soc.*, 91, 162, 1978.
146. Andersen, P. C., Buchanan, D. W., and Albrigo, L. G., Antitranspirant effects on water relations and fruit growth of rabbiteye blueberry, *J. Amer. Soc. Hort. Sci.*, 104, 378, 1979.
147. Spiers, J. M., Influence of peat moss and irrigation on establishment of 'Tifblue' blueberry, *USDA Research Report*, 4, 18, 1980.
148. Spiers, J. M., Root distribution of 'Tifblue' rabbiteye blueberry as influenced by irrigation, incorporated peat moss, and mulch, *J. Amer. Soc. Hort. Sci.*, 111, 877, 1986.
149. Patten, K. D., Neuendorff, E. W., and Peters, S. C., Root distribution of 'Climax' rabbiteye blueberry as affected by mulch and irrigation geometry, *J. Amer. Soc. Hort. Sci.*, 113, 657, 1988.
150. Patten, K. D., Neuendorff, E. W., Nimr, G. H., Peters, S. C., and Cawthon, D. L., Growth and yield of rabbiteye blueberry as affected by orchard floor management practices and irrigation geometry, *J. Amer. Soc. Hort. Sci.*, 114, 728, 1989.
151. Smajstrla, A. G., Haman, D. Z., and Lyrene, P. M., Use of tensiometers for blueberry irrigation scheduling, *Proc. Fla. State Hort. Soc.*, 10, 232, 1988.
152. Haman, D. Z., Smajstrla, A. G., and Lyrene, P. M., Blueberry response to irrigation and ground cover, *Proc. Fla. State Hort. Soc.*, 101, 235, 1988.
153. Brinen, G. H., Bienert, S. A., and Crocker, T. E., Current status of the blueberry industry in Alachua County and delineation of problem areas associated with soil acidity as affected by high pH irrigation water, *Proc. Fla. State Hort. Soc.*, 99, 200, 1986.
154. Haby, V. A., Patten, K. D., Cawthon, D. L., Krejsa, B. B., Neuendorff, E. W., Davis, J. V., and Peters, S. C., Response of container-grown rabbiteye blueberry plants to irrigation water quality and soil type, *J. Amer. Soc. Hort. Sci.*, 111, 1986, 332.
155. Patten, K. D., Neuendorff, E. W., Leonard, A. T., and Haby, V. A., Mulch and irrigation placement effects on soil chemistry properties and rabbiteye blueberry plants irrigated with sodic water, *J. Amer. Soc. Hort. Sci.*, 113, 4, 1988.
156. Patten, K., Neuendorff, E. W., Nimr, G., Haby, V., and Wright, G., Cultural practices to reduce salinity/sodium damage of rabbiteye blueberry plants (*Vaccinium ashei* Reade), *Acta Hort.*, 241, 207, 1989.
157. Beckwith, C. S., Flooding and irrigation—water table, *Cranberries—The National Cranberry Magazine*, 5, 2, 1940.

158. Eck, P., Cranberry growth and production in relation to water table depth, *J. Amer. Soc. Hort. Soc.*, 101, 544, 1976.
159. Hall, I. V., Cranberry growth as related to water levels in the soil, *Can. J. Plant Sci.*, 51, 237, 1971.
160. Dickey, G. L. and Baumer, O. W., Irrigation of cranberries, in *Proc. and Tech. Ref. Man. for the Nat. Cranberry Conf.*, Mass. Coop. Ext. Serv., 1985, 46.
161. Chandler, F. B., Effect of methods of irrigating cranberry bogs on water table and soil moisture tension, *Proc. Amer. Soc. Hort. Sci.*, 51, 65, 1951.
162. Christensen, J. R., Root studies. XI. Raspberry root systems, *J. Hort. Sci.*, 23, 218, 1947.
163. Goode, J. E. and Hyrycz, K. J., The response of 'Malling Jewel' and 'Malling Exploit' raspberries to different soil moisture conditions and straw mulching, *J. Hort. Sci.*, 43, 215, 1968.
164. Goode, J. E., The measurement of sap tension of apple, raspberry and black currant leaves, *J. Hort. Sci.*, 43, 231, 1968.
165. Crandall, P. C. and Chamberlain, J. D., Effects of water stress, cane size, and growth regulators on floral primordia development in red raspberries, *J. Amer. Soc. Hort. Sci.*, 97, 418, 1972.
166. Crandall, P. C., Allmendinger, D. F., Chamberlain, J. D., and Biderbost, K. A., Influence of cane number and diameter, irrigation and carbohydrate reserves on the fruit number of red raspberries, *J. Amer. Soc. Hort. Sci.*, 99, 524, 1974.
167. MacKerran, D. K. L., Growth and water use in red raspberry (*Rubus idaeus* L.). I. Growth and yields under different levels of soil moisture stress, *J. Hort. Sci.*, 57, 295, 1982.
168. DeBoer, D. W., Peterson, R. M., and Evers, N. P., Trickler and sprinkler irrigation of red raspberries, *HortScience*, 18, 930, 1983.
169. Crane, J. H. and Davies, F. S., Periodic and seasonal flooding effects on survival, growth, and stomatal conductance of young rabbiteye blueberry plants, *J. Amer. Soc. Hort. Sci.*, 113, 488, 1988.
170. Abbott, J. D. and Gough, R. E., Growth and survival of the highbush blueberry in response to root zone flooding, *J. Amer. Soc. Hort. Sci.*, 112, 603, 1987.
171. Crane, J. H., Soil temperature and flooding effects on young rabbiteye blueberry plant survival, growth, and ethylene levels, *Proc. Fla. State Hort. Soc.*, 100, 301, 1987.
172. Davies, F. S. and Wilcox, D., Waterlogging of containerized rabbiteye blueberries in Florida, *J. Amer. Soc. Hort. Sci.*, 109, 520, 1984.
173. Crane, J. H. and Davies, F. S., Flooding duration and seasonal effects on growth and development of young rabbiteye blueberry plants, *J. Amer. Soc. Hort. Sci.*, 113, 180, 1988.
174. Herath, H. M. E. and Eaton, G. W., Some effects of water table, pH, and nitrogen fertilization upon growth and nutrient element content of highbush blueberry plant, in *Proc. Amer. Soc. Hort. Sci.*, 92, 274, 1968.
175. Abbott, J. D. and Gough, R. E., Reproductive response of the highbush blueberry to root-zone flooding, *HortScience*, 22, 40, 1987.
176. Davies, F. S., unpublished, 1992.
177. Kawase, M., Anatomical and morphological adaptation of plants to waterlogging, *HortScience*, 16, 30, 1981.
178. Abbott, J. D. and Gough, R. E., Prolonged flooding effects on anatomy of highbush blueberry, *HortScience*, 22, 622, 1987.
179. Crawford, R. M. M., Tolerance of anoxia and the regulation of glycolysis in tree roots, in *Tree Physiology and Yield Improvement*, Cannel, M. G. R. and Last, F. T., Eds., Academic Press, New York, 387, 1976.
180. Bradford, K. J. and Hsiao, T. C., Stomatal behavior and water relations of waterlogged tomato plants, *Plant Physiol.*, 70, 1508, 1982.
181. Kramer, P. J. and Jackson, M. B., Causes of injury to flooded tobacco plants, *Plant Physiol.*, 29, 241, 1954.
182. Coutts, M. P., Effects of waterlogging on water relations of actively growing and dormant 'Sitka' spruce seedlings, *Ann. Bot.*, 47, 747, 1981.
183. Mansfield, T. A., Heterington, A. M., and Atkinson, C. J., Some current aspects of stomatal physiology, *Annu. Rev. Plant Phys. Plant Mol. Biol.*, 41, 55, 1990.
184. Andersen, P., unpublished, 1992.
185. Crane, J. H. and Davies, F. S., Flooding, hydraulic conductivity, and root electrolyte leakage of rabbiteye blueberry plants, *HortScience*, 22, 1249, 1987.

186. Davies, F. S. and Flore, J. A., Root conductivity, CO_2 response and quantum yield of flooded and unflooded highbush blueberry plants, *HortScience*, 20, 89, 1986 (Abstr.).
187. Bergman, H. F., Disorders of cranberries, in *Plant Diseases*, USDA Yearbook of Agric., 1953.
188. Bergman, H. F., The respiratory activity of various parts of the cranberry plant in relation to flooding injury, *Amer. J. Bot.*, 12, 641, 1925.
189. Franklin, H. J., Bergman, H. F., and Stevens, N. E., Weather in cranberry culture, *Mass. Agr. Expt. Sta. Bul.* no. 402, 1943.

Chapter 4

Grape

Larry E. Williams, Nick K. Dokoozlian, and Robert Wample

CONTENTS

0-8493-175-0/94/$0.00+$.50

I. INTRODUCTION

A. GRAPE PRODUCTION

Grapevines are the most widely planted fruit crop worldwide and are cultivated on all continents except Antarctica.[1] Grapevine acreage as of 1988 was greater than 10 million ha.[2] The widespread distribution of grapevines is due to a large genetic diversity of available species and cultivars and a low chilling requirement for the release of buds from dormancy. However, a single species, *Vitis vinifera* L., of which there are currently over 10,000 cultivars accounts for greater than 90% of the annual production worldwide. Sixty percent of the world production of grapes (>63 Tg) is produced in Europe with Spain, Italy, and France each having more than 1 million ha of land devoted to grapevines.

Grapes are primarily used for wine, juice, distilled liquors, dried fruit (raisins) and fresh consumption fruit (table grapes). Italy, France, and Spain produce more than 50% of the world's wine; European countries together with the Commonwealth of Nations (former Soviet Union) account for 80% of the world's production.[1] The production of raisins worldwide is approximately 800 Gg (dried fruit); the United States and Turkey are the top two producing countries. Annual world production of table grapes is approximately 7 Tg. Italy, the Commonwealth of Nations, and Turkey comprise the top table grape producing nations. Fruit juice is concentrated when production exceeds demand.

B. CLIMATIC CONSTRAINTS TO GRAPE PRODUCTION

V. vinifera is a temperate climate species adapted to hot summers with mild winters. The suitability of a given grape cultivar to a local environment is based upon day length, heat summation, rainfall, length of the growing season, and minimum winter temperatures. On a broad scale the main grape production areas are found between 30 and 50°N and 30 and 40°S latitudes, corresponding to the 10 and 20°C yearly isotherms.[1] Grapes can be commercially grown in other areas where climate is moderated due to local geographical conditions (mountains, land masses, and ocean currents). Raisin production is limited to the latitudes of 30 and 39°N in the Northern Hemisphere and between 28 and 36°S in the Southern Hemisphere. This is due to the fact that the best suited raisin cultivars, 'Thompson Seedless' (syn. 'Sultinina') and 'Zante Currant', require warm temperatures for fruit bud differentiation and fruit maturation. In addition, the production of natural raisins (sun-dried grapes) requires high temperatures and lack of rainfall following harvest. Warm, dry weather also favors the production of table grapes as the incidence the fungal diseases is much reduced under these conditions.

C. CYCLE OF VINE GROWTH

Vineyards are planted via vegetative means such as cuttings, rootings, or grafted vines. Vineyards commonly produce a harvestable crop in the third growing season, subsequent to establishment of a root system and training the vines to fit a specific trellis system.[2] Trellis choice depends upon intended use of the grapes (wine, raisin, or table grape production), methods of pruning and harvest (manual or mechanical),[3] and climate and soil conditions.

Current season's vegetative and reproductive growth occurs from compound buds (consisting of a primary, secondary, and tertiary bud) in the spring. The primary bud consists of eight to ten leaf primordia with zero or more cluster primordia. Budbreak is followed by rapid shoot growth. Flower differentiation on the cluster primordia begins prior to budbreak and continues up until anthesis. Anthesis occurs approximately eight weeks after budbreak, therefore, considerable leaf area has developed before pollination, fertilization, and berry set takes place. Berry growth of both seeded and "seedless" cultivars is of the double sigmoid type in which growth occurs in three stages. Vegetative growth commonly continues until veraison (the end of Stage II; characterized by softening of the fruit and change in color for red and black cultivars). Wine and raisin grapes are harvested when the soluble solids (°Brix) concentrations are between 16 and 25 °Brix. Table grapes are generally harvested when the soluble solids levels are 15 to 17 °Brix.

Leaves remain photosynthetically active and will remain such until the first killing freeze. Periderm will have formed on the main axis of the shoot throughout the growing season. Once the leaves abscise the vine goes dormant and the leafless shoots, now called canes, will be pruned during the winter to regulate next year's crop.

Cluster differentiation (for the next year's crop) within the compound buds of spur pruned cultivars (four basal nodes on a shoot) begins around anthesis and is complete prior to veraison for spur-pruned table grape cultivars in California (L.E. Williams, unpublished data). Cluster differentiation is complete at node 15 by late summer. Therefore, environmental conditions and stress during these periods can dramatically affect the next season's yield. Further details on the physiology, anatomy, and morphology of the grapevine can be found in Mullins et al.[1]

II. IRRADIANCE

Solar radiation induces various biological responses through changes in light quality, quantity, direction, and periodicity.[4] Plant responses include thermal effects, photosynthesis, photomorphogenesis, and mutagenesis. Viticulturists have become increasingly aware of the positive effects of light on both the quantity and quality of the harvested fruit due in large part to the pioneering work of Dr. Nelson Shaulis and co-workers at the New York State Agricultural Experiment Station, Geneva, NY.

A. INTERCEPTION BY GRAPEVINE CANOPIES

1. Effect of Canopy Height and Row Direction

Light interception is dependent primarily upon canopy shape and orientation. Smart[5] predicted that sunlight interception by grapevine canopies declines rapidly as canopy height decreases and the distance between walls of foliage (i.e., distance between vine rows) is increased. While a smaller distance between rows increases solar radiation interception on an area basis, cross-row shading becomes a significant factor limiting sunlight interception of individual foliage walls as the distance between rows is reduced. A value of 1:1 for the ratio of canopy height to distance between canopies is recommended to avoid cross shading.[5]

Row direction also has a pronounced effect on solar radiation interception. Greater amounts of direct light are absorbed by the canopy walls in the mid-morning and mid-afternoon in rows oriented north-south compared to east-west.[5] Canopies 1 and 3 m in height and spaced 4 m apart intercepted approximately 10 and 22% more sunlight, respectively, when rows were oriented north-south compared to the east-west row orientation.

2. Effect of Trellis and Canopy Management

The amount of solar radiation intercepted by the grapevine canopy, as well as the light environment within the canopy interior, is largely determined by the training and trellis system employed and vine leaf area.[6] These factors, combined with such cultural practices as shoot positioning and basal leaf removal, determine shoot orientation, canopy surface area, and vine foliage density.[7] Shaulis and co-workers[8] were among the first to recognize the influence of vine training and trellis design on the light environment within grapevine canopies, and its effect on vine productivity and fruit composition. Vegetative growth normally increases as canopy width expands, thereby increasing the total amount of leaf surface available for solar radiation interception.[9] As a result of greater sunlight interception per unit row or canopy length, increasing both canopy height and width via vineyard layout and training/trellising generally increases vine yield.[10–12]

Solar radiation interception and penetration to the canopy's interior has been substantially increased by the separation of the canopy into two vertical curtains.[8] The Geneva double curtain training system has often resulted in greater bud fruitfulness and vine productivity, and improved fruit quality. Carbonneau and Huglin[13] reported that the surface of an open lyre or "U" shaped canopy with two distinct curtains of foliage intercepted 10% more solar radiation per day than a non-separated, single row canopy at wide row spacing. Light measured in the fruiting zone was 21% of ambient for the separated lyre canopy, compared to 6.4% of ambient for the single canopy. Kliewer et al.[14] reported that photosynthetic photon flux density (PFD) within the fruiting zone of a non-divided canopy of 'Sauvignon blanc' was approximately 4% of ambient, while the fruit zone PFD of this cultivar with a divided canopy was greater than 30% of ambient.

Additional canopy management practices may be employed to increase solar radiation interception and penetration into the canopy interior. Shoot positioning, either performed manually or mechanically, can be used to separate tangled foliage within the interior of divided canopies. Shoot positioning prevents shading and improves sunlight interception by maintaining canopy separation throughout the growing season.[7] Basal leaf removal, influences the light microclimate in the canopy interior. Bledsoe et al.[15]

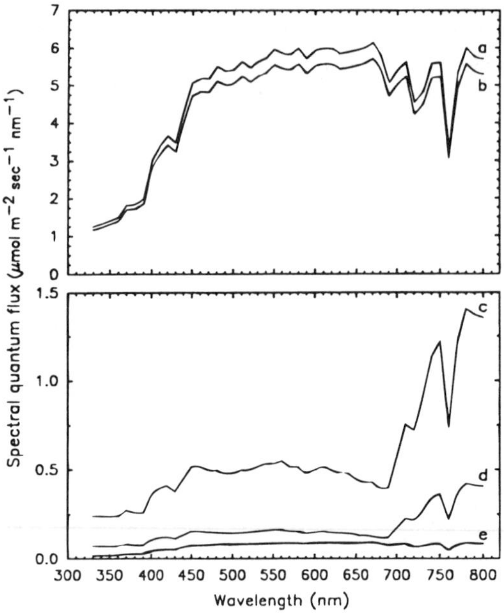

Figure 1 Spectral quantum flux of radiation between the wavelengths of 330 and 800 nm for: ambient sunlight (a), a sunfleck within the interior of a grapevine canopy (b), and for canopy light in the fruiting zone of 'Cabernet Sauvignon' grapevine canopies with 2.1 (c), 5.6 (d) and 12.2 (e) m^2 leaf area m^{-1} canopy length. Each curve represents the mean of five separate scans. Determinations were made at 2-nm wavelength intervals using a scanning spectroradiometer with the sensor positioned horizontally near solar noon under clear skies in Oakville, CA. (N.K. Dokoozlian, unpublished data.)

reported that basal leaf removal following berry set increased PFD within the fruiting region of 'Sauvignon blanc' grapevines by approximately 20% when compared to the untreated control.

B. LIGHT ENVIRONMENT WITHIN GRAPEVINE CANOPIES

Of the PFD arriving perpendicular to the surface of a grape leaf, 80 to 90% is absorbed and the remainder is either transmitted or reflected.[16] The low levels of PFD (5% of ambient or less) commonly measured within the interior of dense grapevine canopies is a result of this high degree of absorption.[17] The spectral quality also is altered when compared to ambient solar radiation in that the blue (400 to 500 nm) and red (650 to 700 nm) wavelengths are absorbed by the vine's canopy to a greater extent than the far-red (710 to 800 nm) wavelengths (Figure 1). This results in a decrease in the amount of blue light in the canopy interior relative to other wavelengths, as well as a decrease in the red (R) to far-red (FR) ratio (R:FR). The spectral composition of sunflecks, which occur when direct solar radiation penetrates through gaps in the canopy, closely resembles that of incident radiation (Figure 1).[18]

The relationship between canopy leaf area and PFD, R:FR, and sunflecks within the canopy of wine grapes grown under the standard (non-separated) training/trellis system used in California is presented in Figure 2. PFD decreased to 10% or less of its ambient value approximately 0.5 m below the canopy surface. PFD was lowest near the fruiting zone, with values of 7.0 and 0.3% of ambient, for vineyards with 2.2 and 12.2 m^2 of leaf area per meter canopy length, respectively. PFD increased below the fruit zone, and ambient values of PFD were found at ground level of canopies with low leaf area densities. Patterns of the R:FR ratio and sunfleck attenuation within low and high leaf area density canopies were similar to those observed for PFD (Figures 2B and 2C, respectively). The R:FR ratio decreased immediately below the canopy surface, and reached its lowest level at or near the fruit zone. The R:FR also increased gradually moving downward along the vertical transect from the fruit zone to the ground. Sunflecks illuminated about 20% of the fruit zone at harvest within the low density canopy, but were nearly absent along the vertical transect in a high density canopy. This study also revealed that the patterns of PFD, R:FR, and sunfleck attenuation changed little during the course of fruit development, despite an approximate doubling of canopy leaf area in various vineyards during this same time period.[19] A close, positive relationship exists between PFD and the R:FR ratio (Figure 3). In very dense canopies PFD may approach 0.1% of ambient sunlight and the R:FR ratio may drop as low as 0.05.

C. EFFECTS ON VINE GROWTH AND METABOLISM

1. Cluster Differentiation

The differentiation of anlagen into either cluster or tendril primordia is dependent upon the irradiance level reaching the compound bud during development as demonstrated by experiments conducted under

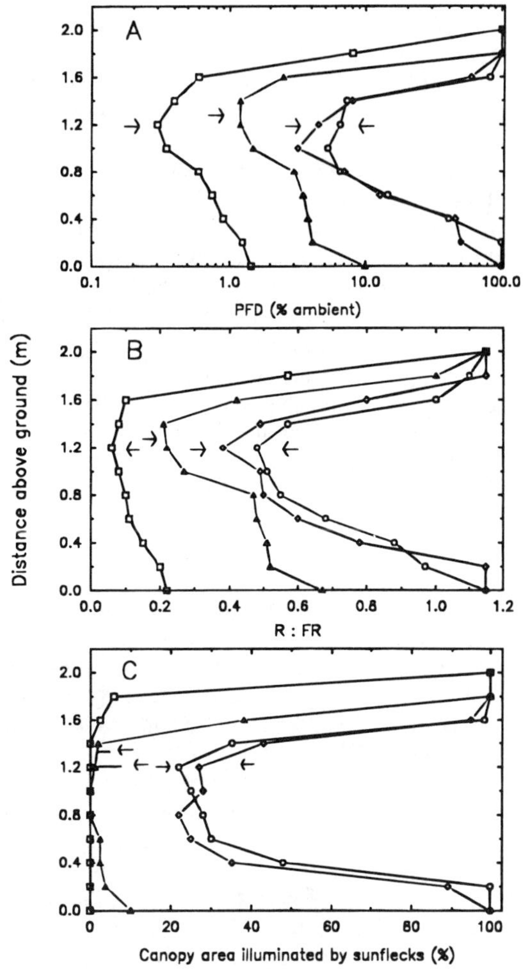

Figure 2 Relationship between photosynthetic photon flux density (PFD) (A), the red:far red (R:FR) ratio (B), and the percent canopy area illuminated by sunflecks (C) as a function of canopy depth for 'Cabernet Sauvignon' canopies at harvest. Symbols correspond to the following canopy leaf areas: (○) 2.2, (◇) 3.3, (△) 8.7, and (□) 12.1 m^2 leaf area m^{-1} canopy length. Readings were taken with sensors positioned vertically upward, at solar noon under clear skies. Arrows indicate the locations of the fruit zone in each canopy. (N.K. Dokoozlian, unpublished data.)

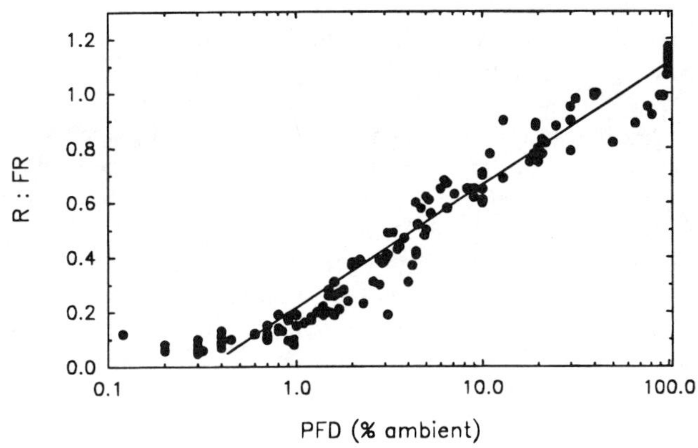

Figure 3 Relationship between the red:far red (R:FR) ratio and photosynthetic photon flux density (PFD) (% ambient, log scale) within grapevine canopies. Values represent a wide range of vineyard sites and canopy densities and depths. Measurements were made with light sensors positioned horizontally at solar noon under clear skies. Data were fitted to a linear function: $y = -0.431 + 2.147x$, $r^2 = 0.96$. (N.K. Dokoozlian, unpublished data.)

controlled environmental conditions[20,21] and in the field.[22,23] Low irradiance favors the differentiation of tendril primordia. The number and size of cluster primordia generally increase with an increase in irradiance level.[20] Bud fruitfulness and subsequent yield are increased by improving the light environment of developing buds by the use of divided canopies[8] or retention of spurs or canes developing at the top of the canopy.[1]

The specific mechanisms responsible for the regulation of bud fruitfulness by light are unknown. However, increased cluster initiation under well-exposed conditions appears to be primarily due to increased light quantity; the R:FR ratio was shown to have no effect on bud fruitfulness.[24,25] Although shading individual buds decreases cluster initiation, some controversy remains regarding the influence of irradiance on the leaves immediately subtending the bud on fruit bud differentiation.

The irradiance required to maximize bud fruitfulness varies among *Vitis vinifera* cultivars.[26] 'Sultana' (syn. 'Thompson Seedless') and 'Ohanez' require relatively high irradiance (approximately one third full sunlight) for notable cluster initiation. In comparison, significant cluster differentiation of 'Rhine Riesling' buds was obtained with only 10% of full sunlight.

It is generally accepted that photoperiod has little effect on cluster differentiation of *V. vinifera.*[1] However, American *Vitis* species will respond to increased day length. For example, the *Vitis* × *labruscana* cultivar 'Delaware' had three times more clusters when grown under long days compared to those grown under short days.[27,28]

2. Leaf Gas Exchange

As for other C_3 species, the relationship between leaf net CO_2 assimilation rate (A) and PFD for grapevine leaves can best be described as a rectangular hyperbole. Light saturation for individual leaves of grapevines may change due to conditions under which the vines are grown.[29,30] However, recent studies using field-grown grapevines indicate that light saturation occurs at approximately 1500 μmol quanta m^{-2} s^{-1}.[1,31,32] The light compensation point for A of grapevines is between 10 and 20 μmol quanta m^{-2} s^{-1}.[32,33]

Stomatal conductance to water vapor (g_s) of well-watered vines showed a hyperbolic response to PFD.[34,35] Maximum stomatal opening of an individual leaf has been recorded at a PFD of 130 to 300 μmol quanta m^{-2} s^{-1}.[30,34,35] Canopy conductance of a grapevine at full canopy, unlike single leaf g_s, is linearly related to PFD (Figure 4). This is expected as individual leaves located throughout the canopy are simultaneously exposed to different PFDs due to shading, leaf angle, zenith angle of the sun, and row and shoot direction. Therefore, maximum canopy conductance is associated with maximum PFD and occurs when the greatest proportion of the leaf canopy is exposed to direct solar radiation.[37]

It has been suggested that there is a high PFD-stress effect on *V. vinifera* leaves that causes an afternoon depression of A independent of leaf temperature.[38,39] This contradicts studies on well-watered field-grown grapevines in which there was no midday depression of A at high irradiance levels.[31,40,41] Data demonstrating the midday depression of A, though, were collected on potted vines either without a measure of vine water status[38] or grown in a glasshouse with measurements taken in the lab.[39] Düring[33] has shown that A of potted glasshouse-grown but not field-grown Riesling vines was slightly depressed at high PFD.

The most extensive research investigating photoinhibition of A in grapevines has been conducted on the native California species *V. californica* Benth. At high PFD both the light and dark reactions of A were more severely inhibited at high (41.5°C) and low (22.7°C) temperatures than at intermediate temperatures.[42] The inhibition of A at high PFD was greater for growth chamber grown vines relative to vines grown outside. Exposure to either high light or high temperature caused reductions in PSII photochemical activity with a subsequent recovery the following day.[43] However, exposure of *V. californica* leaves to both high light and high temperature caused PSII inhibition that was severe and persistent. Finally, field studies using unrestrained and horizontally held leaves of this species confirmed that high PFD (>1800 μmol quanta m^{-2} s^{-1}) had no adverse impact on A.[44]

3. Berry Growth and Composition

Much of the recent information regarding the influence of light on grape berry growth and composition has been obtained from studies investigating the influences of training-trellis systems and other canopy management practices on grapevine yield and fruit composition.[13,14,45,46] In cool climates canopy management practices which improve the exposure of vine foliage and fruit to solar radiation have generally improved grape and wine composition. Fruit of vines in which the canopy interiors are well exposed

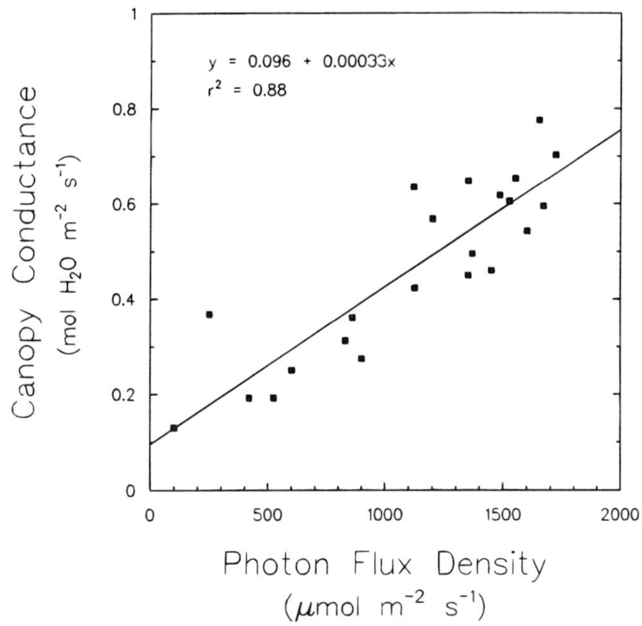

Figure 4 The relationship between canopy conductance and photosynthetic photon flux density. Total vine conductance was calculated according to the method of Grantz and Meinzer.[36] Canopy conductance was calculated as the difference between total vine conductance and aerodynamic conductance.[37] The data were collected on two separate days during the month of August, 1992. (L.E. Williams, unpublished data.)

to sunlight have normally exhibited increased rates of sugar accumulation, greater concentrations of anthocyanins, and total phenols, yet decreased levels of malic acid, potassium, and pH compared to fruits produced from vines with little interior canopy exposure. The above-mentioned practices are used to alter the fruiting zone light environment and have likely influenced other aspects of vine microclimate. For example, training-trellis systems, shoot positioning, and other canopy management practices, alter the light environment of both shoots (leaves) and clusters thus creating uncertainty as to whether improvements in berry composition are attributable to increased fruit exposure, leaf exposure, or both. In addition, increased levels of solar radiation exposure can result in substantial increases in both berry[47] and leaf[8] temperatures.

Recent investigations have shown that a photoreceptor localized in the grape berry is responsible for the detection of the light environment and thus the photoregulation of grape berry growth and composition.[19] The exclusion of light to fruit of field-grown 'Sultana' (syn. 'Thompson Seedless') for a period 2 weeks prior to softening until harvest increased berry weight and soluble solids compared to fruit exposed to natural light in the canopy.[48] Fruit of 'Thompson Seedless' vines exposed to direct solar radiation had similar soluble solids (°Brix), but were lower in weight and acidity and had a higher pH than fruit that ripened in the canopy interior.[49] Anthocyanin concentrations and soluble solids of 'Emperor' berries were decreased when they received 15% compared to berries receiving 54 or 100% of ambient solar radiation.[50] Field-grown 'Cabernet Sauvignon' berries exposed to sunlight had lower berry weights but higher concentrations of tartrate, malate, glucose, fructose, and anthocyanins compared to berries ripened in the canopy interior.[51,52] Morrison[53] reported on the independent effects of cluster shading and leaf shading on the growth and composition of fruit from field-grown vines of 'Cabernet Sauvignon'; cluster shading reduced fruit anthocyanin and total soluble phenolic concentrations, while leaf shading reduced berry weight and decreased the rate of sugar accumulation.

During Stage I of the double sigmoid curve of grape berry growth,[54] berry pericarp growth is rapid due to both cell division and cell expansion while growth during Stage III is due to cell enlargement. Light affects grape berry growth and composition differently during the three stages of fruit development. It was found that when sunlight was excluded from clusters during Stages I and II, berry growth was significantly less compared to those which received 20% of ambient sunlight during these two stages (Figure 5). It is unknown whether the reduction in berry growth was due to a reduction in cell division,

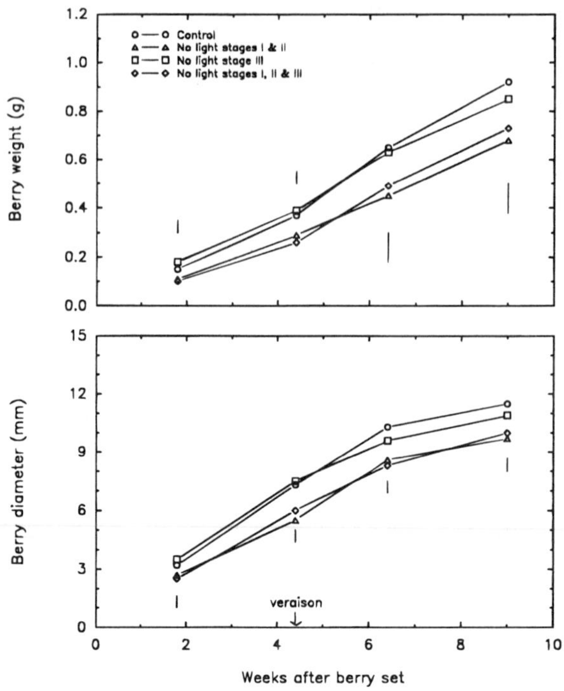

Figure 5 The influence of solar radiation exclusion during various stages of fruit development on the fresh weight (upper) and diameter (lower) of 'Cabernet Sauvignon' grape berries. Sunlight was excluded from the clusters by enclosing the fruit in aluminum lined, paper bags at the appropriate stage of growth. Data represent the mean of four, single-cluster replicates grown on potted vines in a phytotron. Bars represent LSDs (p <0.05) on each sample date. (N.K. Dokoozlian, unpublished data.)

cell enlargement, or both. Berries of these clusters also exhibited reduced rates of solute and color accumulation during Stage III compared to berries exposed to sunlight during Stages I and II. Exclusion of sunlight during Stage III had little effect on berry weight, berry diameter, or solute accumulation. However, berries from these clusters exhibited lower rates of anthocyanin and phenol accumulation indicating that light has the greatest effect on both berry growth and fruit composition during Stages I and II.

The importance of phytochrome in the control of grape berry growth and composition is a topic of interest. Smart et al.[55] reported that 'Cabernet Sauvignon' vines shaded with neutral shade cloth and receiving supplemental red light in the fruiting zone (i.e., exposed to 10% of ambient PFD, and a R:FR ratio of 3.0) produced fruit with greater concentrations of glucose, fructose, and anthocyanins than shaded vines which received no supplemental red light. In a similar study, 'Cabernet Franc' vines grown under artificial (neutral shade cloth) and natural (foliage) shade received similar levels of PFD, but the R:FR ratios of the two treatments were 0.7 and 0.07, respectively.[56] Compared to the controls, both shading treatments reduced berry weight, soluble solids, and anthocyanin concentrations. The effects of natural shade on fruit growth and composition were partially reversed by supplemental red light (R:FR ratio of 3.0) in the fruiting zone indicating the involvement of phytochrome.

By contrast, a study of 'Cabernet Sauvignon' and 'Pinot Noir' fruit exposed to various combinations of light quantity (20 and 1% of ambient PFD) and light quality (R:FR = 1, 0.6, 0.3, or 0.1) indicated that light quantity rather than light quality was of primary importance. A reduction in light quantity from 20 to 1% of ambient PFD, regardless of the R:FR ratio, decreased berry weight and diameter, and delayed the accumulation of sugars, anthocyanins, and phenolics. Also, varying the R:FR ratio under continuous illumination had no influence on anthocyanin accumulation and did not reverse the R to FR mediated anthocyanin synthesis. These results indicate that a photoreceptor which is dependent upon light quantity is most likely involved in the regulation of anthocyanin accumulation in grape berries.

4. Nitrogen Metabolism

The nitrate concentration of grapevine petioles and leaf blades has been shown to be inversely related to the PFD environment.[57] The petiole NO_3 concentrations of 'Chardonnay', 'Zinfandel', and 'Malbec' vines were five-fold higher and nitrate reductase activity was lower in vines grown at 8% of ambient PFD compared to vines grown in full sunlight. Smart et al.[55] reported that shading increased the concentration of NO_3^- in petioles and leaf blades, and the concentration of NH_4^+ in the leaf blades, peduncles and juice of 'Cabernet Sauvignon'. In addition, supplemental red light partially restored leaf nitrate reductase activity of heavily shaded vines, suggesting a phytochrome mediated regulation of nitrate reductase. However, red light has not stimulated leaf nitrate reductase activity of field-grown *V. vinifera* cultivars (Dokoozlian, unpublished data).

III. TEMPERATURE

Temperature is the environmental factor primarily responsible for the distribution of *V. vinifera* throughout the world. Grapes are produced in some of the hottest cultivated areas of the earth. Air temperatures in the Jordan Valley of Israel and the interior valleys of California during the summer often exceed 35°C with maximum temperatures approaching 43°C. Over 40% of the grapes produced in China are grown in the Turpan depression of Xinjiang Province, Northwest China, the mean temperature during July is 33°C, daily temperatures exceed 35°C and the maximum is 48°C.[58,59] While the loss of a grape crop due to high temperatures is rare, partial yield loss in individual vineyards can occur. When the daily high temperature increased from an average of 30 to 47°C over a 3-day period in the Coachella Valley of California, clusters of 'Thompson Seedless' grapevines, which had just been girdled (the removal of a strip of phloem from around the trunk), in several vineyards became desiccated and resulted in crop loss (L.E. Williams, personal observation). It is unknown whether this crop loss was due solely to the high temperatures or a combination of high temperature and improper irrigation applications.

Potential crop loss of grapes due to low temperatures is far greater than that for high temperatures. "Potential crop loss" is used because in many cases, the result of a winter freeze is reported as the percent primary bud loss.[60–62] The interpretation of crop loss is often complicated by the adjustment in pruning practices made by the vineyard managers to compensate for bud injury or compensation by the vine producing fruitful shoots from secondary or tertiary buds, or by increasing total fruit set and berry size.[62,63] Losses due to late spring and especially early fall freezes are less likely to be compensated for by either the vineyard manager or the grapevine. For instance a spring freeze in 1985 caused a reduction in 'Concord' grape yield in the state of Washington from 151,000 tons in 1984 to only 90,000 tons in 1985.[64]

The growth and productivity of many crops are temperature dependent in the range of 5 to 20°C assuming all other factors are non-limiting.[65] One of the first attempts to use this concept was the development of a relationship between air temperatures and the dates of grape harvest.[65] In viticulture, temperature summations (i.e., termed in the literature as degree-days, growing degree days, day degrees, or heat summations) have been used as dependent variables in describing the timing of various grapevine phenological events and growth.[1] Amerine and Winkler[66] used accumulated degree-days above 10°C to formulate recommendations for the growing of wine grape cultivars in California.

While their method has gained wide acceptance, recent studies have indicated that degree-days may not be the most accurate basis for viticultural recommendations.[67,68] A major limitation in calculating degree-days, by taking the mean of the daily maximum and minimum temperatures and subtracting a base temperature, is that periods of fog, cloud cover, or wind, factors which may not affect the daily maximum and minimum temperatures but undoubtedly will affect vine growth. Degree minutes calculated with dataloggers may improve the accuracy of degree-days.[69]

A. EFFECTS ON GAS EXCHANGE

Every aspect of plant growth, such as physical processes, enzyme reactions, ion and carbohydrate transport, and membrane permeability, is controlled by temperature. Vine growth and productivity are dependent upon the assimilation of carbon via photosynthesis and subsequent carbon translocation and allocation. The production of biomass is the result of a balance between carbon gains and losses due to respiration, organ death, and other means where biomass may be lost (i.e., pruning, herbivory).

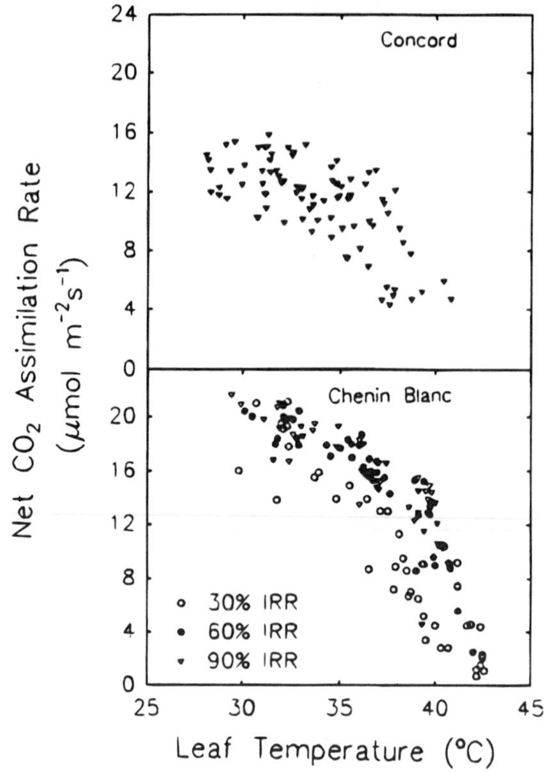

Figure 6 The effect of leaf temperature on net CO_2 assimilation of 'Concord' and 'Chenin blanc' grapevines grown in Washington, USA. The 'Chenin blanc' vines were irrigated with 30, 60, or 90% of daily evapotranspiration. (R. Wample, unpublished data.)

The optimum leaf temperature for photosynthesis of field-grown grapevines is quite broad; generally between 25 and 35°C[1,29] while other studies have demonstrated a more narrow temperature optimum (25–30°C).[29,30] This variation may be due to cultivar, growth conditions or seasonal variation.[70] This could perhaps also explain some of the variation in the data presented in Figure 6 since it is a compilation of data from several seasons and dates of measurement. Photosynthesis generally declines at temperatures above 35°C, for both American and European species of grapevines, yet a positive A occurs even up to 40°C (Figure 6).[1,30,43,44] It is also noteworthy that the leaf temperature of 'Chenin blanc' vines receiving 90% evapotranspiration replacement barely exceeded ambient temperature, i.e., 40°C, as a result of evaporative cooling. The temperature dependency of A of 'Concord' leaves appeared to be similar to 'Chenin blanc' and reports of other *V. vinifera* cultivars.[1] Thus, air temperatures up to 40°C, unless experienced for an extended period of time would not appear to be a major limiting factor in grape production.

Plants grown in thermally contrasting habitats exhibit photosynthetic temperature responses that reflect an adaptation to the temperature regimes of their respective habitat.[71] The photosynthetic temperature response curves of 'Chenin blanc' grapevines grown in two thermally contrasting climates in California are presented in Figure 7. Although fitted curves for the two data sets were similar at temperatures greater than 32°C, the decrease in A occurred more rapidly for vines grown in the San Joaquin Valley as leaf temperature decreased.

Preconditioning temperatures may also influence photosynthetic processes. Balo et al.[72] found that 3 to 6 h of chilling (6 ± 2°C) had little effect on gas exchange, fluorescence kinetics, and water relations of rooted cuttings of 'Merlot' grapevines. However, durations of over 24 h of chilling caused significant reductions in these variables. Chilling stress (6°C) for 4 h under low light reduced A by 10% for leaves of 'Leanyka' and 20 to 70% in 'Zold veltelini'.[73] The effect was reversible and longer exposure resulted in acclimation and improved A rates. Stomata of 'Leanyka' closed in response to chilling while the

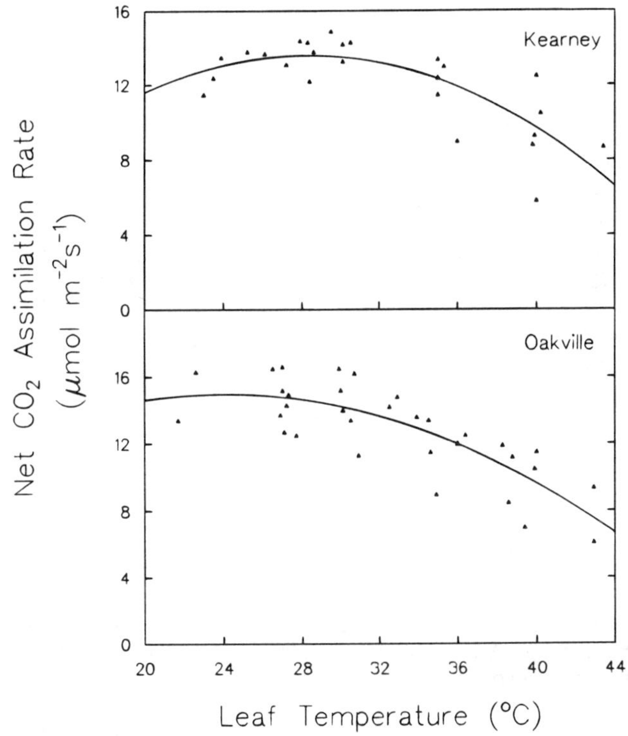

Figure 7 The effect of leaf temperature on net CO_2 assimilation of 'Chenin blanc' grapevines grown either in the Napa Valley (UC-Davis Oakville Field Station) or in the San Joaquin Valley (UC Kearney Ag Center) of California. Leaves were placed inside temperature controlled cuvettes and net CO_2 assimilation was measured after CO_2 exchange reached steady state (approximately 30 to 45 minutes after the temperature was changed). Data were fit to nonlinear functions: Kearney $y = -9.4 + 1.62x - 0.029x^2$, $r^2 = 0.62$; Oakville $y = 2.9 + 1.0x - 0.021x^2$, $r^2 = 0.65$. (L.E. Williams, unpublished data.)

more chilling-sensitive 'Zold veltelini' showed a loss of stomatal control. Both cultivars developed more negative water potentials in response to chilling stress. Sherer[74] found the time for the induction curve of light induced chlorophyll fluorescence to return to a stationary level at 5°C was longer in cold-susceptible cultivars of grapevines than in cold-tolerant cultivars; an indication of a greater effect of chilling on the photosynthetic mechanism of chill-sensitive cultivars.

It is probable that respiration by an entire grapevine commands a large portion of the daily photosynthate as the percentage of photosynthate utilized in respiration by other woody perennial species ranges from 38 to 65%.[75] Respiration can be divided into two components: respiration required for growth and respiration needed for organ maintenance. Carbon requirements for growth and maintenance respiration during the 3 weeks prior to anthesis of an individual grape flower was 0.23 mg CO_2 (3 J) and 0.83 mg CO_2 (10.8 J), respectively.[76] Maintenance respiration costs of mature organs may vary from 0.015 to 0.6 kg CO_2 kg^{-1} dry mass d^{-1}. Even when little growth is occurring, grapevines still demand large amounts of carbon for maintenance respiration and the larger the plant the greater the carbon requirement.

Temperature is the most important abiotic factor affecting respiration under most conditions.[72] The Q_{10} of respiration is approximately 2 in the range of physiological relevant temperatures.[77] For example, respiration rate of mature leaves of Perlette grapevines growing in southern California was close to zero at a leaf temperature of 10°C and doubled with each 10°C increase in temperature (Figure 8). Several factors may affect the Q_{10} and actual rate of respiration such as organ type and age, tissue N content, availability of carbon substrates and growth temperature. For example, leaves of 'Chardonnay' grapevines grown in a cool climate continued to respire down to a leaf temperature of 7°C, at which time respiration rates were too small to quantify (L. E. Williams, unpublished data). It also has been found that the Q_{10} of grapevine leaf respiration may change during leaf ontogeny (H. Schultz, personal communication).

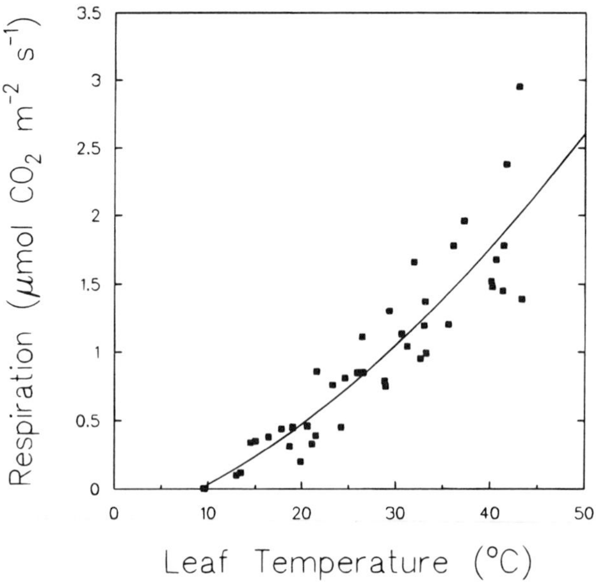

Figure 8 The effect of leaf temperature on respiration rate (CO_2 evolution) of mature leaves of 'Perlette' grapevines grown in the Coachella Valley of California. Measurements were made in temperature controlled cuvettes. (L.E. Williams, unpublished data.)

B. EFFECTS ON ROOT GROWTH

There is generally little information regarding the effects of high root temperatures on grapevine growth and development or physiology due largely to the depth of rooting and the smaller variations in root, compared to shoot, temperatures. In addition, in many of the hottest viticultural areas, irrigation is a normal practice minimizing root zone temperature fluctuations. Furthermore, the shade provided by the grapevine canopy would reduce the heat load on the soil as will the presence of a cover crop. Thus, high temperature effects on roots of grapevine have received relatively little attention.

By contrast, low temperatures may result in root injury or death. Several reports have shown differences in cold hardiness between rootstock selections[78] and for own rooted commercial cultivars.[79,80] In the colder regions of grape production, temperatures as low as −13°C at 20-cm depth have been recorded.[78] Low soil moisture results in lower soil temperatures and increased chance of root injury. Thus, late season irrigation is recommended in viticultural areas that routinely experience low winter temperatures accompanied by low fall and winter precipitation.

C. HIGH TEMPERATURES

1. Species and Cultivar Differences

There has been no comprehensive survey of the high temperature tolerance of *Vitis* species or cultivars, although several species or cultivars have been classified into different groups requiring varying levels of heat units to mature.[81] It was shown that the temperature at which heat injury occurred in leaf discs of 'Venus' was 48°C while that for 'Veeblanc' was only 44°C. The same two hybrids in tissue culture were both injured at 42°C. Damage resulted from irreversible changes in the plasma membrane for both. The loss of membrane function leads to symptoms of water stress.

Fanizza and Ricciardi[83] noted a decline in shoot growth of *in vitro* propagated grape cultivars when subjected to sequential subculture at 35 or 38°C. Four of the seven cultivars examined showed no apex growth or died at 35°C; all cultivars died at 38°C.[83] Growth rates recovered when temperatures were lowered to 25°C. The authors were not certain if the reduction in growth was due to a loss in an unknown heat tolerance mechanism of the apices or ageing of the cultures. Pre-rooted cuttings of 'Muscat of Alexandria' grown in growth chambers showed maximum growth and dry weight at 25/20°C day/night temperatures followed by 30/25°C.[20] During the 13 weeks of this study, leaves represented an increasing percentage of total dry weight; the apparent level of apical dominance with increasing temperature within the temperature range of this study (20 to 30°C).

Buttrose[21] demonstrated that high temperatures increased the number of clusters per shoot. Cluster initiation in 'Muscat of Alexandria' was almost nonexistent at 20°C but increased between 25 and 35°C.[84] Stem dry weight at 20°C increased indicating a change in the sink strength or allocation pattern at this temperature that was not conducive to fruit bud initiation.

The effect of the previous season's temperatures, particularly during the initiation period, on current seasons vine fruitfulness has not been thoroughly researched. Smit[85] studied this in the production of 'Sultana' vines and recognized the potential effect on a commercial level. Unfavorable conditions during the period of bud initiation and early development of the inflorescence primordia lead to a reduction in the crop potential the following year. This reduction in flower initiation and development may be misinterpreted as a reduction in fruit set. Very little is known about the effect of high temperatures on flower buds following initiation, although there appears to be no major detrimental effects on crop production in regions of the world where temperatures often reach 35 to 40°C after flower bud initiation has occurred.

Bud fruitfulness may be affected even during the period just prior to and following budbreak. For example, the number of clusters per shoot was greater at higher temperatures and the number of flowers per cluster was reduced.[86] Hence there is an extended period of time, prior to anthesis, over which the potential productivity of a grapevine can be influenced by temperature.

Cluster development following budbreak of container-grown 'Cabernet Sauvignon', 'Shiraz', 'White Riesling', and 'Clare Riesling' potted vines was increased at 32/27 and 38/33 compared to 14/9, 20/15, and 26/21°C, day/night temperatures.[87] The number of days to flowering was reduced from 70 to 18 over this temperature range. There was a difference of only 2 days between the 26/21 and the 38/33 treatments suggesting a reduction in flowering at the higher temperature. In this same paper, maximum fruit set occurred at 20/15°C with no fruit set occurring at either 14/9 or 38/33°C for the cultivar 'Cabernet Sauvignon'. While the above information may not be directly transferable to field-grown vines, the authors note that in most areas where grapes are grown, the temperature rarely reaches these higher levels during the period from budbreak to bloom.

Numerous reports indicate that fruit set in grapes is inhibited by high temperatures.[88–92] Potential causes of reduced set include a reduction in ovule or pollen viability and/or in pollen tube growth, changes in hormonal status and, indirectly, water stress. Kliewer[88] demonstrated a loss of ovule viability for 'Pinot Noir' and 'Carignane' grapes at 35 and 40°C compared to 25°C. Reduced ovule viability, which results in fewer seeds per berry, could contribute to smaller berries and yield based on the known relationship of seed number, hormones, and berry size.[93] Pollen germination and pollen tube growth in Petri dishes was unaffected in 'Muscat of Alexandria' at 35°C compared to 22 or 25°C[94] while that for 'Delaware' (*V. labruscana*) was reduced to about 30% at 30°C compared to 24°C.[89]

Thirteen years of data collected at over 100 vineyard sites in the Yakima Valley of Washington, showed that high temperatures during bloom did not generally reduce the number of berries per cluster for 'Concord' vines (Figure 9). In fact temperatures above 25°C often gave higher berry set than lower temperatures. This data clearly shows that low temperatures are more detrimental to berry set than high temperatures in this cultivar.

Matsui et al.[95] investigated the effect of plant hormones and high temperature effects on fruit growth of 'Thompson Seedless' and 'Napa Gamay' grapevines. Four-year-old potted vines, with the root temperature being controlled, were subjected to 40/22°C (day/night) for 4 days and fruit development compared to plants held in a greenhouse (temperature range 22 to 32°C). Pre-stress treatment with gibberellic acid (GA_3) or GA_3 plus a cytokinin (benzyladenine) partially overcame the negative effect of high temperature on berry size and total soluble solids accumulation. Only GA_3 overcame the effect on berry weight. Estimates of total endogenous GA_3 levels from 'Thompson Seedless' berries indicated a reduction in the level and a change in the qualitative nature due to heat stress and suggests the reason for recovery with GA_3 applications.

High root temperatures (20 vs. 11°C) have also reduced the number of berries per cluster in 'Cabernet Sauvignon'.[96] High air temperatures do not always have a deleterious effect on fruit development. For example during 1992, the U.S. Pacific Northwest experienced two periods of 10 days or greater when the temperature was 35°C or higher (Figure 10). However, fruit maturation occurred from 2 to 4 weeks earlier than average with normal crop loads for this grape production area.

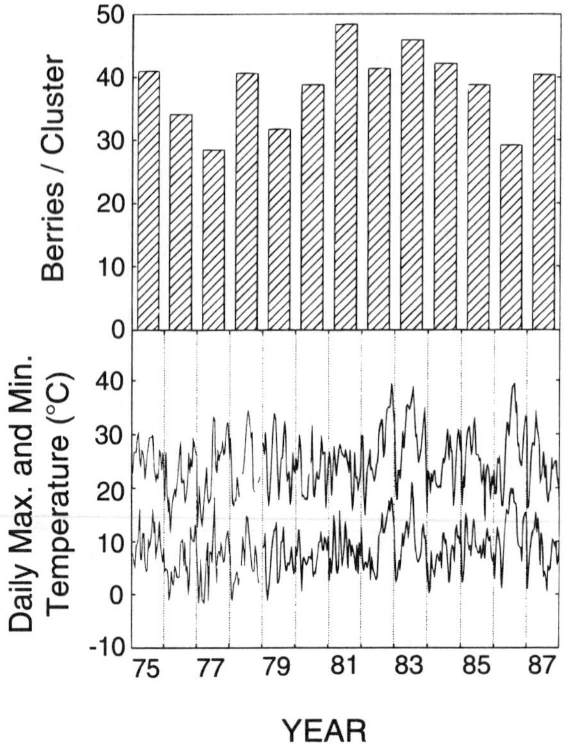

Figure 9 Number of berries per cluster and daily maximum and minimum temperatures for 2 weeks before and 2 weeks after full bloom of 'Concord' grapevines grown in the Yakima Valley of Washington, US, from 1975 to 1987. Data represent the mean of approximately 100 different vineyard sites. (R. Wample, unpublished data.)

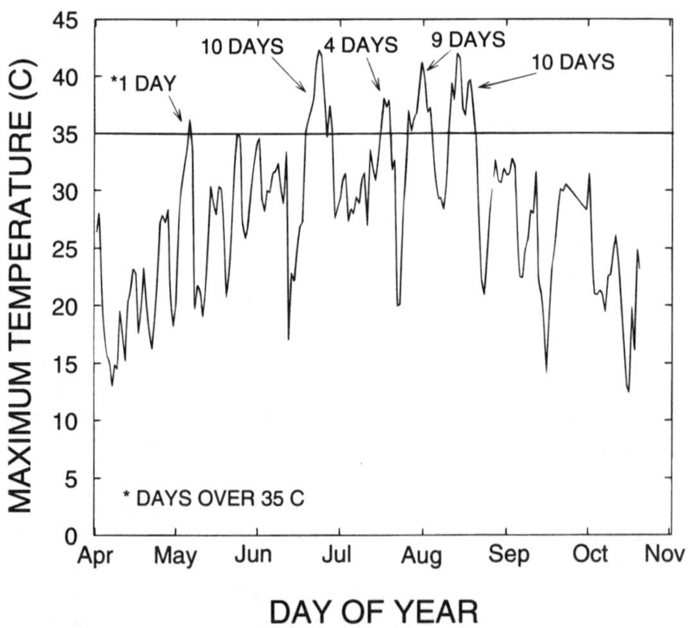

Figure 10 Daily maximum, ambient temperatures during the 1992 growing season at Paterson, Washington, US. The number of days in which the maximum temperature exceeded 35°C are at the top of the graph. (R. Wample, unpublished data.)

2. Effects on Water Relations

Differences have been found in the stomatal response of grape cultivars to temperatures ranging from 34 to 43°C.[97] 'Cardinal' showed the least response to heat stress although the control treatment (25 to 29°C) had a relatively low g_s compared to the other cultivars studied. The response of 'Chardonnay' and 'Chenin blanc' to heat stress was similar whether measured on a diurnal basis or over 4 to 12 days. This work underscores the influence of vapor pressure deficit when evaluating g_s in heat-stressed vines (see Humidity Section).

3. Other Physiological Processes

Growth of grapevines, as well as other plants, at constant high temperatures (>35°C) for extended times (>30 days) is also important as an essential procedure in the elimination of viruses, viroids, mycoplasmas, and perhaps some bacteria. The effect of such treatment appears to have more effect on the microorganism than the grapevine. The response of grapevines to high temperatures and elevated CO_2 concentrations has been examined.[98] Rooted cuttings of 'Cabernet Sauvignon' subjected to heat treatment and elevated (1200–1300 ppm) CO_2 levels manifested a reduced transpiration (E), elevated A and growth rates and a change in the allocation of photosynthates that promoted more root growth. Increased A was apparently due to reduced nonstomatal limitations of carboxylation and a lower level of photorespiration since g_s was lower.

Excessively high temperatures for extended periods of time generally result in a delay of fruit maturation, and a reduction in fruit quality.[99–103] Both of the above characteristics have been associated with a decline in total titratable acidity and increased pH[104] caused primarily by a reduction in the synthesis and the increased catabolism of malic acid;[105] increased mono- and di-basic salts of tartaric acid and di-basic salt of malic acid have also been noted.[101] High temperatures also reduce color development of grape berries.[88,102] For example, at 35°C pigment development was completely inhibited in 'Tokay' and reduced in 'Cardinal' and 'Pinot Noir' compared to 20 or 25°C.[106] In general, cool nights or days improve coloration and a beneficial effect of night cooling by sprinkling has been found[103,107,108] and has become a commercial practice in other fruit crops such as apple. Such a practice in grapes should be carefully managed to prevent disease problems from developing in the fruit and canopy.

High temperatures can also affect the partitioning of photosynthates within the leaf. As temperature increased the concentration of starch within the leaves of 'Cabernet Sauvignon' vines decreased.[109] For example, leaf starch concentrations of vines grown in growth cabinets and exposed to day temperatures of 18, 25, and 35°C were 23.3, 10.9, and 1.3% of dry weight, respectively. Increasing leaf temperature resulted in a shift in lipids from 5.8 to 16% of dry weight over this same temperature range. Interestingly, total chlorophyll content increased from 0.6% of dry weight at 18°C to 1.2% (equivalent to 5.6 mg dm^{-2}) at 35°C. Roper and Williams[40] reported that the starch levels in field-grown grapevine leaves from a warm climate also remained very low. However, it is not clear if changes in lipids and chlorophyll will occur under field conditions.

Translocation of photosynthates, a major factor in fruit development and maturation, may be influenced by high temperatures. Sepulveda et al.[110] exposed a mature leaf of non-bearing 'Chenin blanc' and bearing 'Chardonnay' grapevines to $^{14}CO_2$ for 30 minutes after 4, 8, or 12 days in a greenhouse (29/15°C) or a phytotron (40/20°C). Twenty-four h after $^{14}CO_2$ exposure, heat stress enhanced the transport of ^{14}C photosynthates to the shoot tip at the expense of the roots, trunk, and clusters. High temperatures did not reduce A. Sucrose concentrations increased in all vine organs due to heat stress in both cultivars. Heat-stressed 'Chardonnay' vines had lower concentrations of glucose and fructose in the fruit.

4. Vine Adaptations

Like many other living organisms, the genus *Vitis* is presumed to produce heat-shock proteins (HSP) which apparently play a role in the metabolism of other proteins and in their protection against thermal degradation.[111] Despite the apparent heat tolerance of *Vitis* spp. the authors are not aware of any reports that demonstrate the importance of HSP in grapevines.

D. LOW TEMPERATURES

1. Chilling

Grapevines vary greatly in chilling tolerance. A brief period of chilling (4 h) caused only a 10% reduction in A in the cultivar 'Leanyka' but as much as a 70% reduction in 'Zold veltelini' but in both

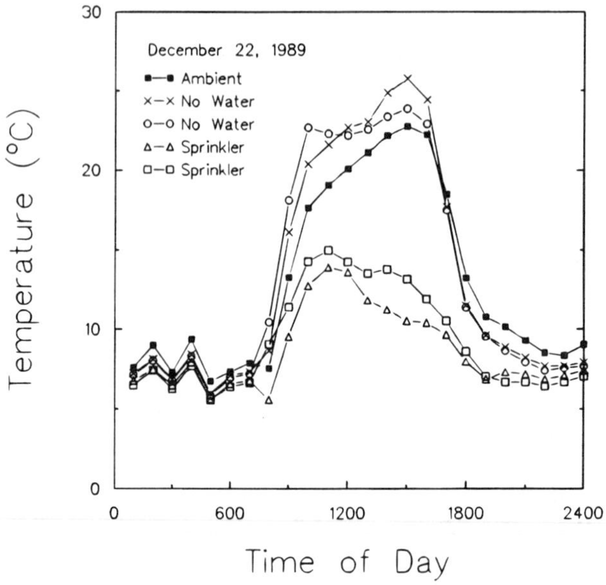

Figure 11 The effect of evaporative cooling on bud temperature of 'Perlette' grapevines grown in the Coachella Valley of California. Vines were either cooled with overhead sprinklers while the controls were not sprinkled with water. Bud temperatures were measured with hypodermic thermocouples connected to a data logger. Individual data points are the hourly means of bud or ambient temperatures measured every minute. (L.E. Williams, R. A. Neja, and E. A. Walker, unpublished data.)

cases this was reversible within 24 h.[76] Some recovery in photosynthetic capacity occurred following chilling periods longer than 12 h. McIntyre et al.[67] working with 100 cultivars of grapes in Davis, CA reported that differences in susceptibility to either spring or fall frosts were dependent upon phenology (i.e., date of budbreak or fruit maturation). This demonstrates the possibility of two levels of tolerance to chilling temperatures. First is the inherent genetic tolerance to low temperatures and the second the effect of environmental and management factors on vine phenology which may confer more or less risk to a low temperature event.

2. Effects on Growth

Although there is some controversy regarding the absolute chilling requirement of grapevines,[112] evaporative cooling and other practices are used to achieve more rapid and uniform budbreak in many warm grape growing areas.[113,114] Evaporative cooling reduced bud temperatures by 10 to 15°C (Figure 11), and resulted in increased yield, and advanced, more uniform berry maturity.[113,114] Chilling has also been linked to decreased catalase enzyme activity in grapevine buds and this has been associated with release from dormancy.[115] Weaver and Iwasaki[116] reported that 4 to 8 weeks of chilling (0°C) were required for rapid and uniform budbreak of 'Zinfandel' grape. They were unable to establish any meaningful relationship between either free or bound ABA and bud response to chilling or calcium cyanamide treatment. Takeno et al.[117] showed that chilling somatic embryos of hybrid grape (*V. vinifera* × *V. rupestris*) for 1 week at 4°C caused an increase in GA-like activity which declined during the second and third week of chilling.

Roubelakis and Kliewer[91] reported cultivar differences in fruit set at day/night temperatures of 15/10°C. Higher light intensities enhanced fruit set at those temperatures. Fruit set in 'Concord' appeared more sensitive to high (32–35°C) than to low (15–18°C) temperatures.[90,118] The temperature range associated with maximum fruit set for most species/cultivars has been between 20 and 30°C.

Several studies have reported an improvement in vine microclimate with evaporative cooling during the growing season.[103,108,109] Evaporative cooling during this stage of vine growth results in changes in vine temperature and vapor pressure deficit (VPD) and may improve vine water relations, photosynthesis, and fruit quality. In Montpellier, France, it was observed that maximum shoot growth occurred at 28°C and that growth ceased at 10°C and below.[120]

3. Effects on Water Relations

Chilling of container-grown *V. vinifera* cvs. in controlled environment chambers for more than a few hours decreased plant water potential (Ψ) despite a significant reduction in g_s indicating that water uptake and/or transport might have been affected.[75,76] However, it is possible that increased water stress was the result of much lower root temperatures (or a greater rate of temperature reduction) in the growth chamber than would have been experienced under field conditions. Cold nights (<10°C) have also induced higher leaf resistances the following day in 'Concord' vines grown in New York. This inhibition of stomatal opening appeared to be independent of Ψ_l and it was speculated that this might be due to a chilling-induced reduction in starch hydrolysis and an inhibition of A.

Thus it seems that chilling may have both direct and indirect effects on plant water relations. Unfortunately, there are few field studies on the effects of chilling on grapevines to help clarify our understanding. This research could be valuable in viticultural areas where low (0–10°C) temperatures are common during the growing season.

4. Other Physiological Processes

Ahmedullah[121] found that exposing 1-year-old 'Cardinal' grapevines to 15°C, as compared to 25°C, led to increased basipetal transport of photosynthates. He also found a higher rate of total recovery of ^{14}C-labelled photosynthates at 15°C indicating a lower respiration rate (see Figure 8). In some viticultural areas, the night time temperatures often, even in mid summer, drop to 15°C or lower. Increased basipetal transport of photosynthates coupled with lower rates of respiration at night may contribute to more rapid fruit maturation and may account for the ability to mature some cultivars of grapes in areas that would not appear to have enough heat units.

To our knowledge there has been very little research designed specifically to evaluate grapevine adaptations to chilling. However, the above-mentioned reports indicate that grapevines can acclimate and adjust to changing temperatures.

E. FREEZING TEMPERATURES

1. Cultivar Tolerance

A wide range of cold hardiness exists in the genus *Vitis* and this genetic variation has been utilized in breeding and cultivar evaluation programs.[122] In a report covering 88 European, 34 American, and 14 French Hybrid cultivars, Clore et al.[123] ranked grapevine cold hardiness and demonstrated the wide range in genetic cold hardiness potential.

Cold hardiness may encompass mid-winter hardiness as well as spring and fall frost hardiness. Mechanisms of cold hardiness may involve tolerance and/or avoidance. For instance, avoidance of spring frost damage due to late budbreak is an important distinction from the ability to survive (i.e., tolerate) frost. Species as well as cultivar differences in timing of budbreak, flowering, and fruit maturation are important in the selection of grapes for a given vineyard. McIntyre et al.[67] found up to 25 days difference in budbreak and more than 100 days difference in the maturity date among 100 cultivars. Often, cultivars with early budbreak are susceptible to spring frost but are also early maturing thereby avoiding crop losses due to fall frost. It appears that vines whose origins are further from the equator are more sensitive to changes in photoperiod and thus less susceptible to fall frost.[124]

An example documenting the need to understand this relationship was reported by Wolf and Cook[125] who found that 'Cabernet Franc' was 1 to 2°C more hardy throughout the winter than 'Cabernet Sauvignon'. However, 'Cabernet Franc' deacclimated more rapidly in the spring and was more susceptible to spring frost or late winter freeze. Comparable results also were found examining the deacclimation of 'Concord', 'White Riesling', and 'Cabernet Sauvignon'.[125] Similarly, Proebsting et al.[126] reported that although cold hardiness of 'Concord' buds was much greater than either 'White Riesling' or 'Cabernet Sauvignon', they deacclimated 1 to 2 weeks earlier thereby explaining the greater crop loss for 'Concords' compared to wine grapes in Washington during a major spring frost on April 28, 1985.

Differences in cold hardiness within a *Vitis* species may be influenced by changes in temperature and water content. Damborska[127] found that warm temperatures induced a cultivar and/or season-specific reduction in cold hardiness. 'Riesling' maintained its hardiness better than 'Muller-Thurgau' in mid-winter after exposure to 10, 12, or 15°C, although 'Riesling' was less cold hardy in the spring. Such cultivar- and species-specific changes were also evident during a winter freeze in the U.S. Pacific Northwest during December 1990. Bud injury following a brief (36 h) period of high temperatures

Table 1 **Percent of primary and secondary bud injury of seven cultivars of *Vitis vinifera* and *V. labruscana* cv. Concord as affected by irrigation**

	Bud injury (%)			
	Irrigated		Stressed	
Cultivar	Primary	Secondary	Primary	Secondary
Chenin Blanc	98.8	64.3	97.5	60.5
Chardonnay	96.0	64.3	87.0	67.0
Cabernet Sauvignon	96.3	73.5	81.3	52.5
Gamay Beaujolais	59.0	16.0	56.0	26.0
Meunier	58.0	27.0	39.0	13.0
Pinot Noir	71.0	39.0	53.0	33.0
Semillon	100.0	97.0	100.0	96.0
Concord	<19.9	<5.0	<10.0	<5.0

Note: Data were collected following a severe winter freeze in December, 1990. Percentages represent the mean of 10 or more vines and 100 or more dissected buds per vine. Stressed vines had received only one irrigation compared to 4 or more for irrigated vines.
From Wample, R., unpublished data.

(10°C) was followed by temperatures as low as −28°C is presented in Table 1. A consistent reduction in primary bud injury was associated with less irrigation the previous season for all but 'Concord' grapevines. Mild water stress during bud development appears to improve winter survival and is associated with smaller cane diameter and shorter internodes. Wolpert and Howell[128] noted the importance of low water content on cold hardiness development during early acclimation. Our understanding of temperature and water interactions on grapevine bud, cane, and trunk cold hardiness is incomplete and in need of additional research.

2. Effects on Vine Growth

Growth following freezing injury may be separated into events that occur pre- and post-budbreak. In a pre-budbreak state, low temperature injury may influence one or more of the following: the primary, secondary, or tertiary bud; the phloem of the trunk and canes or roots; the xylem parenchyma of the trunk and canes or roots; and/or the vascular and cork cambia of the permanent structures. The simplest and perhaps the most frequent case, low temperature injury of the primary bud, results in very few changes in overall growth of the vine, but frequently results in a significant loss of yield for that season. Some cultivars such as 'Tokay' and 'Folle blanche' have fruitful secondary buds and may still produce nearly a full crop.[129,130] Loss of more than the primary bud is often accompanied by damage to other vine organs, frequently resulting in the loss of permanent structures and requires retraining of the vine if it survives. This is a major problem in grafted vines and either regrafting or replanting may be required if the scion is completely killed.

Situations have been recorded where the root system has been injured by low temperature while the majority of the shoot system has been undamaged. This occurred during the winter of 1978–79 in the Pacific Northwest of the United States when a shortage of irrigation water was combined with very little rainfall after harvest. This led to dry soils and resulted in freezing to a depth of 30 to 40 cm. Air temperature varied between 0 and −20°C which resulted in a minimum of bud damage but significant root damage. In the spring, budbreak occurred and shoot growth began normally. After a few weeks of shoot growth, and during a very warm period, the injured root system was unable to meet the water requirements of the shoot system and resulted in the collapse of the green shoots and in many cases vine death. Although not confirmed, there appeared to have been effects of this root damage as much as 3 to 5 years later, as the root system became infected by soil born pathogens.

Low temperature injury may facilitate the development of crown gall [*Agrobacterium tumefaciens* (E.F. Smith and Townsend) Conn, biovar 3] in grapevines which may be more deleterious than low temperature injury itself. Therefore, propagation wood should be from crown gall free vines when possible. Methods of eliminating this bacteria from grapevine cuttings are proceeding.[131–134]

The extent of injury when freezing occurs after budbreak depends upon the severity of the freeze and the subsequent management of the vines. Winkler[129] studied frost injury to 'Thompson Seedless',

'Malaga', and 'Tokay' grapes during the spring of 1932 and 1933. He concluded that if the injury did not extend basipetally beyond the clusters there was little need to adjust the vine. However, if the injury extended below the clusters in cultivars with fruitful secondary and tertiary buds, removal of the shoots to stimulate the growth could result in increased yields. In cultivars with non-fruitful secondary and tertiary buds it made little difference to the present year's crop whether or not the frosted shoots were removed. However, the development of the next years fruiting wood was improved by shoot removal if the injury were such that an excessive number of axillary buds began to grow. Lider[135] in a similar study in the spring of 1964 on 'Folle blanche', 'Cabernet Sauvignon', and 'White Riesling', found that doing nothing was the most economical practice following frost injury. He was unable to confirm the benefit of shoot removal in 'Folle blanche' which reportedly has fruitful secondary buds. A study of 'Chardonnay', 'White Riesling', and 'Cabernet Sauvignon' in Washington (Stimson Lane Wine and Spirits, personal communication) confirmed the results of Lider in that "no treatment" was the most economical practice following frost injury. Despite some significant differences in the number of clusters per vine and cluster weight due to post-frost treatments, there was no difference in yield for 'Cabernet Sauvignon' or 'White Riesling' (Figure 12). For 'Chardonnay', however, shoot removal following frost injury significantly reduced yield.

Pratt and Pool[136] have provided an anatomical description of the recovery of canes of *V. vinifera* from simulated freezing. They found that recovery was dependent upon a sufficient quantity of viable undifferentiated tissues (cork and vascular cambia, and xylem and phloem parenchyma) capable of undergoing cell division. In the case of bud injury, a surviving lower-order bud was required to replace those injured since adventitious buds in grapes have not been reported. The apparent required characteristics for recovery included an apical meristem and at least two vascular traces.

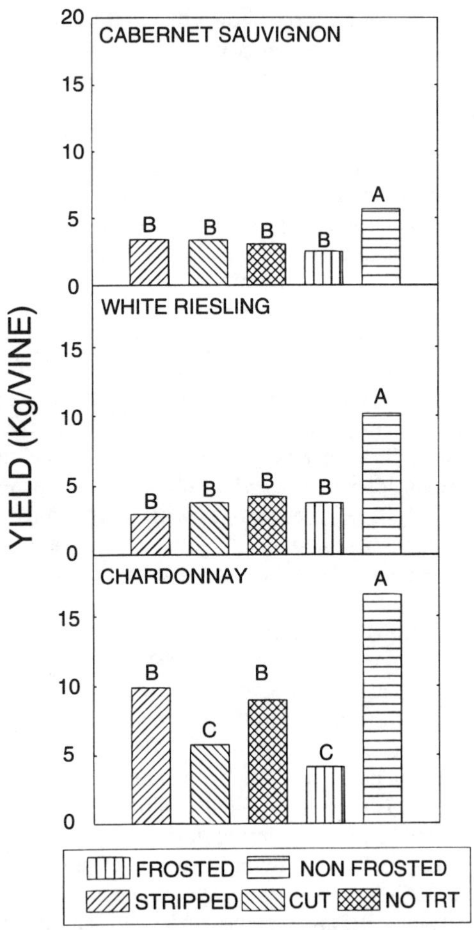

Figure 12 Yield response of 'Cabernet Sauvignon,' 'White Riesling' and 'Chardonnay' vines following spring frost injury during May 1992 in Washington, US. Treatments included removal of the partially frosted shoots by hand (stripped), removal of the frosted portion of the shoot with hand shears (cut), no treatment of partially frosted shoots (no trt), vines that were severely frosted with no additional treatment (frosted), and control (non-frosted) vines. Columns with different letters are significantly different at the p = .05 level or better. (R. Wample, unpublished data.)

3. Effects on Water Relations

Continuous periods of below freezing, non-lethal temperatures often result in the formation of extracellular ice which has the consequence of establishing a strong vapor pressure gradient between the extracellular water and the liquid water in the cells. The slow, continuous diffusion of water out of the cells to the extracellular ice results in an increase in grapevine cold hardiness.[126,127,138–140] This occurs through a combination of reduced cellular water content, a concentration effect of the cellular solutes and additional physiological changes in membrane and protein structure.[158] This "water transfer" process appears to occur in buds but may not take place in cane and trunk tissues.[141] Results similar to these for *V. riparia* have been noted for several cultivars of *V. vinifera* including 'Cabernet Sauvignon', 'White Riesling', and 'Chenin blanc'.

As a part of the acclimation process, the formation of tyloses and other vascular blockages play a role in the reduction of vine water content.[142,143] Vascular blockages also have the effect of inhibiting the rehydration process during the winter and may be important in the maintenance of low water content throughout most of the dormant period. One of the hazards near the end of dormancy is the rehydration of cane and bud tissues which if followed by subfreezing temperatures results in significant vine injury.

4. Other Physiological Processes

Nitrogen nutrition is related to grapevine performance, and may influence cold hardiness.[144] A review of the literature found little evidence to support the contention that high nitrogen nutrition resulted in direct loss of grapevine cold hardiness.[145] Nitrogen metabolism in grapevines during the acclimation and dormant periods is known to be dynamic.[146–148] Total and protein nitrogen levels rise at the onset of acclimation and continues into the second phase of hardening.[147] Higher levels of total nitrogenous substances were found in the more cold hardy cultivars.[147] Higher nitrogen concentration may have resulted from the slower growth rate of these cultivars, and are thus indirectly related to cold hardiness. Similarly, it has been shown that grafting European grapevines onto winter-hardy American rootstocks led to higher mid-winter nitrogen levels and improved cold hardiness.[149] High protein to total nitrogen ratios have also been linked with more winter-hardy cultivars.[148] Despite these reports, two recent publications indicate little or no effect of different nitrogen nutrition levels on the cold hardiness and survival of 'Chardonnay'[150] or 'White Riesling'[151] grapevines.

Cold acclimation and near freezing temperatures have been associated with increases in the soluble carbohydrate levels found in bud and cane samples of grapevines.[151–156] Low temperatures influence the magnitude of the apparent conversion of starch reserves into soluble carbohydrates, but the absolute nature of this relationship is not fully understood. The concentrations of soluble carbohydrates are slightly lower in cane tissues than in buds, but relative seasonal changes were similar.[151,156] Sucrose is the primary soluble carbohydrate with glucose and fructose making up the majority of the balance. Sucrose levels appeared to peak during late winter and early spring while glucose and fructose declined during this time.[155,156] Starch levels showed an inverse relationship to soluble carbohydrates.[151,156]

Increased solute concentrations have been correlated with cold hardiness and deep supercooling (the presence of water in a liquid state below the normal ice nucleation point) of grapevine tissues.[157] Although some reports have associated the rise in soluble carbohydrates with a cause and effect relationship with grapevine cold hardiness, the increase in soluble carbohydrates is probably responsible for only a few degrees freezing point depression and therefore cannot account for the changes in hardiness observed.[151,156,158] Deep supercooling, the primary cold hardiness mechanism in grape,[157,159] is not known to be directly related to the level of soluble carbohydrates.[151,156]

5. Vine Adaptations

Supercooling has been associated with the geographic distribution of some plants.[158] However, because the temperature at which the low temperature exotherms occur varies with different species and cultivars, precise distribution limits have not been established for all *Vitis* species. It may be possible to estimate the limits of distribution for a given cultivar if its minimum exotherm temperature were known and were compared with low temperature isotherms for a geographical area. Other adaptations that exist in some *Vitis* species are the deposition of callose and suberin in the phloem and phellem, which reduces the uptake of water during dormancy and limits mechanical injury due to intracellular ice formation.[142]

The ability of different cultivars of grape to respond to photoperiod is important to survival in cold climates. A synergistic effect of photoperiod and temperature enhanced the development of cold hardiness in 'White Riesling' grapes.[160] Fennell and Hoover[161] reported similar responses for *V. labruscana* and

V. riparia but with distinctions between these species. Interruption of the dark period did not significantly affect the cold hardiness of 'Concord' buds out to the 12th node or in the extent of cane maturation.[162] They did record more actively growing shoots on the night interrupted vines.

IV. WATER STRESS

Grapevines are often cultivated in regions of low rainfall and high evaporative demand and if irrigation is limited vines may experience some water stress during the growing season. Reviews of the effects of water stress on various aspects of grapevine growth and physiology have recently been published.[1,2,163,164] Therefore, in this section we will review much of the basic aspects of vine water stress and whatever new information has been published since 1989. In addition, preliminary data from an irrigation experiment with treatments varying from 0 to 140% of vine water use, determined with a weighing lysimeter, will be included to demonstrate trends between available water and vine response.

A. HUMIDITY

Experimental evidence indicates that the reduction in g_s induced by abscisic acid causes heterogeneous stomatal closure in many plant species[165] including *V. vinifera.*[166] Heterogeneous stomatal closure in response to ABA may be associated with the species' mesophyll anatomy.[165] It appears that plant species having their leaf mesophyll separated into intercellular chambers hermetically sealed from other areas (heterobaric type as compared to homobaric type mesophyll anatomy) will respond to ABA applications or stress with non-uniform stomatal closure[167] although there are exceptions.[168] Heterogeneous stomatal behavior will provide erroneous infrared gas analyzer-calculated values of intercellular CO_2 concentrations (C_i) as a result of the nonlinear relationship of g_s and A.[166]

Stomata are controlled by numerous environmental factors (in addition to internal factors). Generally an increase in VPD above a certain threshold, causes a reduction in g_s in most plant species[169] including *Vitis* spp.[170] The effect of VPD on g_s of grapevines is cultivar dependent.[35,170] Düring[167] recently has shown that high VPD in addition to ABA causes non-uniform stomatal closure in *Vitis* species as determined by the water infiltration technique.

The response of A to VPD may differ from that of g_s in *V. vinifera* depending on where on the curve of the relationship between A and g_s the measurements were taken. The relationship between A and g_s of field-grown 'Thompson Seedless' grapevines is curvilinear with maximum A leveling off at a conductance to water vapor of approximately 500 mmol m^{-2} s^{-1} (Figure 13). Thus, there may be a reduction in g_s due to an increase in VPD without a concomitant decrease in A when measurements are taken beyond the linear portion of the curve. Düring[170] found a linear decrease in both A and g_s with increasing VPD, however, he also found that A and g_s were linearly related up to a g_s of 160 mmol m^{-2} s^{-1}, the maximum g_s measured in that study. This would be equivalent to the linear portion of the curve in Figure 13.

Decreases in g_s due to increases in VPD may also be more pronounced for vines grown under drought conditions.[171,172] Stomatal conductance decreased significantly when 'Müller-Thurgau' and 'Riesling' vines were grown with an aerial environment kept at 50% relative humidity (RH) and soil water content maintained at 60% of field capacity compared to vines grown at 50% RH and a soil water content held at 95% of field capacity.[172] This response can also be measured on field-grown vines. Stomatal conductance decreased as VPD increased throughout the day for vines receiving less than full vineyard evapotranspiration (ET_c) (Figure 14). An increase in VPD from 1 to 3 kPa reduced g_s 50 and 75% for vines irrigated at 60 and 20%, respectively, of vine water use determined with a weighing lysimeter. In semi-arid environments, such as found in the San Joaquin Valley of California, VPD and ambient temperature are highly correlated.[174] Therefore, the relationship between g_s and ambient temperature are similar to the relationship found in Figure 14 for this particular data set.

Investigations into the response of grapevine growth and development to VPD are limited. If carbon assimilation is decreased due to VPD effects on g_s then one would expect a reduction in vine growth. In addition, high evaporative demand may also induce water stress again limiting the uptake of CO_2. A study conducted in growth cabinets demonstrated that vines grown under low RH (50 compared to 95% RH) produced more leaf but less stem (main axis of shoot) biomass than vines grown under the higher humidity.[172] There was no effect on dry matter partitioning to the root. It also was shown that budbreak occurred earlier and more buds broke at 95% RH than at 50% RH in that study for both

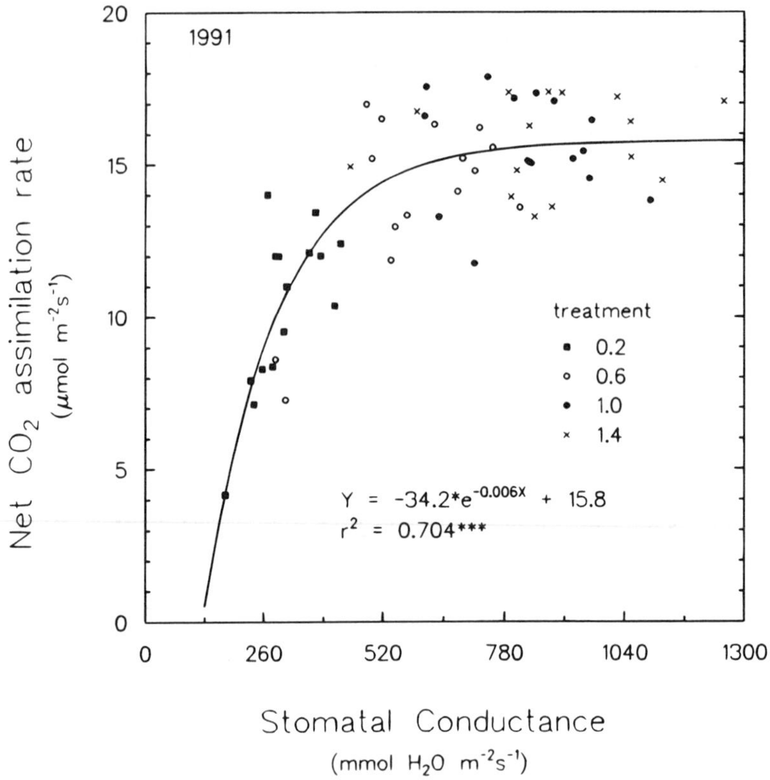

Figure 13 The relationship between net CO_2 assimilation rate and stomatal conductance of 'Thompson Seedless' grapevines measured at solar noon approximately every 2 weeks throughout the 1991 growing season. Vines were irrigated daily at various fractions (treatments 0.2, 0.6, 1.0, and 1.4) of vine water use determined with a weighing lysimeter. A complete description of the weighing lysimeter is found in Phene et al.[173] (L.E. Williams, unpublished data.)

'Müller-Thurgau' and 'Riesling'. The effect of VPD on reproductive development of the current season's crop in grape is unknown.

B. DROUGHT

As mentioned earlier in this chapter the majority of the grape production areas of the world are characterized by Mediterranean type climates having warm to hot temperatures and little rainfall during the summer. Supplemental irrigation, therefore, is necessary if one is to produce a harvestable crop of high quality. However, irrigation generally is not permitted in European Community (EC) vineyards where grapes are destined for "quality wine" while in many other viticultural areas throughout the world the availability of supplemental water is limited. Therefore, vines may undergo a considerable amount of water stress sometime during the growing season in these viticultural production areas.

1. Species and Cultivar Tolerance

It is thought that the cultivation of the grapevine began during the Neolithic era (6000–5000 BC) in the region known as Transcaucasia.[1] By 4000 BC grape growing extended from Transcaucasia to Asia Minor and into the Nile Delta. Many of these regions today are characterized by low summer rainfall and periods of drought. It is probable that many of today's grape cultivars evolved in warm climates with little rainfall during the growing season and therefore may have indirectly been selected early for drought tolerance.

There have been attempts to classify both *V. vinifera*[175] and rootstock cultivars[176] with regards to drought tolerance although the basis for the rankings are not necessarily given. It is thought that *V. vinifera* is very drought-tolerant, and the American species *V. berlandieri* and *V. cordifolia* also are

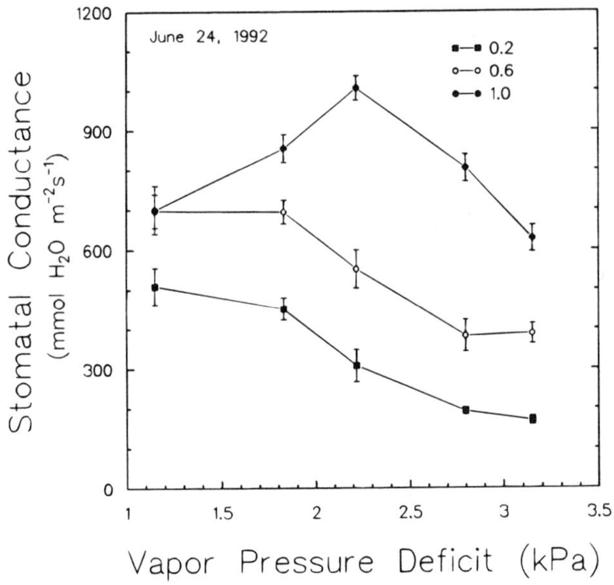

Figure 14 The relationship between stomatal conductance of 'Thompson Seedless' grapevines and vapor pressure deficit (VPD). Measurements were taken at 2-h intervals during the day. Thus, the measurements obtained at the highest VPDs were collected subsequent solar noon (at 1500 and 1700 h) at which time solar radiation may have become a limiting factor. Each value is the mean of nine individual leaf replicates. The numbers in the upper right corner of the figure represent the fraction of full vine-water use the treatments received. (L.E. Williams, unpublished data.)

known for their drought tolerance.[177] Galet[178] has classified the performance of the hybrids of *berlandieri-rupestris* as the most satisfactory in very dry soils. *V. riparia* and *V. rupestris* are thought to be sensitive to soil-water deficits.[177]

2. Effects on Vine Water Relations

Grapevine Ψ_l will undergo diurnal fluctuations[1,2] regardless of the amount of water available to the vine.[179] Vines with adequate soil moisture will have a pre-dawn Ψ_l between -0.01 and -0.1 MPa while those with less available water will have a more negative Ψ_l (Figure 15). The daily minimum Ψ_l potential typically occurs after midday (solar noon) and then increases as solar radiation and evaporative demand decrease in the late afternoon. Pre-dawn Ψ_l of well-watered vines will remain relatively constant throughout the growing season while that of deficit irrigated vines will become less.[179,180] For the data set in Figure 15, vines which had not been irrigated throughout the season had a pre-dawn Ψ_l close to -0.9 MPa on August 18. While midday Ψ_l is dependent upon evaporative demand, this value should not become much less than -1.0 MPa if the vines are irrigated at full vineyard ET even under semi-arid conditions.[181] Midday Ψ_l of deficit irrigated vines will continue to decrease as soil-water content decreases.[179,181]

Stem and cluster water potentials also will fluctuate on a diurnal basis.[34,179] Pre-dawn cluster water potential ($\Psi_{cluster}$) is more negative than Ψ_l and remains such until evaporative demand increases after sunrise with Ψ_l decreasing more rapidly as the day proceeds.[34,179] Clusters will reach their minimum water potential values later in the afternoon than leaves and may ameliorate changes in Ψ_l by supplying water to leaves especially during midday. As with Ψ_l, $\Psi_{cluster}$ is more negative for deficit irrigated vines compared to those receiving adequate water on both a diurnal and seasonal basis.[179,182]

It has been suggested that drought avoidance rather than tolerance is the mechanism by which grapevines respond to soil water deficits.[164] However, there are a few studies that indicate *V. vinifera* cultivars are able to osmoregulate.[181,183,184] Experiments conducted in the lab and field indicate that a decline in the osmotic potential (Ψ_π) of between 0.4 and 0.7 MPa can occur in drought stressed vines.[181,184] The ability to adjust the vine's Ψ_π appears to be cultivar/species dependent.[185]

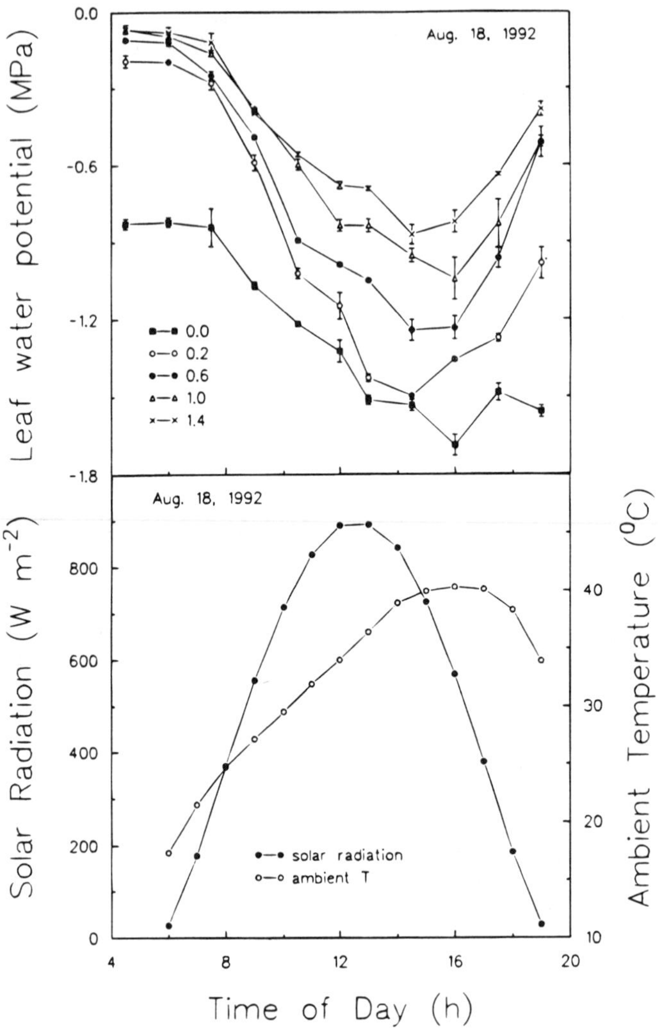

Figure 15 The diurnal time course of leaf water potential for 'Thompson Seedless' grapevines receiving various fractions of full vine water use (top). The diurnal time course of ambient temperature and solar radiation for this day also is included (bottom). Leaf water potential values represent the means of six individual leaf replicates per treatment. Treatments represent irrigation amounts at various fractions of full vine-water use (1.0 treatment). (L.E. Williams, unpublished data.)

3. Effects on Leaf Gas Exchange

Water stress will cause a reduction in g_s of grapevines. The Ψ_l at which stomatal closure begins varies between -0.9 and -1.6 MPa[30,34,179,180,182] depending upon environmental conditions and the rate of water stress imposition (i.e., rapid for potted vines; slow for field-grown vines). The relationship between midday Ψ_l and g_s throughout the 1991 growing season for 'Thompson Seedless' vines irrigated at various fractions of vine water use (ET), however, results in a linear reduction in g_s with a decrease in Ψ_l (Figure 16). The low coefficient of determination indicates that other factors (either internal or environmental) must contribute to the reduction in g_s. It was demonstrated in Figure 14 that vines experiencing soil water deficits are more sensitive to changes in VPD than well-watered vines. A similar r^2 value for the relationship between midday Ψ_l and g_s of 'Colombard' grapevines has been reported by van Zyl.[179]

Studies during the past decade on numerous plant species indicate that the reduction in stomatal conductance and growth of plants due to soil water deficits may be a response to some sort of "root signal".[186] This signal probably arises due to the roots sensing a reduction in soil water content or an

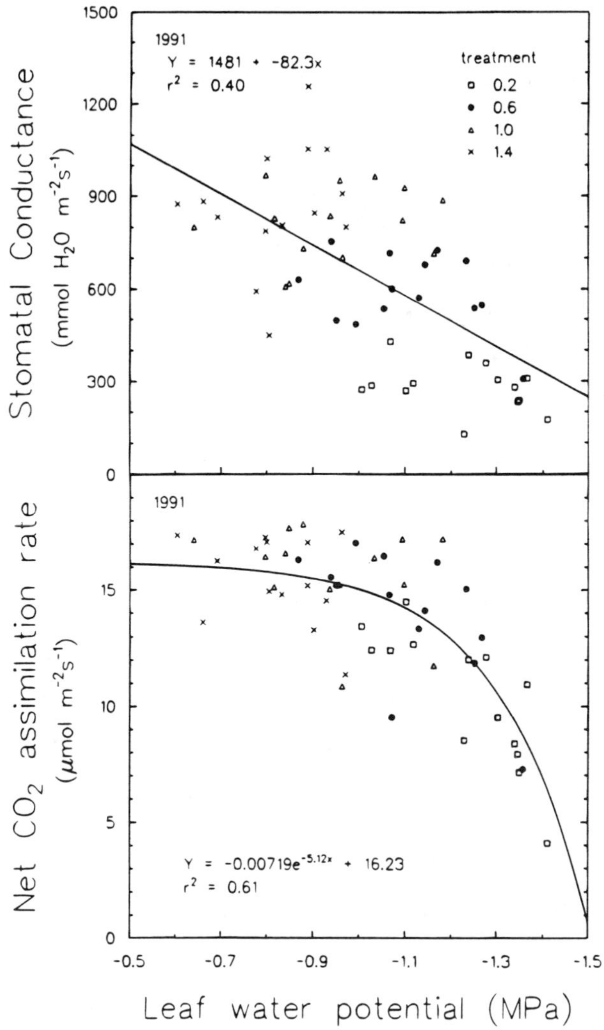

Figure 16 The relationship between midday measurements of net CO_2 assimilation rate and stomatal conductance and leaf water potential of 'Thompson Seedless' grapevines grown in the San Joaquin Valley of California. Data were collected throughout the 1991 growing season. Each data point is the mean of nine individual measurements of A and g_s and six individual Ψ_l measurements. Treatments are as outlined in Figure 13. (L. E. Williams, unpublished data.)

increase in the mechanical impedance as the soil dries out. The most likely candidate for this signal is ABA produced in the roots or an as-yet unidentified regulator.[186,187]

It has been found in grapevines under well-irrigated conditions and exposed to a minimum level of stress that the ABA content of leaves increases after sunrise and peaks at approximately twice the pre-dawn level at midday.[188,189] Decreases in g_s during the day were correlated with ABA accumulation in the leaves. Loveys[190] suggested that ABA is exported from leaves to roots, and then transported back to the leaves via the xylem which ultimately controls g_s optimizing water use efficiency.[189] ABA also will increase in the leaves of water-stressed grapevines.[191–193] The involvement of ABA in stomatal regulation of grapevines is further supported by the fact that water stress causes heterogeneous stomatal closure in grapevine[193] which also has been shown to occur in grapevine leaves supplied with ABA.[166]

van Zyl[179] found that pre-dawn and midday Ψ_l were highly correlated with both soil water content and soil water potential (Ψ_{soil}). Coefficient of determinations were highest between Ψ_{soil} and both pre-dawn and midday Ψ_l. Studies on other plant species indicate that soil water content is the factor responsible for eliciting root sensed responses in the shoot.[194,195] Data in Figure 17 demonstrates that

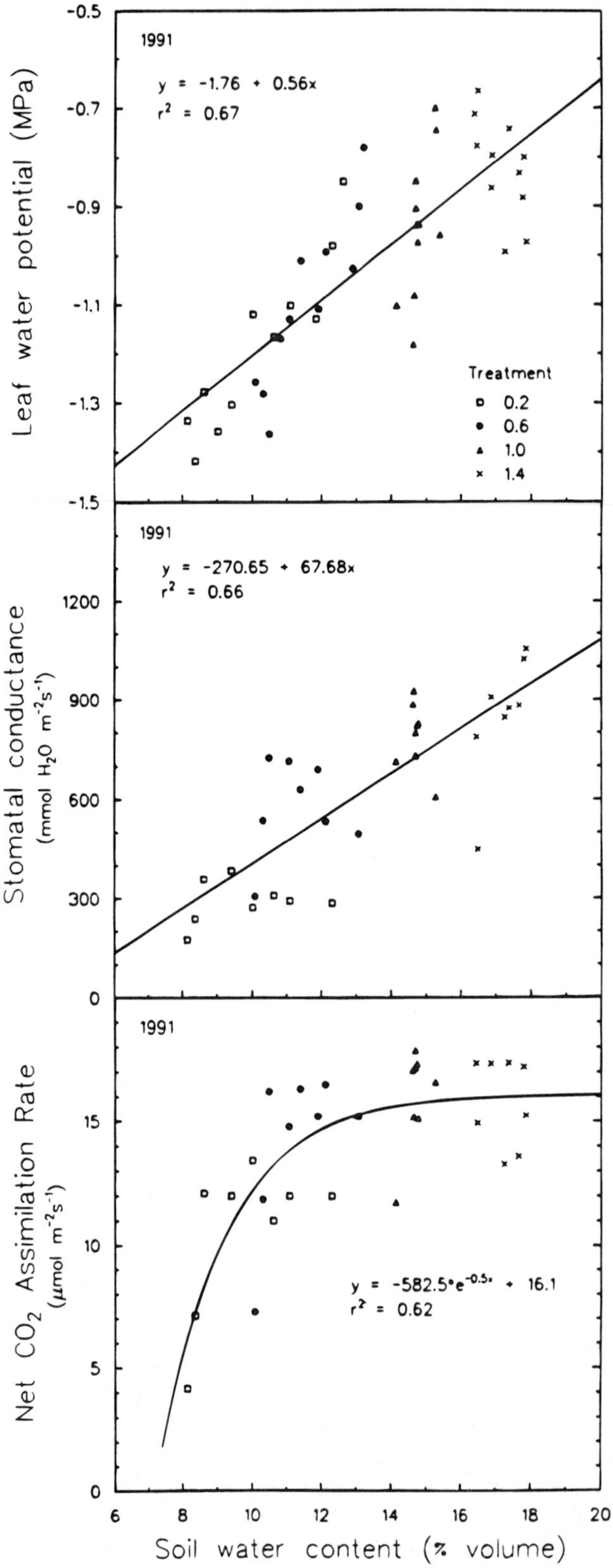

Figure 17 The relationship between midday measurements of net CO_2 assimilation rate, stomatal conductance, and leaf water potential of 'Thompson Seedless' grapevines and soil water content. Treatments are the same as those outlined previously (Figure 13). Soil water content was measured as described in Grimes and Williams.[181] The nine access tubes per individual vine were replicated three times in each irrigation treatment. Soil water content is the mean of those three replicates (nine access tubes measured to a depth of 3 m per individual vine replicate). (L.E. Williams, unpublished data.)

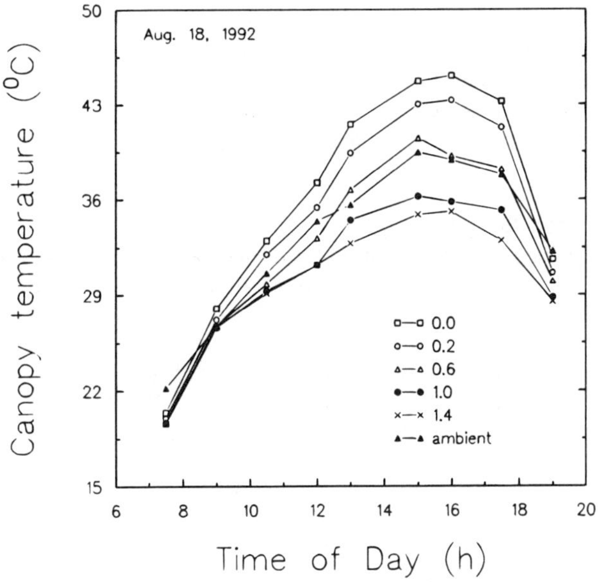

Figure 18 The diurnal time course of 'Thompson Seedless' canopy temperature as a function of irrigation treatment on August 18, 1992. Canopy temperature was measured with an infrared thermometer. Each value is the mean of nine individual measurements per irrigation treatment. Ambient temperature was measured 0.5 m above the canopy. Other information as given in Figure 13. (L.E. Williams, unpublished data.)

during the growing season midday Ψ_l and g_s were highly correlated with soil water content (SWC). Soil water content accounted for up to two thirds of the variation of both plant-based measures of vine water status. It should be pointed out that measurements were taken on vines irrigated at various fractions of full vine ET throughout the season and that only the 0.2 and 0.6 ET_c treatments experienced decreasing soil water content as the season progressed. In a study where water was withheld for 13 days for drip irrigated vines, there also was a linear relationship between SWC and midday Ψ_l (Ψ_l decreased as SWC decreased), and the coefficient of determination was 0.91.[196] The results found in Figure 17 and those of Araujo,[196] would indicate that the rapidity in which the soil dries out affects the degree of coupling between root sensed responses and the aerial portion of the vine.

Water stress is associated with a reduction in A and E of grapevine leaves.[164] The exponential relationships found in Figures 16 and 17 do indicate that A will decrease once the plant or soil water status reaches a particular level. A relationship similar to that shown between Ψ_l and A in Figure 16 previously has been demonstrated on leaves from excised shoots of grapevines.[30] It appears that stomatal control of A occurs during the early stages of drought, perhaps due to ABA's (irrespective of site of origin) effect on stomatal closure. The previously reported non-stomatal limitation to A when Ψ_l exceeds −1.3 MPa may actually be due to non-homogeneous stomatal closure.[193] For example, PSII photochemical efficiency was not a primary target of water stress in *V. californica*.[43]

Canopy temperature (measured via portable infrared thermometers) has often been used to rapidly evaluate plant water status.[181,197,199,200] If plants are well supplied with water and stomata are open, transpiration will proceed at the maximum rate determined by soil and plant hydraulic conductance and by climatic evaporative demand. As VPD increases, transpiration of nonstressed plants will increase with greater evaporative cooling resulting in foliage that is cooler than the surrounding air. As water becomes limiting, transpiration will decrease and the canopy temperature will increase, becoming greater than that of the nonstressed plants. Canopy temperatures of vines irrigated at full vineyard ET or greater always were lower than ambient temperature throughout a hot, summer day in the San Joaquin Valley of California (Figure 18). Vines irrigated at less than full ET had canopy temperatures greater than ambient at least during some portion of the day. Canopy temperature is an average of all of the leaves in the field of view of the infrared thermometer, therefore, individual, sunlit leaf temperature will be warmer than that of the canopy.[197] Canopy-to-ambient temperature differentials of up to 10°C have been measured on grapevines.[198] The concept of a Crop Water Stress Index (CWSI) based upon the difference between canopy and air temperature eventually may be used as a means to schedule irrigations.[201]

Diurnal measurements of A indicate that vines with less soil water availability can be as great as that of well-irrigated vines early in the morning.[202] However, as the day progresses A of stressed vines will decrease as opposed to a near constant rate of A of the nonstressed vines.[202]

4. Effects on Vegetative Growth

Few studies have determined the effects of soil moisture deficits on the growth of the permanent structures (i.e., roots, trunk, and cordons) of field-grown grapevines.[164] A recent study using field-grown 'Chenin blanc' vines demonstrated that dry biomass of the roots was reduced approximately 30% for vines irrigated at 52% of vineyard ET when compared to vines irrigated at full ET_c after the completion of a 4-year experiment. Root growth of container-grown vines was affected less than shoot growth.[203] It has been shown that the number of actively growing root tips diminish due to soil water deficits.[204]

Soil water deficits, due to irrigation at 52% of calculated ET, decreased trunk and cordons biomass by 17 and 30%, respectively. van Zyl[204] concluded that trunk circumference measured annually was a reliable indicator of vine water stress. While there were reductions in biomass of the roots, trunk, and cordons due to deficit irrigation, there was no significant difference in the concentration of non-structural carbohydrates (glucose, fructose, sucrose, and starch).[1]

The effects of soil water deficits on budbreak and subsequent phenological events have been assessed in a few studies. Water stressed vines generally have earlier budbreak than those receiving greater amounts of irrigation water[205,206] whether vines are deficit irrigated throughout the season[205] or subsequent to fruit harvest on vines grown in the desert.[206] In the first year of a timing-of-irrigation study it was shown that differences in vine water status had no significant effect on the duration of vine developmental periods (i.e., between anthesis and veraison) expressed either on calendar days or degree days.[207]

Growth inhibition and final growth cessation due to water stress was shown to be similar among internodes, leaves, and tendrils of container-grown White Riesling.[208] The relative partitioning of growth among these three organs was unaltered when growth was inhibited due to water stress. The pre-dawn tissue Ψ of leaves, internodes, and tendrils which completely inhibited growth of each was -1.0, -1.2, and <-1.2 MPa, respectively. The growth of each organ was inhibited initially at a Ψ_{soil} of -0.065 MPa and ceased completely at a Ψ_{soil} of -0.54 MPa. It was concluded that the sensitivity of growth to water stress increased with ontogeny as some growth was maintained in younger tissues when inhibition was complete in older tissues.

These data contrast with those of field-grown vines. Soil matric potentials (Ψ_m) of -0.05 MPa were insufficient to decrease midseason shoot growth in a cool environment,[209] while the same soil Ψ_m decreased shoot growth in a hot environment with shoot growth ceasing at a $\Psi_m = -0.065$ MPa measured at a depth of 0.3 m.[210] Kliewer et al.[180] found that the rate of shoot elongation of 'Carignane' was reduced by water stress before any differences were detected in pre-dawn Ψ_l and that water stress reduced shoot growth but had no effect on the growth rate of leaves. The differences in results between field- and container-grown vines could possibly be because soil water status in the field was measured only in a limited portion of the root zone and therefore may not reflect the soil Ψ_m of the entire rooting profile. In addition, the study on container-grown vines was conducted on vegetative vines while those in the field had a crop.[208]

A reduction in shoot growth is the first visible symptom of vine water status in the field[211] and may be more sensitive to Ψ_{soil} than physiological processes occurring within the leaf (Figure 17), and recent studies on annual crops indicate that non-hydraulic signals from the roots in drying soil may inhibit leaf elongation without influencing g_s.[212] Such a response may act to conserve water as the soil dries but before the onset of water stress in the aerial portion of the plant. The reduction in shoot elongation is clearly demonstrated on vines irrigated daily at various fractions of vine ET (Figure 19). The soil Ψ_m, for the 1.0, 0.6, and 0.2 irrigation treatments on day-of-year 150 were -0.025, -0.05, and -0.06 MPa, respectively.

Weights of canes pruned from the vine during the dormant portion of the growing season is often used as a measure of shoot growth the previous season. Pruning weights may increase up to 137% with irrigation;[164] the relative increase in pruning weight being largely dependent on the volume and timing of irrigation throughout the season.[213] When vines were irrigated daily at various fractions of full vineyard ET (from 0 to 140%) pruning weights increased linearly with irrigation quantity (L. E. Williams,

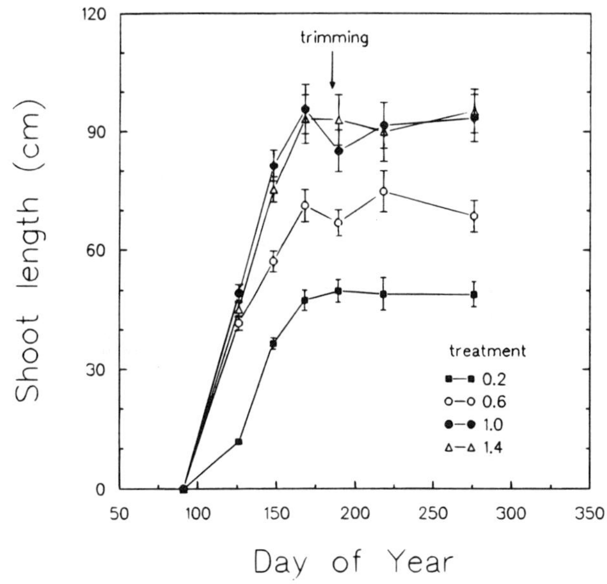

Figure 19 Average shoot length of 'Thompson Seedless' grapevines throughout the 1991 growing season as a function of irrigation treatments. Each value is the mean (± SE) of all shoots on three individual vine replicates per irrigation treatment measured repeatedly during the season. (L.E. Williams, unpublished data.)

unpublished data). Irrigation at full ET, under the conditions of this study, increased pruning weights to almost three times that of the non-irrigated control.

Water use efficiency (WUE) is the amount of plant biomass produced per amount of water transpired and is a useful parameter in assessing the effects of water stress on productivity and drought tolerance.[214] WUE can be constant despite differences in crop water use due to different irrigation regimes.[215] However, the WUE (g dry wt produced per kg water transpired) of container-grown vines ranged from 1.5 to greater than 5 for different cultivars of *V. vinifera*.[216] The authors concluded that WUE increased with increased vegetative growth (vigor) of the vine. The WUE decreased as vine ET increased for 'Thompson Seedless' vines grown in a semi-arid environment over the period from budbreak to veraison (Figure 20) despite increased vine vigor with greater irrigation amounts. WUE decreased from 5.85 for the 0.2 irrigation treatment to 1.84 for the 1.4 treatment. These values are similar to those calculated by Smart and Coombe[164] with data from a study by van Zyl and van Huyssteen.[217] One may expect differences in WUE due to differing amount of applied water with 'Thompson Seedless' grapevines as the relationship between A and g_s is curvilinear (Figure 13). The greatest efficiency between A and transpiration in this study occurred at a g_s of approximately 500 mmol H_2O m^{-2} s^{-1}. Most of the midday values of g_s for vines irrigated at 100 and 140% of vine water use are greater than 500 mmol m^{-2} s^{-1}.

5. Effects on Reproductive Growth

Reproductive growth of grapevines is less sensitive to water stress than vegetative growth.[2,163,164,204,218] Information on the effects of water stress on bud fruitfulness of grapevines is limited due to the inability of separating specific effects of water stress from those of temperature and light intensity in the field.[1] Water stress has decreased bud fruitfulness of container-grown vines under controlled environmental conditions.[219] However, it was suggested that bud fruitfulness is not adversely affected by the levels of water stress experienced in the field,[220] and may even increase bud fruitfulness.[213] Increased fruitfulness may be due to the reduction in vegetative growth which improves the exposure of the buds to light (see Section II.C). However, severe reductions in shoot growth due to water stress will result in fewer buds available for next year's fruiting canes for cane pruned cultivars if there is not sufficient shoot growth.

Using 3-year-old, container-grown vines, it was demonstrated that severe water stress (predawn Ψ_l <0.6 MPa) for the 3-week period after anthesis, induced cluster abscission and reduced berry set.[221] While severe water stress at anthesis in the field is uncommon, cluster abscission did occur early in the season (shoot length 30 cm) for vines irrigated at 0 and 20% of full vineyard ET the previous

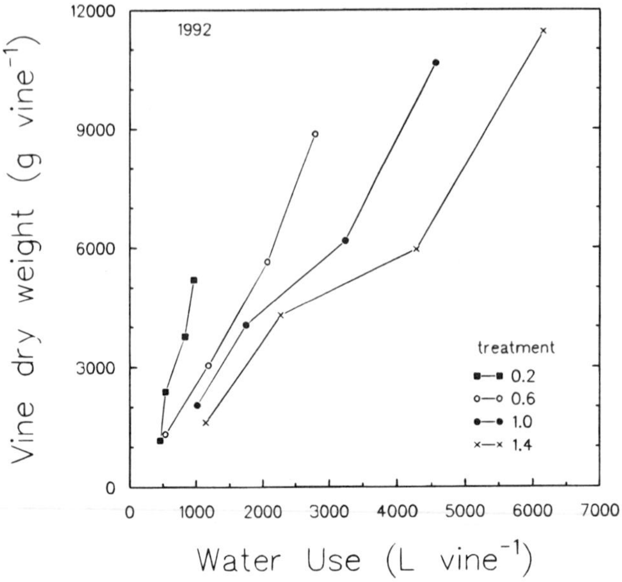

Figure 20 The relationship between the accumulation of dry biomass (leaves, main axis of shoots, and clusters) of 'Thompson Seedless' grapevines and water use throughout the 1992 growing season. Water use of each irrigation treatment was determined by adding the amount of water applied via drip irrigation to each vine and the amount of water depleted in the soil. Each value is the mean of three individual vine replicates. (L.E. Williams, unpublished data.)

growing season (L. E. Williams, unpublished data). Vines in these two treatments lost 70 to 90% of the clusters present on the vines during the period of delayed shoot growth noted in Figure 19.

Water stress will reduce the growth of berries, but does not influence the characteristic double-sigmoid growth curve.[204] A given water deficit during Stage I (when cell division is occurring) will generally reduce final berry size more than water deficits during Stages II and III (growth by cell expansion).[2,54,204,207,221,222] Also, the reduction in berry size due to soil moisture deficits during Stage I cannot be reversed by supplemental irrigation during Stages II and III.[213,223,224]

The sink potential of the fruit, determined during Stage I, appears to dictate the amount of carbon allocated to the cluster regardless of water stress. GA_3 applied at berry set will increase final berry size in seedless cultivars due to increased cell division.[225] An application of GA_3 to non-irrigated 'Thompson Seedless' vines at berry set resulted in comparable berry size and yield to the irrigated control vines (Table 2) underscoring the importance of events occurring during Stage I in determining final berry size. Final yields were similar between the irrigated control vines and the non-irrigated vines that were sprayed with GA_3 at berry set despite large differences in leaf area per vine at harvest. In addition, the ability of the water stressed vines (with reduced leaf area) to mature a crop, similar to that of irrigated vines indicates that alterations in source/sink relationships may be able to overcome the detrimental effects of water stress. Berry growth rate of irrigated and non-irrigated vines is similar subsequent to

Table 2 **The interaction of irrigation amount and gibberellic acid (GA_3) applied at berry set on berry size and yield of 'Thompson Seedless' grapevines**

Irrigation treatment	Berry weight (g)		Yield (kg vine^{-1})	
	Control	GA_3	Control	GA_3
Irrigated	2.0	2.8	16.5	23.4
Nonirrigated	1.2	1.8	9.0	16.1

Note: Mean leaf area per vine at harvest for the irrigated treatment was 20.3 m^2 while that for the nonirrigated treatment was 6.2 m^2. There were no significant differences in berry weight and yield between the irrigated control vines and the nonirrigated vines sprayed with GA_3.

From Williams, L.E., unpublished data.

Stage I despite differences in vine water status.[205] These results indicate that the involvement of plant hormones, other than ABA, should also be studied in plants under water stress.

The effects of water stress on berry growth are reflected in final yield.[164,205] Post-veraison water deficits had less of a detrimental effect on final yield than pre-veraison water deficits when compared to a continuous weekly irrigation treatment.[207,222] Hepner et al.[226] reported no significant differences in yield due to different irrigation amounts for 'Cabernet Sauvignon' vines between May and July 15th or subsequent to July 15th until harvest. There was a linear increase in yield for 'Thompson Seedless' grapevines irrigated daily at fractions of full ET_c, 40 to 100%.[181] Thus, both the timing and degree of water stress will have an important effect on the yield of field-grown grapevines.

6. Effects on Fruit Quality

Vine water status affects fruit solute composition throughout berry development. However, the literature contains conflicting data as to the exact effects of water stress on the berry composition. The accumulation of sugars is less affected by water deficits than is berry growth[2] although severe water stress may decrease sugar accumulation.[205,221] Sugar accumulation may also be delayed by increased water applications or by increased vegetative growth resulting in a less favorable light environment in the fruit zone (see Section II.C).

Organic acids contribute to the quality of the harvested fruit. A decrease in titratable acidity may occur with vine water stress.[204,223,227] Malic acid is the primary acid affected by water deficits; its concentration is dependent upon the specific time the deficit is imposed in relation to veraison.[204,227] The reduction in malate may be due to increased fruit temperature, and therefore increased respiration, as clusters become more exposed because of lack of leaf shading. The losses of malate may be balanced by similar decreases in counter-balancing cations or by accumulation of other acidic moieties such as amino acids.[2] The effects of water deficits on pH of the juice is less clear as some studies report that pH is increased by irrigation while others found no effect of supplemental irrigation.[2]

Water deficits will improve fruit color of red- and black-fruited cultivars.[2] The increase in color is the result of an increase in the production of anthocyanins. It is unknown whether this is a direct effect of water stress or an indirect one due to increased fruit exposure as a result of a reduction in vegetative growth.

Both early and late season water deficits increased juice and skin phenolics in berries of 'Cabernet franc' vines.[228] Wine sensory characteristics also can be manipulated by vineyard irrigation amounts and timings.[229] Wine made from continually irrigated vines differed from those irrigated only before or after veraison while the early season water deficit differed from the late season water deficit in appearance, flavor, taste, and aroma.[229] The wine effects may be associated with reduced berry size and increased skin content.

C. FLOODING

Waterlogging is a serious problem of grapevines.[230] Some species used for rootstocks may have tolerance to excessive soil water. *V. rupestris* is the most sensitive species to low soil 0_2.[177] Less sensitive species are *V. solonis, cinerea, candicans*, and the *riparia* × *rupestris* hybrids. For example, the rootstock (Couderc) 3306 (a *V. riparia* × *rupestris* hybrid) has been shown to tolerate poor soil drainage in Australia.[231]

Sensitivity of grapevines to waterlogging depends upon the time of year. Subsequent to leaf fall, grapevines are little affected by waterlogging conditions. In fact, flooding a vineyard in midwinter for a period of up to 6 weeks has been used as a means to control grape phylloxera (*Daktulosphaira vitifoliae* Fitch).[232] Waterlogging after budbreak will cause reduced shoot growth, leaf chlorosis, and death.[233] Excessive irrigation, resulting in saturated soils reduces new root initiation[234] and inhibits the growth of roots into water saturated soil layers.[204]

Webber and Jones[235] have recently summarized the indirect effects of waterlogged conditions on vine growth. Many of the American *Vitis* species used for rootstocks are intolerant of lime and they will suffer from chlorosis,[231] which is aggravated by waterlogging. In addition, waterlogging can change soil pH and affect the availability of nutrients.

While waterlogged soils may have deleterious effects on vine performance, over-irrigation (water applications slightly greater than vineyard ET) has more subtle effects on growth, productivity and fruit quality. Such vines have reduced bud fruitfulness and yields but increased pruning weights (L. E. Williams, unpublished data). Excessive irrigation generally will reduce fruit sugar concentration, titratable acidity and delay color development.[204] This is thought to be due to competition for photosynthates between the fruit and post-veraison vegetative growth.[164] However, much of the negative effects of over-irrigation may actually be due to shading effects due to excessive vegetative growth (see Section II.C).

V. MISCELLANEOUS ENVIRONMENTAL FACTORS

A. SALINITY

Grapes are grown in areas where salinity is a problem, most notably in areas of Australia, Israel, and portions of southwestern United States. Grape has been classified as moderately sensitive to salinity (chloride) based upon vegetative growth measurements.[239] Most annual crops are affected by the reduction in Ψ_π of the soil solution due to salinity while woody perennial crops are primarily affected by specific ion toxicities.[236] Grapevines accumulate chloride readily either via the root system[237] or through the leaf.[238]

The visible symptoms of salt stress on grapevines first appear as marginal chlorosis on the leaves followed by necrosis progressing towards the center of the leaf blade. These toxicity symptoms are probably due to the uptake of the chloride ion as grapevines grown on sodic soils rarely exhibit these symptoms.[240] The maximum permissible chloride in soil water without leaf injury was shown to range from 60 to 80 mol m^{-3} for three container-grown, commercial rootstocks.[236] However, Prior et al.[241] found that symptoms of leaf damage (marginal necrosis) in the field were more closely related to the onset of hot dry weather than with reaching a particular Cl or Na concentration in the lamina. Under severe salt stress the entire vine may defoliate.

There is variability in the uptake of salt among *Vitis* species, cultivars, and rootstocks.[1] Downton[242] categorized *V. rupestris* as the most salt tolerant species followed in order of descending tolerance by *berlandieri, riparia, candicans, champinii, longii, cinerea, cordifolia*, and *vinifera*. Antcliff et al.[243] found the order of *V. berlandieri, champinii*, and *cinerea* similar to that of Downton[242] but that the only clone of *V. rupestris* used by Downton, 'Rupestris du Lot' (syn. 'Rupestris St. George'), was atypical of that species. Their data indicated *V. rupestris* salt tolerance as comparable to that of *V. cinerea*. The variation in Cl exclusion both among and within species indicates there may be a genetic basis for this characteristic.[244] The ability to exclude Cl by the *V. champinii* species (the 'Ramsey' rootstock, syn 'Salt Creek') is probably due to the action of many genes.[245] The genetic basis for Cl exclusion in *V. berlandieri* may be due to a single dominant gene.[246] In that study, Cl exclusion was independent of both yield and berry weight.

V. champinii rootstocks 'Dogridge' and 'Ramsey'[242,243,247,248] and the *berlandieri* × *rupestris* hybrids '110 Richter', '140 Ruggeri', and '1103 Paulsen'[243,247] are effective Cl excluders.[243,247] Salt tolerant *V. vinifera* cultivars include 'French Colombard' (most tolerant) > 'Grenache', 'Chenin blanc', 'Thompson Seedless' > 'Barbera', 'Muscat of Alexandria', or 'Ribier' (susceptible).[1,240]

Reductions in growth and A have been observed to occur in response to soil salinity before any toxic symptoms appear in grapevines.[249] Prior et al.[241] found that the response of A to salinity of field-grown grapevines was almost identical to that found by Downton[249] on container-grown vines and was more strongly correlated with leaf Cl than leaf Na. The reduction in A is due to a uniform decrease in g_s up to a tissue concentration of 165 m*M* Cl.[250] It also was found that at tissue concentrations above 165 m*M* Cl non-stomatal inhibition of A was actually due to non-uniform stomatal closure (determined visually using $^{14}CO_2$ uptake and autoradiograms). It has been observed that ABA levels in the leaves of salt-stressed grapevines increase rapidly and remain such for several weeks.[251] Therefore, in grapevines the salt-induced reductions in A (mediated by the increase in ABA levels) are a result of heterogeneous stomatal closure.

Container-grown grapevines exposed to saline water are able to osmotically adjust shortly after exposure.[252] Osmotic adjustment is due to the accumulation of Na, K, and Cl ions and an increase in reducing sugars.[251] It was concluded that the maintenance of turgor pressure and osmotic adjustment during salt stress prevents immediate damage to PSII activity. However, the continued accumulation of chloride eventually causes membrane permeability changes, cell damage, and the loss of turgor.

As mentioned above, grapevine tissue ion content will change with the use of saline water. Prior et al.[253] found that the accumulation of Na and Cl in leaves and petioles of field-grown vines was not linear but tapered off at high irrigation water electrical conductivity (EC) values, indicating that the tissue was becoming saturated with salt. K concentrations in petioles and leaf blades generally are reduced due to elevated levels of NaCl in the rooting medium.[240,241] However, it was shown that leaves of container-grown grapevines began to accumulate K^+ and Cl within 6 h of exposure to high levels of NaCl while the uptake of Na was not evident until the next day.[251] The accumulation of K^+ and Na generally balanced the accumulating Cl ion. The effect of saline water on the accumulation of other mineral nutrients in vine tissues is less clear and differs depending whether the study is conducted on container-grown vines over a short period or on field-grown vines over a long-term period.

Vegetative growth (shoot length, pruning weights, and leaf weight) of 'Sultana' (syn. 'Thompson Seedless') was reduced by salinity to a much greater extent than yield.[241] In that study, all vegetative growth parameters measured were reduced by salinity with the effects of salinity were more severe on heavier soils. Growth reductions occurred despite the fact that there were no differences in Ψ_l among salinity treatments. Root density was reduced to a greater extent than the reduction in total leaf area.[253] Root growth was severely restricted on container-grown 'Cabernet Sauvignon' even in concentrations of NaCl as low as 20 m*M* in the rooting medium.[254] Soil salinity has also reduced starch levels in the canes and increased the concentrations of reducing sugars.[241,250] It was concluded that the reduction in vegetative growth and photosynthetic area was the primary reason for a reduction in yield in the study of Prior et al.[241,253,255]

A 6-year study on own-rooted 'Sultana' vines conducted in the River Murray flood basin of Australia[241,253,255] showed that vine performance declined with the duration of exposure to salinity and was strongly influenced by soil texture.[255] Yield response to salinity levels in the lightest textured soil resembled the model developed by Maas and Hoffman,[239] although more severe losses than predicted occurred in the heavier soils. Yield was correlated with EC[253] of the soil extract, but that relationship was not as good as that derived between yield and plant-based measures (lamina content) of salinity.[241] For own-rooted 'Sultana': (1) there was no evidence of a yield threshold in response to soil salinity when averaged across all soil texture types; (2) there was a 10% yield loss when EC at the end of the winter exceeded 1.0 dS m^{-1}; (3) vine-based measurements of salinity effects were better than soil-based measurements as they were able to integrate the rootzone salinity over space and time; (4) the best relationship between yield and vine tissue concentrations was obtained using petiole Cl and Na values; and (5) petiole Cl and Na values should be kept below 420 and 191 mmol kg^{-1} dry weight, respectively, to avoid the detrimental effects of salinity on vine productivity.

There are instances when salt in the irrigation water may be present only for a short period of time. A study investigating transient (two-month period), soil salinization of a Colombard vineyard on 'Ramsey' rootstock found that the treatments increased EC values and petiole Na and Cl concentrations and decreased Ψ_l[256]; however, vegetative growth or yield were not affected.[256] The authors concluded that the lack of growth or yield response in this vineyard was not due the use of the 'Ramsey' rootstock as other studies have found that growth of various scion/rootstock combinations will decline over a range of saline water.[257] Thus, it appears that transient exposure to high EC levels (3 to 6 dS m^{-1}) for 2-month periods may not necessarily reduce vine productivity. Of significant importance in minimizing the effects of salinity is to ensure that adequate leaching of salts take place sometime during the growing season.[253]

Fruit composition may be affected by salinity. The accumulation of sugars in the fruit was not affected for the first 3 years of soil salinization,[255,258] but declined subsequently. Juice titratable acidity increases with salt treatments as does the concentrations of Na, K, and Cl ions; wines made from the fruit were similarly affected.[242] The use of rootstocks, known to exclude Cl, reduced the levels of Cl in the wine.[242]

B. WIND

Wind can affect the physiology and growth of plants in numerous ways.[259] Wind speed is of significant importance as it affects heat and mass transfer of individual leaves and the vine canopy as a whole. High wind velocities can lead to structural damage of plant tissue while constant winds at low to medium velocities can lead to deformation and disruption of physiological processes. Vineyards in many regions may be exposed to chronic low- to medium-wind velocities; the effects of chronic wind exposure on vine physiology and productivity have not been quantified.

Vineyards are considered aerodynamically rough as their surfaces are covered by discontinuous canopies. Wind will determine the depth of the boundary layer which ultimately affects the exchange of CO_2 and water vapor between the plant and the atmosphere. Weiss and Allen[260] did not find a constant flux zone over the vineyard in which their measurements were made as has been found for other crops. They concluded that the vineyard boundary layer actually consisted of an inner and outer zone. Wind direction also influences the degree of roughness of a vineyard. The drag coefficient, which is a nondimensional measure of the roughness of a surface, is higher for cross-row flow than for down-row flow of air in a vineyard.[260,261] Wind speed also has been shown to be more important than wind direction in maintaining between-row circulations.[262]

An increase in wind speed will increase the boundary layer conductance which generally increases the rate of transpiration from leaves and plant canopies.[215] Hicks[261] found that vineyard ET increased 10 to 20% with cross-row flow as compared with down-row flow of air. This indicated that ET increased due to an increase in the drag coefficient (see above). However, studies conducted both in the lab and the field on individual leaves of grapevines have shown that g_s and transpiration is decreased when wind speed exceeds a given threshold;[263–265] wind velocities greater than 3 m s^{-1} were required to significantly decrease g_s and transpiration. Preliminary results from Australia indicate that ABA increases in leaves of grapevines exposed to chronic, low- to medium-wind velocities (B.R. Loveys, personal communication). The reduction in g_s of grapevines exposed to wind may be due to the accumulation of ABA as others have found that ABA increases in plant tissues that are exposed to wind.[266]

Wind has been reported to have little effect on the water relations of various plant species[215,259] including *V. vinifera.*[264] However, in studies examining the differences in water relations between sheltered and non-sheltered, field-grown grapevines in windy locations, Ψ_l of the sheltered vines always was more negative than that of the controls.[263,265] This may be expected as g_s was always greater for the sheltered vines.

The only report documenting a wind-induced reduction of A of grapevines did not include g_s measurements.[267] Many of the authors who have studied the effects of wind on grapevines suggest that the reduction in g_s due to increased wind speeds also will reduce A. The degree in which A is reduced due to increased wind speeds is largely dependent (although not linearly dependent) upon the extent that g_s is reduced. However, preliminary assessment of wind-breaks on vine physiology and growth indicates that there may not always be a large reduction in A when g_s is reduced due to chronic wind exposure (Figure 21).

Kobringer et al.[264] reported a carryover effect on g_s by wind velocities greater than 10 m s^{-1} but not at 3.6 m s^{-1}. Stomatal conductance remained depressed for up to 4 days after the treatment stopped, however, there were cultivar differences. A similar type of response has been observed in the field using windbreaks in a ‘Chardonnay’ vineyard (L. E. Williams, unpublished data). Wind has been shown to both increase and decrease the density of stomata on leaves of other plant species.[259] The carryover effect on g_s reported by Kobringer et al.[264] was not due to a reduction in stomatal densities as measurements were made only on mature leaves over a short period. It is uncertain whether this was true for the field study as the leaves had been exposed to chronic wind stress throughout the growing season.

While extremely high winds can cause physical damage in vineyards,[268] constant exposure to medium velocity winds (i.e., 1 to 2 m s^{-1}) also may affect vine growth and productivity.[269] At vineyard sites planted perpendicular to prevailing winds the vines will have an asymmetric growth habit (i.e., growth is much reduced on the windward side). The reduction in shoot growth is primarily due to a reduction in internode length (Table 3). There also is a reduction in individual leaf size and it is this parameter which is principally responsible for the reduction in total vine leaf area. The reduction in total vine leaf area is probably a more important determinant of reduced vine growth due to wind than the small decreases in leaf A. The reduction in growth may be the result of physiological and/or mechanical effects.[215] The production of ethylene in response to mechanical perturbation may be responsible for thigmomorphogenic responses of plants.[270]

Yields are greater for vines grown within windbreaks (Table 3) or close to windbreaks in vineyards when compared to vines grown further away or upwind from the break.[268,269] Simon[271] found that growth and productivity of vines increased when grown within a distance of five times the height of the windbreak. The reduction in yield was due both to reduced berry weight and cluster numbers per vine. Fruit soluble solids were lower for exposed vines when compared to the sheltered vines.[269] Other differences in fruit composition measured between sheltered and non-sheltered vines may be the result of differences in maturity between the treatments.

C. AIR POLLUTION

Extensive reviews of the effects of airborne pollutants on the grapevine were published in the early 1980s.[272-274] Therefore, this section will briefly review the effects of air pollutants on vines and present some more recent data on this subject.

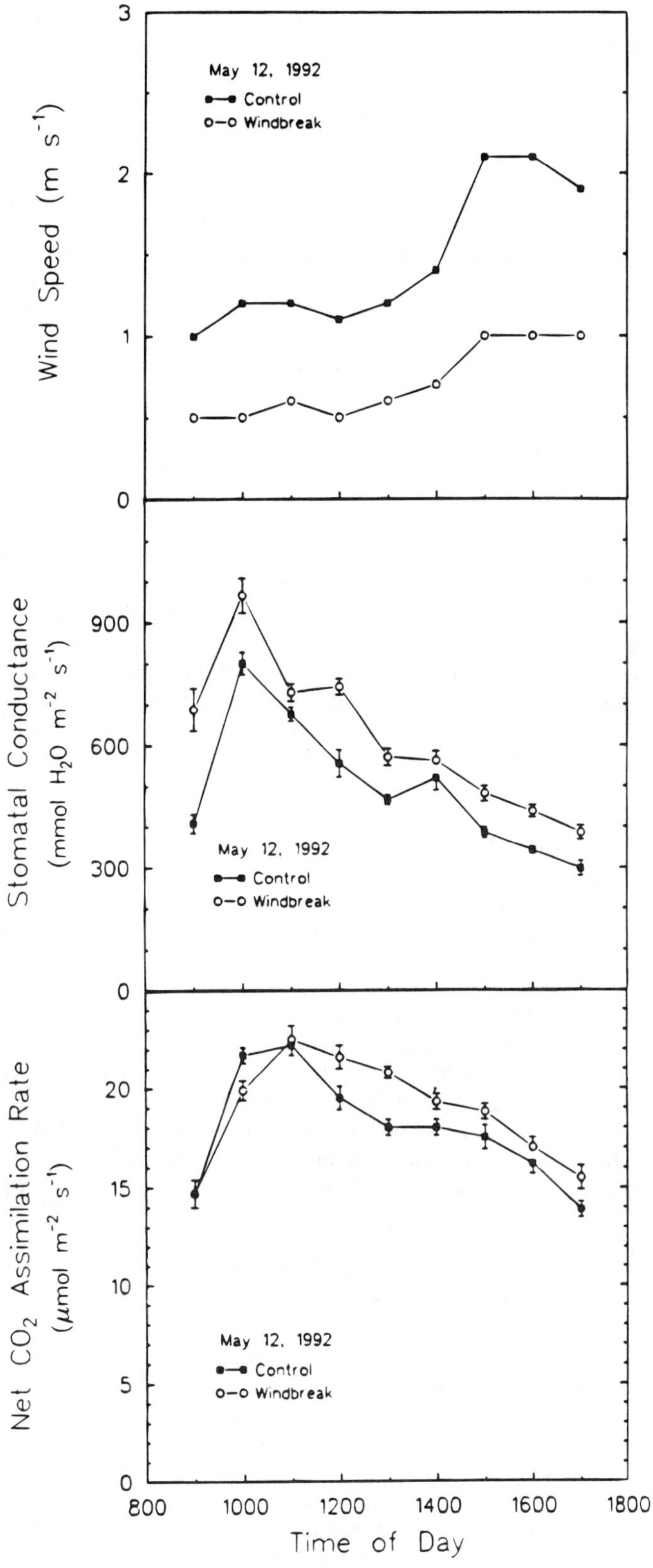

Figure 21 The effect of a windbreak on the diurnal time course of net CO_2 assimilation rate, stomatal conductance, and wind speed of 'Chardonnay' grapevines grown in the Salinas Valley of California. Control vines were exposed to ambient wind velocities. For an explanation of the windbreak see Table 3. Each value of A and g_s are the means of eight individual leaf measurements. Wind speed values are the hourly means of measurements taken every minute by a datalogger. (L.E. Williams, N.K. Dokoozlian, and L. Bettiga, unpublished data.)

Table 3 **The influence of windbreaks on the growth and yield of 'Chardonnay' grapevines grown in the Salinas Valley**

Treatment	Internode length (cm)	Shoot length (cm)	Individual leaf area[a] (cm^2)	Total vine leaf area (m^2)	Vine yield (kg)
Control	3.3	62	77	8.5	6.8
Windbreak	4.6	99	112	11.7	7.6

Note: The windbreak consisted of 50% density shade cloth surrounding seven vines. The shade cloth extended from the soil surface up to 3.05 m into the air. Each plot was replicated eight times.

[a]These values represent the area of individual leaves from the primary shoot. Leaves from lateral shoots are not included.

From Dokoozlian, N. K., Bettiga, L., and Williams, L. E., unpublished data.

Airborne pollutant injury to grapevines was first reported in France early in the Twentieth Century.[272] Pollution injury in the Western Hemisphere was first reported to occur in the 1950s in Southern California.[275] Since those times there have been numerous reports of air pollution affecting grapevines in vineyards throughout the industrialized world. Exposure of grapevines to air pollutants results in foliar symptoms as these molecules are taken up through the stomata.[276] Symptoms of pollution injury may vary among cultivars and might be confused with mineral nutrient deficiencies or toxicities. Exposure of grapevine leaves to ozone results in small patches of necrotic tissue surrounded by healthy green tissue and is termed 'Oxidant stipple'.[275] These lesions are localized in the palisade tissue of the leaf. Severe injury will result in chlorosis, bronzing, and premature leaf senescence and abscission. Older leaves are more susceptible to O_3 than younger leaves.

There is evidence of tolerance to airborne pollutants by native *Vitis* species, various cultivars of *V. vinifera*, and some French and American hybrids.[272] Tolerance is based upon foliar symptoms. For example, it was found that grape oxidant stipple (symptom of ozone exposure) was more prevalent on 'Carignane' and 'Grenache' than on 'Zinfandel' and 'Thompson Seedless'.[275] Similar categories have been established for susceptibility to hydrogen fluoride and sulfur dioxide.[272] Unfortunately, foliar symptoms of grapevines in response to air pollutants are dependent upon concentration and dose of the particular pollutant, stage of shoot development, leaf age, and cultural practices. Thus, there is some disagreement among studies with regard to the tolerance classification of individual cultivars.

Ambient partial pressures of O_3 (12-h mean 50 to 60 μPa Pa^{-1}) will reduce grapevine A anywhere from 5 to 13%.[41] The authors concluded that reductions in A were due to an O_3-induced decline in g_s which resulted in decreased C_i. This differs from the findings of Shertz et al.[277] who found that stomata of grapevine were opened by exposure to O_3. Studies on other plant species have shown ambient partial pressures of O_3 will reduce g_s to a greater extent than mesophyll conductance.[278,279] Even O_3 partial pressures greater than ambient levels significantly reduced g_s while having no significant effect on grapevine A over a 5-h fumigation period.[41] The exact mode of action of O_3 on grapevine g_s is unknown. ABA may be produced in response to pollutant exposure.[280,281] As discussed previously, the reduction in grape A when ABA is taken up is due solely to its effect on g_s and results in heterogeneous stomatal closure.[166] Heterogeneous stomatal closure in response to ozone fumigation has been shown on other plant species.[282] The symptoms of O_3 damage in grape leaves, oxidant stipple, would be consistent with heterogeneous stomatal closure upon exposure to this air pollutant. Stomata which remain open allow ozone to diffuse into the mesophyll and may result in acute damage to cells; stomata which remained closed would ostensibly protect the tissue from damage.

Most studies characterizing the effects of air pollutants on growth and productivity of grapevines have been conducted on container-grown plants.[272] However, there have been three studies conducted in the field assessing the effects of ambient oxidants (O_3) using open-top chambers. The chambers are constructed around mature vines and then either charcoal-filtered air or ambient air is forced through the chambers and out the top. Studies conducted on 'Zinfandel'[283,284] and 'Thompson Seedless'[285] grapevines in California, indicate that ambient pollution reduced pruning weights ca. 25 and 12%, respectively, compared to the charcoal-filtered controls. A study conducted on 'Concord' vines in New York found no significant effects of ambient oxidants on vegetative growth although there was a trend for less growth in the ambient air chambers.[286,287]

Thompson and co-workers[283,284] found that yields of 'Zinfandel' vines exposed to ambient air were reduced 12 and 61% when compared to vines in the charcoal-filtered chambers in the first and second years of the study, respectively. Yields of 'Thompson Seedless' were reduced 28 and 17% after the second and third years of exposure to ambient oxidants, respectively.[285] The number of clusters per vine was the yield component affected most for 'Thompson Seedless'.[285] It appears that O_3 exposure affects the differentiation of clusters within the compound bud; however, it is unknown whether O_3 directly affects the differentiation of cluster primordia or indirectly affects the process via a reduction in available vine carbohydrates due to a reduction in A.

Berry weight and the accumulation of soluble solids within the fruit of 'Zinfandel' were reduced due to exposure of vines to ambient oxidants.[283,284] Apparently, the reduction in A on both a single leaf basis[41] and whole vine basis (foliage levels estimated) reduced the amount of photosynthate available for growth of the fruit. In addition to a reduction in A, premature leaf abscission and loss of effective leaf area due to toxic levels of O_3 can cause a reduction in plant biomass.[288]

Sulfur dioxide injury results in grayish-brown lesions along the margins or tip of the leaf with middle-aged leaves most susceptible.[272] The first symptom of fluoride injury is a gray-green color at the margins of young leaves which then becomes brown or reddish-brown in color.[272]

The effects of sulfur dioxide and hydrogen fluoride on grapevine productivity is less clear.[272] Sulfur dioxide and ozone may exert a synergistic effect on vine physiology.[289] Reduced yields due to exposure to hydrogen fluoride may depend less on the degree of foliar injury than on characteristics of exposure over one or more growing seasons and stage of vine development.[272]

There have been no studies assessing the effects of air pollutants on the post-harvest storage of fruit used for table grapes (fresh market). This is crucial as appearance and storeability are important quality characteristics of table grapes. Crisosto et al.,[290] found that elevated partial pressures of O_3 decreased wax deposition and cuticle thickness of plum fruit (*P. salicina*, cv. Casselman) and the highest O_3 partial pressures resulted in a greater weight loss. Unfortunately, no data are available for grapevines.

VI. SUMMARY

The world-wide distribution of grapes (*Vitis* sp.) attests to the large genetic diversity both across the genus and within a species. The diverse climates, under which grapevines are grown, has resulted in a remarkable selection of cultivars that meet a variety of uses. Fortunately, there has been a minimum of "gene pool" reduction within the *Vitis* genus which will provide opportunities for grape breeding programs in the future.

Although there are several commercially-grown species, *V. vinifera* constitutes the majority of the acreage in the United States and around the world. The major limitations in the distribution of *Vitis* are low temperatures, seasonal heat accumulation, and water availability. Despite the fact that low temperature stress is the predominant limitation to the distribution of grapevines, little is known about the acclimation and deacclimation processes that are important for continued production in geographically and climatically extreme areas. For *V. vinifera*, high temperatures are clearly not a limiting factor and this species may have some unique characteristics as a crop plant in this regard. This species is also very tolerant to drought stress although irrigation may improve vine productivity.

A major theme present throughout this chapter is the mechanism by which grapevine leaves respond to various environmental stresses. It has been documented that heterogeneous stomatal closure in the leaves of grapevines occurs due to applied ABA, increases in VPD, soil water deficits, and salinity. A similar response may occur when grapevines are exposed to air pollutants and wind. It is tempting to suggest a causal relationship between the accumulation of ABA in grape leaves and the vine's response to environmental stresses as ABA causes heterogeneous stomatal closure.[166] Interestingly, a recent study has shown that ABA accumulates in the leaves of grapevines and grape callus tissue in response to high temperatures.[291] The authors suggested that ABA may be a factor in high-temperature acclimation and heat-tolerance induction in grape.

While the involvement of ABA in mediating environmental stresses at the leaf level in grapevines is apparent, other phytohormones may also play a role.[292] This is exemplified in the data from Table 2. It was demonstrated that the application of GA_3 at berry set of the seedless cultivar, 'Thompson Seedless', was able to overcome the negative effects of soil water deficits on berry growth. Further studies are needed to clarify how stress affects phytohormones in grape and other horticultural fruit crops.

Temperature extremes and fluctuations and availability of water will determine in large part the global distribution of species and cultivars of grapevines used for commercial purposes. However, it is possible to manipulate a vine's microclimate with the use of various training and trellising systems. While temperature, VPD and wind are ameliorated to some degree by the use of these cultural practices, it is the light microclimate within and at the surface of the canopy that is most impacted by training and trellising systems. Beginning with the pioneering work of Shaulis and co-workers, great advances have been made regarding our understanding of the effects of irradiance on vine growth, productivity, and fruit composition. Much attention has also been given to canopy management practices such as basal leaf removal and shoot positioning, for the improvement of the canopy's light environment.

Although some uncertainty remains regarding the location of the photoreceptor responsible for the regulation of bud fruitfulness, it is generally believed that irradiance received directly by the bud governs cluster initiation. Existing evidence suggests that light quantity rather than quality regulates fruit bud differentiation. Recent work also has revealed that many aspects of grape berry growth and composition are regulated by light and that the photoreceptors are located in the fruit. Much speculation has centered around the involvement of phytochrome for the photoregulation of berry metabolism. Again, it has been shown that light quantity rather than light quality is responsible, suggesting that chlorophyll or the putative blue light photoreceptor cryptochrome, may be involved in the regulation of grape berry growth. The elucidation of the exact location and nature of the photoreceptor(s) regulating vine growth and metabolism should be of high priority for future research.

Further improvements in vine productivity and fruit quality may only be possible by increasing our basic understanding of the interactions of solar radiation interception by the vine and other environmental factors. Therefore, future studies in viticultural research must be conducted under a wide variety of vine training and trellising systems in contrasting mesoclimates. With this knowledge and a subsequent expanded database, modelling vine performance will be enhanced, and the use of expert systems in viticulture will become common place.

REFERENCES

1. Mullins, M. G., Bouquet, A., and Williams, L. E., *Biology of the grapevine*, Press Syndicate of the University of Cambridge, Cambridge, 1992, 239.
2. Williams, L. E., and Matthews, M. A., Grapevine, *Irrigation of agricultural crops*, Agronomy Monograph No. 30, Stewart, B. A., and Nielsen, D. R., Eds. ASA-CSSA-SSSA, Madison, 1990, 1019.
3. Clingeleffer, P. R., Response of Riesling clones to mechanical hedging and minimal pruning of cordon trained vines (MPCT)—implications for clonal selection, *Vitis*, 27, 87, 1988.
4. Hart, J. W., *Light and plant growth* (Topics in plant physiology, No. 1), Unwin Hyman Ltd., London, 1988, 204.
5. Smart, R. E., Sunlight interception by vineyards, *Am. J. Enol. Vitic.*, 24, 141, 1973.
6. Kliewer, W. M., Vineyard canopy management—A review, in *Grape and Wine Centennial Symposium Proceedings*, Webb, A.D., Ed., Univ. California, Davis, 1982, 342.
7. Shaulis, N. J., Responses of grapevines and grapes to spacing of and within canopies, in *Grape and Wine Centennial Symposium Proceedings*, Webb, A. D., Ed., Univ. California, Davis, 1982, 353.
8. Shaulis, N., Amberg, H., and Crowe, D., Response of concord grapes to light, exposure, and geneva double curtain training, *Proc. Am. Soc. Hort. Sci.*, 89, 268, 1966.
9. Weaver, R. J., Kasimatis, A. N., Johnson, J. O., and Vilas, N., Effect of trellis height and crossarm width and angle on yield of 'Thompson seedless' grapes, *Am. J. Enol. Vitic.*, 35, 94, 1984.
10. Kasimatis, A. N., Lider, L. A., and Kliewer, W. M., Trellising and training practices to influence yield, fruit composition, and growth of 'Chenin blanc' grapes, in *Grape and Wine Centennial Symposium Proceedings*, Webb, A. D., Ed., Univ. California, Davis, 1982, 386.
11. Weaver, R. J., and Kasimatis, A. N., Effect of trellis height with and without crossarms on yield of 'Thompson Seedless' grapes, *J. Am. Soc. Hort. Sci.*, 100, 252, 1975.
12. May, P., Clingeleffer, P. R., and Brien, C. J., Sultana (*Vitis vinifera* L.) cane and their exposure to light, *Vitis*, 14, 278, 1976.
13. Carbonneau, A., and Huglin, P., Adaptation of training systems to French regions, in *Grape and Wine Centennial Symposium Proceedings*, Webb, A. D., Ed., Univ. California, Davis, 1982, 376.

14. Kliewer, W. M., Marois, J. J., Bledsoe, A. M., Smit, S. P., Benz, M. L., and Silvestroni, O., Relative effectiveness of leaf removal, shoot positioning, and trellising for improving winegrape composition, in *Proceedings Second International Cool Climate Viticulture and Oenology Symposium*, Smart, R. E., Thornton, R. J., Rodriques, S. B., and Young, J. E., Eds., NZ Soc. Vitic. Oenol., Auckland, 1988, 123.
15. Bledsoe, A. M., Kliewer, W. M., and Marois, J. J., Effects of timing and severity of leaf removal on yield and fruit composition of Sauvignon blanc grapevines, *Am. J. Enol. Vitic.*, 39, 49, 1988.
16. Smart, R. E., Principles of grapevine canopy management microclimate manipulation with implications for yield and quality. A review, *Am. J. Enol. Vitic.*, 36, 230, 1985.
17. Smart, R. E., Shaulis, N. J., and Lemon, E. R., The effect of Concord vineyard microclimate on yield. I. The effects of pruning, training, and shoot positioning on radiation microclimate, *Am. J. Enol. Vitic.*, 33, 99, 1985.
18. Kriedemann, P. E., Torokfalvy, E., and Smart, R. E., Natural occurrence and utilization of sunflecks by grapevine leaves, *Photosynthetica*, 7, 18, 1973.
19. Dokoozlian, N. K., Light quantity and light quality within *Vitis vinifera* L. grapevine canopies and their relative influence on berry growth and composition, Ph.D. Diss., University of California, Davis, 1990.
20. Buttrose, M. S., Some effects of light intensity and temperature on dry weight and shoot growth of grapevines, *Ann. Bot.*, 32, 753, 1968.
21. Buttrose, M. S., Climatic factors and fruitfulness in grapevines, *Hort. Abstr.*, 44, 319, 1974.
22. Baldwin, J. G., The effect of some cultural practices on nitrogen and fruitfulness in the Sultana vine, *Am. J. Enol. Vitic.*, 17, 58, 1966.
23. May, P., The effect of direction of shoot growth on fruitfulness and yield in Sultana vines, *Aust. J. Agric. Res.*, 17, 479, 1966.
24. May, P., Reducing inflorescence formation by shading individual Sultana buds, *Aust. J. Biol. Sci.*, 18, 463, 1965.
25. Morgan, D. C., Stanley, C. J., and Warrington, I. J., The effects of simulated daylight and shadelight on vegetative and reproductive growth in kiwifruit and grapevine, *J. Hort. Sci.*, 60, 476, 1985.
26. Buttrose, M. S., Fruitfulness in grapevines: The response of different cultivars to light, temperature, and day length, *Vitis*, 9, 121, 1970.
27. Kobayashi, A., Sigiura, A., Watanabe, H., and Yamamura, H., On the effects of daylength on the growth and flower bud formation of grapes, *Mem. Res. Inst. Food Sci. Kyoto Univ.*, 27, 15, 1966.
28. Sigiura, A., Utsunomiya, N., and Kobayashi, A., Effects of daylength and temperature on growth and bunch differentiation of grapevines, *Jap. J. Hort. Sci.*, 43, 387, 1975.
29. Kriedemann, P. E., Photosynthesis in vine leaves as a function of light intensity, temperature, and leaf age, *Vitis*, 7, 213, 1968.
30. Kriedemann, P. E., and Smart, R. E., Effects of irradiance, temperature, and leaf water potential on photosynthesis of vine leaves, *Photosynthetica*, 5, 6, 1971.
31. Downton, W. J. S., Grant, W. J. R., and Loveys, B. R., Diurnal changes in the photosynthesis of field-grown grape vines, *New Phytol.*, 105, 71, 1987.
32. Downton, W. J. S., and Grant, W. J. R., Photosynthetic physiology of spur pruned and minimal pruned grapevines, *Aust. J. Plant. Physiol.*, 19, 309, 1992.
33. Düring, H., CO_2 assimilation and photorespiration of grapevine leaves: Responses to light and drought, *Vitis*, 27, 199, 1988.
34. Liu, W. T., Wenkert, W., Allen Jr., L. H., and Lemon, E. R., Soil-plant water relations in a New York vineyard: Resistances to water movement, *J. Am. Soc. Hort. Sci.*, 103, 226, 1978.
35. Winkel, T. and Rambal, S., Stomatal conductance of some grapevines growing in the field under a Mediterranean environment, *Agric. For. Meteorol.*, 51, 107, 1990.
36. Grantz, D. A., and Meinzer, F. C., Regulation of transpiration in field-grown sugarcane: Evaluation of the stomatal response to humidity with the Bowen ratio technique, *Agric. For. Meteorol.*, 53, 169, 1991.
37. Williams, L. E., Williams, D. W., and Phene, C. J., Modeling grapevine water use, in *Proceedings Eighth Australian Wine Technical Conference*, Stockley, C. S., Johnstone, R. S., Leske, P. A., and Lee, T., Eds. Winetitles, Adelaide, 29, 1993.
38. Chaves, M. M., Harley, P. C., Tenhunen, J. D., and Lange, O. L., Gas exchange studies in two Portuguese grapevine cultivars, *Physiol. Plant.*, 70, 639, 1987.

39. Correia, M. J., Chaves, M. M. C., and Pereira, J. E., Afternoon depression in photosynthesis in grapevine leaves—evidence for a high light stress effect, *J. Exp. Bot.*, 41, 417, 1990.
40. Roper, T. R., and Williams, L. E., Net CO_2 assimilation and carbohydrate partitioning of grapevine leaves in response to trunk girdling and gibberellic acid application, *Plant Physiol.*, 89, 1136, 1989.
41. Roper, T. R., and Williams, L. E., Effects of ambient and acute partial pressure of ozone on leaf net CO_2 assimilation of field-grown *Vitis vinifera* L., *Plant Physiol.*, 91, 1501, 1989.
42. Gamon, J. A., and Pearcy, R. W., Photoinhibition in *Vitis californica*: The role of temperature during high-light treatment, *Plant Physiol.*, 92, 487, 1990.
43. Gamon, J. A., and Pearcy, R. W., Photoinhibition in *Vitis californica*: interactive effects of sunlight, temperature and water status, *Plant Cell. Environ.*, 13, 267, 1990.
44. Gamon, J. A., and Pearcy, R. W., Leaf movement, stress avoidance and photosynthesis in *Vitis californica, Oecologia*, 79, 475, 1989.
45. Smart, R. E., Vine manipulation to improve grape quality, in *Grape and Wine Centennial Symposium Proceedings*, A.D. Webb, Ed., Univ. California, Davis, 1982, 362.
46. Smart, R. E., and Smith, S. M., Canopy management: Identifying the problems and practical solutions, in *Proceedings Second International Cool Climate Viticulture and Oenology*, R.E. Smart, R.J. Thornton, S.B. Rodriques, and J.E. Young, Eds., NZ Soc. Vitic, Oenol., Auckland, 1988, 109.
47. Smart, R. E., and Sinclair, T. R., Solar heating of grape berries and other spherical fruits, *Agric. Meteorol.*, 17, 241, 1976.
48. Kliewer, W. M., and Antcliff, A. J., Influence of defoliation, leaf darkening, and cluster shading on the growth and composition of Sultana grapes, *Am. J. Enol. Vitic.*, 21, 26, 1970.
49. Kliewer, W. M., and Lider, L. A., Influence of cluster exposure to the sun on the composition of 'Thompson Seedless' fruit, *Am. J. Enol. Vitic.*, 19, 175, 1968.
50. Kliewer, W. M., Influence of temperature, solar radiation, and nitrogen on coloration and composition of Emperor grapes, *Am. J. Enol. Vitic.*, 28, 96, 1977.
51. Crippen, D. D., and Morrison, J. C., The effects of sun exposure on the compositional development of 'Cabernet Sauvignon' berries, *Am. J. Enol. Vitic.*, 37, 235, 1986.
52. Crippen, D. D., and Morrison, J. C., The effects of sun exposure on the phenolic content of 'Cabernet Sauvignon' berries during development, *Am. J. Enol. Vitic.*, 37, 243, 1986.
53. Morrison, J. C., The effects of shading on the composition of 'Cabernet Sauvignon' grape berries, in *Proceedings Second International Cool Climate Viticulture and Oenology Symposium*, R.E. Smart, R.J. Thornton, S.B. Rodriques, and J.E. Young, Eds., NZ Soc. Vitic. Oenol., Auckland, 144, 1988.
54. Coombe, B. G., The development of fleshy fruits, *Annu. Rev. Plant Physiol.*, 27, 507, 1976.
55. Smart, R.E., Smith, S. M., and Winchester, R. V., Light quality and quantity effects on fruit ripening for 'Cabernet Sauvignon', *Am. J. Enol. Vitic.*, 39, 250, 1988.
56. Kliewer, W. M., and Smart, R. E., Canopy manipulation for optimizing vine microclimate crop yield and composition of grapes, in *Manipulation of Fruiting*, C. J. Wright, Ed., Butterworth, London, 1989, 275.
57. Perez, J., and Kliewer, W. M., Influence of light regime and nitrate fertilization on nitrate reductase activity and concentrations of nitrate and arginine in tissues of three cultivars of grapevines, *Am. J. Enol. Vitic.*, 33, 86, 1982.
58. Huang, S. B., Agroclimatology of the major fruit production in China: A review of current practice, *Agric. For. Meteor.*, 53, 125, 1990.
59. Qinghua, W. A. N. G., Turpan: China's prime grape producer, *Fruit Var. J.*, 45, 187, 1991.
60. Forsline, P., Winter cold acclimation and deacclimation. Eastern Grape Grower & Winery News, April/March, 16, 1984.
61. Clore, W. J., and Brummund, V. P., Cold injury to grapes in eastern Washington 1964–65, *Washington State Hort. Assoc. Proceedings*, 61, 143, 1965.
62. Strik, B., and Lombard, P., Assessment of Winter Injury of Grapevines in Oregon, 1991, *Agric. Exp. Stn. Oregon Special Report 901*, 1992, 29.
63. Anonymous, *Washington agricultural statistics 1991–92*, Washington State Department of Agriculture, 1986.
64. Anonymous, *Washington agricultural statistics 1985–86*, Washington State Department of Agriculture, 1986.
65. Grace, J., Temperature as a determinant of plant productivity, in *Plants and Temperature*, S. P. Long and F. I. Woodward, Eds., The Company of Biologists, London, 1988, 91.

66. Amerine, M. A., and Winkler, A. J., Composition and quality of musts and wines of California grapes, *Hilgardia*, 15, 493, 1944.
67. McIntyre, G. N., Lider, L. A., and Ferrari, N. L., The chronological classification of grapevine phenology, *Am. J. Enol. Vitic.*, 33, 80, 1982.
68. McIntyre, G. N., Kliewer, W. M., and Lider, L. A., Some limitations of the degree day system as used in viticulture in California, *Am. J. Enol. Vitic.*, 38, 128, 1987.
69. Williams, L. E., The effect of cyanamide on budbreak and vine development of 'Thompson Seedless' grapevines in the San Joaquin Valley of California, *Vitis*, 26, 107, 1987.
70. Stoev, K., and Slavtcheva, T., La photosynthese nette chez la vigne et les facteurs ecologiques, *Con. Vigne Vin.*, 16, 171, 1982.
71. Berry, J., and Bjorkman, O., Photosynthetic response and adaptation to temperature in higher plants, *Annu. Rev. Plant Physiol.*, 31, 491, 1980.
72. Balo, B., Lajko, F., and Garab, G., Effects of chilling on photosynthesis of grapevine leaves, *Photosynthetica*, 25, 227, 1991.
73. Balo, B., Mustardy, L. A., Hideg, E., and Faludi-Daniel, A., Studies on the effect of chilling on the photosynthesis of grapevine, *Vitis*, 25, 1, 1986.
74. Sherer, V. A., Delayed fluorescence of grape leaves in relation to frost resistance, *Fiziol. I. Biokhimiya Kul'Turnykh Rastenii*, 19, 170, 1987.
75. Kramer, P. J., and Kozlowski, T. T., *Physiology of woody plants*, Academic Press, Orlando, 1979.
76. Blanke, M. M., Carbon economy of the grape inflorescence. 5. Energy requirement of the flower bud of grape, *Vitic. Enol. Sci.*, 45, 33, 1990.
77. Amthor, J. S., *Respiration and Crop Productivity*, Springer-Verlag, New York, 1989, 215.
78. Guo, X. W., Fu, W. H., and Wang, G. J., Studies on cold hardiness of grapevine roots, *Vitis*, 26, 161, 1987.
79. Carrick, D. B., Resistance of the roots of some fruit species to low temperature, *Cornell Univ. Agric. Exp. Sta. Mem.* 36, 1920.
80. Ahmedullah, M., and Kawakami, A., Evaluation of laboratory tests for determining the lethal temperature of *Vitis labruscana* Bailey Concord roots exposed to subzero temperatures, *Vitis*, 25, 142, 1986.
81. Jackson, D. I., and Cherry, N. J., Prediction of a district's grape-ripening capacity using a latitude-temperature index (LTI), *Am. J. Enol. Vitic.*, 39, 19, 1988.
82. Abass, M., and Rajashekar, C. B., Characterization of heat injury in grapes using H Nuclear Magnetic Resonance methods, *Plant Physiol.*, 96, 957, 1991.
83. Fanizza, G., and Ricciardi, L., The response of a range of genotypes of *Vitis vinifera* to sequential shoot tip cultures at high temperatures, *Euphytica*, 39, 19, 1986.
84. Buttrose, M. S., Fruitfulness in grapevines: effects of light intensity and temperature, *Bot. Gaz.*, 130, 166, 1969.
85. Smit, C. J., Flower differentiation of Sultana vines as influenced by cumulative effects of low temperature during the preceding season, *Dried Fruit*, 2, 6, 1970.
86. Pouget, R., Action de la temperature sur la differenciation des inflorescences et des fleurs durant la phase de predebourrement et de post-debourrement des bourgeons latents de la vigne, *Conn. Vigne Vin*, 15, 65, 1981.
87. Buttrose M. S., and Hale, C. R., Effect of temperature on development of the grapevine inflorescence after budburst, *Am. J. Enol. Vitic.*, 24, 14, 1973.
88. Kliewer, W. M., Effect of high temperatures during the bloom-set period on fruit-set, ovule fertility, and berry growth of several grape cultivars, *Am. J. Enol. Vitic.*, 28, 215, 1977.
89. Kobayashi, A., Yukinaga, H., Fukushima, T., and Wada, K., Studies on the thermal conditions of grapes. II. Effects of night temperatures on the growth, yield, and quality of Delaware grapes, *Bull. Res. Inst. Food Sci.*, Kyoto Univ., 24, 29, 1960.
90. Tukey, L. D., Effects of controlled temperatures following bloom on berry development of the Concord grape (*Vitis labrusca*), *Amer. Soc. Hort. Sci.*, 71, 157, 1958.
91. Roubelakis, K. A., and Kliewer, W. M., Influence of light intensity and growth regulators on fruit-set and ovule fertilization in grape cultivars under low temperature conditions, *Am. J. Enol. Vitic.*, 27, 163, 1976.
92. Ewart, A., and Kliewer, W. M., Effects of controlled day and night temperatures and nitrogen on fruit set, ovule fertility and fruit composition on several wine grape cultivars, *Am. J. Enol. Vitic.*, 28, 88, 1977.

93. Coombe, B. G., The regulation of set and development of the grape berry, *Acta Hort.*, 34, 261, 1973.
94. Kobayashi, A. H., Yukinaga, H., and Matsunaga, E., Berry growth, yield and quality of Muscat of Alexandria as affected by night temperature, *J. Jap. Soc. Hort. Sci.*, 34, 151, 1965.
95. Matsui, S., Ryugo, K., and Kliewer, W. M., Inhibition of Napa Gamay and 'Thompson Seedless' berry development by heat stress and its partial reversibility by applications of gibberellic acid and promalin, *Acta Hort.*, 179, 425, 1986.
96. Kliewer, W. M., Effect of root temperature on budbreak, shoot growth and fruit set of 'Cabernet Sauvignon' grapevines, *Am. J. Enol. Vitic.*, 26, 82, 1975.
97. Sepulveda, G., and Kliewer, W. M., Stomatal response of three grapevine cultivars (*Vitis vinifera* L.) to high temperatures, *Am. J. Enol. Vitic.*, 37, 20, 1986.
98. Kriedemann, P. E., Sward, R. J., and Downton, W. J. S., Vine response to carbon dioxide enrichment during heat therapy, *Aust. J. Plant Physiol.*, 3, 605, 1976.
99. Kliewer, W. M., Influence of environment on metabolism of organic acids and carbohydrates in *Vitis vinifera.* I. Temperature, *Plant Physiol.*, 39, 869, 1964.
100. Kliewer, W. M., Effect of day temperature and light intensity on coloration of *Vitis vinifera* L. grapes, *J. Amer. Soc. Hort. Sci.*, 95, 693, 1970.
101. Kliewer, W. M., Effect of day temperature and light intensity on concentrations of malic and tartaric acids in *Vitis vinifera* L. grapes, *J. Amer. Soc. Hort. Sci.*, 96, 372, 1971.
102. Kliewer, W. M., and Lider, L. A., Effects of day temperature and light intensity on growth and composition of *Vitis vinifera* L. fruits, *J. Amer. Soc. Hort. Sci.*, 95, 766, 1970.
103. Kliewer, W. M., and Schultz, H. B., Effect of sprinkler cooling of grapevines on fruit growth and composition, *Am. J. Enol. Vitic.*, 24, 17, 1973.
104. Kliewer, W. M., Effect of temperature on the composition of grapes grown under field and controlled conditions, *Proc. Amer. Soc. Hort. Sci.*, 93, 797, 1968.
105. Possner, D., Ruffner, H. P., and Rast, D. M., Regulation of malic acid metabolism in berries of *Vitis vinifera, Acta Hort.*, 139, 117, 1983.
106. Kliewer, W. M., and Torres, R. E., Effect of controlled day and night temperature on grape coloration, *Am. J. Enol. Vitic.*, 23, 771, 1972.
107. Fukushima, M., Iwasaki, N., Gemma, H., and Oogaki, C., Effect of night cooling at high temperature season on vine growth and berry ripening of grape 'Kyoho' (*Vitis vinifera* L. × *V. labrusca* L.), *Acta Hort.*, 279, 321, 1990.
108. Matthias, A. D., and Coates, W. E., Wine grape vine radiation balance and temperature modification with fine-mist nozzles, *HortScience*, 21, 1453, 1986.
109. Buttrose, M. S., and Hale, C. R., Effects of temperature on accumulation of starch or lipid in chloroplasts of grapevine, *Planta*, 101, 166, 1971.
110. Sepulveda, G., Kliewer, W. M., and Ryugo, K., Effect of high temperature on grapevines (*Vitis vinifera* L.). I. Translocation of 14-C-photosynthates, *Am. J. Enol. Vitic.*, 37, 13, 1986.
111. Vierling, E., The roles of heat shock proteins in plants, *Annu. Rev. Plant Physiol. Plant Mol. Biol.*, 42, 579, 1991.
112. Antcliff, A. J., and May, P., Dormancy and bud burst in Sultana vines, *Vitis*, 3, 1, 1961.
113. Nir, G., Klein, I., Lavee, S., Spieler, G., and Barak, U., Improving grapevine budbreak and yields by evaporative cooling, *J. Amer. Soc. Hort. Sci.*, 113, 512, 1988.
114. Natali, S., Guerriero, R., Di Ciano, A., and Bianchi, M. L., Three years' results of winter evaporative cooling on grapevines: effects on bud opening and crop characteristics, *Adv. Hort. Sci.*, 1, 20, 1987.
115. Nir, G., Shulman, Y., Fanberstein, L., and Lavee, S., Changes in the activity of catalase (EC 1.11.1.6) in relation to the dormancy of grapevine (*Vitis vinifera* L.) buds, *Plant Physiol.*, 81, 1140, 1986.
116. Weaver, R. J., and Iwasaki, K., Effect of temperature and length of storage, root growth and termination of bud rest in 'Zinfandel', *Am. J. Enol. Vitic.*, 28, 149, 1977.
117. Takeno, K., Koshioka, M., Pharis, R. P., Rajasekaran, K., and Mullins, M. G., Endogenous gibberellin-like substances in somatic embryos of grape (*Vitis vinifera* × *Vitis rupestris*) in relation to embryogenesis and the chilling requirement for subsequent development of mature embryos, *Plant Physiol.*, 73, 803, 1983.
118. Haeseler, C. W., and Fleming, H. K., Response of Concord grapevines to various controlled day temperatures, *The Penn. State Univ. Bull.*, 739, 1967.

119. Aljibury, F. K., Brewer, R., Christensen, P., and Kasimatis, A. N., Grape response to cooling with sprinklers, *Am. J. Enol. Vitic.*, 26, 214, 1975.
120. Nigond, J., The effect of temperature on growth and development of vines at Montpellier, *Stat. Bioclimat. Agric.*, Montpellier, 1957, 20.
121. Ahmedullah, M., Translocation, storage and utilization of ^{14}C photosynthates in grapevine (*Vitis vinifera*), in the year of application and in the following season, Diss. Abst. Inter., 39, 254, 1979.
122. Isaenko, V. V., Study of certain aspects of winter hardiness and development of a biological method of increasing it, *Soviet Plant Phys.*, 24, 919, 1977.
123. Clore, W. J., Wallace, M. A., and Fay, R. D., Bud survival of grape varieties at sub-zero temperatures in Washington, *Am. J. Enol. Vitic.*, 25, 24, 1974.
124. He, P., and Niu, L., Study of cold hardiness in the wild *Vitis* native to China, *Acta Hort. Sin.*, 16, 81, 1989.
125. Wolf, T. K., and Cook, M. K., Comparison of 'Cabernet Sauvignon' and 'Cabernet franc' grapevine dormant bud cold hardiness, *Fruit Var. J.*, 45, 17, 1991.
126. Proebsting, E. L., Ahmedullah, M., and Brummund, V. P., Seasonal changes in low temperature resistance of grape buds, *Am. J. Enol. Vitic.*, 31, 329, 1980.
127. Damborska, M., The effect of higher winter temperatures on changes of the frost resistance of grapevine buds, *Vitis*, 17, 341, 1978.
128. Wolpert, J. A., and Howell, G. S., Cold acclimation of Concord grapevines. III. Relationship between cold hardiness, tissue water content, and shoot maturation. *Vitis*, 25, 151, 1986.
129. Winkler, A. J., The treatment of frosted grapevines, *Proc. Amer. Soc. Hort. Sci.*, 30, 253, 1933.
130. Lider, J. V., Some responses of grapevines to treatment for frost in Napa Valley, *Am. J. Enol. Vitic.*, 16, 231, 1965.
131. Burr, T. J., Ophel, K., Katz, B. H., and Kerr, A., Effect of hot water treatment on systemic *Agrobacterium tumefaciens* Biovar 3 in dormant grape cuttings, *Plant Dis.*, 73, 242, 1989.
132. Ophel, K., Nicholas, P. R., Magarey, P. A., and Bass, A. W., Hot water treatment of dormant grape cuttings reduces crown gall incidence in a field nursery, *Am. J. Enol. Vitic.*, 41, 325, 1990.
133. Wample, R. L., Bary, A., and Burr, T. J., Heat tolerance of dormant *Vitis vinifera* cuttings, *Am. J. Enol. Vitic.*, 42, 67, 1991.
134. Wample, R. L., and Bary, A., Harvest date as a factor in carbohydrate storage and cold hardiness of 'Cabernet Sauvignon' grapevines, *J. Amer. Soc. Hort. Sci.*, 117, 32, 1992.
135. Lider, J. V., Some responses of grapevines to treatment for frost in Napa Valley, *Am. J. Enol. Vitic.*, 16, 231, 1965.
136. Pratt, C., and Pool, R. M., Anatomy of recovery of canes of *Vitis vinifera* L. from simulated freezing injury, *Am. J. Enol. Vitic.*, 32, 223, 1981.
137. Damborska, M., Adaptation of grapevine buds to winter frost. Effect of temperature near 0°C at 10- to 30-day intervals, *Rostlinna Vuroba*, 25, 77, 1979.
138. Kondo, I. N., The frost resistance of grapevine under the conditions of central Asia, *Trudy Vses Nauchno-Issled. Vinodel. Vinogradar. Physiol.* (Magarach), 1, 1, 1960.
139. Pogosyan, K. S., Some peculiarities of the hardening of grapevine shoots, *Agrobilogia*, 5, 688, 1960.
140. Reuther, G., Physiological criteria of resistance to climatic stress as variety-specific characteristics, *Angewandte Botanik*, 49, 75, 1975.
141. Pierquet, P., and Stushnoff, C., Relationship of low temperature exotherms to cold injury in *Vitis riparia* Michx., *Am. J. Enol. Vitic.*, 31, 1, 1980.
142. Paroschy, J. H., Meiering, A. G., Peterson, R. L., Hostetter, G., and Neff, A., Mechanical winter injury in grapevine trunks, *Am. J. Enol. Vitic.*, 31, 227, 1980.
143. Parsons, L. R., Water relations, stomatal behavior and root conductivity of red osier dogwood during acclimation to freezing temperatures, *Plant Physiol.*, 62, 64, 1978.
144. Rantz, J., Ed., Proceedings International Symposium on Nitrogen in Grapes and Wine, *Am. Soc. Enol. Vitic.*, Davis, 1991, 323.
145. Wample, R. L., Spayd, S. E., Evans, R. G., and Stevens, R. G., Nitrogen fertilization and factors influencing grapevine cold hardiness, in Proceedings International Symposium on Nitrogen in Grapes and Wine, Rantz, J. M., Ed., *Am. Soc. Enol. Vitic.*, Davis, 1991, 120.
146. Kirillou, A. F., Levitt, T. K., Grozova, V. M., and Koz'mik, R. A., Soluble proteins in grapevine shoots in relation to winter hardiness, *Referativnyi Zhurnal*, 20, 1979.

147. Kikvidze, M. V., Chanishvili, S. H., and Gvamichava, N. E., Effect of cold hardening on content of protein and forms of nitrogen in the shoots of grapevine, *Bull. Acad. Sci. Georgian SSR*, 106, 377, 1982.
148. Chanishvili, S. H., Georgobiani, E. L., Kikvidze, M. V., Datukishvili, N. M., Doledze, M. D., and Purtseladze, T. D., Change in the content of forms of phosphorus and nitrogen compounds in varieties of grape plants differing in resistance, *Fiziol. Morozoustoich, Vinograd. Lozy*, 98, 1987.
149. Mikeladze, E. G., Abramidze, S. P., and Razmadze, N. G., Change in nitrogenous substances in grapevine grafts in relations to frost resistance, *Fiziol. Morozoustoich. Vinograd. Lozy*, 124, 1986.
150. Wolf, T. K., and Pool, R. M., Nitrogen fertilization and rootstock effects on wood maturation and dormant bud cold hardiness of cv. 'Chardonnay' grapevines, *Am. J. Enol. Vitic.*, 39, 308, 1988.
151. Wample, R. L., Spayd, S. R., Evans, R. G., and Stevens, R. G., Nitrogen fertilization of 'White Riesling' grapes in Washington: Nitrogen and seasonal effects on bud cold hardiness and carbohydrate reserves, *Am. J. Enol. Vitic.*, 44, 159, 1993.
152. Grozova, V. M., Carbohydrate metabolism of introduced grapevine cultivars during the period of winter hardening, *Referativnyi Zhurnal*, 21, 1978.
153. Pickett, T. A., and Cowart, F. F., Carbohydrate changes in Muscadine grape shoots during the growing season, *Am. Soc. Hort. Sci.*, 38, 393, 1941.
154. Richey, H. W., and Bowers, H. A., Correlation of root and top growth of the Concord grape and translocation of elaborated plant food during the dormant season, *Am. Soc. Hort. Sci.*, 21, 33, 1924.
155. Winkler, A. J., and Williams, W. O., Starch and sugars of *Vitis vinifera, Plant Physiol.*, 20, 412, 1945.
156. Wample, R. L., and Bary, A., Harvest date as a factor in carbohydrate storage and cold hardiness of 'Cabernet Sauvignon' grapevines, *J. Amer. Soc. Hort. Sci.*, 117, 32, 1992.
157. Andrews, P. K., Sandidge, C. R., and Toyama, T. K., Deep supercooling of dormant and deacclimating *Vitis* buds, *Am. J. Enol. Vitic.*, 35, 175, 1984.
158. Levitt, J., *Responses of Plants to Environmental Stresses*, Vol. 1, 2nd ed., Academic Press, New York, 1980.
159. Pierquet, P., Stushnoff, C., and Burke, M. J., Low temperature exotherms in stem and bud tissues of *Vitis riparia* Michx., *J. Amer. Soc. Hort. Sci.*, 102, 54, 1977.
160. Schnabel, B. J., and Wample, R. L., Dormancy and cold hardiness in *Vitis vinifera* L. cv. 'white riesling' as influenced by photoperiod and temperature, *Am. J. Enol. Vitic.*, 38, 265, 1987.
161. Fennell, A., and Hoover, E., Photoperiod influences growth, bud dormancy, and cold acclimation in *Vitis labruscana* and *V. riparia, J. Am. Soc. Hort. Sci.*, 116, 270, 1991.
162. Wolpert, J. A., and Howell, G. S., Effect of night interruption on cold acclimation of potted 'Concord' grapevines, *J. Am. Soc. Hortic. Sci.*, 111, 16, 1986.
163. Nagarajah, S., Physiological responses of grapevines to water stress, *Acta. Hort.*, 240, 249, 1989.
164. Smart, R. E., and Coombe, B. G., Water relations of grapevines, in *Water Deficits and Plant Growth*, Vol. 7, Kozlowski, T. T., Ed., Academic Press, New York, 1983, 137.
165. Terashima, I., Wong, S. C., Osmond, C. B., and Farquhar, G. D., Characterization of non-uniform photosynthesis induced by abscisic acid in leaves having different mesophyll anatomies, *Plant Cell Physiol.*, 29, 385, 1988.
166. Downton, W. J. S., Loveys, B. R., and Grant, W. J. R., Stomatal closure fully accounts for the inhibition of photosynthesis by abscisic acid, *New Phytol.*, 108, 263, 1988.
167. Düring, H., Low air humidity causes non-uniform stomatal closure in heterobaric leaves of *Vitis* species, *Vitis*, 31, 1, 1992.
168. Gunasekera, D., and Berkowitz, G. A., Heterogenous stomatal closure in response to leaf water deficits is not a universal phenomenon, *Plant Physiol.*, 98, 660, 1992.
169. Schulze, E.-D., and Hall, A. E., Stomatal responses, water loss and CO_2 assimilation rates of plants in contrasting environments, in *Encyclopedia of Plant Physiology, New Series*, Vol. 12B; Physiological Plant Ecology II Water Relations and Carbon Assimilation, Lange, O. L., Nobel, P. S., Osmond, C. B., and Ziegler, H., Eds., Springer-Verlag, Berlin, 1982, 181.
170. Düring, H., Stomatal responses to alterations of soil and air humidity in grapevines, *Vitis*, 26, 9, 1987.
171. Düring, H., Studies on the environmentally controlled stomatal transpiration in grape vines. I. Effects of light intensity and air humidity, *Vitis*, 15, 82, 1976.
172. Düring, H., Effects of air and soil humidity on vegetative growth and water relationships of grapevines, *Vitis*, 18, 211, 1979.

173. Phene, C. J., Hoffman, G. J., Howell, T. A., Clark, D. A., Mead, R. M., Johnson, R. S., and Williams, L. E., Automated lysimeter for irrigation and drainage control, in *Lysimeters for Evapotranspiration and Environmental Measurements*, IR DIV/ASCE, St. Joseph, 1991, 28.
174. Grimes, D. W., Yamada, H., and Hughes, S. W., Climate-normalized cotton leaf water potentials for irrigation scheduling, *Agric. Water Man.*, 12, 293, 1987.
175. Düring, H., and Scienza, A., Studies on drought resistance of *Vitis* species and cultivars, in *Proceedings 3rd Intern. Symp. Grape Breeding*, Univ. California, Davis, 1980, 179.
176. Carbonneau, A., The early selection of grapevine rootstocks for resistance to drought conditions, *Am. J. Enol. Vitic.*, 36, 195, 1985.
177. Pongracz, D. P., *Rootstocks for Grape-Vines*, David Philip, Cape Province, 1983, 150.
178. Galet, P., *A Practical Ampelography* (transl. L. T. Morton), Cornell University Press, Ithaca, 1979, 248.
179. van Zyl, J. L., Diurnal variation in grapevine water stress as a function of changing soil water status and meteorological conditions, *S. Afr. J. Enol. Vitic.*, 8, 45, 1987.
180. Kliewer, W. M., Freeman, B. M., and Hossom, C., Effect of irrigation, crop level and potassium fertilization on Carignane vines. I. Degree of water stress and effect on growth and yield, *Am. J. Enol. Vitic.*, 34, 186, 1983.
181. Grimes, D. W., and Williams, L. E., Irrigation effects on plant water relations and productivity of 'Thompson Seedless' grapevines, *Crop Sci.*, 30, 255, 1990.
182. Smart, R. E., Aspects of water relations of the grapevine (*Vitis vinifera*), *Am. J. Enol. Vitic.*, 25, 84, 1974.
183. Düring, H., and Loveys, B. R., Diurnal changes in water relations and abscisic acid in field grown *Vitis vinifera* cvs. I. Leaf water potential components and leaf conductance under humid temperate and semiarid conditions, *Vitis*, 21, 223, 1982.
184. Düring, H., Evidence for osmotic adjustment to drought in grapevines (*Vitis vinifera* L.), *Vitis*, 23, 1, 1984.
185. Düring, H., Osmotic adjustment in grapevines, *Acta Hort.*, 171, 315, 1985.
186. Davies, W. J., and Zhang, J., Root signals and the regulation of growth and development of plants in drying soil, in *Annual Review of Plant Physiology and Plant Molecular Biology*, Vol. 42, Briggs, W. R., Jones, R. L., and Walbot, V., Eds., Annual Review, Palo Alto, 1991, 55.
187. Trejo, C. L., and Davies, W. J., Drought-induced closure of *Phaseolus vulgaris* L. stomata precedes leaf water deficit and any increase in xylem ABA concentration, *J. Exp. Bot.*, 42, 1507, 1991.
188. Loveys, B. R., and Düring, H., Diurnal changes in water relations and abscisic acid in field-grown *Vitis vinifera* cultivars. II. Abscisic acid changes under semi-arid conditions, *New Phytol.*, 97, 37, 1984.
189. Loveys, B. R., Diurnal changes in water relations and abscisic acid in field-grown *Vitis vinifera* cultivars. III. The influence of xylem-derived abscisic acid on leaf gas exchange, *New Phytol.*, 98, 563, 1984.
190. Loveys, B. R., Abscisic acid transport and metabolism in grapevine (*Vitis vinifera* L.), *New Phytol.*, 98, 575, 1984.
191. Loveys, B. R., and Kriedemann, P. E., Rapid changes in abscisic acid-like inhibitors following alterations in vine leaf water potential, *Physiol. Plant*, 28, 476, 1973.
192. Loveys, B. R., and Kriedemann, P. E., Internal control of stomatal physiology and photosynthesis. I. Stomatal regulation and associated changes in endogenous levels of abscisic and phaseic acids, *Aust. J. Plant Physiol.*, 1, 407, 1974.
193. Downton, W. J. S., Loveys, B. R., and Grant, W. J. R., Non-uniform stomatal closure induced by water stress causes putative non-stomatal inhibition of photosynthesis, *New Phytol.*, 110, 503, 1988.
194. Passioura, J. B., Root signals control leaf expansion in wheat seedlings growing in drying soil, *Aust. J. Plant Physiol.*, 15, 687, 1988.
195. Gollan, T., Passioura, J. B., and Munns, R., Soil water status affects the stomatal conductance of fully turgid wheat and sunflower leaves, *Aust. J. Plant Physiol.*, 13, 459, 1986.
196. Araujo, F. J., The response of three year-old 'Thompson Seedless' grapevines to drip and furrow irrigation in the San Joaquin Valley, M. S. Thesis, University of California, Davis, 1988.
197. van Zyl, J. L., Canopy temperature as a water stress indicator in vines, *S. Afr. J. Enol. Vitic.*, 7, 53, 1986.
198. Freeman, B. M., Lee, T. H., and Turkington, C. R., Interaction of irrigation and pruning level on grape and wine quality of Shiraz vines, *Am. J. Enol. Vitic.*, 31, 124, 1980.
199. Jackson, R. D., Kustas, W. P., and Choudhury, B. J., A reexamination of the crop water stress index, *Irrig. Sci.*, 9, 309, 1988.

200. Wanjura, D. F., Hatfield, J. L., and Upchurch, D. R., Crop water stress index relationships with crop productivity, *Irrig. Sci.*, 11, 93, 1990.
201. Idso, S. B., Pinter, Jr., P. J., and Reginato, R. J., Non-water-stressed baselines: the importance of site selection for air temperature and air vapour pressure deficit measurements, *Agric. For. Meteor.*, 53, 73, 1990.
202. Downton, W. J. S., Grant, W. J. R., and Loveys, B. R., Diurnal changes in the photosynthesis of field-grown grapevines, *New Phytol.*, 105, 71, 1987.
203. Eibach, R., and Alleweldt, G., Influence of water supply on growth, gas exchange and substance production of fruit-bearing grapevines. III. Substance production, *Vitis*, 24, 183, 1985.
204. van Zyl, J. L., Response of Colombar grapevines to irrigation as regards quality aspects and growth, *S. Afr. J. Enol. Vitic.*, 5, 19, 1984.
205. Williams, L. E., and Grimes, D. W., Modelling vine growth-development of a data set for a water balance subroutine, in *Proceedings of the Sixth Australian Wine Industry Technical Conference*, (14–17 July, 1986, Adelaide), T. Lee, Ed., Australian Industrial Publishers, Adelaide, Australia, 1987, 169.
206. Williams, L. E., Neja, R. A., Meyer, J. L., Yates, L. A., and Walker, E. L., Postharvest irrigation influences budbreak of 'Perlette' grapevines, *Hort. Sci.*, 26, 1081, 1991.
207. Matthews, M. A., Anderson, M. M., and Schultz, H. R., Phenologic and growth responses to early and late season water deficits in 'Cabernet Franc', *Vitis*, 26, 147, 1987.
208. Schultz, H. R., and Matthews, M. A., Vegetative growth distribution during water deficits in *Vitis vinifera* L., *Aust. J. Plant Physiol.*, 15, 641, 1988.
209. Neja, R. A., Wildman, R. S., Ayers, R. S., and Kasimatis, A. N., Grapevine response to irrigation and trellis treatments in the Salinas Valley, *Am. J. Enol. Vitic.*, 28, 16, 1977.
210. Christensen, P., Response of 'Thompson Seedless' grapevines to the timing of preharvest irrigation cut-off, *Am. J. Enol. Vitic.*, 26, 188, 1975.
211. Kasimatis, A. N., Grapes and berries, Part I. Grapes. Irrigation of agricultural lands, In R. M. Hagan et al. (Eds.), *Agronomy*, 11, 719, 1967.
212. Saab, I. N., and Sharp, R. E., Non-hydraulic signals from maize roots in drying soil: Inhibition of leaf elongation but not stomatal conductance, *Planta*, 179, 466, 1989.
213. Smart, R. E., Turkington, C. R., and Evans, J. C., Grapevine response to furrow and trickle irrigation, *Am. J. Enol. Vitic.*, 25, 62, 1974.
214. Howell, T. A., Relationships between crop production and transpiration, evapotranspiration, and irrigation, *Irrigation of Agricultural Crops*, Agronomy Monograph No. 30, Stewart, B. A., and Nielsen, D. R., Eds., ASA-CSSA-SSSA, Madison, 1990, 391.
215. Jones, H. G., *Plants and Microclimate*, Cambridge University Press, Cambridge, 1992, 428.
216. Bravdo, B., Lavee, S., and Samish, R. M., Analysis of water consumption of various grapevine cultivars, *Vitis*, 10, 279, 1972.
217. van Zyl, J. L., and van Huyssteen, L., Comparative studies on wine grapes on different trellising systems: I. consumptive water use, *S. Afr. J. Enol. Vitic.*, 1, 7, 1980.
218. Ruhl, E., and Alleweldt, G., Investigations into the influence of time of irrigation on yield and quality of grapevines, *Acta Hort.*, 171, 457, 1985.
219. Buttrose, M. S., Fruitfulness in grapevines: effect of water stress, *Vitis*, 12, 299, 1974.
220. Freeman, B. M., Lee, T. H., and Turkington, C. R., Interaction of irrigation and pruning level on growth and yield of Shiraz vines, *Am. J. Enol. Vitic.*, 30, 218, 1979.
221. Hardie, W. J., and Considine, J. A., Response of grapes to water-deficit stress in particular stages of development, *Am. J. Enol. Vitic.*, 27, 55, 1976.
222. Matthews, M. A., and Anderson, M. M., Reproductive development in grape (*Vitis vinifera* L.): Response to seasonal water deficits, *Am. J. Enol. Vitic.*, 40, 52, 1989.
223. van Rooyen, F. C., Weber, H. W., and Levin, I., The response of grapes to a manipulation of the soil-plant-atmosphere continuum. II. Plant-water relationships, *Agrochemophysica*, 12, 69, 1980.
224. Vaadia, Y., and Kasimatis, A. N., Vineyard irrigation trials, *Am. J. Enol. Vitic.*, 12, 88, 1961.
225. Sachs, R. M., and Weaver, R. J., Gibberellin and Auxin-induced berry enlargement in *Vitis vinifera* L., *J. Hort. Sci.*, 43, 185, 1968.
226. Hepner, Y., Bravdo, B., Loinger, C., Cohen, S., and Tabacman, H., Effect of drip irrigation schedules on growth, yield, must composition and wine quality of 'Cabernet Sauvignon', *Am. J. Enol. Vitic.*, 36, 77, 1985.

227. Bravdo, B., Hepner, Y., Loinger, C., Cohen, S., and Tabacman, H., Effect of irrigation and crop level on growth, yield and wine quality of 'Cabernet Sauvignon', *Am. J. Enol. Vitic.*, 36, 132, 1985.
228. Matthews, M. A., and Anderson, M. M., Fruit ripening in *Vitis vinifera* L.: Responses to seasonal water deficits, *Am. J. Enol. Vitic.*, 39, 313, 1988.
229. Matthews, M. A., Ishii, R., Anderson, M. M., and O'Mahony, M., Dependence of wine sensory attributes on vine water status, *J. Sci. Food Agric.*, 51, 321, 1990.
230. Northcote, K. H., Soils and Australian viticulture, in *Viticulture*, Vol. 1, *Resources in Australia*, Coombe, B. G. and Dry, P. R., Eds., Australian Industrial Publishers Pty., Underdale, 1988, 61.
231. Hardie, W. J., and Cirami, R. M., Grapevine rootstocks, *Viticulture*, Vol. 1, *Resources in Australia*, Coombe, B. G., and Dry, P. R., Eds., Australian Industrial Publishers Pty., Underdale, South Australia, 1988, 154.
232. Winkler, A., Cook, J. A., Kliewer, W. M., and Lider, L. A., *General Viticulture*, University of California Press, Berkeley, 1974, 710.
233. Hale, C. R., Response of grapevines to prolonged flooding of the soil, M.S. Thesis, University of California, Davis, 1959.
234. Freeman, B. M., and Smart, R. E., A root observation laboratory for studies with grapevines, *Am. J. Enol. Vitic.*, 27, 36, 1976.
235. Webber, R. T. J., and Jones, L. D., Drainage and soil salinity, in *Viticulture*, Vol. 2, *Practices*, Coombe, B. G., and Dry, P. R., Eds., Winetitles, Underdale, 1992, 129.
236. Maas, E. V., Salt tolerance of plants, in *Handbook of Plant Science in Agriculture*, Vol. II, B. R. Christie, Ed., CRC Press, Boca Raton, FL, 1987, 57.
237. Downton, W. J. S., and Loveys, B. R., Abscisic acid content and osmotic relations of salt-stressed grapevine leaves, *Aust. J. Plant Physiol.*, 8, 443, 1981.
238. Francois, L. E., and Clark, R. A., Accumulation of sodium and chloride in leaves of sprinkler-irrigated grapes, *J. Amer. Soc. Hort. Sci.*, 104, 11, 1979.
239. Maas, E. V., and Hoffman, G. J., Crop salt tolerance-current assessment, *Journal of the Irrigation and Drainage Division*, ASCE, 103(IR2), Proc. Paper 12993, 115, 1977.
240. Ehlig, C. F., Effects of salinity on four varieties of table grapes grown in sand culture, *Proc. Am. Soc. Hort. Sci.*, 76, 323, 1960.
241. Prior, L. D., Grieve, A. M., Slavich, P. G., and Cullis, B. R., Sodium chloride and soil texture interactions in irrigated field grown Sultana grapevines. II. Plant mineral content, growth and physiology, *Aust. J. Agric. Res.*, 43, 1067, 1992.
242. Downton, W. J. S., Salinity effects on the ion composition of fruiting 'Cabernet Sauvignon' vines, *Am. J. Enol. Vitic.*, 28, 210, 1977.
243. Antcliff, A. J., Newman, H. P., and Barrett, H. C., Variation in chloride accumulation in some American species of grapevine, *Vitis*, 22, 357, 1983.
244. Sykes, S. R., and Newman, H. P., The genetic basis for salt exclusion in grapevines, *Aust. Grapegrower and Winemaker*, Apr., 75, 1987.
245. Sykes, S. R., Variation in chloride accumulation by hybrid vines from crosses involving the cultivars Ramsey, Villard blanc, and Sultana, *Am. J. Enol. Vitic.*, 36, 30, 1985.
246. Newman, H. P., and Antcliff, A. J., Chloride accumulation in some hybrids and backcrosses of *Vitis berlandieri* and *Vitis vinifera, Vitis*, 23, 106, 1984.
247. Sauer, M. R., Effects of vine rootstocks on chloride concentration in Sultana scions, *Vitis*, 7, 223, 1968.
248. Bernstein, L., Ehlig, C. F., and Clark, R. A., Effect of grape rootstocks on chloride accumulation in leaves, *J. Amer. Soc. Hort. Sci.*, 94, 584, 1969.
249. Downton, W. J. S., Photosynthesis in salt-stressed grapevines, *Aust. J. Plant Physiol.*, 4, 183, 1977.
250. Downton, W. J. S., Loveys, B. R., and Grant, W. J. R., Salinity effects on the stomatal behavior of grapevine, *New Phytol.*, 116, 499, 1990.
251. Downton, W. J. S., and Loveys, B. R., Abscisic acid content and osmotic relations of salt-stressed grapevine leaves, *Aust. J. Plant Physiol.*, 8, 443, 1981.
252. Downton, W. J. S., and Millhouse, J., Turgor maintenance during salt stress prevents loss of variable fluorescence in grapevine leaves, *Plant Sci. Lett.*, 31, 1, 1983.
253. Prior, L. D., Grieve, A. M., Slavich, P. G., and Cullis, B. R., Sodium chloride and soil texture interactions in irrigated field grown Sultana grapevines, III. Soil and root system effects, *Aust. J. Agric. Res.*, 43, 1085, 1992.

254. Hawker, J. S., and Walker, R. R., The effect of sodium chloride on the growth and fruiting of 'Cabernet Sauvignon' vines, *Am. J. Enol. Vitic.*, 29, 172, 1978.
255. Prior, L. D., Grieve, A. M., and Cullis, B. R., Sodium chloride and soil texture interactions in irrigated field grown Sultana grapevines. I. Yield and fruit quality, *Aust. J. Agric. Res.*, 43, 1051, 1992.
256. Stevens, R. M., and Harvey, G. C., Grapevine responses to transient soil salinization, in Proc. Symp. Management of Soil Salinity in South East Australia (18-20 September, 1989, Albury, Australia), Humphreys, E., Muirhead, W. A., and van der Lelij, A., Eds., *Aust. Soc. Soil Sci.*, 1990.
257. Downton, W. J. S., Growth and mineral composition of the Sultana grapevine as influenced by salinity and rootstock, *Aust. J. Agric. Res.*, 36, 425, 1988.
258. Stevens, R., Coombe, B., and Aspinall, D., Salinity increases juice total acids, sodium, potassium, and pH, *Proc. 4th Inter. Symp. on Grapevine Physiology*, 1992.
259. Grace, J., *Plant Response to Wind*, Academic Press, London, 1977, Chapter 2.
260. Weiss, A., and Allen Jr., L. H., Vertical and horizontal air flow above rows of a vineyard, *Agric. Meteor.*, 17, 433, 1976.
261. Hicks, B. B., Eddy fluxes over a vineyard, *Agric. Meteor.*, 12, 203, 1973.
262. Weiss, A., and Allen Jr., L. H., Air-flow patterns in vineyard rows, *Agric. Meteor.*, 16, 329, 1976.
263. Freeman, B. M., Kliewer, W. M., and Stern, P., Influence of windbreaks and climatic region on diurnal fluctuation of leaf water potential stomatal conductance, and leaf temperature of grapevines, *Am. J. Enol. Vitic.*, 33, 233, 1982.
264. Kobringer, J. M., Kliewer, M. W., and Lagier, S. T., Effects of wind on stomatal behaviour of several grapevine cultivars, *Am. J. Enol. Vitic.*, 35, 164, 1984.
265. Gates Jr., D. S., Influence of wind on growth, productivity, and stomatal response of 'Chenin blanc' grapevines, M.S. Thesis, University of California, Davis, 1984.
266. Hiron, R. W. P., and Wright, S. T. C., The role of endogenous abscisic acid in the response of plants to stress, *J. Exp. Bot.*, 24, 769, 1973.
267. Takahashi, K., Kuranaka, M., Myogavra, A., and Taheshita, O., The effects of wind on grapevine growth, windbreaks for vineyards, *Bull. Shimane Agri. Exp. Station*, 704, 39, 1976.
268. Hamilton, R. P., Wind effects on grapevines, in *Proceedings of the second international symposium for cool climate viticulture and oenology*, Smart, R. E., Thornton, R. J., Rodriques, S. B., and Young, J. E., Eds., NZ Soc. Vitic. and Oen., Auckland, 1988, 65.
269. Dry, P. R., Reed, S., and Potter, G., The effect of wind on the performance of 'Cabernet franc' grapevines, *Acta Hort.*, 240, 143, 1989.
270. Telewski, F. W., and Jaffe, M. J., Thigmomorphogenesis: The role of ethylene in the response of *Pinus taeda* and *Abies fraseri* to mechanical perturbation, *Physiol. Plant*, 66, 227, 1986.
271. Simon, J. C., Etude des influences agronomiques des brisevent dans les perimetres irriges de CentreQuest de l'Argentine. 1. Effects des brisevent sur la croissance et le developpement d'une culture type: la vigne, *Ann. Agron.*, 28, 75, 1977.
272. Weinstein, L. H., Effects of air pollution on grapevines, *Vitis*, 23, 274, 1984.
273. Viessman, S. M., Knudson, D. A., and Streets, D. C., *The potential effects of sulfur pollutants on grape production in New York State*, ANL/EES-TM-213, Argonne National Laboratory, Argonne, IL, 1982.
274. Greenhalgh, W. J., Brown, G. S., and McFayden, L., The effect of airborne fluorides on the growth and cropping of grapes—A literature review. Unpublished manuscript, Dept. of Agronomy and Horticultural Sciences, Univ. of Sydney, Sydney, 1980.
275. Richards, B. L., Middleton, J. T., and Hewitt, W. B., Air pollution with relation to agronomic crops. V. Oxidant stipple of grape, *Agron. J.*, 50, 559, 1958.
276. Tingey, D. T., and Taylor Jr., G. E., Variation in plant response to ozone: A conceptual model of physiological events, in *Effects of gaseous air pollution in agriculture and horticulture*, Unsworth, M. H., and Ormrod, D. P., Eds., Butterworth Scientific, London, 1982, 113.
277. Shertz, F. D., Kender, W. J., and Musselman, R. C., Effects of ozone and sulfur dioxide on grapevines, *Sci. Hortic.*, (Amsterdam), 13, 37, 1980.
278. Martin, B., Bytnerowicz, A., and Thorstenson, Y. R., Effects of air pollutants on the composition of stable carbon isotopes, $\delta^{13}C$, of leaves and wood, and on leaf injury, *Plant Physiol.*, 88, 218, 1988.
279. Greitner, C. S., and Winner, W. E., Increases in $\delta^{13}C$ values of radish and soybean plants caused by ozone, *New Phytol.*, 108, 489, 1988.

280. Kondo, N., and Sugahara, K., Changes in transpiration rate of SO_2-resistant and -sensitive plants with ozone. SO_2 fumigation and the participation of abscisic acid, *Plant Cell Physiol.*, 19, 365, 1978.
281. Olszyk, D. M., and Tinsey, D. T., Joint action of O_3 and SO_2 in modifying plant gas exchange, *Plant Physiol.*, 82, 401, 1986.
282. Omasa, K., Hashimoto, Y., and Aiga, I., Image instrumentation of plants exposed to air pollutants. 2. Relationships between SO_2 or NO_2 sorption and their acute effects on plant leaves, *Res. Rep. Natl. Inst. Environ. Stud.*, Jan., 66, 81, 1984.
283. Thompson, C. R., Hensel, E., and Kats, G., Effects of photochemical air pollutants on 'Zinfandel' grapes, *Hort. Sci.*, 4, 222, 1969.
284. Thompson, C. R., and Kats, G., Antioxidants reduce grape yield reductions from photochemical smog, *Calif. Ag.*, 24, 12, 1970.
285. Brewer, R. F., and Ashcroft, R., The effects of ambient oxidants on 'Thompson Seedless' grapes, *Final Report on ARB Contract AL-132-33*, "The effects of present and potential air pollution on important San Joaquin Valley Crops: Grapes," 1983.
286. Musselman, R. C., Kender, W. J., and Crowe, D. E., Determining air pollutant effects on the growth and productivity of 'Concord' grapevines using open top chambers, *J. Amer. Soc. Hort. Sci.*, 103, 645, 1978.
287. Olszyk, D. M., Takemoto, B. K., Kats, G., Dawson, P. J., Morrison, C. L., Preston, J. W., and Thompson, C. R., Effects of open-top chambers on 'Valencia' orange trees, *J. Environ. Qual.*, 21, 128, 1992.
288. Retzlaff, W. A., Williams, L. E., and DeJong, T. M., Photosynthesis, growth, and yield response of 'Casselman' plum to various ozone partial pressures during orchard establishment, *J. Amer. Soc. Hort. Sci.*, 117, 703, 1992.
289. Musselman, R. C., Air pollution injury to grapevines, *Proc. Annu. Meet. N.Y. State Hortic. Soc.*, 125, 129, 1980.
290. Crisosto, C. H., Retzlaff, W. A., Williams, L. E., DeJong, T. M., and Zoffoli, J. P., Postharvest performance evaluation of plum (*Prunus salicina* Lindel., 'Casselman') fruit grown under three ozone partial pressures, *J. Amer. Soc. Hort. Sci.*, 118, 497, 1993.
291. Abass, M., and Rajashekar, C. B., Abscisic acid accumulation in leaves and cultured cells during heat acclimation in grapes, *HortScience*, 28, 50, 1993.
292. Seeley, S., Hormonal transduction of environmental stresses. *HortScience*, 25, 1369, 1990.

Chapter 5

Kiwifruit

Garth S. Smith and James G. Buwalda

CONTENTS

I. INTRODUCTION

The kiwifruit of commerce [*Actinidia deliciosa* (A. Chev.) C. F. Liang et A. R. Ferguson var. *deliciosa*] is a warm-temperate deciduous vine that has been subjected to little genetic selection and is essentially indistinguishable from the plant which grows wild in its natural habitat in China.[1] Only a small part of the gene pool of this species has been exploited. The majority of commercial plantings throughout the world are of the one cultivar, 'Hayward'.[2] This cultivar was selected in New Zealand just over 50 years ago from an extraordinarily small group of about 40 seedlings, which themselves can be traced back to a small quantity of seed introduced from China in 1904.[3] The first commercial planting of 'Hayward' was in New Zealand in the late 1930s. However, commercial success of the fruit did not

0-8493-175-0/94/$0.00+$.50

Table 1 **World production of kiwifruit from 1985 to 1991**

	Proportion of total production (%)	
Country	**1985**	**1991**
New Zealand	57.5	40.4
Italy	16.7	31.5
USA	8.7	3.6
Japan	7.4	6.0
France	7.1	6.8
Greece	1.2	1.4
Australia	0.9	1.0
Chile	0.5	6.0
Spain	0.0	0.6
Portugal	0.0	0.3
Others	0.0	2.4
Total production	194.6	665.7

Note: Data given in thousands of tonnes.
From New Zealand Kiwifruit Marketing Board, 1992.

take place internationally until much later in the late 1960s. Although kiwifruit vines were introduced from New Zealand into many countries in the 1970s, it was not until after 1980 that the area planted in any single country other than New Zealand exceeded 1000 ha.[4] The dramatic increase in the quantity of kiwifruit being produced worldwide has seen the total annual trade in fruit increase from approximately 195,000 t in 1985 to the current level of production of 666,000 t (Table 1). New Zealand and Italy are now the main kiwifruit producing countries.

The rapid increase in plantings of kiwifruit in many parts of the world has resulted in vines being grown under a diverse range of soils and climatic conditions. In their natural habitat in southern and central China, kiwifruit vines are confined to the sides of forest-covered hills and mountains where they grow in well-illuminated environments at forest margins.[1] Annual rainfall in these areas is evenly distributed and ranges from 1045 to 1950 mm, maintaining a relatively high humidity of 75–85%.[5] The local soils are very acidic (pH < 5) with high contents of organic matter (6–18%) and moderate reserves of nitrogen and potassium.[5] The climate is continental in nature with winter temperatures often falling well below 0°C. There are long frost-free periods during the growing season of 210–280 d, while summer temperatures occasionally reach 40°C.[6] By contrast, conditions under which kiwifruit vines are now grown commercially range from the cool humid environment of New Zealand where rainfall during the growing season often exceeds 800 mm and the average temperature rarely exceeds 17°C,[7] to the hot dry environment of the Central Valley of California where the average rainfall is often less than 300 mm and the average temperature usually exceeds 23°C.[8] Associated with the increased diversity of these growing areas has been a range of physiological disorders, of which many can be linked to environmental stresses.[9]

As the history of research into the physiology of the kiwifruit vine is very recent compared to that for most other fruit crops, little attempt has been made to develop an integrated understanding of the functioning of the entire vine. Hence, solutions to many disorders affecting the vine have been considered in isolation. In this chapter, growth and development of the kiwifruit vine are described to provide a basis for examining the physiological responses of the vine to environmental stress and opportunities for minimizing the effects of these stresses on fruit production and quality. Unless stated otherwise, the growth patterns described in this chapter are typical of vines grown under New Zealand conditions. While much of the research reviewed in this chapter was carried out in New Zealand, reflecting the influence this country has had on the culture of the kiwifruit vine, the conclusions drawn have direct relevance for kiwifruit culture elsewhere.

Table 2 **Typical distribution of leaf area and fruit number in "replacement cane" and "fruiting" zones of the canopy of kiwifruit vines (planted at 400 vines ha^{-1}) in New Zealand**

	Leaf area (m^2 vine^{-1})	Fruit number (vine^{-1})
Replacement cane zone	64.5	338
Fruiting zone	25.7	789
Total	90.2	1127

From Buwalda, J. G. and Smith, G. S., *Sci. Hortic.*, 42, 29, 1990.

II. VINE GROWTH AND DEVELOPMENT

A. VINE ARCHITECTURE

Kiwifruit vines are dioecious and up to 25% of the vines in an orchard may be staminate.[10] In New Zealand, the planting density is typically 400 vines ha^{-1} (25 m^2 vine^{-1}). The components of the pistillate vine have been classified as perennial (trunk, cordon or leader, and structural roots), transient (laterals and shoots), and deciduous (fruit, leaves, and fibrous roots).[11] This classification reflects both the natural growth pattern and conventional management practices. Usually the trunk is about 1.8 m high. A single cordon is retained and laterals (fruiting canes) trained at right angles from the cordon. Laterals are usually retained for only 1 year, and are replaced using shoots (replacement canes) emerging from or near the cordon during the growing season. Such management of the canopy enables a further division of the canopy into "replacement cane zone" and "fruiting zone"[12] to reflect the predominant locations of the vegetative and reproductive components of the canopy (Table 2).

Vines, either wild or cultivated, can live for more than 50 years and develop trunks of more than 0.2 m in diameter.[13] However, the vines do not become self-supporting and require a strong framework on which to grow. Almost all commercial kiwifruit vines are supported on one of two trellis types (Figure 1); (1) a horizontal and continuous trellis known as a "Pergola" in New Zealand and a "Tendone" in Italy (which will be subsequently referred as Pergola in this chapter), and (2) a discontinuous trellis with a central horizontal plane and inclined planes on either side usually facing east and west known as a "T-bar" in New Zealand[10] and a "Pergola" in Italy (subsequently referred to as T-bar in this chapter). The ratio of canopy surface area to apportioned orchard area is 1.0 for the horizontal pergola, and typically 1.0–1.2 for the T-bar trellis.[9] Leaves typically comprise 92% of the total surface area of the canopy components, with wood comprising 5% and fruit 3%.[14]

Two main types of shoot can be identified. "Terminating" shoots are variable in length but have no terminal buds and the shoot tips usually wither and die within 40 d of emergence, leaving shoots with three to six full-sized leaves.[15] Short-terminated shoots are sometimes called spurs. "Non-terminating" shoots continue growing, and can reach 3–5 m in length during a single growing season. A large proportion of non-terminating shoots develops within the replacement cane zone of the canopy, although both types of shoot arise from apparently identical buds. The cause of the differences in the pattern of

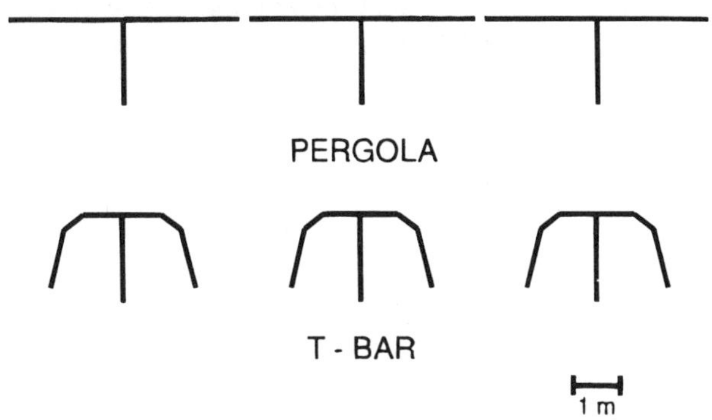

Figure 1 Schematic representation of the major trellis types used to support kiwifruit vines. The "Pergola" comprises a continuous horizontal trellis so that the ratio of the canopy surface area to appointed ground area (surface area ratio) is 1.0. The "T-Bar" comprises a discontinuous trellis including in each row a central horizontal portion and two inclined portions usually facing east and west, respectively.

shoot growth is not known, although the level of carbohydrate reserve laid down the previous growing season may be involved.[9,16]

The term "water shoots" has been used to describe a sub-group of non-terminating shoots that are especially vigorous.[1] These shoots generally arise close to the cordon, and are relatively unfruitful in the season following their development. Morphologically, water shoots are similar to shoots on juvenile vines. In commercial culture, these shoots are often pruned during the growing season, or removed during dormant pruning.[10] However, in their natural habitat water shoots represent a means of moving kiwifruit vines to new areas of the forest canopy, where subsequent generations of shoots would again become fruitful with the better light conditions. Such cycling between juvenile and mature forms appears to be very similar to that reported for another vine, *Syngonium* spp., in response to increases and decreases in soluble carbohydrate levels in the plant.[17]

Root development and distribution are much more variable than the growth characteristics of the above-ground components of the vine[13] due to the influences of the soil conditions.[18–21] The distribution of roots of noncultivated kiwifruit vines growing in the wild in China indicate that the roots are largely found in the top 1.0 m of soil, with the greatest densities (m m^{-3}) about 0.4 m below the surface.[5,13] In New Zealand, roots of mature cultivated kiwifruit vines, grafted to seedling rootstocks[10] have been found to depths of at least 4.0 m in deep porous soils and extend well beyond the area occupied by the leaf canopy.[22] However, it is more typical to find over 50% of the total root within 0.5 m of the soil surface[23–25] and more than 90% of the roots within the top 1.0 m.

The expansion of the root system as a vine matures determines the soil volume for water and nutrient supply, and occurs by a combination of root extension and branching. The root system of immature vines occupies a bowl-shaped soil volume and the root density (m m^{-3}) decreases with both depth and radial distance from the crown.[26] Roots from adjacent vines usually meet within 4–6 years of planting. For mature vines, the root density decreases with depth but at any soil depth is unaffected by distance from the crown.

B. VINE DIMENSIONS

Annual fruit yields for mature vines (>6 years) are 10–25 t ha^{-1} but can exceed 30 t ha^{-1}, with a fruit water content at harvest of 81–85%.[11] The biomass (dry weight) of mature vines with a fruit yield (fresh weight) equivalent to 30 t ha^{-1} is about 21.5 t ha^{-1}.[9] For maturing vines (aged 3–6 years), the distribution of dry matter among the component organs, especially for the deciduous and transient organs, remains almost constant.[25,27] The root system accounts for about 40% of the total plant biomass. This contrasts with apple, where the root system usually comprises 20% or less of the total tree biomass.[28,29] The biomass of the current season's growth (leaves, shoots, fruit, and fibrous roots) may constitute more than 50% of the total vine biomass (Table 3). For mature vines, the proportion of total biomass in perennial organs may be greater than in developing vines, but the relative dry weights of deciduous and transient organs appear to be very similar to those for younger vines.[30]

The mean root density in the surface 0.5 m of soil is usually about 5–10 km m^{-3}.[22,23,31] This density is comparable to (or greater) than that of other perennial fruiting crops including apple,[32] cherry[32] and grape.[33] The fibrous roots (<2 mm diameter) comprise more than 98% of the total root length, but less than 20% of the total root biomass.[25,27] Over a wide range of soil types, the distribution of roots of mature vines has been shown to be extremely clumped.[26]

Table 3 **Typical distribution of biomass for mature kiwifruit vines with fruit yield (fresh weight) of 30 t ha^{-1}**

	Biomass (t ha^{-1})	Total (%)
Leaves	2.52	10.8
Fruit	4.97	21.2
Shoots	2.66	11.4
Laterals	1.94	8.3
Cordon	0.73	3.1
Stem	0.94	4.0
Structural roots	7.60	32.9
Fibrous roots	1.92	8.2

From Buwalda, J. G. and Smith, G. S., *Tree Physiol.*, 3, 295, 1987.

For cultivated vines, the leaf area depends on dormant and summer pruning practices, but a total leaf area index (m^2 leaf area m^{-2} apportioned orchard area) of about 3.0 is common.[12] Location of the leaves in space depends on the growth and management of the shoots. While individual shoots may extend more than 2 m from the trellis on which the vine is supported, more than 80% of the leaf area is located within 0.6 m of the trellis.[14] Leaf area distribution within the canopy is highly heterogeneous (Figure 2). More than 70% of the total leaf area for a vine is located within the replacement cane zone.[12] The leaf area index can vary greatly (i.e., more than four-fold) within distances of less than 0.2 m, anywhere within the canopy (Figure 2). Leaf orientation within the canopy usually shows no significant azimuthal preference.[14,34] The distribution of leaf inclination angles is non-spherical and is consistent with an ellipsoidal model, and mean inclination angles ranging from 29° to 33° have been reported.[14,34]

Flowers are borne on current season's terminate and non-terminate shoots arising from axillary buds on lateral canes that developed the previous growing season.[35] The inflorescence is potentially a compound dichashium containing a terminal flower and primary and secondary lateral flowers.[35] Many of the lateral flowers of 'Hayward' abort, but in some years they complete their development resulting in inflorescences with two or three flowers. The number of fruit on commercially managed vines often ranges from 25–40 m^{-2} of apportioned ground area.[10] Fruit location within the canopy also depends strongly on vine management, although more than 65% are usually located within the fruiting zone. The disparate arrangement of leaves and fruit within the canopy results in leaf:fruit ratios varying greatly between the two canopy zones, ranging from 2:1 in the fruiting zone to 10:1 in the replacement cane zone.[12]

C. ANNUAL GROWTH

1. Shoots and Leaves

Shoots of the current season's growth develop from axillary buds on laterals (canes) of the previous season's growth, although some growth can occur from the buds on older wood. Vines are pruned during winter to retain about 25–40 axillary buds m^{-2} of apportioned orchard area. Under New Zealand conditions, fewer than 50% of these axillary buds develop into shoots.[15] However, the proportion of buds that break can be greater in other countries such as Italy.[36] Many of the buds on the long vigorous laterals or on the undersides of the laterals remain dormant. The proportion of buds that break can be modified through the use of the dormancy-breaking chemical, hydrogen cyanamide,[37–39] which increases the uniformity of shoot development, advances the date of bud break, and compresses the bud break period. A large increase in the concentration of proline in the buds of vines treated with hydrogen cyanamide has been associated with the onset of flowering and increased fruitfulness.[40] Bud break usually occurs over a period of about 20 d.[41] Shoots that develop early tend to be more fruitful and they also appear to suppress growth from other buds.[41,42]

Shoot extension proceeds rapidly after bud break, with the relative growth rate peaking at 0.25–0.30 m m^{-1} d^{-1} during the first 20 d.[31] The relative growth rate of the shoots then declines rapidly between 20–60 d after bud break, followed by a much lower growth rate of <0.01 m m^{-1} d^{-1} for the remaining 150–170 d of the growing season.

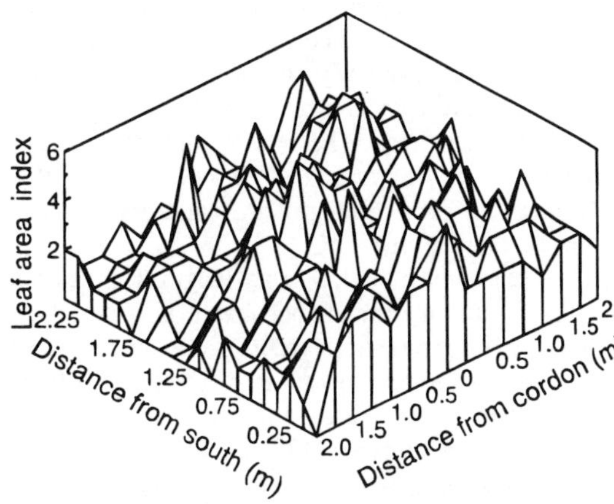

Figure 2 Spatial distribution of leaf area index for a kiwifruit vine. Leaf area index was measured at 320 canopy positions by counting the number of interceptions with leaves as a long needle passes through the canopy at an angle normal to that of the trellis surface. The total leaf area index for the whole vine was 2.74.

Leaf numbers increase rapidly during the first 60 d after bud-break to more than 200 leaves m^{-2} of apportioned orchard area, with only a slight increase thereafter.[12] The leaves are arranged in a spiral phyllotaxis of 2/5 or 2 + 3; the spiral may be either clockwise or anticlockwise.[43] The total leaf area on an individual shoot can be linearly related to shoot length.[44] Increase in leaf area follows a sigmoid pattern with time. Maximum rate of expansion occurs about 20 d after appearance, with final leaf size occurring about 60 d later.[16] As the leaves enlarge they change shape, with more growth in width than in length.[13] The average final area of individual leaves can range from 0.013 m^2 for leaves in the fruiting zone to 0.020 m^2 for leaves in the replacement cane zone.[12]

The specific leaf weight changes with leaf age, increasing from 0.09 kg m^{-2} about 30 d after leaf emergence to 0.20 kg m^{-2} after the leaves have fully expanded.[45] Leaves on non-terminating shoots accumulate dry weight for an extended period compared to those on terminating shoots.[46] A transitory decline in dry weight of the leaves occurs during the period of rapid fruit growth. This decline coincides with a marked reduction in the starch content of the leaves.[47] The weight of leaves in shaded positions within the canopy is usually much lower than that of leaves in exposed positions.[48] Leaf chlorophyll (chlorophyll $a + b$) increases during the initial 100–150 d after leaf emergence, reaching maximum values of 1.2 to 1.4 g m^{-2}.[45] The lower surface of the leaf is covered by a dense layer of stellate trichomes which are found largely along the veins.[13] Trichomes on the upper surface are also found along the primary and secondary veins, but they do not persist for any length of time. Stomata are found only on the lower leaf surface.[49] Large raphide cells are found in the palisade and spongy mesophyll, alongside the veins, and in the trichomes.[1,50] Styloids are also found in the palisade tissue and in the spongy mesophyll.

Leaves abscise in autumn, usually after the first frost.

2. Flowers and Fruit

Flower primordia are not present in the buds during dormancy, but begin to differentiate at bud swell, approximately 10–15 d prior to bud break.[15] Flowering occurs about 50–70 d after bud break. The fruit develops from a multicarpellate ovary borne on a pedicel.[1] The pattern of fruit enlargement can be related to three distinct growth phases within the fruit.[51] During the first 70 d after anthesis cell division in the central core, inner and outer pericarp, followed by cell enlargement in all tissues, results in rapid growth and weight gain. During the next 20–30 d, cell enlargement slows in both the inner pericarp and central core so that growth and weight gain are reduced. Cell enlargement in the inner pericarp and central core results in a second period of growth and weight gain in the final 70–85 d until harvest. The dry weight of the fruit increases linearly with time.[16,51–53] The double sigmoid curve for gains in fruit volume and fresh weight therefore implies that assimilates are incorporated into the fruit with variable quantities of water.[9]

The relative growth rate of fruit volume is greatest (>0.1 $mm^3\ mm^{-3}\ d^{-1}$) during the first 20–30 d after anthesis, then declines rapidly during the following 40 d to a rate less than 0.005 $mm^3\ mm^{-3}\ d^{-1}$ for the remaining 90–100 d until harvest.[9] Fruit are harvested in autumn, about 160–180 d after anthesis.

3. Roots

The root system is highly dynamic, with continual growth and senescence (turnover) of the roots. Seasonal turnover appears to be limited mainly to the fibrous roots.[9] The seasonal pattern of root growth, described as incidence of non-suberised (white) root expressed as a fraction of the total root length, shows an increase in early spring to a peak in late summer, before declining during late autumn and winter.[31] The presence of white root throughout the year indicates that there is at least some root growth occurring at all times. The relative growth rate of new roots ranges from 0.5×10^{-3} m m^{-1} d^{-1} in spring to 3.5×10^{-3} m m^{-1} d^{-1} in late summer. Root growth was especially low during the period of flowering and early fruit growth,[54] and appears to be very sensitive to changes in assimilate supply within the whole vine.[12,47,54]

In mature vines, the total root length declines during spring, and increases during late summer and autumn, so that the initial root length is restored by the following winter.[54] These changes imply that considerable turnover of the roots occurs during the season. An assumption that new roots remain white for 4 weeks led to an estimate that more than 30% of the total root length of mature vines was replaced annually.[31] However, more rapid suberization of roots during summer[32] would imply that root turnover rates are greater, perhaps 100% $year^{-1}$.[9] If turnover is limited to the fine root fraction (<2 mm diameter), this would involve 6–8% of the total biomass annually. It is further possible that rapid death of a large proportion of new roots may mask considerable root growth. For apple, pear, and sour cherry seedlings, 2–4.8% of the root tips are often found dead at sampling time.[55] Assuming that dead roots are sloughed off quickly (i.e., less than 1 week), this incidence suggests a rate of replacement of root tips greater than 100% per year.

III. IRRADIANCE

The radiation environment of the orchard, the interception of available radiation by leaves within the canopy, and the irradiance response of photosynthesis (net CO_2 assimilation, A) and hence carbon acquisition, for individual leaves are the main factors determining the effects of irradiance on the whole vine.

A. RESPONSE OF SINGLE LEAVES

An early report, based on spot measurements of A with variable photosynthetically active radiation (PAR), suggested that A was radiation saturated at 500–700 μmol PAR m^{-2} s^{-1}.[56] Later measurements of A at variable ambient PAR,[48] fitted to a rectangular hyperbola, indicated that an irradiance greater than 1400 μmol PAR m^{-2} s^{-1} was required to saturate A.[9] However, interactions between the influences of irradiance and other environmental variables (e.g., temperature, leaf-to-air vapor pressure gradient) limit the value of such descriptions of the irradiance response of A.[57] More recently, an examination of the irradiance response of A for single leaves, made using a climate controlled mini-cuvette system, led to the description of this response with an asymptotic exponential equation[45]

$$A = A_{sat} - \beta \rho^{PAR} \tag{1}$$

where A_{sat} is the radiation saturated rate of photosynthesis (μmol CO_2 m^{-2} s^{-1}), and β and ρ are empirical parameters ($\rho < 1$). Other leaf gas exchange characteristics such as the dark respiration rate (R_d) and the quantum yield of photosynthesis (ϕ_i), can be estimated from the fitted parameters. For kiwifruit, the asymptotic exponential curve (Equation 1) describes the irradiance response of A for single leaves more accurately than rectangular hyperbola or non-rectangular hyperbola curves.[58]

Measurements made throughout the season have shown that kiwifruit leaves exhibit considerable temporal variation in the irradiance response of A.[45] The stage of growing season was found to be less important than the age of the leaves at the time of measurement (Figure 3). A_{sat} increased from about 10 μmol CO_2 m^{-2} s^{-1} at 30 d after leaf emergence to a peak of about 17 μmol CO_2 m^{-2} s^{-1} at 120–180 d after leaf emergence, and declined with leaf senescence (Figure 3). The A_{sat} levels so far defined are within the range reported for other fruit crop species, such as grape,[59] apple,[60] and peach.[61] Factors that enhance leaf senescence may accelerate the rate of decline of A_{sat} at the end of the growing season. An early study indicated that A_{sat} declined from mid-summer to the end of the growing season.[48] The radiation level required to saturate A, PAR_{sat}, increased from about 600 μmol m^{-2} s^{-1} at 30 d after leaf emergence to peak at about 800–900 μmol m^{-2} s^{-1} at about 90 d after leaf emergence, before declining again at the end of the season. These estimates of PAR_{sat} are consistent with those reported for a wide range of agricultural and horticultural crops.[57,62] Photosynthesis for the leaves of kiwifruit vines is

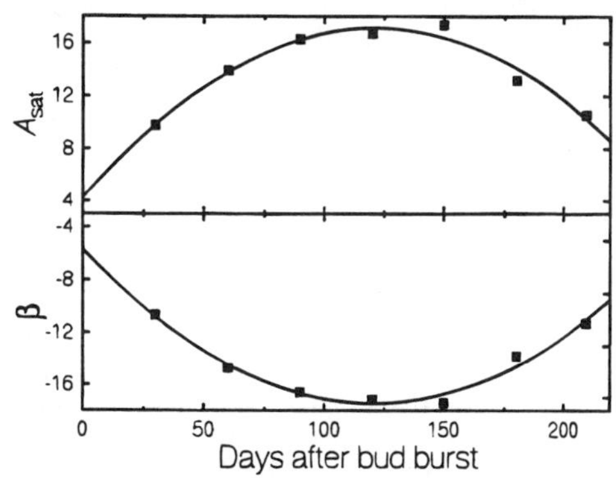

Figure 3 Variation with leaf age in parameter values for the asymptotic exponential curves used to describe irradiance response of single leaf photosynthesis.

therefore likely to be radiation limited for much of the growing season. Relatively low PAR_{sat} and A_{sat} levels have been reported for kiwifruit vines grown in controlled environments.[63–65] However, the irradiance levels used in these studies were usually less than the equivalent of one-third of full sun, so the results probably reflect acclimation to low light. R_d showed no significant temporal variation, averaging 0.4 μmol CO_2 m^{-2} s^{-1}. However, expressed on a leaf dry weight basis, R_d deceased with increasing leaf age. This decline coincided with a decline in the concentration of leaf nitrogen expressed on a dry weight basis. ϕ_i increased slightly with leaf age from about 0.05 mol CO_2 mol^{-1} PAR 30 d after leaf emergence to about 0.07 mol CO_2 mol^{-1} PAR by 120–180 d after leaf emergence (Figure 3). As R_d and ϕ_i were affected little by leaf age, the compensation point was also little affected, averaging 22 μmol PAR m^{-2} s^{-1}.

Leaves within the canopy show spatial variation in photosynthetic capacity at any time. The radiation environment within which leaves grow can affect photosynthetic capacity. Leaves growing in shaded parts of the canopy have been shown to be acclimated to lower irradiance conditions with lower A_{sat} and PAR_{sat}, and changes in chlorophyll a/b ratios, although ϕ_i is little affected.[34,48] These responses are typical of those reported for a wide range of C-3 species.[66] Exposure of previously shaded leaves to high levels of irradiance can lead to photoinhibition of photosynthesis, especially at low temperatures.[67] The low A_{sat} rates for leaves growing in shaded positions in the canopy are unlikely to limit canopy photosynthesis. Radiation levels in the shaded positions of the canopy are usually very low, and photosynthesis will be affected more by ϕ_i than by A_{sat}.

The slow development of A_{sat} after leaf emergence limits whole-plant photosynthesis during canopy expansion. For many species, maximum photosynthetic capacity for any leaf is attained at or prior to full leaf expansion; e.g., within 15–17 d for strawberry,[68] and 11 d for bean.[69] While A_{sat} generally develops more slowly for leaves of perennial plants compared to leaves of annual plants,[70] the development of A_{sat} for leaves of kiwifruit vines is still slower than that reported for other perennial species such as cherry.[71] Stepwise multiple regression analysis indicated that the slow development of A_{sat} for kiwifruit vines was related more closely to leaf nitrogen (per unit of leaf area) than to other leaf attributes such as leaf chlorophyll and stomatal conductance.[45] In fact, the limited photosynthetic capacity in young leaves is physiologically similar to that usually associated with nitrogen deficiency.[72] It is possible that the ontogenetic trend in leaf nitrogen content regulates photosynthetic capacity for kiwifruit leaves. Nitrogen deficiency has been shown to affect directly the radiation saturated rate of A.[63]

B. INTERCEPTION OF RADIATION

Canopy shape strongly affects the radiation environment for leaves at different positions. The irradiance on the surface of canopies trained on a horizontal Pergola shows no spatial variability, so temporal variability depends mainly on incoming radiation.[73] For T-bar vines, however, irradiance at the surface also varies according to the variable angle of incidence of the surface and the solar beam. The east-facing surface typically experiences relatively high irradiance during the morning, while the west-facing surface experiences relatively high irradiance for a period during the afternoon (Figure 4). However, the east- and west-facing surfaces of the T-bar vines are shaded by other parts of the canopy for long periods of the day, so diurnal integrals of incident radiation decrease with distance down the inclined surfaces, and can be only 66% of those for the central horizontal surface (Figure 4).

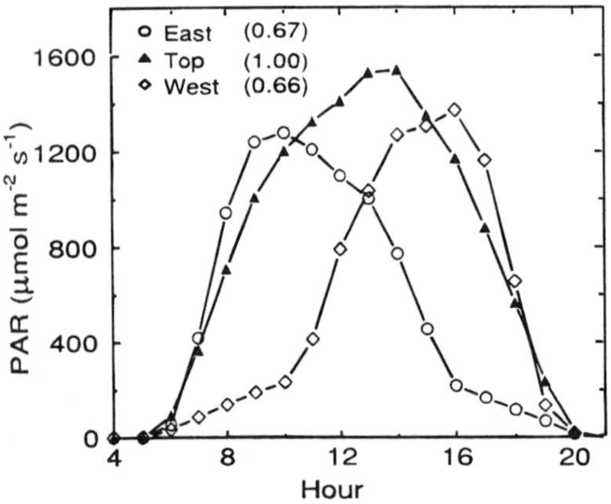

Figure 4 Average diurnal trend of surface irradiance for eastern, central (top), and western faces of a kiwifruit vine supported on a T-bar trellis. Data represent mean hourly irradiance over a 5-week period during mid-summer. Diurnal integrals for the east and west facing surfaces were 67 and 66%, respectively of the integral for the central (top) horizontal surface.

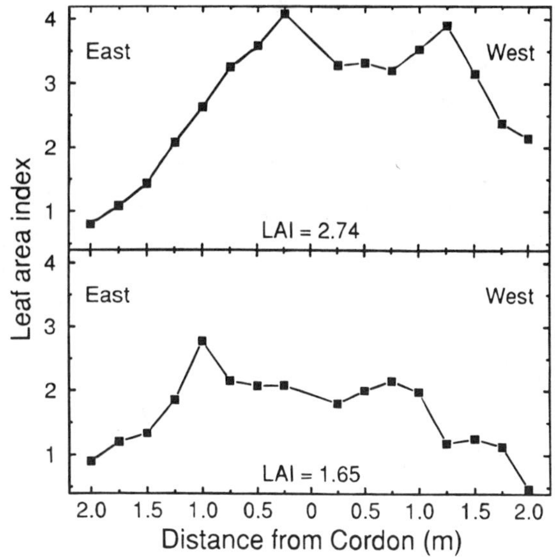

Figure 5 Mean leaf area index at different distances from the central cordon for vines with total leaf area index (LAI) of (a) 2.74 and (b) 1.61. The LAI was measured as described in Figure 2.

Spatial variation in leaf area for canopies of kiwifruit vines also affects radiation interception. Typical distributions of leaf area index across a surface of vines trained on a T-bar trellis are shown in Figure 5. Such vines are characterized by continuity of canopy within a row, but discontinuity between adjacent rows (Figure 1). The leaf area index is generally highest in the inner region of the vine near the cordon, and may be up to 50% greater in that position than the average for the whole vine. Conversely, the leaf area index tends to be lowest near the ends of the laterals, at the edges of the canopy. Within any part of the canopy, the leaf area index can show substantial spatial variability, often ranging from 1.0 to 6.0 within distances of 0.3 m. This variability results in localized areas of heavy shade or light transmission within the canopy.

C. CANOPY PHOTOSYNTHESIS

Together, spatial variation in leaf arrangement and temporal variation in irradiance conditions lead to highly dynamic irradiance of individual leaves within the canopy. Furthermore, at any time the canopy includes leaves of widely varying photosynthetic capacity, due to varying leaf age within the canopy and age-related changes in the photosynthetic capacity of individual leaves. Consequently the photosynthetic integral for all leaves within the canopy is highly dynamic. The complex plant-environment interactions and the seasonal dynamics of canopy photosynthesis for a kiwifruit vine have been described using a mathematical model of carbon acquisition and utilization.[44] This model considers radiation attenuation through the leaf canopy (assuming Beer's Law for exponential attenuation),[74] and computes integrated canopy photosynthesis according to the irradiance of leaves at different layers. Leaf arrangement within the canopy influences radiation attenuation. An ellipsoidal distribution of leaf inclination angles, with a mean inclination angle of 29° is assumed.[34] The asymptotic exponential curve describing the irradiance response of A for single leaves must be integrated numerically, but "Gaussian 3-point" integration[75] enables rapid and convenient computation of canopy A.[58]

Canopy A for vines trained on a horizontal trellis can be simulated by assuming uniform incident radiation at all points on the canopy surface. The simulated irradiance response of canopy photosynthesis depends on the irradiance response of single leaf photosynthesis (Figure 6a) and the leaf area index (Figure 6b). Typically, A_{sat} and PAR_{sat} for the canopy increase with increasing leaf area.

A modified canopy photosynthesis model enabling estimation of the incident radiation at any point on surfaces with complex shapes, and distribution of photosynthesis within canopies with non-homogeneous distribution of leaf area, has recently been developed.[73] For vines on a T-bar trellis with the leaf area shown in Figure 5, A near the cordon is usually greater than that near the ends of the laterals (Figure 7). On a clear day (e.g., PAR > 1800 μmol m^{-2} s^{-1} at midday), this difference was about 60% for vines with a high leaf area and 85% for vines with low leaf area. On cloudy days A at different parts of the canopy was more uniform, but still 10–20% greater near the center of the vine than near the ends of the laterals.

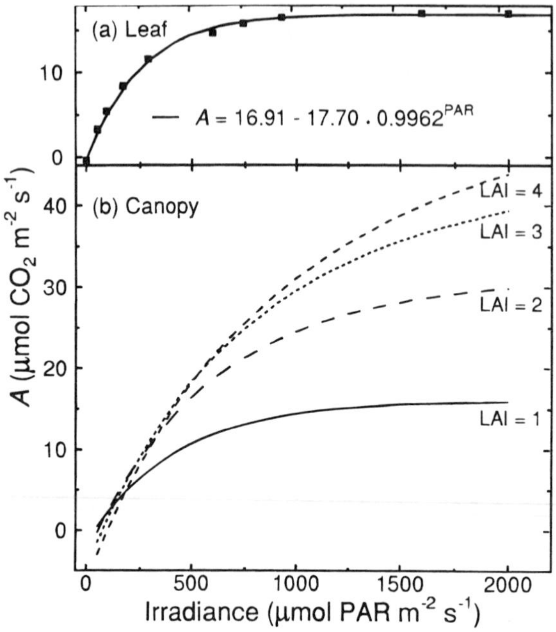

Figure 6 Irradiance response of photosynthesis (A) for (a) single leaves, and (b) entire canopies. The single leaf response was described with an asymptotic exponential equation which was then used to estimate canopy photosynthesis for vines with different total leaf area index using a mathematical model.[44]

Canopy A has been measured directly for kiwifruit vines using whole-canopy gas exchange cuvettes.[76,77] The irradiance response of a canopy on a T-bar trellis was described with an asymptotic exponential curve similar to that used for single leaves. For measurements made during late summer, diel trends of canopy A indicated that R_d averaged -8.24 µmol CO_2 m^{-2} s^{-1}, ϕ_i averaged 0.091 mol CO_2 mol^{-1} PAR, the compensation point averaged 105 µmol PAR m^{-2} s^{-1}, PAR_{sat} averaged 1312 µmol PAR m^{-2} s^{-1}, and A_{sat} averaged 28.2 µmol CO_2 m^{-2} s^{-1}.[76] The R_d rate reflects respiration of non-photosynthetic components (fruit and wood) of the canopy as well as respiration of the leaves. The higher PAR_{sat} and A_{sat} for the canopy compared to that for single leaves illustrate the capacity for increased photosynthesis where radiation is attenuated through a canopy with overlapping leaf layers. The higher apparent ϕ_i for the canopy compared to that for single leaves is difficult to explain. The

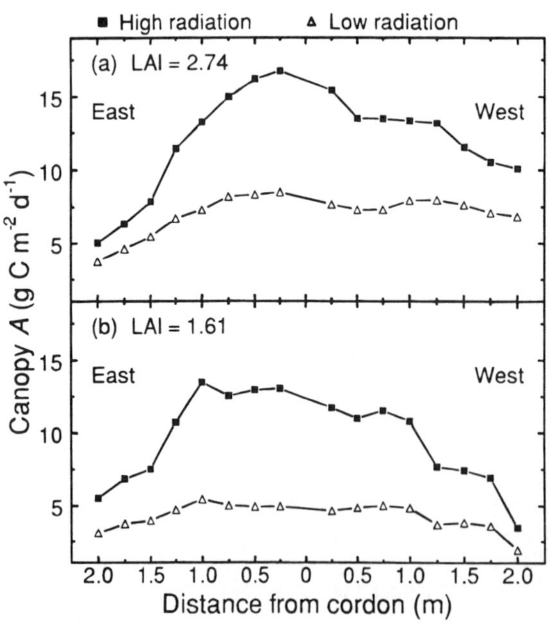

Figure 7 Diurnal integrals of simulated photosynthesis (A) at different distances from the cordon of kiwifruit vines with (a) high and (b) low total leaf area index. These simulations used a modified canopy photosynthesis model,[73] and radiation data for a sunny and cloudy day in late summer.

irradiance response for the canopies was described according to diurnal measurements of canopy A as irradiance varied. It is possible that the diffusing effect of the film used to clad the gas-exchange cuvette and the canopy shape affected the diurnal trend for radiation interception. Expressing ϕ_i in terms of intercepted radiation, rather than incident PAR would enable better comparisons to be made between canopies and single leaves. A more recent study of whole-canopy gas exchange illustrated clearly increased canopy A for vines with a leaf area index of 2.7 compared to vines with leaf area index of 1.6.[78] Direct measurement of canopy A also illustrated mid-day declines in A, typically after 09:00–10:00 h on days with high radiation levels.[78] Such declines led to canopy A rates in the afternoon of only 70–80% of those recorded at similar radiation levels in the morning. Increased respiration with increasing temperature was insufficient to account for the decline in canopy A. Feedback inhibition of A has been suggested as a possible contributor to this phenomenon.[78]

The cyclic behavior of the stomata of kiwifruit leaves may also be an expression of feedback inhibition.[79] Typically the period of stomatal cycling for most plants studied has been about 10–60 min.[80] However, an average cycle found for kiwifruit vines was of 4–6 d duration and corresponded closely to the fluctuations in PAR.[79] The strongly negative relationship between stomatal conductance and PAR is consistent with the suggestion that accumulation of photosynthate in the leaves may be responsible for the partial closure of the stomata.[81]

D. EFFECTS OF REDUCED IRRADIANCE ON VINE GROWTH AND DEVELOPMENT

A feature of kiwifruit cultivation in New Zealand is the use of dense networks of windbreaks (shelter trees or shelter cloth) to protect the vines from wind damage.[82] Radiation interception by vines growing adjacent to shelter belts has been shown to be reduced by up to 40% where the shelter trees were 20 m in height.[83] For vines growing in a controlled environment where the radiation was reduced from 600 to 220 μmol PAR m^{-2} s^{-1}, there was a marked reduction in shoot growth and leaf weight, while bud break and flowering in the following year were also reduced.[84] Overhead shading of vines to 45% of ambient radiation for three consecutive seasons reduced the average fruit weight by 14 g and the return bloom by 56%.[85] During cold storage, fruit from shaded vines had lower soluble solids concentrations, and were only slightly less firm, than fruit from unshaded vines.[85] By contrast, shading had no effect on the starch content or the internal color of the fruit. Shading vines for short periods during the growing season can also affect fruit growth. Shading the canopy during the most rapid phase of growth of the fruit immediately after anthesis, had a greater effect on final fruit size than shading later in the growing season.[85]

Low irradiance within the canopy can limit carbon acquisition by the leaves and may lead to premature leaf senescence, particularly those leaves in the denser parts of the canopy. Low irradiance within the canopy may result from low irradiance at the periphery of the vine and/or high leaf area. However, the effects of these two factors can be difficult to discern. Simulation modelling of canopy photosynthesis[44] has been applied to examine the interaction between incoming radiation and leaf area index on net carbon acquisition of the canopy. While light interception and hence total photosynthesis continue to increase with increasing leaf area (Figure 6b), marginal gains in photosynthesis at high leaf area are small relative to marginal costs of tissues synthesis and maintenance. Accordingly, there is an optimum leaf area beyond which the carbon cost of tissue synthesis and maintenance exceeds the carbon acquired by photosynthesis. The optimum leaf area reduces with decreasing radiation (e.g., cloudy weather or shading from shelter). Net carbon acquisition by the canopy is reduced slightly at excessive leaf area, but is reduced strongly by low radiation (Figure 8a). For kiwifruit in New Zealand conditions, the optimum leaf area index appears to be about 3.0–3.5 m^2 m^{-2}.[34] This leaf area is typically attained about 100 days after bud break (Figure 8b). During canopy expansion, canopy photosynthesis is limited more by low leaf area (i.e., low radiation interception) than by irradiance.

IV. TEMPERATURE

A. PHENOLOGY AND GROWTH

Estimations of the chilling requirements of kiwifruit vines to "break" dormancy have been based on both laboratory and field observations. A significant delay in bud break was found for vines that received less than 950–1000 h of chilling at 4°C.[86,87] These authors also showed an increase in flower numbers per shoot with an increase in the duration of chilling during the early to mid stages but not in the later stages of dormancy. Constant temperature conditions were less effective in breaking dormancy than where there was a change in the temperature between day and night.[87,88]

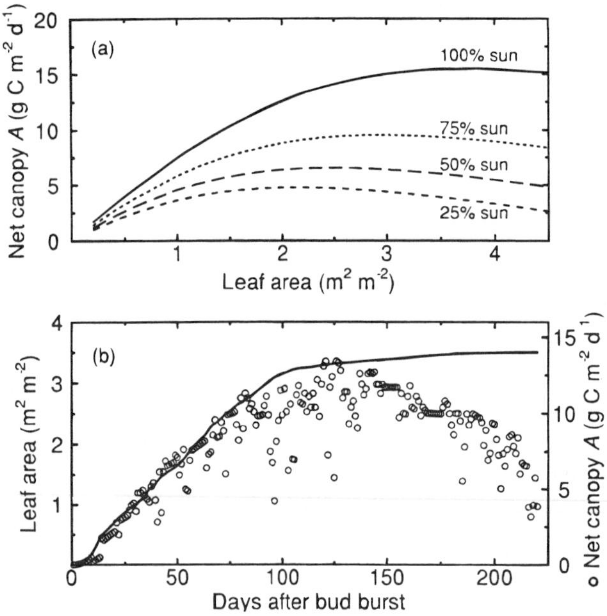

Figure 8 (a) Interaction between radiation and leaf area (LAI) for diurnal integral of simulated net canopy photosynthesis (*A*); simulations assumed fractions of radiation recorded for a clear day in mid-summer. (b) Seasonal trend of canopy leaf area and simulated canopy photosynthesis (*A*).

Field observations initially indicated that kiwifruit vines required approximately 700–800 Richardson Chill Units (RCUs) to break dormancy.[89] However, a later report claimed that only 500 RCUs were needed to produce a crop.[90] In New Zealand, RCUs calculated for the winter period typically range from 1033 in the warmest kiwifruit growing district to 1892 in the coolest district.[91] Bud break was reported to be approximately 10% greater in the cooler regions than in the warmer regions.[91] However, recent research has shown that RCUs cannot be used to reliably predict bud break and flowering in kiwifruit vines as the response of vines to winter chilling occurs over a wide range of accumulated RCUs.[92] The main effects of increased winter chilling are to improve the proportion of dormant buds that break in spring and to reduce the number of developing flowers that abort.

The rate of leaf appearance (leaves per shoot per day) in spring increases approximately linearly with temperatures above 10°C to a maximum rate at 20°C, above which the rate declines.[93] The time from bud break to flowering also appears to be temperature dependent. Full bloom was shown to occur approximately 360 growing degree days (calculated from a threshold temperature of 7.5°C) after bud break.[94] Controlled environment studies using container-grown vines have shown that bud break was 16 d later when the temperature was maintained at 10°C than it was at 19°C.[41] The duration from bud break to flowering increased approximately 10 d for every 1°C reduction in temperature. In the same experiment, flowering was found to be spread over 34 d with a mean temperature of 12°C, but only over 13 d where the temperature was 19°C.

Temperature is also thought to have an important influence on fruit maturity. Harvest maturity is determined in kiwifruit by the concentration of soluble solids (Brix) of expressed juice.[95] In New Zealand, the required minimum Brix for harvest is 6.2°.[96] Cool temperatures during autumn enhance the rate of increase in the Brix level of the fruit.[97] At a mean temperature of 11°C, starch in the fruit degrades very quickly with a coincidental increase in total sugar.[98] In contrast, at a mean temperature of 17°C there is little degradation of starch and a proportionally smaller increase in total sugar. Changes in total sugar are similar to those recorded for Brix values.[98]

The seasonal pattern of new root growth has been shown to be broadly related to the annual changes in soil temperature.[31] Experiments where the temperature of the root and shoot environments were controlled independently have shown that vine growth ceased when the temperature of the root zone was maintained at 10°C, whereas maximum root growth occurred at 20°C.[99] Root growth of kiwifruit vines therefore appears to have a temperature optimum lower than that for other deciduous fruiting plants. Root growth of apple, for example, has an optimum root temperature of 25°C.[100] Soil temperatures could have a major influence on the commencement of root growth in spring, as soil temperatures in New Zealand, for example, do not exceed 10°C until late spring or early summer.[99] By this stage of the season the vine has undergone several major growth phases, including bud break, leaf growth, anthesis, and early fruit growth.[9]

B. LOW-TEMPERATURE STRESS

The minimum temperature tolerated by kiwifruit vines depends on the stage of growth. In spring, flower buds, flowers, and young actively growing shoots are seriously damaged by temperatures of −1.5°C or less for durations as short as 30 min.[101] In autumn, following leaf fall, temperatures as low as −5.0°C have been found to cause no obvious shoot damage or reduction in the number of buds that developed the following spring, but temperatures of −7.0°C were damaging.[102] By mid-winter, vines can tolerate temperatures of −7.0°C, but exposure to −10.0°C for 1 h has been shown to damage dormant shoots.[101]

Studies in controlled environments have shown that leaves of kiwifruit vines suffer severe damage when exposed to moderate light (650 μmol m^{-2} s^{-1}) at 10°C, including photobleaching, yellowing of the leaf margins, reduced expansion rates of the leaf, and death of the stem apices.[93] However, these symptoms were not apparent when vines were grown at lower light (280 μmol m^{-2} s^{-1}) at 10°C, suggesting that the vines were affected more from high-light stress at low temperatures than directly from low temperature stress. Photoinhibition of photosynthesis has been cited as the cause of this leaf damage.[67,103] A reduction in the photon yield of photosynthesis and changes in the chlorophyll fluorescence emission at 77K occurred when shade-grown leaves were exposed to high light (1500 μmol m^{-2} s^{-1}). Photoinhibition occurred at all temperatures examined (5–35°C), but was greatest at low temperatures. Recovery from the effects of photoinhibition was also found to be temperature dependent with little or no recovery occurring below 20°C, but rapid recovery at 30–35°C.[67,104] Visual leaf symptoms typical of high-light low-temperature stress have been observed during early spring in New Zealand. Gas exchange and fluorescence analysis have also shown that exposed leaves can exhibit symptoms of partially impaired photosystem II activity, particularly in spring and autumn, which are consistent with high-light low-temperature stress.[105] Photoinhibition in the field appears to be at a much lower level than would be expected from studies on plants grown in controlled climate facilities.

C. EFFECTS OF HIGH TEMPERATURE

While optimal temperatures for the growth of various components of the kiwifruit vine have not been defined, temperatures greater than 35°C have been shown to cause significant reductions in photosynthesis and excessive rates of transpiration.[106] For vines growing in the field, increasing the average air temperature during late winter and early spring from 12.2 to 17.1°C increased the rate of growth of the apical shoots, which in turn reduced the number of subtending shoots that produced flowers.[107] Elevated temperatures during spring, however, advanced the date of flowering. At harvest fruit from the vines grown at the higher temperature were significantly larger and had higher soluble solids concentrations than fruit from the vines grown at the ambient temperature.[107]

Root temperatures in excess of 25°C reduced both shoot and root growth.[99] The effects on the shoots however, were less severe so that the shoot/root ratio increased from 4.6 at 20°C to 6.4 at 30°C. Increased root respiration with increasing temperature,[108] and hence increased carbon losses from the root system, may partly explain this reduction in root growth.

V. WATER

A. WATER RELATIONS

Stomatal conductance to water vapor (g_s) at saturating irradiance (2000 μmol PAR m^{-2} s^{-1}) varies with leaf age, increasing from 0.2–0.3 mol H_2O m^{-2} s^{-1} at about 30 d after leaf emergence to 0.6–0.9 mol H_2O m^{-2} s^{-1} at 90–120 d after leaf emergence.[45] The temporal changes in g_s and A influence the water-use efficiency. The slow increase in A_{sat} with leaf age results in the transpiration ratio at saturating irradiance increasing from about 3–5 mmol CO_2 mol^{-1} H_2O at 30–60 d after leaf emergence to 9–10 mmol CO_2 mol^{-1} H_2O at 120–150 d after leaf emergence.

The typical diurnal pattern of g_s for leaves of unstressed vines shows a rapid rise in the early morning with little change during the day until radiation levels decline again during the evening.[48,77,109,110] However, partial stomatal closure about midday has recently been reported, which resulted in a reduction in g_s by 20–30% and was associated with a reduction in leaf and canopy A.[77] Relatively high values for g_s have been recorded at night, especially for vines growing in advective conditions,[110] and are consistent with nocturnal water use measured for whole canopies.[78] As a consequence, up to 20% of the water use by the vine may occur at night.[111]

Because g_s is high for kiwifruit leaves, water use is influenced strongly by the boundary layer conductance. Any additional limitations due to trapping of a boundary layer by the stellate hairs could therefore be significant.

B. DROUGHT STRESS

Like most other plants, drought stress reduces g_s of the leaves of kiwifruit vines, with a g_s of 0.016 mol H_2O m^{-2} s^{-1} being recorded for severely droughted vines in the field.[48] Effects of water stress appear to be closely related to the leaf water potential (Ψ_l). Typically Ψ_l for vines without water stress is lowest at about midday, but generally no lower than −0.6 MPa.[48,109,110,112] Stomatal conductance has been shown to be insensitive to Ψ_l greater than −0.6 MPa, but declines once Ψ_l falls below −0.6 MPa. Leaves may become severely wilted when Ψ_l falls below −0.9 MPa, but the stomata do not close fully even where Ψ_l as low as −2.9 MPa has been measured.[48]

Measurements of Ψ_l at dawn have been found to be useful for predicting drought stress of kiwifruit vines. The degree of wilting during peak evaporative demand at midday has been related to dawn measurements of Ψ_l.[109] Vines with Ψ_l of less than −0.12 MPa at dawn readily wilt during peak periods of evaporative demand during sunny days, but were shown to regain turgor at night. Wilting was also shown to be irreversible where the Ψ_l at dawn was less than −0.65 MPa.[109]

The effects of drought stress on *A* are larger than those on g_s, so that the transpiration ratio increases as the stomata close.[48] In fact, an early report suggesting that the water-use efficiency increased with leaf age[48] probably reflected restricted stomatal conductance due to water stress.

1. Effects on Growth

Studies of kiwifruit vines grown in a polyethylene lined trench have shown that drought stress reduces dry matter accumulation of all vine components of kiwifruit.[27] In an experiment where the supply of irrigation water was restricted for 3 years, root length increased from 19.8 KM $vine^{-1}$ for the irrigated kiwifruit vines to 27.5 KM $vine^{-1}$ for the non-irrigated vine.[118] The corresponding volume of soil explored by the roots increased from 6.9 m^3 $vine^{-1}$ to 11.8 m^3 $vine^{-1}$ respectively. This result is consistent with an increased root:shoot ratio, typical of responses of many plant species to drought stress.[113]

Drought stress has immediate and significant effects on fruit growth, with a severe stress level halting growth. While reducing water supply per se may affect fruit growth directly, it is also possible that an indirect effect, through reduced assimilate supply as a result of partial closure of the stomata and altered allocation to the root system, may also contribute to reduced fruit growth during periods of drought stress. Relieving the vine of drought stress restores the growth rate of the fruit to that of fruit on non-stressed vines but growth lost during the period of drought stress is never recovered.[114,115] Although fruit size can be restricted by drought stress, beneficial effects on the post-harvest storage characteristics of the fruit can result. Fruit from non-irrigated vines were found to be firmer and have greater concentrations of soluble solids at harvest and during storage than fruit from irrigated vines.[116] Storage life of fruit has also been found to be enhanced when vines were drought stressed in early summer, during the cell division phase of fruit growth.[117]

2. Effects on Root Activity

Measurements using a neutron probe indicated that kiwifruit roots extract water from deeper soil horizons as the water deficit increases in the soil.[109] For vines with large and extensive root systems, the increased soil volume explored and the capacity to extract water from soils at lower water potentials reduces the dependence of the vine on irrigation.[118] Results from a recent study of vines growing on deep volcanic loam soils in the Bay of Plenty in New Zealand, where root systems are usually large and extensive,[22,23] indicate no significant response to irrigation over a 6-year period.[119]

C. ROOT ANOXIA

Kiwifruit vines are not well adapted to tolerate low concentrations of oxygen in their root zone. The reported death of over 30,000 vines in New Zealand as a result of excessive rainfall (six times the seasonal average) during early summer, graphically demonstrates the sensitivity of kiwifruit vines to root anoxia.[120] The death of large numbers of vines in Italy has been associated with high winter water tables of 0–0.5 m.[21] Fruit yields were significantly reduced on the surviving vines growing on these soils as a result of a reduction in the number of fruit per vine.

There are a number of physiological and morphological features of kiwifruit vines which make them especially vulnerable to root anoxia. The rate of consumption of oxygen (9.1×10^{-10} mol $m^{-1}s^{-1}$ at 20°C) by the roots is relatively high, and it has been estimated that the time taken to deplete the oxygen supply in a well aerated soil to an anaerobic condition would be approximately 5 h assuming no replenishment of oxygen.[99] Generally, plant species with root porosities of less than 5% are highly sensitive to anaerobic conditions in the soil.[121,122] The average air space between the cells of kiwifruit roots accounts for 2% or less of the total root volume of aerated as well as oxygen stressed vines.[99] Unlike plant species more tolerant of root anoxia, kiwifruit roots do not develop aerenchyma tissue in response to decreasing oxygen concentrations in the root zone.[79,99] Aerenchyma formation has been shown to be triggered by enhanced internal concentrations of ethylene.[123] Anoxia is known to cause a stimulation of the ethylene precursor, 1-amino-cyclopropane-1-carboxylic acid (ACC) in the roots of many plants.[121,124] A key co-factor in the production of ACC is methionine[125] and it is interesting to speculate that the absence of free methionine in the tissues of kiwifruit vines[126] may be responsible in part for the lack of formation of aerenchyma in the roots.

1. Effects on the Roots

Within 3 d of the roots being exposed to anaerobic conditions the fibrous roots change from an opaque cream color to a translucent grey color.[79] Substantial losses of up to 50% of dry weight of the fibrous roots have been recorded for vines growing under anaerobic conditions in the field and under controlled conditions.[18,19,79] The loss of tissue has been shown to be caused by the detachment of the cortex from the central stele.[79] The point of separation occurs at the endodermis where a layer of cortical cells, usually packed with starch, collapses.

The speed with which the roots of kiwifruit vines die under anoxic conditions greatly limits their ability to resume growth once the supply of oxygen to the roots has been restored. However, in situations where partial recovery has occurred, growth of new fibrous roots from the base of the trunk is one of the first signs of compensation for the loss of the original root system.[20,79] The extent of the growth of new roots was found to be inversely related to the exposure time to anoxia.

2. Effects on the Leaves and Fruit

A rapid reduction in g_s of the leaves occurs in response to a decrease in the concentration of oxygen in the root zone.[79,99,122,127] Exposing as little as 25% of the total root system to anaerobic conditions has been found to result in substantial reductions in g_s after only 2 d.[128] While leaf turgor of the oxygen-stressed vines has been shown to be maintained in the short term, the stomata eventually close completely after the roots have been exposed to anoxic conditions for as little as 4 d.[79] Characteristic leaf symptoms and reduction in gas exchange by the leaves develop once the oxygen concentration in the root zone falls below 3%. One of the first symptoms to appear is the temporary wilting of the older leaves. These symptoms are quickly followed by the sudden appearance of interveinal necrotic patches, which on some leaves coalesce to form large areas of affected tissue.[99,127] In some cases an interveinal chlorosis precedes the necrosis. The rapid desiccation of the leaves associated with the closure of the stomata may be partly explained in terms of temperature stress. A marked increase in the maximum daily temperature (from 28 to 42°C) has been found for leaves attached to oxygen stressed vines.[99]

Marked increases in the concentration of abscisic acid (ABA) in the leaves, but not the roots, of oxygen stressed kiwifruit vines growing under controlled conditions have been found to accompany the closure of the stomata.[129] While elevated concentrations of ABA have more recently been found in the leaves of waterlogged kiwifruit vines growing in the field, the greatest accumulation of ABA was found in the fruit (Figure 9). Elevated concentrations of ABA were detected in the fruit almost immediately after the roots were flooded with water, whereas ABA did not begin to accumulate in the leaves until 12 d after the vines were flooded and the stomata began to close.

Field studies have also shown that the damaging effects of root anoxia on the leaves of kiwifruit vines were greater in early summer when the ambient temperature was higher and the evaporative demand was greater than in late summer or early autumn.[128] Naturally low concentrations of oxygen in the soil during mid-summer following increased consumption of oxygen by soil micro-organisms as a result of the higher temperatures in the soil[130] are also likely to have contributed to the susceptibility of the vine to root anoxia during this period. By contrast with the effects on the leaves, a greater percentage of fruit was found to be soft at harvest on those vines that had been stressed during late summer or early autumn than on non-stressed vines or those that had been stressed for a similar length

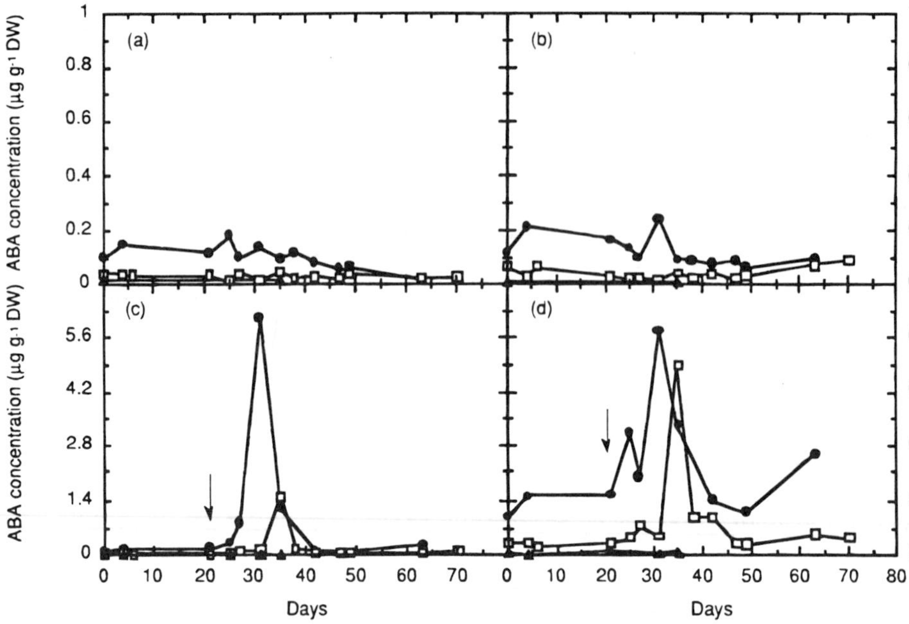

Figure 9 ABA concentrations in the root (▲—▲), leaves (□—□), and fruit (●—●) of (a) low nitrogen control vines, (b) high nitrogen control vines, (c) low nitrogen vines subjected to root anoxia, and (d) high nitrogen vines subjected to root anoxia. The arrow indicates when leaf symptoms were first observed. The anoxic vines were flooded on day 1. The treatments were applied in early summer. Note the different scale for ABA concentrations in (c).

of time in early summer (Table 4). The nitrogen status of the vines was also shown to influence the extent to which root anoxia damaged the vine and the speed with which the vine recovered from the stress. Vines that were nitrogen deficient were much more susceptible to root anoxia than vines that had received adequate quantities of nitrogen fertilizer.[128] Although nitrogen fertilizer was applied well before the imposition of oxygen stress, there was a much greater production of new roots, and hence recovery, by those vines that received nitrogen fertilizer than those which received no nitrogen fertilizer (Table 4).

VI. MINERAL NUTRITION

A. NUTRIENT UPTAKE

For young kiwifruit vines growing in soils with non-limiting nutrient supply the annual uptake of nutrients is determined by the combined requirements for deciduous growth and the expansion of the perennial components.[11] At maturity, the annual uptake by the vine is proportional to productivity and reflects the nutrient composition of the deciduous and transient components rather than that of the perennial components, which by this stage of development show little net increase in size. Differences

Table 4 **Effect of waterlogging on root growth and fruit quality**

	Proportion of new root (%)		
Time when vines waterlogged	**Low nitrogen**	**High nitrogen**	**Soft fruit at harvest (%)**
Control	24	21	3
December	36	68	2
January	2	14	14
March	6	23	28

From Smith, G. S. and Miller, S. A., *Acta Hortic.*, 297, 401, 1991.

have been found, however, among the vine components in their ability to accumulate nutrients. For example, 49% of the total quantity of nitrogen in the vine accumulates in the deciduous components, 16% in the transient components, and 35% in the perennial components; whereas 63% of the total quantity of potassium is in the deciduous components, 11% in the transient components, and only 25% in the perennial components.[25] The relatively high concentrations of nitrogen in the perennial tissues results in a disproportionate accumulation during vine development, such that developing vines may take up 50% more nitrogen over a year than mature vines with the same fruit yield.[11] Likewise developing vines may take up 22% more potassium than mature vines with similar yields.

At maturity, annual uptakes are greatest for nitrogen, potassium, and calcium (between 125 and 180 kg ha^{-1}), while smaller quantities of chlorine (60 kg ha^{-1}) phosphorus, magnesium, and sulfur (<25 kg ha^{-1}) are taken up.[25,131–133] The quantity of nutrient recovered from fertilizer by mature kiwifruit vines is usually less than 50% for most elements.[11,133] A recent study using ^{15}N-labelled nitrogen fertilizer applied in a single dose at 100 or 200 kg ha^{-1} in early spring showed that 48 to 53% of the applied nitrogen had been recovered by the vines.[134] The proportion of added ^{15}N utilized by the vines tended to be slightly greater at the lower application rate. Removal of ^{15}N in harvested fruit was small, at 5 to 6% of the total applied in the first year and 8% over three years. After 2–3 years, over 60% of the ^{15}N in the vine resided in the roots. Estimates of ^{15}N removal in harvested produce of other perennial crops have also been relatively low, ranging from 3–4% for asparagus,[135] 11% for citrus,[136] and up to 10–25% for almonds.[137]

For most nutrients, over 65% of the annual accumulation by the leaves occurs during the first 10 weeks of growth after budbreak,[46] although accumulation by the whole vine continues until harvest.[133] Uptake studies using ^{15}N confirm that the accumulation of nitrogen by kiwifruit vines is rapid and almost complete within 10 weeks of application in either late spring or early summer.[138] Initially most of the ^{15}N was present in the leaves and the roots, but this declined due to the translocation to other components of the vine. Analysis of xylem sap approximately 4 weeks after bud break showed that more than 50% of the nitrogen absorbed by the roots and transferred in the xylem, remained as nitrate.[138] The other half of the nitrogen in the sap which had been derived from soil and fertilizer nitrogen was first reduced in the roots and translocated predominantly as glutamine.[126,139] During summer and autumn, the nitrogen required for developing fruit and shoots was met to a large extent by the redistribution of the recently absorbed nitrogen from the leaves and roots. Hence, the decline in leaf ^{15}N coincided with an equivalent accumulation of ^{15}N in the fruit.[138]

While a large fraction of the annual uptake of nutrients occurs during the first 10 weeks after bud break, mobilization of nutrients stored in the vine from the previous season also makes a contribution during this period of growth. Sufficient quantities of nitrogen, potassium, phosphorus, and magnesium can be mobilized from the laterals of the vine to support the growth of 20–40% of the leaves during the first 30 d after bud break.[46] An analysis of the xylem sap has indicated that remobilization of nitrogen stored in the roots contributes about 60% of the total nitrogen for new growth at 4 weeks after bud break and about 30% approximately 4 weeks later.[138]

In addition to the buffering effects of the nutrient reserves in the vine, the rate of uptake is also influenced by environmental variables. For example, a strong association has been found between soil temperature and the concentration of potassium in the leaves in spring, suggesting that the uptake of potassium by kiwifruit vines is temperature dependent.[46] While there are no published results on direct measurements of the uptake of nutrients per unit length of root of vines growing in the field, estimated rates for macronutrients range from 13.7×10^{-12} mol m^{-1} s^{-1} for nitrogen to 0.82×10^{-12} mol m^{-1} s^{-1} for phosphorus.[11] These values are similar to the uptake rates measured in the field for annual *Graminaceous* sp.,[140,141] and estimated for apple.[11]

An unusual feature of kiwifruit vines is their relatively high requirement for chlorine. Kiwifruit vines require at least 2–6 g kg^{-1} dry weight of chlorine in their leaves to maintain healthy growth.[142,143] These levels are at least 10 times those required by other non-halophytic plants.[144,145] Typically the concentration of chloride in the leaves of high-producing kiwifruit vines in New Zealand ranges from 8–20 g kg^{-1} dry weight.[132] Chlorine toxicity is not usually observed in kiwifruit until the concentrations in the leaf exceed 25 g kg^{-1}.[132] A close association between the availability of nitrate and the absorption of chloride ions by kiwifruit vines has been established.[143] The high requirement for chlorine was considered to be an adaptation which minimizes the expenditure of energy on the generation of osmotic pressure, and probably reflects the high availability of chloride and ammonium ions in the habitat in which this species has evolved.[143]

Recently the effects of potassium fertilizer applied with chloride or sulfate as the accompanying anion on the potassium nutrition of kiwifruit vines have been assessed in the field. The concentrations of potassium in the leaves during the first 6 weeks after bud break were significantly greater for those vines receiving potassium chloride compared to those receiving potassium sulfate.[146] Subsequently there was no significant difference in the concentration of potassium in the leaves for the vines receiving potassium in the chloride or sulfate form. This transient effect of the accompanying anion on the potassium status of the leaf was associated with large effects on flowering with the resulting fruit yields being about 28% greater for the vines receiving potassium chloride rather than potassium sulfate. Applying potassium in the chloride form increased the concentration of chloride in the leaf, especially in spring, while applying potassium in the sulfate form had no significant effect on the concentration of sulfur in the leaf at that time. Kiwifruit vines are well adapted to using chloride ions rather than organic-acid anions such as malate for charge balance.[142,143] Hence, it is likely that the requirement for organic-acid anions for charge balance and maintenance of potassium uptake was greater where potassium sulfate rather than potassium chloride was the source of potassium. The lower concentrations of potassium in the leaves of kiwifruit vines in spring following the application of potassium sulfate, suggests that organic-acid anion availability at this time limited potassium uptake, as has been reported for other plant species.[147,148] For kiwifruit vines this effect may be associated with the slow canopy development in spring,[9,34] which in turn limits the availability of carbon skeletons for synthesis of organic-acid anions.

The developing fruit are undoubtedly responsible for many of the seasonal changes in the distribution of mineral nutrients within the vine during the growing season. Generally it has been found that leaves close to the fruit are the main suppliers of nutrients accumulated during early fruit growth.[46] The losses of potassium and nitrogen from the leaves, especially those close to the fruit, reflect the large demand of the developing fruit for these two elements; whereas the much smaller losses of phosphorus, sulfur, magnesium, calcium, and most micronutrients reflect their lower mobility in the vine and the smaller demand of the developing fruit for these elements. A mass-flow vascular transport model has been developed to predict the relative importance of the phloem and xylem supply for nutrient accumulation by the fruit.[52] The model indicates that transport to the fruit is principally by the phloem, except for calcium, manganese, and zinc, which are transported almost exclusively by the xylem.

B. NUTRIENT DISORDERS

1. Potassium Deficiency

The most significant nutrient disorder affecting kiwifruit vines in New Zealand in recent years has been a deficiency of potassium.[132,149] This deficiency appears most commonly in orchards approaching maturity and beginning to produce large quantities of fruit. The high incidence of potassium deficiency reflects the large quantity of this element that is removed from the orchard in fruit and inadequate applications of potassium in fertilizer.[132,149] As potassium deficiency can be detected readily by leaf analysis and corrected with fertilizer,[149] this disorder is now much less common in New Zealand. Maximum fruit yields were associated with potassium concentrations in the leaves in excess of 25 g kg^{-1} dry weight 6 weeks after bud break.[149] The average fruit weight decreased from 107 g on healthy vines to 90 g on the severely deficient vines.

2. Nitrogen Deficiency

Recent studies suggest that nitrogen deficiency of kiwifruit vines is more common than has been thought previously.[134,138,150] Maximum fruit yields were associated with concentrations of nitrogen in the leaves of 42 g kg^{-1} dry weight at 6 weeks after bud break and 25 g kg^{-1} dry weight at 12–20 weeks after bud break. A 50% reduction in photosynthetic rate has been measured for leaves from nitrogen-deficient kiwifruit vines.[63] One reason for the incidence of this deficiency is the large requirement for nitrogen in the developing framework of the kiwifruit vine.[133,143]

Minor nitrogen deficiencies are difficult to detect unless leaf samples for analysis are taken very early in the growing season.[132] Detection of nitrogen deficiency early in the season enables correction of the disorder while the canopy and fruit growth can still respond to an improvement in the nitrogen status of the vine.[46,150]

3. Other Nutrient Deficiencies

Production losses have also been associated with deficiencies of magnesium[151] and manganese.[152] In New Zealand, manganese deficiency has been most easily overcome through reducing the pH of the soil below 6.8.[152] In California, zinc deficiency is reported to be common,[153] while iron and manganese deficiencies are widespread in parts of Italy.[154,155] However, the effects of these disorders on fruit production have not been measured.

4. Nutrient Toxicities

Instances of excess nutrients affecting productivity have also been measured. Kiwifruit vines are especially sensitive to excess boron. Large reductions in fruit yield have been reported where the concentrations of boron in the leaves exceeded 80 $\mu g\ g^{-1}$ dry weight and associated concentrations of hot water-soluble boron in the topsoil were greater than 0.5 $\mu g\ g^{-1}$ dry weight.[156] Typically the concentration of boron in the leaves of high producing kiwifruit vines in New Zealand ranges from 40 to 50 $\mu g\ g^{-1}$ dry weight.[132]

Suggestions that kiwifruit vines are sensitive to chlorine toxicity[157] have not been supported by direct experimentation. Where in excess of 1.2 t ha^{-1} of chlorine has been applied in one application in spring with equivalent quantities of nitrogen to kiwifruit vines in the field, there was no indication that chloride ions adversely affected the vines despite the concentrations in the soil being up to 10–15 times those usually found in commercial kiwifruit orchards in New Zealand.[146,158] In a long-term study with field-grown kiwifruit vines, nitrogen deficiency has been associated with the appearance of leaf chlorosis and necrosis and increased concentrations of chloride in the leaves despite the concentrations of chloride in the soil remaining relatively constant at about 60 mg kg^{-1}.[150] These changes in the composition of kiwifruit leaves and appearance of symptoms are typical of those as nitrogen concentrations vary in the vine.[146] Hence the suggestion that these leaf symptoms indicate chlorine toxicity[157] does not appear to be valid.

Kiwifruit vines are generally tolerant of relatively high concentrations of soluble salts in the root zone, with the exception of sodium. Severe damage to kiwifruit vines has resulted from the use of irrigation water containing sodium concentrations that ranged from 230–860 mg l^{-1}.[132] This sensitivity to sodium, rather than chlorine, largely accounts for their sensitivity to saline irrigation water.[132,159] Where exceptionally large quantities of soluble fertilizer (up to 10 t ha^{-1} of a 12:10:10 N:P:K mixture) have been used to increase the soluble salt level in the root zone to high levels (the electrical conductivity of the soil was increased to over 2 d sm at 25°C), there was a marked increase in the number of buds that developed in spring, along with flower numbers, and root growth.[158] Measurements at harvest showed that fruit yields were doubled for the high fertilizer treatment compared to the control vines which received typical annual inputs of fertilizer. The results from this study suggest that osmotic effects may play an important part in the bud-breaking process in spring.

A feature common to all mineral nutrient disorders of kiwifruit vines is that the loss of productivity results primarily from a reduction in fruit numbers.[11] The average weight of individual fruit from affected and unaffected vines has been found to be similar in most cases.

C. EFFECTS OF MINERAL NUTRITION ON FRUIT QUALITY

A feature of fruit from kiwifruit vines is their ability to be stored for remarkably long periods (6–8 months) at low temperatures (−0.5 to 0.5°C) before being ripened to an edible state.[1] Various claims have been made, however, about the extent to which mineral nutrition influences the post-harvest storage of kiwifruit. In the specific case of nitrogen, results from the long-term study of responses to nitrogen fertilizer have shown no significant effect on fruit quality with fertilizer inputs ranging from 0–200 kg ha^{-1} $year^{-1}$.[150] Applying even greater quantities of nitrogen (1400 kg ha^{-1}) just before bud break, had no effect on the long-term storage of the fruit.[158] In a survey of the relationship between vine nitrogen status and post-harvest fruit quality for kiwifruit growing in Italy, no relationship could be found between the concentration of nitrogen in the leaves sampled mid-season and flesh firmness of the fruit after harvest.[160] These results fail to support an earlier claim that nitrogen may be linked to a more rapid deterioration of fruit in storage and with a higher incidence of *Botrytis* infection.[161,162] It would appear that the detrimental effects of nitrogen fertilizers on the post-harvest storage characteristics of kiwifruit have been confounded by other factors. For example, the small but significant decrease in fruit firmness attributed to an increase in the nitrogen concentration in the fruit was confounded by the use of ammonium sulfate as the source of nitrogen.[163] An even greater change in the concentration of manganese than nitrogen was measured in these fruits, as resulted from the acidification of the root zone through the use of this form of nitrogen. Consequently, the extent to which nitrogen fertilizer was having a direct effect on fruit quality or an indirect effect through changes in the acidity of the root zone remains

Table 5 **Relationship between mineral composition and postharvest fruit disorders**

Nutrient element	Botrytis infection (%)	Soft fruit (%)	Surface pitting (%)
Potassium	0.133	−0.013	0.048
Nitrogen	0.076	−0.019	−0.122
Chlorine	0.048	−0.130	0.058
Calcium	−0.077	−0.045	−0.009
Phosphorus	0.053	−0.028	0.114
Sulfur	0.077	−0.021	−0.084
Magnesium	0.098	0.085	0.064
Sodium	0.038	−0.074	−0.086
Iron	−0.027	0.051	0.054
Boron	0.151	0.059	−0.047
Copper	0.100	0.021	−0.085
Zinc	0.011	0.145	−0.057
Manganese	0.017	0.029	0.078

Note: The values in the table are correlation coefficients (r). Significant values for r were p (0.05) = 0.156.

uncertain. Similarly an earlier report of a reduction in storage life of kiwifruit,[164] was associated with the abnormal practice of applying nitrogen in late summer/early autumn when vine growth had largely ceased.[9] In New Zealand it is general practice to apply most, if not all of the annual requirement of nitrogen before or during the first 10 weeks after bud break in spring.[11]

Recently the relationship between the mineral composition and post-harvest disorders of kiwifruit was investigated using a large sample of fruit (1660) from 166 commercial orchards in the major kiwifruit growing districts of New Zealand. Included were examples of fruit that had softened prematurely and/or were infected with *Botrytis* while in storage and fruit with no storage disorders.[165] There were no clear relationships between the mineral composition and the incidence of soft fruit or *Botrytis* infection. The correlation coefficients rarely exceeded 0.1 for any of the individual nutrient elements (Table 5). There was also little evidence to suggest that nitrogen or calcium was more strongly associated with post-harvest fruit disorders than any other mineral nutrient in the fruit. Principal component analysis of the results indicated that nutrient elements could be separated into two distinct groups. The first group, comprised of nitrogen, phosphorus, sulfur, potassium, and copper, behaved in the opposite manner to the second group of elements, which were equally strongly linked together and were comprised of calcium, manganese, and zinc. Hence the effect of an individual element cannot be separated from that of the other elements present in these groupings.

The absence of any obvious relationship between calcium and post-harvest fruit disorders of commercially grown fruit was consistent with earlier findings where no relationship was found between total calcium concentrations in the fruit or young leaves and the firmness of the fruit during storage.[166,167] Calcium concentrations in kiwifruit (0.5–3.7 g kg^{-1} dry weight) are up to 10 times greater than those in the flesh of other fruits which often develop calcium related disorders, such as apples.[168,169] However, exceptionally low concentrations of calcium (<0.2 g kg^{-1} dry weight) have been measured in kiwifruit from China.[165] These fruit had large surface pits and necrotic tissue surrounding the vascular tissues to the seeds. A large fraction of calcium in kiwifruit has been found to be present as calcium oxalate,[1] but the concentration of the more soluble forms of calcium in the flesh can be at least three times greater than the total calcium concentration in apples.[170] The suggestion that soluble forms of calcium may reflect more closely the behavior of kiwifruit during storage than total calcium[162] has not been supported by direct measurement.[170]

No significant differences in fruit quality have been associated with potassium deficiency,[132] magnesium deficiency,[151] or manganese deficiency,[152] but excess boron has been associated with a significant reduction in the firmness of the fruit after only a short period in storage.[156]

VII. WHOLE-PLANT EXPRESSION OF STRESS

In an earlier review,[9] the interactions between roots and shoots were described in terms of the acquisition of resources for growth and how these relationships changed in response to environmental stresses.

Carbon acquisition has been shown to be central to the responses of kiwifruit vines to most environmental stresses. As shown in the sections above, the gross effects of stresses on the vine are relatively easy to identify and describe. However, it is the subtle effects of these stresses that are less understood, frequently overlooked, and often these effects are attributed to other factors and confused with natural variation within the vine itself. Hence understanding the growth patterns and extent of the variation of "stress free" vines is important in examining the responses of the vine to environmental stresses.

Spatial heterogeneity of the physical, chemical, and post-harvest attributes of fruit appears to reflect effects of at least some "subtle" environmental stresses. Recent studies of this spatial heterogeneity suggest that a large fraction of the variation in fruit quality encountered in the New Zealand kiwifruit industry originates at the level of the individual vine.[171] For many species, differences in the physical, chemical, and post-harvest attributes of fruit can be detected when the entire population of a single plant is analyzed.[172,173] The recent development of a rapid method for representing in 3-dimensional space the location of each fruit[174] has highlighted relationships between spatial heterogeneity of fruit quality and the position of the fruit in the canopy. Fruit with above-average physical and post-harvest attributes and flavor were from parts of the canopy with high leaf area density, while fruit with less desirable attributes were from the less dense parts of the canopy close to the ground.[171,175] Such patterns suggest that, at least in parts of the canopy, resources for fruit growth are limited.

Of the limited number of studies where the post-harvest characteristics of the fruit from vines exposed to an environmental stress (root anoxia) have been assessed according to their position in the canopy of the vine, the results show an increase in the number of fruit with inferior attributes from the less dense parts of the canopy.[128] However, the apparent relationship between the temporal and spatial variation in photosynthetic activity within the canopy and fruit growth[171] and quality indicates that the dynamics of acquisition and allocation of resources within the vine are likely to be very complex and require further clarification.

VIII. PROSPECTS FOR MINIMIZING STRESS

Understanding the mechanisms and environmental factors regulating the allocation of resources within the whole vine is central to identifying options for minimizing adverse effects of environmental stresses. The importance of recognizing the complex patterns that exist in the spatial arrangement of the leaves and fruit in the canopy lies in developing strategies for producing fruit of a more uniform quality. Differences in canopy architecture and management practices such as leaf pruning and fruit thinning can substantially modify the interception of radiation and allocation of resources within the vine.

For kiwifruit in New Zealand, the effect of environmental stress is usually inversely related to vine vigor. In particular, high producing vines growing on deep, free-draining soils in the Bay of Plenty region are often considered to be excessively vigorous.[1] These vines usually have extensive root systems,[22,23] but are rarely affected by water stress[109,119] or mineral nutrient stress.[132] It is likely that this "vigorous" growth of such vines is central to high rates of carbon acquisition and hence carbon allocation for fruit and root growth and tolerance of below-ground stresses. Options for minimizing stress therefore include:

- Ensuring a uniform bud break in spring. In locations where natural bud break is poor, the use of the bud breaking chemical, hydrogen cyanamide, increases the number of dormant buds that break and uniformity of shoot development.[38] The suppression of shoot growth by shoots that develop early can be prevented by removing the terminal bud of any early developing shoot. Such treatment can increase fruit yields by 20–40%.[176]
- Reducing the period over which the flowers open in spring. Hydrogen cyanamide is effective in compressing the flowering period in locations where there is a history of flowering problems.[38,39] A prolonged flowering period results in a population of fruit on a single vine with a wide range of physiological ages at harvest.[171,177] This diversity in fruit age impacts on the post-harvest storage potential of these fruit.
- Increasing the rate of leaf canopy development in spring by avoiding leaf pruning during the first 100 d after bud break and ensuring that the vine is adequately supplied with mineral nutrients, especially nitrogen. For New Zealand conditions the optimum leaf index appears to be about 3.0–3.5 m^2 m^{-2}.[34] Although the direct effects of limited carbon acquisition may not necessarily be immediate, fruit production in the long term can be very sensitive to even subtle changes in the leaf area index. Often these changes are reflected in the production of flowers for the following season.[12] However, where the limitation of

carbon acquisition within any one season is sufficiently severe, especially during the first few weeks after fruit set, fruit growth can be affected directly as illustrated by the reduction in average fruit size following a reduction in total leaf area.[12,178]

- Increasing radiation interception by the leaf canopy. The relative importance of limitations to carbon acquisition due to radiation environment, leaf area, and the photochemical efficiency of the canopy is difficult to measure directly. Mathematical modelling, however, suggests that variations in radiation interception are relatively more important than variations in photochemical efficiency.[44] Minimizing shading, such as from wind shelter, provides the greatest opportunity for manipulating the radiation environment in an orchard.
- Maximizing starch reserves in the vine by avoiding overcropping or removal of too many leaves during the previous season.
- Ensuring that a large root system is maintained. Because the root system of kiwifruit vines is a poor competitor for resources of growth, indirect effects of limited supplies of these resources increase the sensitivity of the vine to below-ground stresses such as drought, anoxia, and mineral nutrient deficiencies. Increasing the size of the root system effectively increases the volume of soil explored by the roots thereby decreasing the concentration of resources required in the soil to support a given rate of uptake.[141]
- Reducing the dependence of the vine on irrigation by encouraging the growth of a large root system. The greater volume of soil explored by a large root system means that the transpiration requirements of the vine can be met at a greater soil moisture tension because of the overall increase in the quantity of water available to the vine.[9]
- Avoiding the risks of root anoxia and the adverse effects which excess water has on the storage properties of the fruit, by refraining from flood irrigation or the application of large quantities of water after infrequent intervals to slowly draining soil types.
- Improving the carbon allocation to the roots by ensuring a complete canopy and hence radiation interception at an early stage of the growing season and restricting crop load, especially on young developing vines.
- Selecting rootstocks that are better adapted to local environments. Kiwifruit production has traditionally involved 'Hayward' grafted on to seedling rootstocks or 'Hayward' grown on its own roots.[10] Neither root system confers any particular advantage to the 'Hayward' scion. However, the recent selection of the clonal rootstock of *Actinidia hemsleyana* ('Kaimai') for 'Hayward',[179] has shown what can be achieved in terms of improvements in production and fruit quality.

REFERENCES

1. Ferguson, A. R., Kiwifruit: a botanical review, *Horticultural Reviews,* 6, 1, 1984.
2. Ferguson, A. R. and Bollard, E. G., Domestication of the kiwifruit, in *Kiwifruit Science and Management,* Warrington, I. J. and Weston, G. C., Eds., Richards Publisher, Auckland, 1990, 165.
3. Ferguson, A. R., The genus *Actinidia,* in *Kiwifruit Science and Management,* Warrington, I. J. and Weston, G. C., Eds., Richards Publisher, Auckland, 1990, 15.
4. Warrington, I. J., Areas and trends of kiwifruit production in New Zealand and around the world, in *Kiwifruit Science and Management,* Warrington, I. J. and Weston, G. C., Eds., Richards Publisher, Auckland, 1990, 511.
5. Li, R., Huang, C., Liang, M. and Huang, Z., Investigation of germplasm resources of *Actinidia* in Guangxi, *Guihaia* 5, 253, 1985.
6. Ferguson, A. R., Botanical nomenclature: *Actinidia chinensis, Actinidia deliciosa and Actinidia setosa,* in *Kiwifruit Science and Management,* Warrington, I. J. and Weston, G. C., Eds., Richards Publisher, Auckland, 1990, 36.
7. Sale, P. R. and Lyford, P. B., Cultural, management and harvesting practices for kiwifruit in New Zealand, in *Kiwifruit Science and Management,* Warrington, I. J. and Weston, G. C., Eds., Richards Publisher, Auckland, 1990, 247.
8. Walton, E. F. and De Jong, T. M., Growth and compositional changes in kiwifruit berries from three California locations, *Annals of Botany,* 66, 285, 1990.
9. Buwalda, J. G. and Smith, G. S., Acquisition and utilisation of carbon, mineral nutrients, and water by the kiwifruit vine, *Horticultural Reviews,* 12, 307, 1990.
10. Sale, P. R., *Kiwifruit Growing,* GP Books, Wellington, 1990, 84 pp.

11. Smith, G. S., Buwalda, J. G. and Clark, C. J., Nutrient dynamics of a kiwifruit ecosystem, *Scientia Horticulturae,* 37, 87, 1988.
12. Buwalda, J. G. and Smith, G. S., Effects of partial defoliation at various stages of the growing season on fruit yields, root growth and return bloom of kiwifruit vines, *Scientia Horticulturae,* 42, 29, 1990.
13. Ferguson, A. R., Stem, branches, leaves, and roots of the kiwifruit vine, in *Kiwifruit Science and Management,* Warrington, I. J. and Weston, G. C., Eds., Richards Publisher, Auckland, 1990, 58.
14. Morgan, E. R. and McNaughton, K. G., Architecture of a kiwifruit canopy, *New Zealand Journal of Crop and Horticultural Science,* 19, 237, 1991.
15. Brundell, D. J., Flower development of the Chinese gooseberry (*Actinidia chinensis* Planch.), 1. The development of the flowering shoot, *New Zealand Journal of Botany,* 13, 473, 1975.
16. Lai, R., Leaf-fruit relationship in kiwifruit [*Actinidia deliciosa* (A. Chev), Liang, C. F. and Ferguson, A. R.], PhD Thesis, Massey University, Palmerston North, 1987.
17. Ray, T. S., Cyclic heterophylly in *Syngonium* (Araceae), *American Journal of Botany,* 74, 16, 1987.
18. Hughes, K. A. and Wilde, R. H., The effect of poor drainage on the root distribution of kiwifruit vines, *New Zealand Journal of Crop and Horticultural Science,* 17, 239, 1989.
19. Reid, J. B. and Petrie, R. A., Effects of soil aeration on root demography in kiwifruit, *New Zealand Journal of Crop and Horticultural Science,* 19, 423, 1991.
20. d'Andria, R., Magliulo, V., Giorio, P., Sorrentino, G. and Quagilietta, C., Root distribution of kiwifruit as influenced by winter water-table level and irrigation regime, *New Zealand Journal of Crop and Horticultural Science,* 19, 369, 1991.
21. Magliulo, V., d'Andria, R., Giorio, P., Martorella, A. and Quagilietta, C., Cumulative effects of winter shallow water-table conditions on stem growth and yield of kiwifruit, *New Zealand Journal of Crop and Horticultural Science,* 19, 375, 1991.
22. Greaves, A. J., Root distribution of kiwifruit (*Actinidia deliciosa*) in a deep sandy loam soil of the Te Puke district, New Zealand, *New Zealand Journal of Agricultural Research,* 28, 433, 1985.
23. Hughes, K. A., Gandar, P. W., Menalda, P. H. and Snow, V. O., A survey of kiwifruit root systems, Tech. Rept. 22, DSIR Plant Physiology Division, Palmerston North, 1986.
24. Lemon, C. W., The root system of *Actinidia deliciosa* (A. Chev.), Liang, C. F. et Ferguson, A. R. var. *deliciosa* (kiwifruit vine), MSc Thesis, University of Auckland, Auckland, 1986.
25. Buwalda, J. G. and Smith, G. S., Accumulation and partitioning of dry matter and mineral nutrients in developing kiwifruit vines, *Tree Physiology,* 3, 295, 1987.
26. Gandar, P. W. and Hughes, K. A., Kiwifruit root systems, 1. Root length densities, *New Zealand Journal of Experimental Agriculture,* 16, 35, 1988.
27. Wilson, K. S., Water and energy relationships of kiwifruit, MSc Thesis, University of Waikato, Hamilton, 1988.
28. Haynes, R. J. and Goh, K. M., Distribution and budget of nutrients in a commercial apple orchard, *Plant and Soil,* 56, 445, 1980.
29. Buwalda, J. G. and Lenz, F., Effects of cropping, nutrition and water supply on accumulation and distribution of biomass and nutrients for apple trees on 'M9' root systems, *Physiologia Plantarum,* 84, 21, 1992.
30. Ferguson, A. R. and Bank, R. J., The great DSIR demolition derby, *New Zealand Kiwifruit,* March 1986, 25.
31. Buwalda, J. G. and Hutton, R. C., Seasonal changes in root growth of kiwifruit, *Scientia Horticulturae,* 36, 251, 1988.
32. Atkinson, D., The distribution and effectiveness of the roots of tree crops, *Horticultural Reviews,* 2, 424, 1980.
33. Freeman, B. M. and Smart, R. E., A root observation laboratory for studies with grapevines, *American Journal of Enology and Viticulture,* 27, 36, 1976.
34. Buwalda, J. G., Meekings, J. S. and Smith, G. S., Radiation and photosynthesis in kiwifruit canopies, *Acta Horticulturae,* 297, 307, 1991.
35. Hopping, M. E., Floral biology, pollination, and fruit set, in *Kiwifruit Science and Management,* Warrington, I. J. and Weston, G. C., Eds., Richards Publisher, Auckland, 1990, 71.
36. Costa, G., Succi, F., Biasi, R. and Miserocchi, O., Vegetative and cropping performance of kiwifruit (cv. Hayward) as related to pruning length and bud number, *Acta Horticulturae,* 282, 1990, 113.
37. Walton, E. F., Winter-chilling requirements of kiwifruit, in *1983/84 Annual Report,* Agricultural Research Division, Ministry of Agriculture and Fisheries, Wellington, New Zealand, 1985, 30.

38. Henzell, R. F. and Briscoe, M. R., Hydrogen cyanamide: a tool for consistently high kiwifruit production, *New Zealand Kiwifruit Special Publication,* 1. National Research Conference, 1986, 8.
39. Linsley-Noakes, G. C., Improving flowering of kiwifruit in climatically marginal areas using hydrogen cyanamide, *Scientia Horticulturae,* 38, 247, 1989.
40. Walton, E. F., Clark, C. J. and Boldingh, H. L., Effect of hydrogen cyanamide on amino acid profiles in kiwifruit buds during budbreak, *Plant Physiology,* 97, 1256, 1991.
41. McPherson, H., Stanley, J., Warrington, I. and Jansson, D., Dynamics of budbreak and flowering, *New Zealand Kiwifruit Special Publication,* 2. National Research Conference, 1988, 9.
42. Grant, J. A. and Ryugo, K., Influence of developing shoots on flowering potential of dormant buds of *Actinidia chinensis, HortScience,* 17, 977, 1982.
43. Brundell, D. J., Flower development of the Chinese gooseberry (*Actinidia chinensis* Planch.) and some factors influencing it, MHort Sci. Thesis, Massey University, Palmerston North, New Zealand, 1973.
44. Buwalda, J. G., A mathematical model of carbon acquisition and utilisation by kiwifruit vines, *Ecological Modelling,* 57, 43, 1991.
45. Buwalda, J. G., Meekings, J. S. and Smith, G. S., Seasonal changes in photosynthetic capacity of leaves of kiwifruit (*Actinidia deliciosa*) vines, *Physiologia Plantarum,* 83, 93, 1991.
46. Smith, G. S., Clark, C. J. and Henderson, H. V., Seasonal accumulation of mineral nutrients by kiwifruit, 1. Leaves, *The New Phytologist,* 106, 81, 1987.
47. Smith, G. S., Clark, C. J. and Boldingh, H. L., Seasonal accumulation of starch by components of the kiwifruit vine, *Annals of Botany,* 70, 19, 1992.
48. Dick, J. K., Gas exchange in orchard grown kiwifruit vines, MSc Thesis, University of Waikato, Hamilton, 1987.
49. Nuzzo, V., Lain, O. and Giulivo, C., Techniques and experimental approaches in water relations studies of kiwi, *International Symposium on Kiwi Abstracts,* Edizioni Libreria Progetto Padova, 1987, 64.
50. Clark, C. J., Smith, G. S. and Walker, G. D., The form, distribution and seasonal accumulation of calcium in kiwifruit leaves, *The New Phytologist,* 105, 477, 1987.
51. Hopping, M. E., Structure and development of fruit and seeds in Chinese gooseberry (*Actinidia chinensis* Planch.), *New Zealand Journal of Botany,* 14, 63, 1976.
52. Clark, C. J. and Smith, G. S., Seasonal accumulation of nutrients by kiwifruit, 2. Fruit, *The New Phytologist,* 108, 399, 1988.
53. Walton, E. F., Estimation of the bioenergetic cost to grow a kiwifruit (*Actinidia deliciosa*) berry, PhD Thesis, University of California, Davis, 1988.
54. Buwalda, J. G., Root growth of kiwifruit vines and the impact of canopy manipulations, in *Plant Roots and Their Environment; Developments in Agricultural and Managed Forest Ecology,* 24, McMichael, B. L. and Persson, H., Eds., Elsevier, Amsterdam, 1991, 431.
55. Kolesnikov, V. A., The interrelationship between the aerial portion and the root system of fruit trees (in Russian), *Vestnik Sel'koh,* Nauki 11, 46, 1966.
56. Grant, J. A. and Ryugo, K., Influence of within-canopy shading on net photosynthetic rate, stomatal conductance, and chlorophyll content of kiwifruit leaves, *HortScience,* 19, 834, 1984.
57. Flore, J. A. and Lakso, A. N., Environmental and physiological regulation of photosynthesis in fruit crops, *Horticultural Reviews,* 11, 111, 1989.
58. Buwalda, J. G. and Meekings, J. S., Parameter estimation and validation of a canopy photosynthesis model for kiwifruit vines, *Acta Horticulturae,* 313, 29, 1992.
59. Kriedemann, P. E. and Smart, R. E., Effects of irradiance, temperature, and leaf water potential on photosynthesis of vine leaves, *Photosynthetica,* 5, 6, 1971.
60. Avery, D. J., Maximum photosynthesis rate—a case study in apple, *The New Phytologist,* 78, 55, 1977.
61. De Jong, T. M., CO_2 assimilation characteristics of five *Prunus* tree fruit species, *Journal of American Society of Horticultural Science,* 108, 303, 1983.
62. Jones, H. G., *Plants and Microclimate,* Cambridge University Press, 1983.
63. De Jong, T. M., Tombesi, A. and Ryugo, K., Photosynthetic efficiency of kiwi (*Actinidia chinensis* Planch.) in response to nitrogen deficiency, *Photosynthetica,* 18, 139, 1984.
64. Morgan, D. C., Warrington, I. J. and Halligan, E. A., Effect of temperature and photosynthetic photon flux density on vegetative growth of kiwifruit (*Actinidia chinensis*), *New Zealand Journal of Agricultural Research,* 28, 109, 1985.
65. Laing, W. A., Temperature and light response curves for photosynthesis in kiwifruit (*Actinidia chinensis*) cv. Hayward, *New Zealand Journal of Agricultural Research,* 28, 117, 1985.

66. Björkman, O., Responses to different quantum flux densities, in Physiological plant ecology I: Responses to the physical environment, *Encyclopedia of Plant Physiology,* New Series, Vol. 12A, Lange, O. L., Nobel, P. S., Osmond, C. S. and Ziegler, H., Eds., Springer-Verlag, Berlin, 1981, 57.
67. Greer, D. H., Effect of temperature of photoinhibition and recovery in *Actinidia deliciosa, Australian Journal of Plant Physiology,* 15, 195, 1988.
68. Jurik, T. W., Chabot, J. F. and Chabot, B. F., Ontogeny of photosynthetic performance, in *Fragaria virginiana* under changing light regimes, *Plant Physiology,* 63, 542, 1979.
69. Catsky, J. and Ticha, I., Ontogenetic changes in the internal limitations to bean-leaf photosynthesis, *Photosynthetica,* 14, 392, 1980.
70. Harper, J. L., The value of a leaf, *Oecologia,* 80, 53, 1989.
71. Sams, C. E. and Flore, J. A., Net photosynthetic rate of sour cherry (*Prunus cerasum* L. 'Montmorency') during the growing season, with particular reference to fruiting, *Photosynthesis Research,* 4, 307, 1983.
72. Field, C. B. and Mooney, H. A., The photosynthesis-nitrogen relationship in wild plants, in *On the Economy of Plant Form and Function,* Givnish, T. J., Ed., Cambridge University Press, Cambridge, 1986, 25.
73. Curtis, J. P. and Buwalda, J. G., Computer simulation of radiation interception and photosynthesis for kiwifruit canopies in the field, *Acta Horticulturae,* 313, 37 1992.
74. Monsi, M. and Saeki, T., Über der lichtfaktor in den Pflanzengesellschaften und seine Bedeutung für die Stoffproduktion, *Japan Journal of Botany,* 14, 22, 1953.
75. Goudriaan, J., A family of saturation type curves, especially in relation to photosynthesis, *Annals of Botany,* 43, 783, 1979.
76. Buwalda, J. G., Green, T. G. A. and Curtis, J. P., Canopy photosynthesis and respiration of kiwifruit (*Actinidia deliciosa* var deliciosa) vines growing in the field, *Tree Physiology,* 10, 327, 1992.
77. Succi, F., Studio di fotosintesi e transpirazione in *Actinidia deliciosa* nel corso della stagione vegetativa, PhD Thesis, Università degli Studi di Bologna, Bologna, 1992, 112 pp.
78. Buwalda, J. G., Green, T. G. A., Meekings, J. S. and Coneybear, D. J., Measurement of canopy gas exchange of kiwifruit vines using a suite of whole-canopy cuvettes, *Environmental and Experimental Botany,* 32, 425, 1992.
79. Smith, G. S., Judd, M. J., Miller, S. A. and Buwalda, J .G., Recovery of kiwifruit vines from transient waterlogging of the root system, *The New Phytologist,* 115, 325, 1990.
80. Raschke, K., Stomatal action, *Annual Review of Plant Physiology,* 26, 309, 1975.
81. Azcon-Bieto, J., Inhibition of photosynthesis by carbohydrates in wheat leaves, *Plant Physiology,* 73, 681, 1983.
82. McAneney, K. J., Judd, M. J. and Trought, M. C. T., Wind damage to kiwifruit (*Actinidia chinensis* Planch.) in relation to windbreak performance, *New Zealand Journal of Agricultural Research,* 27, 255, 1984.
83. Manson, P., Snelgar, B. and Palmer, J., Live shelter: trim, top or take it out?, *New Zealand Kiwifruit,* June, 9, 1989.
84. Morgan, D. C., Stanley, C. J. and Warrington, I. J., The effects of simulated daylight and shade-light on vegetative and reproductive growth in kiwifruit and grapevine, *Journal of Horticultural Science,* 60, 473, 1985.
85. Snelgar, W. P. and Hopkirk, G., Effect of overhead shading on yield and fruit quality of kiwifruit (*Actinidia deliciosa*), *Journal of Horticultural Science,* 63, 731, 1988.
86. Brundell, D. J., The effect of chilling on the termination of rest and flower bud development of the Chinese gooseberry, *Scientia Horticulturae,* 4, 175, 1976.
87. Lionakis, S. M. and Schwabe, W. W., Some effects of day length, temperature and exogenous growth regulator application on the growth of *Actinidia chinensis* Planch., *Annals of Botany,* 54, 485, 1984.
88. Lawes, G. S., Winter temperatures and kiwifruit bud development, *The Orchardist of New Zealand,* 57, 110, 1984.
89. Lötter, J. de V., An evaluation of the climatic suitability of various areas in Southern Africa for commercial production of Hayward kiwifruit (*Actinidia chinensis*), *Deciduous Fruit Grower,* 34, 112, 1984.
90. Lötter, J. de V., Potential of kiwifruit production in the Western Cape, *Deciduous Fruit Grower,* 34, 311, 1984.
91. Davison, R. M., The physiology of the kiwifruit vine, in *Kiwifruit Science and Management,* Warrington, I. J. and Weston, G. C., Eds., Richards Publisher, Auckland, 1990, 127.

92. McPherson, H., Warrington, I. J. and Stanley, J., Winter and spring temperature effects on budbreak and flowering, *Kiwifruit Cropping with Hi-Cane,* Nufarm and MAFTech North, 1989, 5.
93. Morgan, D. C., Warrington, I. J. and Halligan, E. A., Effect of temperature and photosynthetic photon flux density on vegetative growth of kiwifruit (*Actinidia chinensis*), *New Zealand Journal of Agricultural Research,* 28, 109, 1985.
94. Morley-Bunker, M. J. and Salinger, M. J., Kiwifruit development—the effect of temperature on bud burst and flowering, *Weather and Climate,* 7, 26, 1987.
95. Beever, D. J. and Hopkirk, G., Fruit development and fruit physiology, in *Kiwifruit Science and Management,* Warrington, I. J. and Weston, G. C., Eds., Richards Publisher, Auckland, 1990, 97.
96. Harman, J. E., Kiwifruit Maturity, *The Orchardist of New Zealand,* 54, 126, 1981.
97. Seager, N. G., Hewett, E. W., Warrington, I. J. and MacRae, E. A., The effect of temperature on the rate of kiwifruit maturation using controlled environments, *Acta Horticulturae,* 297, 247, 1991.
98. Seager, N. G., Hewett, E. W. and Warrington, I. J., How temperature affects Brix increase, *New Zealand Kiwifruit,* 75, 17, 1991.
99. Smith, G. S., Buwalda, J. G., Green, T. G. A. and Clark, C. J., Effect of oxygen supply and temperature at the root on the physiology of kiwifruit vines, *The New Phytologist,* 113, 431, 1989.
100. Voorhees, W. B., Allmaras, R. R. and Johnson, C. E., Alleviating temperature stress, in *Modifying the Root Environment to Reduce Crop Stress,* Arkin, G. F. and Taylor, H. M., Eds., American Society of Agricultural Engineers, Michigan, 1981, 217.
101. Hewett, E. W. and Young, K., Critical freeze damage temperatures of flower buds of kiwifruit (*Actinidia chinensis* Planch.), *New Zealand Journal of Agricultural Research,* 24, 73, 1981.
102. Pyke, N. B., MacRae, E. A., Hogg, M. G. and Stanley, C. J., Frost damage in many growing areas, *New Zealand Kiwifruit—February,* 16, 1985.
103. Greer, D. H., Laing, W. A. and Kipnis, T., Photoinhibition of photosynthesis in intact kiwifruit (*Actinidia deliciosa*) leaves: Effect of temperature, *Planta,* 174, 152, 1988.
104. Greer, D. H. and Laing, W. A., Photoinhibition of photosynthesis in intact kiwifruit (*Actinidia deliciosa*) leaves: Effect of growth temperature on photoinhibition and recovery, *Planta,* 180, 32, 1989.
105. Greer, D. H. and Laing, W. A., Photoinhibition of photosynthesis in intact kiwifruit (*Actinidia deliciosa*) leaves: Changes in susceptibility to photoinhibition and recovery during the growing season, *Planta,* 186, 418, 1992.
106. Tombesi, A., Ryugo, K. and De Jong, T. M., L'influenza della intensita di illuminazione, della temperatura e della concentrazione di CO_2 sulla fotosinteri di *Actinidia deliciosa* (Liang e Ferguson), *Rivista di Frutticoltura—N,* 11, 63, 1988.
107. Snelgar, W. P., Bayley, G. S. and Manson, P. J., Temperature studies on kiwifruit vines using relocatable greenhouses, *New Zealand Journal of Experimental Agriculture,* 16, 329, 1988.
108. Buwalda, J. G., Fossen, M., and Lenz, F., Carbon dioxide efflux from roots of calamodin and apple, *Tree Physiology,* 10, 391, 1992.
109. Van Oostrum, A. J., Kiwifruit (*Actinidia chinensis* Planch.) water use characteristics, and physiological responses to the presence and absence of irrigation in an Ohinepanea loamy sand, MSc Thesis, University of Waikato, Hamilton, 1985.
110. Judd, M. J., McAneney, K. J. and Trought, M. C. T., Water use by sheltered kiwifruit vines under advective conditions, *New Zealand Journal of Agricultural Research,* 29, 83, 1986.
111. Green, S. R. and Clothier, B. E., Water use of kiwifruit vines and apple trees by the heat-pulse technique, *Journal of Experimental Botany,* 39, 115, 1988.
112. Judd, J. J., McAneney, K. J. and Wilson, K. S., Influence of water stress on kiwifruit growth, *Irrigation Science,* 10, 303, 1989.
113. Klepper, B., Root-Shoot relationships, in *Plant Roots—The Hidden Half,* Waisel, Yoau., Eshel, Amram. au Kafkaf, Uzi, Eds., 1991, 948 pp.
114. Prendergast, P., McAneney, K. J., Astill, M. S., Wilson, A. D. and Barber, R. F., Water extraction and fruit expansion by kiwifruit, *New Zealand Journal of Experimental Agriculture,* 15, 345, 1987.
115. Judd, M. J. and McAneney, K. J., Economic analysis of kiwifruit irrigation in a humid climate, *Advances in Irrigation,* 4, 307, 1987.
116. Brown, N. S. and Brash, D. W., Effect of various irrigation regimes on kiwifruit storage quality, *The Second International Symposium on Kiwifruit,* Massey University, Palmerston North, New Zealand, 1991, 127 pp.

117. Reid, J. P., personal communication, 1991.
118. Buwalda, J. G., The carbon costs of root systems of perennial fruit crops, *Environmental and Experimental Botany,* 33, 131, 1993.
119. Henzell, R. F. and Allison, P. A., Is it worth irrigating mature kiwifruit vines in the Western Bay of Plenty?, *New Zealand Kiwifruit—February,* 9, 1992.
120. McAneney, K. J., Clough, A., Green, A., Harris, B. and Richardson, A., Waterlogging and vine death at Kerikeri, *New Zealand Kiwifruit,* 56, 15, 1989.
121. Jackson, M. B. and Drew, M. C., Effects of flooding on growth and metabolism of herbaceous plants, in *Flooding and Plant Growth,* Kozlowski, T. T., Ed., Academic Press, Orlando, 1984, 47.
122. Justin, S. H. F. W. and Armstrong, W., The anatomical characteristics of roots and plant response to soil flooding, *The New Phytologist,* 106, 465, 1987.
123. Drew, M. C., Jackson, M. B. and Giffard, S., Ethylene-promoted adventitious rooting and development of cortical air spaces (aerenchyma) in roots may be adaptive responses to flooding, in *Zea mays* L, *Planta,* 147, 83, 1979.
124. Bradford, K. J. and Yang, S. F., Physiological responses of plants to waterlogging, *HortScience,* 16, 25, 1981.
125. Yang, S. F. and Hoffman, M. E., Ethylene biosynthesis and its regulation in higher plants, *Annual Review of Plant Physiology,* 35, 155, 1984.
126. Clark, C. J. and Smith, G. S., Seasonal variation of nitrogenous compounds in components of the kiwifruit vine, *Annals of Botany,* 68, 441, 1991.
127. Savé, R. and Serrano, L., Some physiological and growth responses of kiwifruit (*Actinidia chinensis*) to flooding, *Physiologia Plantarum,* 66, 75, 1986.
128. Smith, G. S. and Miller, S. A., Effects of root anoxia on the physiology of kiwifruit vines, *Acta Horticulturae,* 297, 401, 1991.
129. Caldwell, J. D., Peterlunger, E. and Flore, J. A., Flooding stress effects on gas exchange and growth of kiwifruit, *XXIII International Horticultural Congress—Abstracts of Contributed Papers,* Firenze, Italy, 1990, 481.
130. Glinski, J. and Stepniewski, W., *Soil Aeration and Its Role for Plants,* CRC Press, Florida, 1985.
131. Ferguson, A. R., Turner, N. A. and Bank, R. J., Management and nutrition of kiwifruit vines, *Journal of Plant Nutrition,* 10, 1531, 1987.
132. Smith, G. S., Asher, C. J. and Clark, C. J., *Kiwifruit Nutrition: Diagnosis of Nutritional Disorders,* AgPress Communications, Wellington, 1987, 61 pp.
133. Clark, C. J. and Smith, G. S., Seasonal dynamics of biomass and mineral nutrient partitioning in mature kiwifruit vines, *Annals of Botany,* 70, 229, 1992.
134. Ledgard, S. F., Smith, G. S. and Sprosen, M. S., Fate of ^{15}N labelled nitrogen fertiliser applied to kiwifruit (*Actinidia deliciosa*), 1. ^{15}N recovery in plant and soil, *Plant and Soil,* 147, 49, 1992.
135. Ledgard, S. F., Douglas, J. A., Sprosen, M. S. and Follett, J. M., Influence of time of application on the efficiency of nitrogen fertiliser use by asparagus, estimated using ^{15}N, 2. Long term, *Plant and Soil,* 147, 41, 1992.
136. Feigenbaum, S., Brelorai, H., Erner, Y. and Dasberg, S., The fate of ^{15}N labelled nitrogen applied to mature citrus trees, *Plant and Soil,* 97, 179, 1987.
137. Weinbaum, S. A., Klein, I., Broadbent, F. E., Micke, W. C. and Mutraoka, T. T., Effects of time of nitrogen application and soil texture on the availability of isotopically labelled fertiliser nitrogen to reproductive and vegetative tissue of mature almond trees, *Journal of American Society of Horticultural Science,* 109, 339, 1984.
138. Ledgard, S. F. and Smith, G. S., Fate of ^{15}N-labelled nitrogen fertiliser applied to kiwifruit (*Actinidia deliciosa*) vines, 2. Temporal changes in ^{15}N within vines, *Plant and Soil,* 147, 59, 1992.
139. Clark, C. J., Holland, P. T. and Smith, G. S., Chemical composition of bleeding xylem sap from kiwifruit vines, *Annals of Botany,* 58, 353, 1986.
140. Mengel, D. B. and Barber, S. A., Rate of nutrient uptake per unit of corn root under field conditions, *Agronomy Journal,* 66, 399, 1974.
141. Nye, P. H. and Tinker, P. B., Solute movement in the soil-root system, *Studies in Ecology,* Vol. 4, Blackwell Scientific Publications, Oxford, 1977.
142. Smith, G. S., Clark, C. J. and Holland, P. T., Chlorine requirement of kiwifruit, *The New Phytologist,* 106, 71, 1987.

143. Smith, G. S., Clark, C. J., Buwalda, J. G. and Gravett, I. M., Influence of light and form of nitrogen on chlorine requirement of kiwifruit, *The New Phytologist,* 110, 5, 1988.
144. Johnson, C. M., Stout, P. R., Broyer, T. C. and Carlton, A. B., Comparative chlorine requirements of different plant species, *Plant and Soil,* 8, 337, 1957.
145. Flowers, T. J., Chloride as a nutrient and osmoticum, in *Advances in Plant Nutrition,* Vol. 3, Tinker, B. and Läuchli, A., Eds., 1988, 55.
146. Buwalda, J. G. and Smith, G. S., Influence of anions on the potassium status and productivity of kiwifruit, (*Actinidia deliciosa*) vines, *Plant and Soil,* 133, 209, 1991.
147. Hiatt, A. J., Electostatic association and Donnan phenomena as mechanisms of ion accumulation, *Plant Physiology,* 43, 893, 1968.
148. Kochian, L. V. and Lucas, W. J., Potassium transport in roots, *Advances in Botanical Research,* 15, 93, 1988.
149. Smith, G. S., Clark, C. J. and Buwalda, J. G., Effect of potassium deficiency on kiwifruit, *Journal of Plant Nutrition,* 10, 1939, 1987.
150. Buwalda, J. G., Wilson, G. J., Smith, G. S. and Littler, R. A., The development and effects of nitrogen deficiency in field grown kiwifruit (*Actinidia deliciosa*) vines, *Plant and Soil,* 129, 173, 1990.
151. Clark, C. J. and Smith, G. S., Magnesium deficiency of kiwifruit (*Actinidia deliciosa*), *Plant and Soil,* 104, 281, 1987.
152. Asher, C. J., Smith, G. S., Clark, C. J. and Brown, N. S., Manganese deficiency of kiwifruit (*Actinidia chinensis* Planch.), *Journal of Plant Nutrition,* 7, 1497, 1984.
153. Warfield, D., Courtney, K. and Gouthier, G., Nutrient deficiency symptoms in kiwifruit, *Avocado Grower,* July 1979, 48.
154. Zuccherelli, G. and Zuccherelli, G., *L'Actinidia Pianta da Frutto e da Giardino,* Bologna, Edargricole, 1981.
155. Mauro, B., Rita, B. and Frorenzo, P., *L'Actinidia Note tecniche, esperienze e considerazion,* IRFATA, Centro Per Lo Suiluppo Della Progressionalita in Agricoltura Forli, 1986, 84 pp.
156. Smith, G. S. and Clark, C. J., Effect of excess boron on yield and postharvest storage of kiwifruit, *Scientia Horticulturae,* 38, 105, 1989.
157. Limmer, A., Leaf necrosis in kiwifruit, *New Zealand Kiwifruit Special Publication,* 2, National Research Conference, 24, 1988.
158. Smith, G. S. and Miller, S. A., Osmotic effects on vine performance and fruit quality, *Acta Horticulturae,* 297, 331, 1991.
159. Lionakis, S. M., Toxicity of kiwifruit plant (*Actinidia chinensis* Planch.) by chloride and/or sodium ions, *Fruits,* 40, 261, 1985.
160. Testoni, A., Monastra, F. and Turii, E., Mineral content in leaves and quality of kiwi fruit after harvest and after storage, *International Symposium on Kiwi Abstracts,* Edizioni Libreria Progetto Padova, 1987, 10.
161. Prasad, M., Speirs, T. M. and Lill, R. E., A rapid sap nitrate test for kiwifruit, *Journal of Plant Nutrition,* 10, 1689, 1987.
162. Prasad, M., Speirs, T. M. and Fietje, G., Effect of calcium on fruit softening and rot during storage, *New Zealand Kiwifruit Special Publication,* 3, National Research Conference, 24, 1990.
163. Testoni, A. and Granelli, G., Quality, storability, and yield in kiwifruit with different fertilisation, *XXIII International Horticultural Congress Abstract Contribution Papers,* 1. Oral, 1990, 245.
164. Prasad, M., King, G. and Speirs, T. M., Nutrient influence on storage quality, *New Zealand Kiwifruit Special Publication,* 2. National Research Conference, 1988, 22.
165. Smith, G. S., Karl, C. M., Clark, C. J. and Gravett, I. M., Is mineral nutrition really the problem? Look for other factors concerning soft fruit, *New Zealand Kiwifruit—February,* 18, 1991.
166. Hopkirk, G., Nelson and Te Puke fruit are much the same, *New Zealand Kiwifruit—August,* 19, 1986.
167. Hopkirk, G., Harker, F. R. and Harman, J. E., Calcium and the firmness of kiwifruit, *New Zealand Journal of Crop and Horticultural Science,* 18, 215, 1990.
168. Ferguson, I. B. and Watkins, C. B., Bitter pit in apple fruit, *Horticultural Reviews,* 11, 289, 1989.
169. Kirkby, E. A. and Pilbeam, D. J., Calcium as a plant nutrient, *Plant, Cell and Environment,* 7, 397, 1984.
170. Clark, C. J. and Smith, G. S., Seasonal changes in the form and distribution of calcium in fruit of kiwifruit vines, *Journal of Horticultural Science,* 66, 747, 1991.
171. Smith, G. S., Gravett, I. M. and Curtis, J. P., The position of fruit in the canopy influences fruit quality, *New Zealand Kiwifruit Special Publication,* 4, National Research Conference, 38, 1992.

172. Jackson, J. E., Sharples, R. O. and Palmer, J. W., The influence of shade and within tree position on apple fruit size, colour and storage quality, *Journal of Horticultural Science,* 46, 277, 1971.
173. Marini, R. P. and Trout, J. R., Sampling procedures for minimising variation in peach quality, *Journal of American Society of Horticultural Science,* 109, 361, 1984.
174. Smith, G. S., Curtis, J. P. and Edwards, C. M., A method for analysing plant architecture as it relates to fruit quality using three-dimensional computer graphics, *Annals of Botany,* 70, 265, 1992.
175. McMath, K. L., Fruit taste—the world's finest, *New Zealand Kiwifruit Special Publication,* 4, National Research Conference, 41, 1992.
176. Manson, P. J., Snelgar, W. P. and Seelye, R. J., Relationship between timing of cane tipping and yield of kiwifruit (*Actinidia deliciosa*), *Journal of Horticultural Science,* 66, 415, 1991.
177. Lai, R., Woolley, D. J. and Lawes, G. S., The effect of inter-fruit competition, type of fruiting lateral and time of anthesis on the fruit growth of kiwifruit (*Actinidia deliciosa*), *Journal of Horticultural Science,* 65, 87, 1990.
178. Cooper, K. M. and Marshall, R. R., Croploading and canopy management, *Acta Horticulturae,* 297, 501, 1991.
179. Lowe, R. G., Wang, Z. Y., Patterson, K. J. and Gould, K. S., Rootstocks—enhancing the vine's potential, *New Zealand Kiwifruit Special Publication,* 4, National Research Conference, 33, 1992.

Chapter 6

Olive

Guido Bongi and Alberto Palliotti

CONTENTS

I. INTRODUCTION

A. ORIGINS, EVOLUTION, AND GENETIC VARIABILITY OF *OLEA EUROPAEA* L.

The genus *Olea* includes at least 30–35 species that belong to the family Oleaceae, subfamily Oleoideae (n = 23), which has been recognized as a distinct family since 1809.[1,2] The cultivated olive (*Olea europaea* L.) is an evergreen tree that evolved from tropical and subtropical species. The palaeontological remains of many extinct *Olea* spp. have also been found in France and Italy.[3–5] The area of origin of

0-8493-175-0/94/$0.00+$.50

the olive was the Laurophyllus forest around the Thetys.[6] The Tertiary Mediterranean basin was inside a tropical zone, but partial basin drying out and Pleistocene glaciation were strong natural selection criteria for a sclerophyllous plant with moderate freezing avoidance characteristics.[6,7] Glaciation events probably restricted the original olive population. Only populations with a capacity for freezing avoidance from −5 to −12°C survived; minima below −12°C greatly limited the natural distribution if *Olea* species.[6]

Evidence of early olive cultivation at around 4800 BC were found in Cyprus.[8] The age of the species and its vast prehistoric range may explain it's very high intraspecific variability. It has even been speculated that 'forms' hybridized under different climatic events now constitute the species *Olea europaea*.[3,4] According to recent results, measurable genetic variability exists among individuals of the same variety. It has been reported that some clonal characteristics persisted in trees obtained from self-rooting cuttings.[5,9,10]

Frost-resistant varieties are required in northern and central Italy, since damaging frosts occurred several times in the 19th and 20th centuries.[11] Although dry summers and mild winters provide the ideal climate for olive cultivation, summer drought limits the overall productivity. In most cases, it is necessary to improve resistance to adverse climates by breeding new varieties. Resistance is normally found in land-races with poor pomological traits.

The use of clonal rootstocks is of particular interest for stress resistance.[12] For example, the cv. Oblonga confers resistance to *Verticillium* wilt and may also influence fruit shape,[13] while *O. cuspidata* and cv. Zhongshan 80 are more tolerant to root hypoxia than other olive species and cultivars.[14,15]

B. GROWTH, MORPHOLOGY, AND PHENOLOGY

Olive trees may grow to a considerable size, but a slow growth rate and longevity are characteristics of the species. Olive plants reach full size in 10 years in the southern growing area. However, because of alternate bearing and rainfall fluctuations, 30 years are required to estimate overall productivity. Shoot elongation occurs mostly in spring and correlates well with rising temperatures.[16] The terminal growing apices are dominant and repress axillary bud development in a basipetal direction; however apical dominance is influenced by irradiance, soil fertility and growth regulators.[11]

The strong dominance of a leader shoot inside a branch can originate 'decurrent' or 'excurrent' forms. When dominance control is exerted for only one season, the lateral shoots become dominant and, in the following year, the plant crown tends to an irregular sphere with basitonic multiple heads (Figure 1). When concurrent apices are removed by pruning or when the apical dominance lasts more than one season, stable excurrent forms may be maintained with a single or multiple axes. It seems that natural spindle forms, characterized by strong dominance and small basitony, have higher yields.[5]

Olive trees bare flowers on the lateral buds of 1-year-old wood. Apical flowering is rarely found in *O. europaea* (Figure 2) and flowering on 2-year-old shoots is uncommon. Pruning keeps the fruiting parts not too far from the roots, avoiding embolism and transport limitations, but some one-year-old wood must be preserved for flowering. Vegetative and fruiting buds are mostly located at the leaf axils. Flowering, fruit-set and fruit development processes are extremely climate-dependent.[17]

The flowers of olive are hermaphroditic and grouped in dichotomous panicles with 12–40 flowers, but generally set only one fruit per inflorescence. Only a few cultivars with small fruits (e.g., cv. Koroneiki) may set more than three fruits per panicle. Within each inflorescence, flowering is not synchronous (Figure 2) and early drop reflects imperfect fertilization. Olive cultivars are generally self-incompatible and pollination is wind dependent.[11] Mature olive trees may produce 500,000 or more flowers; however, fruit set is normally below 5%.

Successful pollination by wind requires a continuous source of pollen over the period of stigma receptivity. Some cultivars are better pollinizers than others; male sterility or high ovary abortion is present in some cultivars.[11,17–19] The high degree of heterosis in olive is primarily caused by its self-incompatibility,[8] but is also dependent upon climate.[20,19]

Growth and fruit development in olive also depend on climate. Environmental variability may be more important than genetic variability for several traits, including oil content and quality. The best quality oil is obtained from districts of the northern and eastern parts of the Mediterranean basin and near the margins of 'maquis vegetation'.[6] However, frost damage restricts further expansion northward. The chemical components (e.g., polyphenols, aliphatic alcohols, triterpenic alcohols, sterols, and triterpenic diols) that determine oil quality vary according to cultivar and year of cultivation.[21] Rainfall is one of the predominant environmental factors affecting oil quality. Irrigation may also affect the variability in oil quality.[22]

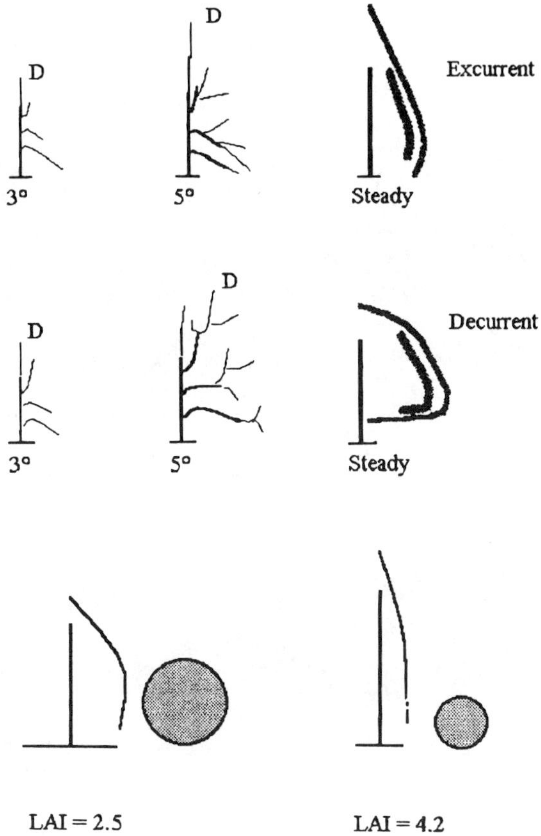

Figure 1 Evolution of olive tree training systems. The first row presents an excurrent form with moderate basitony; D depicts the dominant shoot that maintains dominance over years, from the third to the fifth year after planting; on years 1–3 a tutor is needed. The second row reports a decurrent form with the same basitony. The last row shows how excurrent forms may give rise to an increased LAI.

II. LIGHT

A. LIGHT RESPONSES

Light intensity is the radiant flux of a source per steradian of spherical volume, including the source at the center (Watt per steradian). At the physiological level, the radiant flux density per leaf surface unit (Watt per unit surface, mmol photon m^{-2} s^{-1}) is more relevant. The term 'light fluence' will be used here to indicate 'radiant flux density on the leaf surface'. The use of the term irradiance rather than fluence is technically incorrect as it apportions the fluence from all directions on a surface. A discrepancy arises because sensors used for measuring fluence density and irradiance read spherical space of 2π or 4π steradians, respectively, and differ in their responses. Only collimated beams at 90° of incidence give the same reading in both instruments. If radiation is substantially diffused, the irradiance level may reach 4 times the fluence level.

1. Light Fluence Density

Light fluence is a very important factor for olive flower induction. Shading reduces flower differentiation.[23,24] However, shading after the period of flower induction does not influence flowering,[24] but can induce high morphological sterility (e.g., ovary abortion). Leaf shading or leaf removal may totally inhibit flowering.[25]

Light fluence incident on leaves decreases through the olive canopy and light extinction may vary according to the shape of the trees and the density of the canopy. The olive canopy is moderately dense and evergreen. The amount of leaf area per projected ground surface (LAI) often approaches 2.5. The influence of cultivars, training systems, and orchard management practices (irrigation, fertilization, pruning, etc.) on LAI is unknown. The potential production, e.g., the number of flowers per hectare, is related to the average number of new nodes per hectare. The distance between internodes is under genetic control,[5] and the presence of evergreen leaves inhibits growth of new nodes in shaded positions.

To evaluate the effects of plant shape and size on light extinction, it is essential to know both the LAI and the harvest-index (HI), which is the ratio between fruit production and above ground dry

Figure 2 Apical inflorescence (a), and 'normal' vegetative apical development at full bloom (b) on cv. Koroneiki. The plants, kept in 25-l pots, were subjected to an induction period of 8 weeks at 12/7°C (day/night). These pictures were taken after 60 days, with a day/night cycle of 22–15°C. A significant proportion of the apices (5%) were completely flowers. This is indicative of the biological presence of this flowering topology in olive plants which, if fixed in a new cultivar, may drastically change the schemes of olive growing.

matter. There is a positive effect of high LAI on olive production. The new cultivar, 'Barnea', with a tall natural spindle form produces more fruits per projected ground surface area than other cultivars.[5] In this case, the increased productivity is associated with the increase of LAI while HI is constant. In olive, flowering is associated with growth and the HI is inversely proportional to current season vegetative growth. Production during the subsequent year tends to be reduced as flowers originate only from one-year-old leaf axils.

Both canopy density and the light response of individual leaves also require further discussion in olive. Maximum net CO_2 assimilation (A_{max}) of olive, under optimal conditions, is lower than that of most C_3 species. At high CO_2 concentrations (15% CO_2, 2% O_2, 83% N_2, vol vol^{-1}) A_{max} can be measured by O_2 evolution with polarographic gas phase electrodes.[26] Stomatal limitations to net CO_2 assimilation (A) are obviated and the system becomes CO_2 saturated (i.e., respiration in light is minimized). In these conditions, A_{max} is achieved and is satisfactorily described by the term 'photosynthetic capacity'. The photosynthetic capacity has been positively correlated with the rate of uncoupled electron transport from water to NADP in spinach chloroplasts.[26] However, similar experiments have not been performed on olive.

Olive leaves have a photosynthetic capacity not higher than 18 μmol CO_2 $m^{-2}s^{-1}$ at a fluence of 900 μmol quanta $m^{-2}s^{-1}$. The low photosynthetic capacity in olive is related to leaf thickness and a low density of the photosynthetic reaction centers. Maximal rates of both electron transport in thylakoid suspensions and chlorophyll a/b ratios are typical of shade-adapted chloroplasts.[27] A substantial proportion of olive chloroplasts are dispersed in inner palisade layers and are exposed to low light levels. Up to 6 palisade layers may develop in olive leaves.[27]

The photosynthetic capacity of olive may be drastically reduced if leaves develop in shade, as normally exists in the canopy understory (Figure 3). The photosynthetic capacity of shade-grown leaves may be less than 30% of that of sun-grown leaves. Thus, it may be advisable to prune to enhance light penetration in the canopy. The olive LAI of 2.5 is low compared to that of other fruit crops (in *Citrus*

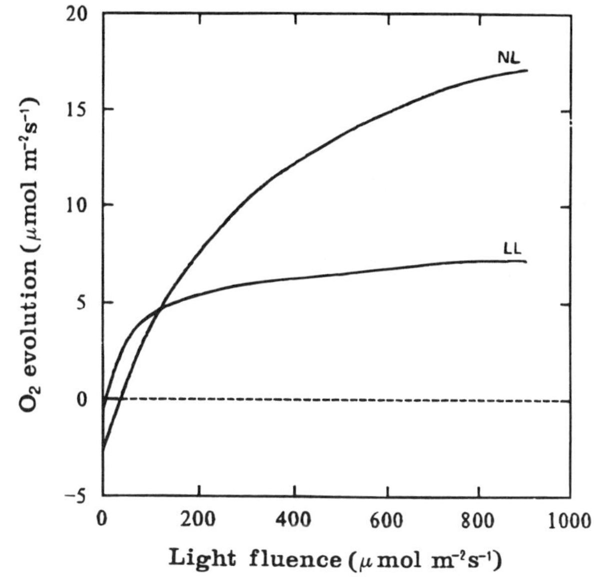

Figure 3 Response of O_2 evolution to light fluence in olive leaves, cv. Frantoio, developed in normal light (NL, 1200 μmol $m^{-2} s^{-1}$), and in low light [LL, under a screen (i.e., neutral filter) of 20% light transmission]. Net exchange was determined as described by Walker (1989)[26] with a CO_2 partial pressure of 15,000 Pa.

it may reach 10), perhaps as a consequence of the rapid reduction of the photosynthetic capacity of olive leaves kept in dim light.

The quantum efficiency (ϕe, i.e., mol O_2 absorbed quanta^{-1}) in olive is 0.089,[27] compared to 0.107 for most higher plants.[28] The lower ϕe for olive may be caused by respiration of inner palisade layers. Basic mechanisms of photochemistry are similar across species of higher plants; 8 photons are required for each evolved oxygen. In environmental conditions characterized by high fluence and variable temperature, a reduction in ϕe, known as photoinhibition, may occur. Structural damage of functional components of photosystem 2 (PS2) (mostly D1 protein)[28] and adverse effects on oxidative processes (e.g., Calvin and glycollate cycles and peroxidases), all reduced ϕe. A long duration may be required to achieve pre-photoinhibition values of ϕe.

Chlorophyll fluorescence is a useful method for determining ϕe of intact leaves in the field. The ratio of variable (Fv) to maximal (Fm) fluorescence has been linearly correlated with ϕe and is a good determinant of photoinhibition.[29,30] By use of chlorophyll fluorescence measurements, ϕe can be monitored in olive leaves conditioned to very long-term stress periods.[31–34] Broad-leaf evergreens are sometimes able to avoid the short-term effects of combined environmental stresses that may be lethal to the leaves of deciduous species.[32] However, the combination of high fluence and temperatures lower than 5°C for a week or more, typically induce photoinhibition in leaves located in the outer canopy (Figure 4).[33] Olive should not be cultivated in cool areas with a high frequency of chilling, largely due to the long time necessary to reconstitute the original ϕe. A marked loss of productivity (50%) of olives occurs between 41° and 36° latitude north.[35,36]

An increase in fluence is often concomitant with a drop in absolute humidity around midday. Stomata of olive leaves close in response to low humidity which may decrease CO_2 concentration in the chloroplast. Consequently, A may increase up to a fluence of 1500 μmol quanta $m^{-2}s^{-1}$, but may decrease at higher fluence as a result of increases in vapor pressure deficit (VPD).[37]

2. Light Quality

Olive leaves are bifacial and morphologically heterogeneous. The adaxial surface of the leaf is noticeably darker than the abaxial surface. Baldy et al.[38] found that the adaxial surface absorbs more photosynthetically active radiation (PAR, 400–700 nm) than the abaxial surface. The abaxial surface reflected 20 to 40% of the PAR. Near-infrared radiation (i.e., wavelengths > 700 nm) are poorly absorbed and highly reflected by both leaf surfaces. The transmission of direct radiation is very low in olive leaves (i.e., less than 1% at 440 nm). The quality of transmitted light is very different from the incident light quality and has a much higher infrared/PAR ratio. The poor PAR transmission limits the photosynthetic capacity of the lower part of the olive canopy. During aging, olive leaf thickness increases, further reducing

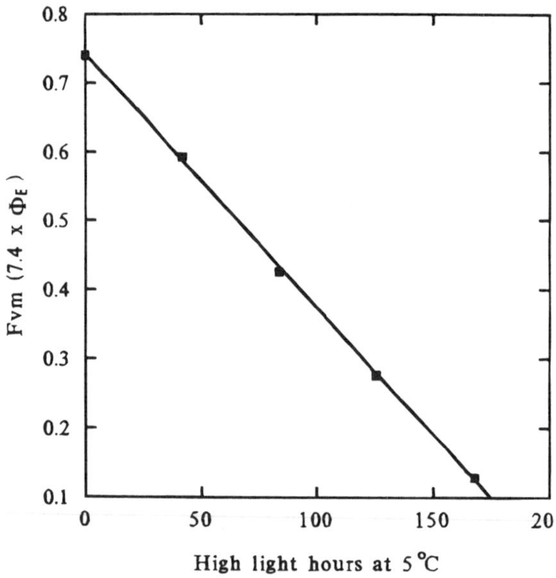

Figure 4 Effect of high light and low temperature on the intensity of chlorophyll fluorescence in olive leaves. Temperature was maintained steady at 5°C during the experimental period and leaves were subjected to hardening in natural light. The F_{vm} is $F_v F_m^{-1}$, and ϕ_E is obtained in olive leaves dividing F_{vm} by 7.4.[33]

light transmittance and photosynthetic capacity of leaves in the lower canopy. Diffuse light is not thought to contribute significantly to leaf photosynthesis in olive.[39]

III. TEMPERATURE

A. THERMAL REQUIREMENTS

Temperature is the most important environmental factor limiting olive cultivation. Olive plants do not survive temperatures lower than −12°C,[6] but require a chilling period for flower bud induction.[40] Thus, temperature limits olive growth to areas mostly between 30° and 45° latitude.[35]

Olive oil production is qualitatively improved by cold winters. This may explain why 10 to 15% of olive production occurs in areas with an inherent risk of frost injury. The yearly primary production per square meter of leaf area at 34° latitude north is 50% more than at 45° due to winter growth limitations,[39] and to photoinhibition.

Olea europaea is an obligate thermoperiodic species that requires at least 10 weeks below 12.2 to 13.3°C for full expression of flowering.[41,42] The total number of chilling hours required for flowering is cultivar dependent. Cultivars acclimated to warm winters require fewer chilling hours.[43] Flowering is induced best when temperature fluctuates between 2 and 15°C for 70–80 days.[40] Under these conditions, flowering can be induced year-round and independently in different parts of the canopy. Temperatures constantly below 7°C or above 15°C may inhibit flower induction,[40] while winter temperatures above 20°C for 2–3 weeks may impede the release of flower buds from dormancy.[44] The hormonal status of trees also regulates the chilling requirement for flower induction. High-yield years or late harvest may increase the chilling requirement of olive.[44] Flower differentiation occurs at temperatures higher than those required for induction. Abnormally high temperatures in spring can impede flower development.[17] Self-pollination may also be limited by high temperatures during the flowering period, due to the inhibition of pollen tube growth.[45] Low spring temperatures prolong olive flowering up to 3 weeks,[17] and can also temporarily interrupt flowering or interfere with cross-pollination.

The temperature requirements of olive vary with the tissue and the phenological stage. Olive seed germination is optimized by keeping the seeds at 10°C for a month before seeding at 20°C.[46] However, a seeding temperature of 25°C seems to reverse the effect of a chilling treatment of less than 5 weeks. Rooting of olive stem cuttings is greatly improved when the basal temperature is kept high (30°C) for 15 days, and then gradually decreased from 30 to 18°C.[47]

Photosynthesis of olive trees saturates at different temperatures on different cultivars, and the role played by cultivar acclimation is likely to be important.[48]

Concentrations of metabolites in different parts of the olive tree are both seasonal and temperature dependent.[49,50] Starch accumulation in leaves is associated with periods of vegetative growth, principally during the spring, whereas leaf starch, soluble carbohydrates, and mannitol concentrations decrease in summer and their depletion may be related to low metabolic activity during this period.[49] Leaf starch concentration also decreases in winter, but concentration of soluble sugars increases. Export of mannitol from bark to xylem is reduced by low temperatures.[49]

B. THERMAL STRESS

1. High Temperature

The southern limit for the reproduction of olive is determined by winter temperatures, which must be lower than 14°C for 2 weeks in order to induce some flowering. Only a few cultivars, such as 'Rubra' and 'Arpa', have lower winter chilling requirement,[43] and these cultivars could be used in breeding programs to extend olive cultivation to warmer areas. Olive productivity is not limited by high temperatures, but very high temperatures immediately after flowering may totally eliminate the crop, as occurred in California in the spring of 1984.[51] Olive leaf tissue can withstand 52°C before the dissociation of protein antennas of PS2.[52]

Photosynthesis is generally inhibited by temperatures higher than 35°C. However, olive cultivars acclimated to high temperatures maintain 70–80% of their photosynthetic rate at 40°C.[48] The olive leaf cuticles underwent irreversible structural changes at temperatures higher than 55°C; in many other species this damage occurred in the interval between 40 and 50°C.[53]

High temperatures are frequently associated with high VPDs between leaf and air, water stress, and high light fluence; this situation is very common in the Mediterranean climate. Vapor pressure deficits between fruit and air also rises with increasing temperature and peduncle withering may produce drastic late July fruit drop.

2. Low Temperature

When winter temperature drops below 5°C, plants begin an acclimation process known as hardening. Hardened plants stop growing and undergo important metabolic changes. In particular, low temperatures cause the starch to be converted into soluble sugars. Starch is also reallocated and concentrated in frost-protected organs such as the roots.[50] A winter temperature of −12°C damaged all the aerial parts of olives. Buds and inflorescences were also particularly sensitive to spring-frost,[11] and during flowering, temperatures below 10°C reduced pollination.[45]

In all plants, ice formation first takes place in the apoplast since it is about 100 times more dilute than the symplast; the matric forces created by large molecules also contribute to a freezing point depression of the symplast. The matric forces exerted by cell wall microfibrils may inhibit the freezing of apoplastic water down to −10°C. When an ice crystal forms in the apoplast, the cell remains permeable to water and a high difference in water potential causes water loss from the symplast. For example, in Eucalyptus at −2°C, ice formation resulted in an osmotic potential (Ψ_π) of −2.5 MPa, while at −10°C the Ψ_π was −11.8 MPa.[54,55]

It is known that olive and palm undergo deep supercooling, a freeze avoidance mechanism.[56,57] However, if ice formation occurs, 'water soaking' and 'leaf bending' take place upon thawing (Figure 5), due to the relative difficulty of reversing cellular plasmolysis and initiating and maintaining water uptake through membranes.[58] Cold-hardening for a period of 4 weeks at 5°C reduced the exotherm temperature by 2.3°C compared to unhardened leaves (Figure 6).[59]

At least two classes of carbohydrates, rhamnose glycosides (i.e., verbascoside)[60] and sugar alcohols (i.e., mannitol)[49] are present in the xylem fluid of olives. Their concentration mainly increases when the winter pseudo-dormancy of roots results in a massive amount of starch deposition in xylem tissue.

Many independent traits must be combined to improve freezing resistance (e.g., fruit size, waxes, photoinhibition tolerance, etc.).[61] However, Roselli et al.[62] reported that stomatal density was lower in olive cultivars that were tolerant of cold temperatures. Thus, stomatal density may be used as a screening criterion for cold tolerance in olives.

IV. WATER

A. WATER REQUIREMENTS OF OLIVE

Olive is considered a drought-resistant species because it thrives in areas where water stress is frequent (Mediterranean climate). It has been postulated that the minimum water requirement for olive is 200 mm year^{-1}.[63]

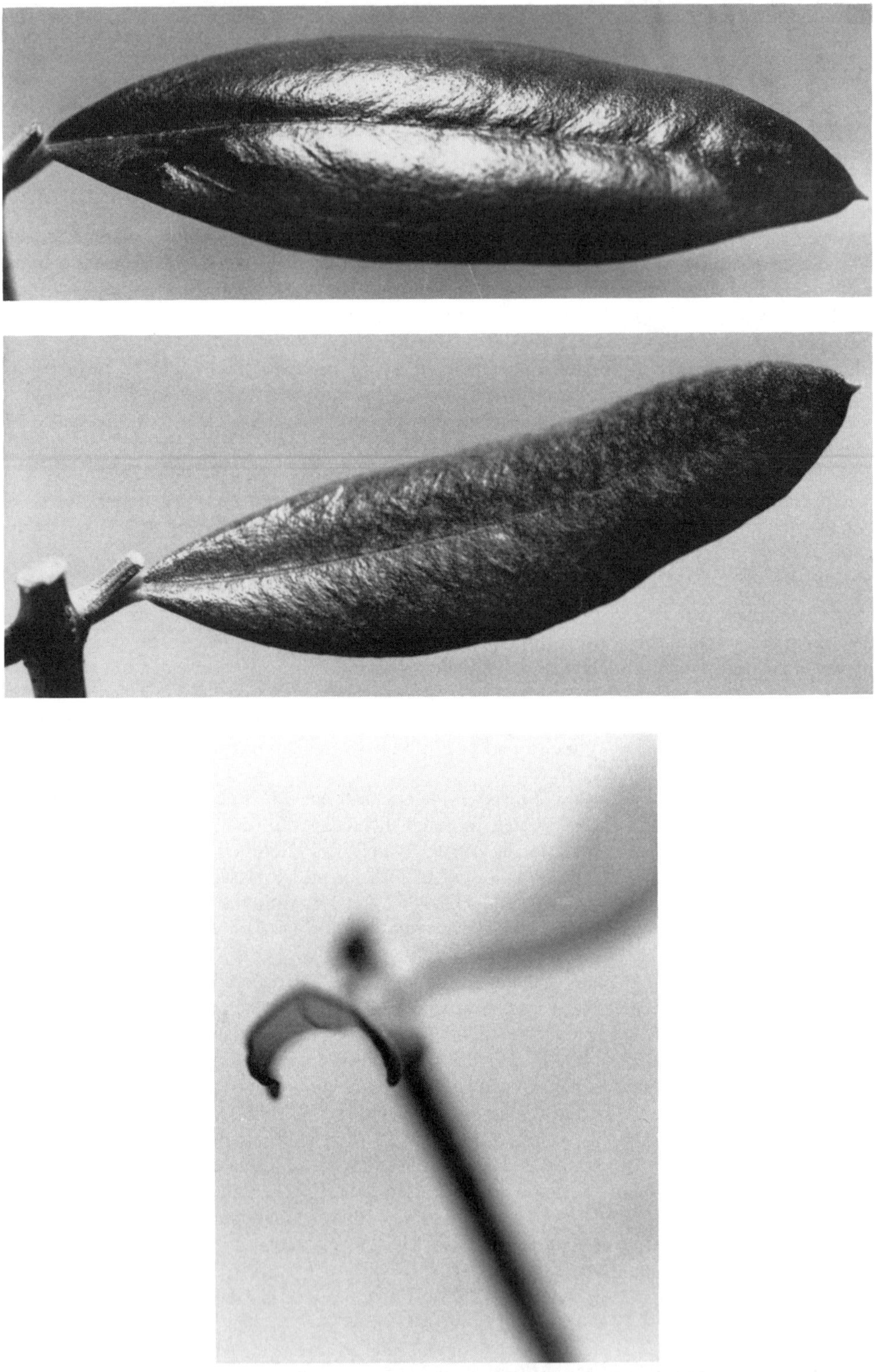

Figure 5 The top picture shows the leaf of cv. Frantoio before freezing damage. If freezing occurs, upon thawing the leaf appears as in the middle picture, becoming 'water soaked' with superficial pit development and a color change which resembles a leaf kept in water for several days. The bottom picture is a side view of the middle leaf, taken several hours later. Leaf abscission occurred after 2 weeks.

1. Root Level

Olive trees do not have a dominant tap root. Even in small plants, the main roots almost invariably follow soil fertility channels, where lateral roots grow rapidly.[11] The root system is mainly confined to the top meter of soil and exceeds the projection area of the aerial parts. Olive xylopode (i.e., the tissue between the trunk and the root crown) is characterized by a stellate section and by the presence of ovular hyperplastic bodies, which contain sphaeroblasts (i.e., wartlike protuberances containing meristematic and conductive tissues).[11] Up to 30% of the total carbon fixed by the plant is invested in xylopodes, which are the origin of the olive adventitious root system.[11] This rooting habit is probably the result of sensitivity to hypoxia[61] and may allow for more efficient water absorption in response to light and intermittent rainfall than a deep rooting system. A drought avoidance response, on the other hand, not displayed by olive is the development of a deep root system to reach water during dry periods.

Interesting data on root distribution have been obtained from 20-year-old 'Manzanillo' trees.[64] The plant crown radius was 2 m, while roots extended less than 3 m and more roots were present at distances greater than 1 m from the trunk. Mean root density of irrigated and non-irrigated trees was measured in mini-rhizotrons. Using this data set the total length of living roots was calculated to be 2283 m on non-irrigated and 975 m on irrigated trees, a ratio of 2.3 between treatments. The increase in linear weight versus root length indicates that irrigation may decrease the carbon investment in roots.[64]

If water is supplied to trees maintained in non-irrigated conditions for a long period, care must be taken to supply water to the entire system. When the roots of the olive plants were split and irrigated at two different soil potentials (Ψ_{soil}, -0.2 and -1.1 MPa), the plants grew less than plants maintained with both split roots irrigated at a Ψ_{soil} of -0.2 MPa. This effect occurred even when only a small fraction of the root system was exposed to a Ψ_{soil} of -1.1 MPa (Figure 7).[65] The leaves of the 'split-root olive plants' maintained full turgor and optimal water content, but stomatal conductance was reduced. Zhang and Davies,[66] obtained similar results for maize which they attributed to a large increase in the abscisic acid content in fine roots and in the transpiration stream. Thus, olive trees in the field may adapt to non-uniform soil potentials. If irrigation does not reach the entire root system of olive plants developed in non-irrigated conditions, a stress signal coming from the non-irrigated portion of the root may inhibit growth. Irrigation should therefore be a regular practice that confines the root system within a projected canopy area.[67]

2. Water Relations: Fruit and Leaf Gas Exchange

Feedback reactions between leaf conductance (g_s) and VPD exist in leaf and fruit. We can measure water losses by g_s, since leaf porosity is much more constant than transpiration (E). For example, in the fruits of cvs. Leccino and Dolce Agogia, g_s decreased from 38.4 mmol $m^{-2}s^{-1}$ at 20 days after full

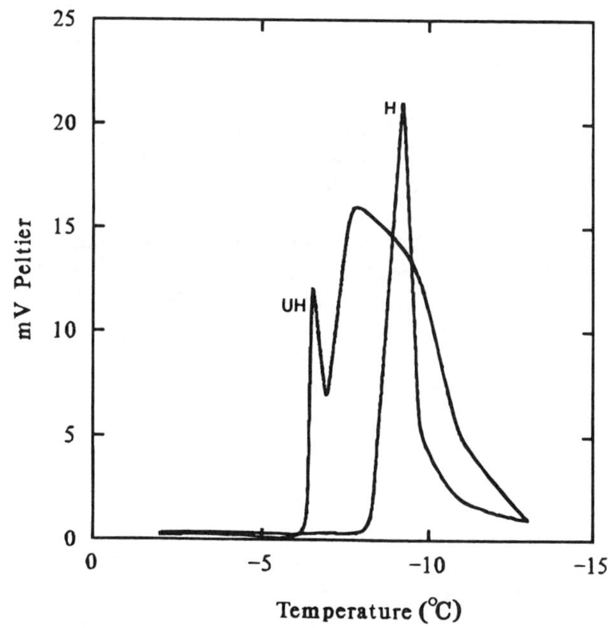

Figure 6 Microcalorimetric differential traces of freezing in hardened leaves (H, 4 weeks at 5°C under natural light) and unhardened leaves (UH) of the same plant (cv. Frantoio) before the hardening. H trace shows a single peak at -9.0°C, whereas in UH, a double heat emission with the primary peak at -6.7°C was detected. The relative intensities are the potentials generated on a differential microcalorimeter with a matrix of 144 Peltier thermoelements. The temperature program was -0.05°C min^{-1}.[59]

Figure 7 Development of small olive plants (cv. Frantoio) grown with a small portion of the roots in drought and the rest in optimal conditions (split-roots). The roots in the left plant were maintained in uniform soil water potential of −0.2 MPa, whereas, in the right one, a small portion of the roots was maintained at −1.1 MPa. Turgor loss was never observed in the leaves of split-root plants, but instead stomatal closure occurred which appeared to be mediated by a translocatable signal.

bloom to 22.4 mmol $m^{-2}s^{-1}$ at 170 days after full bloom. The leaf g_s ranged from 120 to 190 mmol $m^{-2}s^{-1}$.[68,69] The stomatal density of olive fruit is much less than that of the leaves; during ripening fruit stomata or pores gradually cover with wax and sink below the fruit surface.[70] Leon and Bukovac,[71] found 470 stomata mm^{-2} in leaves of cv. Manzanillo. Blanke and Lenz,[72] reported that the frequency of stomata or pores per surface unit in apple fruit was at least 30 times lower than in leaves. Large fruit with a lower surface area:volume ratio retain water more efficiently than small fruit. The following calculations are based upon an olive tree with a crown soil-projected area of 12.5 m^2, a leaf area of 31.4 m^2, and 22 kg of fruits per tree. Assuming g_s = 120 mmol H_2O m^{-2} s^{-1} for leaves, about 58.6 kg of water is lost daily through leaf stomata compared to about 6.5 kg from the fruit which represents 11% of the total water transpired.

3. From Gas Exchange to Orchard Management

Orchard irrigation schedules (called CRIS) were obtained after integrating water losses for a month under typical climatic conditions[73] in olive, *Prunus* and *Citrus.*[65] Temperatures lower than 40°C did not significantly affect g_s. Under this temperature, A_{max} is maintained by irrigating every 3 days. Using the CRIS model and data from measurements of leaf gas exchange, an estimate of irrigation requirements of olive orchards has been obtained.[65] These requirements are similar to those based on lysimetric measurements[74] and favorable leaf water potentials (Ψ_l).[75–78] Evapotranspiration (ET) was comparatively low in olives (603 mm $year^{-1}$) compared to peach (with winter weeds in the orchard, 1088 mm $year^{-1}$) and *Citrus* (940 mm $year^{-1}$). However, olive requires 638 mm $year^{-1}$ water in order to avoid frequent water stress.

According to the CRIS model, in the southern Mediterranean district the volume of supplemental irrigation required to maintain the leaf relative water content deficit (RWCD) below 15% for *Prunus, Citrus,* and *Olea* were 561, 418, and 173 mm, respectively. Thus, olives required 3.2 times less irrigation than *Prunus* and 2.4 less than *Citrus* spp. While *Citrus* and *Prunus* may reach 10.3 tons ha^{-1} of fruit dry matter, the 'on-year' limit for olives is 5.5 tons ha^{-1}. Dividing the production rates by irrigation volumes, 3.17, 2.46, and 1.78 g of dry fruit matter was calculated per kilogram of water in olive, *Citrus*, and *Prunus*, respectively. Under the same conditions, the energy yields of the crop is 103.4 MJ ha^{-1} for *Olea*, 90.5 MJ ha^{-1} for *Prunus*, and 62.3 MJ ha^{-1} for *Citrus*.[79]

When an olive orchard in a northern Mediterranean district is considered, we find that a very limited amount of irrigation is needed during summer (180 m^3 ha^{-1}). The absence of supplemental water (about 288 kg $tree^{-1}$ $year^{-1}$) may induce fruit drop. Since water turn-over time in fruit is about 1 day, only a few days with a dry soil are sufficient to induce fruit drop.

B. RESPONSES TO WATER STRESS IN OLIVE

Water deficits are tolerated by olive for prolonged periods. Olive has evolved a series of adaptations which confer tolerance to water stress. One adaptation of olive leaves is the drastic increase in Ψ_π of water-stressed leaf cells.[80]

1. Mannitol and Osmotic Potential

Sucrose is a major solute among many higher plants, but mannitol, a polyol, predominates in olives and represents from 1/2 to 2/3 of the total soluble sugars in leaves and bark.[49,50] The vacuolar concentration of sucrose can reach 2 *M*. Above this range, the water binding capacity of sucrose may have deleterious effects. At concentrations higher than 0.5 *M*, carbohydrates start to bind water chemically, and the "bound" water is sequestered from physiological equilibria. This leads to substantial reductions in Ψ_π far below van't Hoff's calculations of Ψ_π. In fact, a 2 *M* solution of sucrose in cells may exert a Ψ_π of −13.6 MPa instead −4.9 MPa.[81] Many species use sugar alcohols, called polyols, instead of sucrose. Polyols are more similar to water than sucrose in terms of polarity, electroneutrality and redox potentials and tend to bind water less than sucrose. Cells with high mannitol concentrations absorb water from apoplast at very low water potentials.

Olive fruits are unable to carry out a large degree of osmotic adjustment because little mannitol is present (less than 10% of soluble sugars);[82] in fruits, mannitol is readily re-synthesized into fructose-6P. Mannitol has also been found in xylem sap after winter pseudo-dormancy,[49] and may play a role in cold hardening.

2. Water Content and Capacitance Effects

Olive leaves can tolerate Ψ_l near −9 to −10 MPa with full rehydration ability,[80] and it is difficult to establish a critical Ψ_l for growth or physiological processes. However, a stress parameter able to accurately establish if a leaf is under stress is essential. Olive leaves have been reported to absorb water from humid air (poikilohydricity).[83] However, poikilohydricity can only occur under very humid conditions (i.e., mean relative humidity ca. 100%) and when air temperature exceeds leaf temperature. In addition stomata must be open and Ψ_l must be near zero.

The instantaneous water potential of leaves can be obtained from the relationship $\Psi_l = A + B\ e^{(-C/t)}$, where A is a soil potential parameter, B is a constant associated with Ψ_l, C is a kinetic constant derived by the resistance offered by the hydraulic connections to water transport, and t is the time.[84] The transpiration rate affects both A and B parameters, and any changes in stomatal opening modifies Ψ_l.[84] In steady conditions, hydraulic resistances determine the Ψ_l across segments of the vascular system. If E is prevented, the plant equilibrates to the root water potential.

The large midday water potential drop found from leaves to root in olive may reflect a strong resistance to water movement. A specific hydraulic resistance of 250 MPa s m kg^{-1} has been found in potted olive cv. Nocellara plants, whereas in many deciduous species the hydraulic resistance ranged between 35 and 4 MPa s m kg^{-1}.[85] The linear pressure gradient necessary to move water was 65, 170, and 560 kPa m^{-1} in olive primary, secondary, and tertiary stems, respectively. Values for intact field trees are not known.[85] Maintaining a low Ψ_π in olive leaves may be essential to drive water through the apoplast given such large hydraulic resistances.

3. Volumetric Elastic Modulus

Olive leaf cells have a high volumetric modulus of elasticity (ϵ).[86] This parameter represents an applied pressure divided by a fractional change in volume size,[87] according to:

$$\epsilon = \frac{\Delta P}{\Delta V/V} \quad (1)$$

where P is turgor pressure and V is cell volume. Note that ϵ is a parameter that is inversely proportional to tissue elasticity.

In olive, ϵ tends to increase under drought. At 87.5% of maximal cell volume, ϵ of 7 and 22.5 MPa have been recorded in hazelnut and mature olive leaves, respectively.[88] Turgor loss for the more rigid cells of olive leaves occurred at 80% of maximal cell volume; however, positive turgor was maintained in hazelnut leaves at 66% of maximal volume. The ϵ of young olive leaves was 8.4 MPa, indicating that tissue elasticity was reduced with leaf aging. The negative adaptation of ϵ is typical of Mediterranean evergreens.[86] By contrast, succulent plants, like *Agave deserti,* displayed a decrease in ϵ (i.e., less rigid cells) and tended to retain higher turgor pressure (Ψ_p) under drought conditions.[86] Olive apparently relies more on reducing Ψ_π than on maintaining a low ϵ in response to drought. A reduced Ψ_π (and hence Ψ_l) facilitates the extraction of water from the soil at low soil water potentials (Ψ_{soil}). As a consequence of the low Ψ_π and Ψ_l generated in the symplast, apoplastic water uptake is a rapid phenomenon in olive leaves.[86,89] Water-stressed olive leaves are strong sinks for water at very low Ψ_π values. By contrast, olive fruits are succulent bodies that have very low g_s and may even contribute to a "capacitance effect" by supplying water to leaves during midday.

4. Growth Rate and Carbon Allocation

Leaf surface-based photosynthetic capacity and growth have recently been linked by an allocation function, the Carbon Molar Density (ρ)[90] where ρ represents the ratio of total plant carbon to total leaf area (mol C m^{-2}).

The relative growth rate (RGR) has been related to photosynthesis,[90] by

$$RGR = \frac{Al\,(1 - \phi)}{\rho} \quad (2)$$

where ϕ is the respiratory loss, l is the light period as a proportion of the day (i.e., for 12 h of day light, $l = 0.5$), A is the net CO_2 assimilation rate and ρ is the ratio of total carbon to photosynthetic area.

In wheat a value of ρ = 2.5 mol m^{-2} has been obtained.[90] In well-watered small olive plants (cv. Frantoio) ρ was 12.2 mol m^{-2}, whereas in water-stressed olive plants ρ was 14.9 mol m^{-2}.[65] The RGR was 72×10^{-3} d^{-1} for wheat, 10×10^{-3} d^{-1} for well-watered olive plants and 3.6×10^{-3} d^{-1} for olive plants, which were kept at one half of soil capacity. Therefore, the RGR for irrigated olives is about 7 times less than for wheat, due to a combination of two effects: the reduction of A because of stomatal limitation and an increase of ρ.

5. Stress Caused by Waterlogging

Excessive irrigation or rainfall may result in waterlogging. Hypoxic conditions associated with waterlogged soil conditions alter root metabolism,[91] and, when hypoxia conditions are persistent, metabolic functions, including water and nutrient absorption and transport, are impaired.[92]

Olive is considered a sensitive species to root hypoxia and the sensitivity varies among cultivars.[93] Plants of cv. Mission died after a 30-day period of waterlogging, whereas about 50% of the cv. Kalamata plants survived a 60-day period of waterlogging.[93] Resistance to waterlogging in olives has been correlated with the capacity to avoid root hypoxia by producing adventitious roots close to the soil surface.[93] Waterlogging damage can be reduced by using resistant rootstocks, such as *Olea oblonga* and *O. cuspidata,* or resistant cultivars such as 'Zhongshan 80'.[14,15] The darkening of heartwood or "duramen" has been used as sensitivity index for hypoxia in olive.[15] When the duramen of small potted plants darken after 10 days of waterlogging they were classified as sensitive.[15]

V. NUTRIENTS

A. NUTRIENT REQUIREMENTS

The nutrient requirements of olive are lower than that of many other fruit trees such as peach, grape, or citrus. However, when the requirement is not met by soil supply, nutrient stress can cause major physiological changes in the plant. Olive macronutrient requirements may be satisfied by soils containing more than 0.1% N and more than 0.03% P_2O_5 and K_2O.[94] Other macronutrients are normally present in sufficient amounts in the soil. When nutrients in the leaves are adequate for optimal growth, the rate of fertilizer depletion is the product between RGR and the percentage of nutrients in the leaves.[95]

The optimal macronutrient composition of an olive leaf is 2.1% N, 0.35% P, and 1.05% K in leaf dry matter.[96] The optimum total (N-P-K) concentration, however, vary significantly according to soil, cultivar, plant age, and leaf age.[94] Bouat[97] reported that the total nutrient concentration decreased from 5.03% in young leaves to 2.45% in senescent leaves. However, few large scale experiments have investigated the relationship between nutrients and RGR.

B. NUTRIENT STRESS

High concentrations of starch and soluble carbohydrates have been detected in olive leaves during spring and autumn, whereas in summer there is a reduction in starch, soluble carbohydrates and mannitol.[49] In winter, bark and xylem tissues are used to store mannitol and starch.[49] Carbohydrate concentration is more stable in roots than in shoots.[98] Flowering stems are stronger sinks for carbohydrates than vegetative stems.[99] It has been proposed that floral differentiation requires high carbohydrate and nitrogen concentrations in the leaves.[100,101]

When nutrients are not sufficient to support vegetative and reproductive flushes of olive trees, a series of physiological events are triggered. Vegetative growth is reduced by nitrogen deficiency. However, nitrogen excess (frequent in nurseries) favors lush growth, increases susceptibility to pathogens and impairs the physiological balance with other elements, particularly phosphorus.[94,102] Heavy cropping depletes the carbohydrate concentration and reduces the vegetative growth of olive trees.[17] Leaf concentrations of N, P, and K are high before heavy cropping.[94] Lateral shoot development is influenced by the N/P ratio. A high N/P ratio induced a significant reduction in lateral shoot development.[103] A balanced fertilization program can significantly reduce the problem of alternate bearing. Boron deficiency (below 15 ppm in leaf tissue) results in leaf chlorosis and necrosis, followed by leaf drop, bud abortion, lack of flowering, and the development of abnormal fruits.[17,94] Stress caused by antagonistic interactions of elements or other microelement deficiency is rare.

VI. SALT

A. TOLERANCE TO SALT

Olive is moderately salt tolerant, although salinity may be a problem, due to the high salt concentration in the irrigation water used in the Mediterranean area. Olive orchards growing along the coast must also cope with salt deposited by sea winds. Useful parameters for a quantitative and qualitative analysis of salinity are the total salt concentration in the irrigation water and the sodium absorption ratio (SAR, the ratio between the cation Na^+ and the cations Ca^{2+}, and Mg^{2+} in the irrigation water). Olives can be irrigated with water containing up to 3200 ppm of salt with an SAR of 26 or lower.[94] Salt accumulates in the soil if the irrigation to evaporation ratio is low. Robinson[104] observed that on sand an irrigation to evaporation ratio of 1 supports olive growth when water with 1430 ppm of salt is used. When the irrigation/evaporation ratio is lower than 1, however, salt accumulates in the soil and, eventually, olives may succumb to salt toxicity. Tolerance to salt appears to be cultivar-dependent. For example, an NaCl concentration of 80 meq l^{-1} may increase or decrease the percentage of perfect flowers in different cultivars.[105] The growth of all cultivars tested was reduced under salt stress to varying degrees.

B. SALINITY DAMAGE

Salinity decreases the Ψ_π of the soil solution and makes water absorption from the soil more difficult for olive plants.[105] Increasing the ionic concentrations of the nutrient solution from 43 meq l^{-1} to 69 meq l^{-1} has reduced shoot length, leaf area, root length, and rooting ability of olive cuttings.[106]

A reduction in growth, shoot and leaf differentiation, and total leaf area has also been induced by increasing concentrations of NaCl in the water.[106] The photosynthetic characteristics of salt-stressed olives did not change provided that Cl^- concentration was lower than 80 m*M* in total tissue water.[107] Above this threshold, however, a reduction in photosynthesis and stomatal closure occurred and plant morphology was altered.[108] It has been found that, following salt stress, stomata were also less responsive to environmental changes, particularly air humidity. Leaf thickness and plant growth was reduced; a leaf drop of 50% has also been reported.[108] When the Cl^- concentration approached 200 m*M* in tissue water, leaves became permanently dessicated.[108] Accumulation of Cl^- and Na^+ has been reported to decrease distally in olive leaves and to be lower in lateral shoots, stems, and roots than in leaves. There is four times less chloride accumulation in the apical leaves of a shoot compared to the basal leaves.[108] The salt tolerance mechanism in olives may be related to the capacity to accumulate salt in the leaf vacuoles.[108] By accumulating salt in the basal leaves, rather than in the apex or stems, olive plants retain their capacity to recover and grow by shedding leaves containing excess salt.[108]

VII. CO_2

At a high fluence level, leaf photosynthesis is limited by CO_2 partial pressure. In C_3 plants, the enzyme ribulose-bisphosphate carboxylase-oxygenase (RUBISCO) fixes CO_2 inside the chloroplast stroma. The steady state concentration of CO_2 depends on CO_2 in air and CO_2 conductance from the stomata. Conductance of CO_2 in intercellular spaces across membranes into cells until the CO_2 reaches the site of fixation may also affect the inner CO_2 levels. CO_2 concentration progressively diminishes from the ambient air to the chloroplast stroma. The chemical CO_2 gradient is the diffusive force which allows CO_2 to move from the environment to the stroma.[109]

A comprehensive treatment of biochemical and biophysical principles of gas exchange parameters is discussed in the Citrus chapter of Volume II.

A. BIOCHEMICAL REGULATION LIMITED BY STOMATAL DIFFUSION

The partial pressure of CO_2 in the intercellular spaces in the leaf (C_i) can be estimated from CO_2 and H_2O gas exchange measurements.[110] In non-stressed leaves of C_3 plants, the ratio between intercellular and ambient CO_2 concentration ($C_i\ C_a^{-1}$) is about 0.75, with a small variance, within 0.15; this variance is associated with relative water efficiency among C_3 types. Note that C_i is a volume-averaged value and CO_2 is reduced in concentration continuously as it diffuses from the stomata to the site of fixation in the chloroplast. Similarly, the accuracy of calculating C_i from gas-exchange values determined with infrared gas analyzers is contingent upon homogeneous stomatal behavior. However, heterogeneous stomatal closure has been reported in olive leaves.[111]

Net CO_2 assimilation in high light and in ambient air is linked to CO_2 partial pressure in the chloroplast by:[110]

$$A = Vc\,\frac{C_i - \Gamma^*}{C_i + kap} - Rd \qquad (3)$$

where Vc is the maximal capacity of RUBISCO in the leaf exposed to saturating CO_2, Γ^* is the compensation point in the absence of photorespiration, kap is the apparent Michaelis constant of RUBISCO (including the competitive inhibition of oxygen) and Rd is the respiration rate. In ambient air (0.198 mol mol^{-1} O_2) at 28°C, in 3-month-old olive leaves the Vc was 75–95 mmol m^{-2} s^{-1}, Γ^* was 46 mmol mol^{-1}, kap was 720 mmol, and Rd was about 1.2 mmol m^{-2} s^{-1}.[112]

Within these assumptions, the differences of A in C_3 plants, at constant C_i, are mostly due to changes in Vc; the amount of RUBISCO is proportional to total leaf nitrogen, since at least 20% of the leaf protein is in the form of RUBISCO. Therefore, if a spinach leaf with $A = 35$ μmol m^{-2} s^{-1} and an olive leaf with $A = 11$ μmol m^{-2} s^{-1} have the same C_i, then the Vc of the olive leaf is three times lower than the Vc of spinach. This conclusion is supported by the low nitrogen concentration of olive leaves. Photosynthesis of non-stressed olive leaves with a C_i of 200 mmol mol^{-1} is about 11 mmol m^{-2} s^{-1}, but the most accurate C_i determinations never gave less than 240 mmol mol^{-1}, with an ambient value of 330 mmol mol^{-1}.[112] Most of the $^{14}CO_2$ in olive is fixed by the Calvin cycle and employs the glycolate pathway, with a significant proportion (10%) of photorespiration through alanine.[112]

B. INTERACTIONS WITH WATER CONTENT, HUMIDITY, AND NITROGEN

The effect of drought on leaf gas exchange of 4-month cv. Verdiale has been studied (Figure 8). The CO_2 response of leaves that reached turgor point loss at 20% RWCD (i.e., $\Psi_l = -1.9$ MPa) was slightly different from the controls at 4% of RWCD. There was a limited loss (24%) in negative biochemical capacity associated with an osmotic adjustment (i.e.,Ψ_π varied from 1.3 to 1.9 MPa). Field values of A were reduced by 76% and stomatal conductance to CO_2 (g_s') fell from 0.212 to 0.086 mol m^{-2} s^{-1} and C_i diminished to 120 mmol mol^{-1}. Thus, stomata may play a major role in controlling photosynthesis

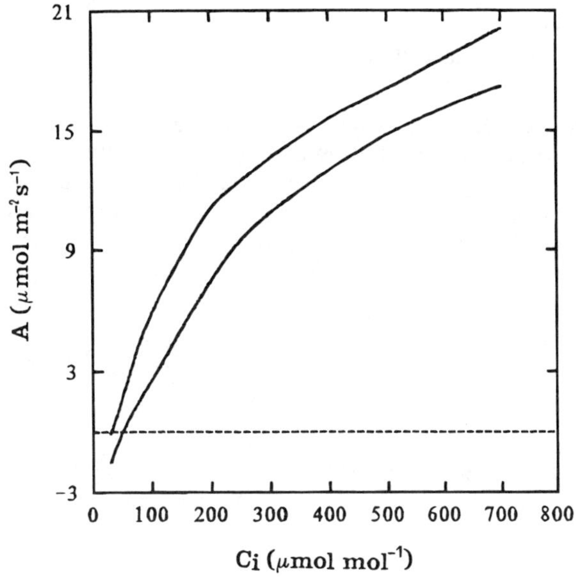

Figure 8 CO_2 response curves corrected for stomatal conductance, and hence expressed in substomatal CO_2 molar fraction (C_i). The upper curve is for leaves in full turgor, whereas the lower one is for leaves which lost 20% of maximal water content at a turgor pressure near zero. The conspicuous reduction in photosynthesis (A) is due mostly to a C_i drop, from 260 to 150 μmol mol^{-1} and, to a lesser extent, to biochemical factors associated with a reduced Ψ_π.

under drought,[112] but these data should be viewed with caution due to the possibility of stomatal heterogeneity.

An increase in VPD caused a marked reduction in A and C_i in greenhouse olive plants (Figure 9).[107] The reduction was also associated with reduction of g_s'. As reducing the VPD neither restored the original A and g_s' nor increased C_i, the responses to VPD appear to be irreversible. Extensive stomatal heterogeneity with closed patches of stomata was also observed after a high VPD treatment,[111] but heterogeneity ceased if the treatment duration was longer than 30 min.

C. DOWN REGULATION CASES

The relationship between A and Vc, may be obviated in many cases by phosphate limitations caused by export of photoassimilates from chloroplasts to cytoplasm, and by starch accumulation inside chloroplasts.[113] The photosynthetic rate is reduced also after increasing CO_2 partial pressure, light, or decreasing O_2 concentration.[113] Such reduction is often termed "down regulation" because it does not depend on stomatal closure, but instead may depend on photoinhibition, phosphate starvation, low temperature inactivation of phosphate chloroplast turnover or photosynthate accumulation inside leaf cells.[113] Treat-

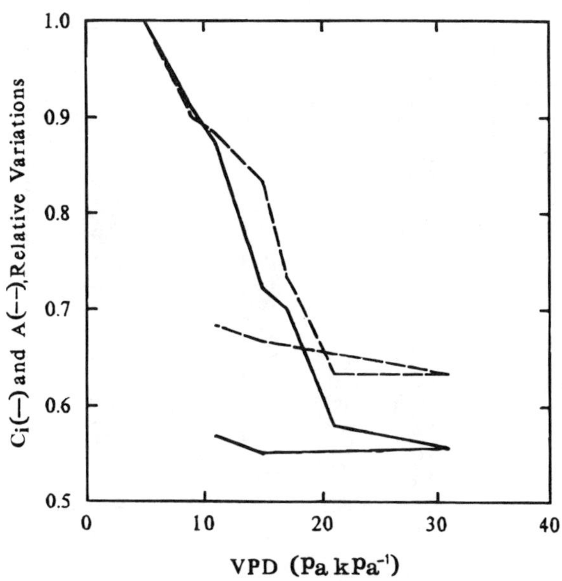

Figure 9 Leaf-air vapor pressure difference (VPD) noticeably affects photosynthesis (A). The time course was 20 min for each VPD, starting from A and C_i equal to unity: when VPD was again lowered, instability of C_i was generated. A subsequent $^{14}CO_2$ experiment confirmed the presence of heterogeneous stomatal closure.[111]

ments, such as girdling, that limit photosynthate translocation in olives have the side effect of blocking photosynthesis.[114]

D. CO_2 UPTAKE AND WATER LOSS

1. Biochemical Models of Water Use Efficiency

The ratio of ^{13}C to ^{12}C in plant tissue has been employed for assessing water use efficiency. Isotope discrimination of $^{13}C/^{12}C$ (Δ) is a measure of the difference in the ratio of $^{13}C/^{12}C$ in plant tissue compared to air.[115,116] Carbon isotope discrimination is measured in a stable isotope ratio analyzer from isotopic ratios in leaf cellulose and air as:

$$\Delta = \frac{R_{air}}{R_{leaf}} - 1; \quad R_{air,leaf} = \frac{^{13}C(mol)}{^{12}C(mol)} \tag{4}$$

R_{air} and R_{leaf} are the molar ratios of carbon isotopes ($^{13}C/^{12}C$) in the air and in the leaf. About 1.1% of the carbon in air is naturally ^{13}C. Farquhar et al.[116] found that in C_3 plants the observed variances of Δ were linked to $C_i\ C_a^{-1}$ variance. Discrimination originates from a diffusional effect produced by the laminar flow within stomata ($\Delta = 4.4 \times 10^{-3}$) and by the smaller reactivity of RUBISCO for $^{13}CO_2$ ($\Delta = 27 \times 10^{-3}$) compared to $^{12}CO_2$. The total "isotope" effect depends on $C_i\ C_a^{-1}$ according to:

$$\Delta\ 10^3 = 4.4 + (27 - 4.4)\frac{Ci}{Ca} \tag{5}$$

Cellulose is the material used most frequently for Δ determinations.[115,117] Leaf cellulose does not change in composition with respect to the Calvin cycle metabolites, and may be conveniently purified by nitric acid digestion.[117] Another useful characteristic is that cellulose synthesis occurs predominantly when the leaf is at its maximal photosynthetic capacity and is minimal under stress. Consequently, Δ is an indicator of the differences in the stomatal regulation during plant growth. Figure 10 reports the leaf cellulose Δ values recently obtained in five olive cultivars in an environment with a mean VPD of 18.5 Pa kPa^{-1}. Fruit-bearing olive shoots had a Δ of 21.64×10^{-3} in 'Frantoio' and of 21.76×10^{-3} in 'Ascolana'; Δ of non-fruiting shoots were 20.29×10^{-3} and 21.28×10^{-3}, respectively.[118] The presence of fruits results in significant differences in Δ.

From the relationship given for the C_i calculation,[119] the following parametric link among A, E, Δ, and $C_i\ C_a^{-1}$ results:

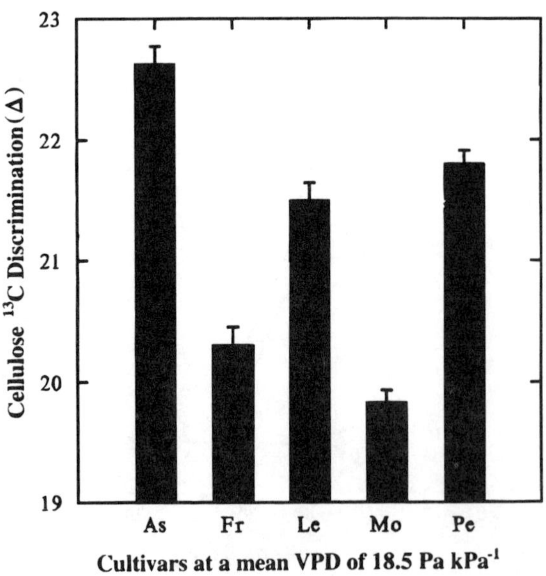

Figure 10 Field experiment of olive cultivars, with a midday mean VPD of 18.5 Pa kPa^{-1} when new shoot formation took place, it clearly showed genetic differences in cellulose ^{13}C discrimination against air CO_2 ($\Delta\ 10^3$). The acronyms are: As, Ascolana; Fr, Frantoio; Le, Leccino; Mo, Moraiolo, and Pe, Pendolino. Bars represent 1 SE.

$$\text{WUE} = \frac{27 - \Delta\ 10^3}{22.6} \frac{\text{Ca}}{1.6\ \text{VPD}} \quad (6)$$

The estimated WUE (water use efficiencies) were 2.16×10^{-3} and 3.48×10^{-3} mol C mol H_2O^{-1} for 'Ascolana' and 'Moraiolo', respectively (Figure 10).

2. Complete Gas Diffusion Paths in Leaves

The CO_2 concentration in chloroplast stroma (C_c) in olive is noticeably different from C_i calculated from gas-exchange measurements.[112] The inverse notation of conductance, resistance, is more appropriate here because, in 'in-series' resistance circuits, like the one we are now introducing, total resistance is the sum of resistances to CO_2 transfer. It should be kept in mind that up to this point the parameter C_i that results from the difference Ca − [A r_s], was being applied.[110] The relationship between WUE, C_i, and Δ, which considered C_i not as a volume-averaged CO_2 concentration in leaf intercellular spaces but as the actual partial pressure of CO_2 in the chloroplast stroma, has been successfully used on glycophytic plants.[116] In olive and in most "hypostomatous plant species", a corrective term must be incorporated due to a conspicuously high diffusive resistance between stomata and stroma. Several measurements indicate that internal diffusive resistance may be equivalent in magnitude to stomatal resistance in mature leaves of all woody species studied.[120,121] For hypostomatous leaves, the total resistance to CO_2 (rtc) pathway from chloroplast to stroma is rtc = rs + rb + rw, where rs is stomatal resistance, the rb is microboundary layer resistance, and rw is the sum of remaining diffusion resistances between stomata and stroma.

The diffusion equilibrium becomes Cc = Ca − [A rtc]; total resistance rtc is expressed for CO_2 transfer. The reason for this correction arose after applying on-line isotope discrimination to olive.[122,123] On-line discrimination, a technique to measure the instantaneous values of Δ, involves determination of Δ of the air surrounding the leaf in a gas exchange cuvette.[120] By contrast, carbon isotope discrimination for structural carbohydrates corresponds to integrated or long-term values.[122] For mature olive leaves the cellulose discrimination was within the limits for a typical C_3 plant, whereas the on-line Δ value fell below 12×10^{-3}, a value more typical for a CAM plant.[122] Cellulose was formed while leaves were young, indicating a change in Δ with ageing in olive. Stomata were invariably closed at night and CO_2 fixation in malate was never detected in olives. Cellulose discrimination and on-line Δ were similar for young leaves, but there was a significant difference in C_i values when they were calculated from leaf gas exchange determinations compared to C_i determinations predicted from the relationship between Δ and C_i.[123] Bongi and Hubick[123] calculated that rb + rw was three times higher in young olive leaves than in wheat, and proposed that internal restriction to CO_2 diffusion into the chloroplast was responsible for high rw. Packing of palisade cell layers called 'scala' are present in olive leaves,[124] that may produce a longer path for CO_2 from stomata to the chloroplast envelope. Further restrictions to the CO_2 pathway are caused by dense thick cell walls and the presence of trichomes on the abaxial leaf surfaces.[71,125] In summary, diffusive resistances, other than stomatal, may considerably limit photosynthetic capacity in olive leaves.

Using the carbon isotope discrimination method, the rb + rw has been estimated to be 3.2 mol^{-1} m^2 s in peach and 7.6 mol^{-1} m^2 s in *Citrus paradisi* and *Macadamia* spp.[120] In mature olive leaves a rb + rw of 11.2 mol^{-1} m^2 s has been calculated.[123]

When trichomes of mature olive leaves were removed, rb + rw decreased from 11.2 to 4.5 mol^{-1} m^2 s (Figure 11). Therefore, rb makes a large contribution to rtc. When rb was calculated as the difference between rtc with and without trichomes, it was 6.7 $mol^{-1}m^2$ s. The advantage of disposing of a barrier outside the stomata in olive leaves, can be derived from the diffusion theory.[119] In turbulent air, the diffusivity ratio for H_2O vapor and CO_2 is 1.37, whereas the laminar flow produced by the chemical gradients inside the stomata results in a diffusivity ratio of 1.6. The 0.23 difference in diffusivity allows 14% more CO_2 uptake than H_2O loss; hence WUE increases.[119] Ying et al.[14] found a large genetic variation on rb in olive leaves; cv. B-114 had the highest rb and more than 8 layers of trichomes, but only 25% of the abaxial surface of cv. ZS-586 was covered with trichomes. The abaxial surface of the leaves was silver-white on cv. B-114, but pale-green on cv. ZS-586. Other genotypes displayed a range of 17 color gradiations.[14]

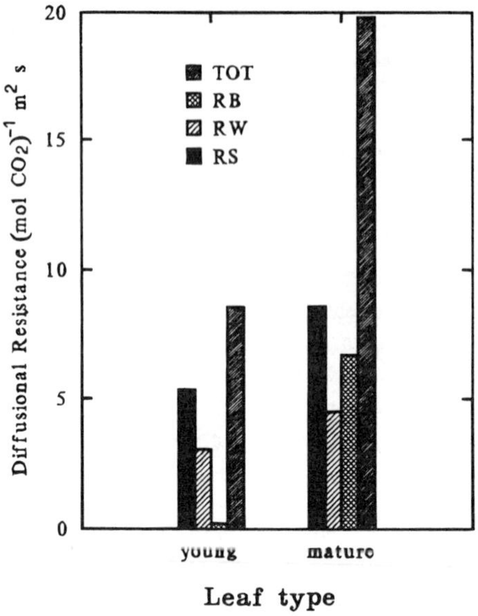

Figure 11 The evolution of strong diffusive resistance in young (2-months-old) and mature (1-year-old) leaves of cv. Verdiale. The use of on-line discrimination recorded a great increase in rb, the resistance to transfer of CO_2 through external boundaries among leaf types due to the build-up of multiple trichome layers. TOT indicates the total resistance to CO_2 transfer, which should be employed to calculate the partial pressure in the chloroplasts (C_c), according to $C_c = C_a - [A\ TOT]$; C_a is the molar fraction of CO_2 outside the boundary layer.

VIII. CONCLUSIONS

Olive has developed adaptations to environmental stress consistent with evolution in a Mediterranean climate. The mechanisms are, however, poorly understood. The performance of a genotype is a cumulative response to various environmental variables and the interaction with environmental variables. The simultaneous occurrence of two or more stresses is generally more deleterious than the occurrence of an isolated stress and makes interpreting stress effects more complex. The low growth rate and photosynthetic capacity of the olive should be considered an adaptation to the environment. Moreover, most olive cultivars which are considered resistant to environmental stresses also exhibit poor horticultural performance.

Morphological features of the olive leaf (size, thickness, anatomy of the mesophyll), limit the exchange of water vapor and CO_2 between leaf and atmosphere, which reduces evapotranspiration and may substantially reduce photosynthesis because of carbon starvation in the stroma. Water stress induces a change in carbon partitioning used for root formation in water stressed plants, thereby reducing the carbon available for photosynthetic end-product formation and canopy growth.

The tolerance of olive to salt is associated with salt accumulation in leaf cells, which inhibits photosynthesis and eventually causes leaf drop only in the leaf population proximal to the roots.

Photoinhibition is limited by low efficiency of the PS2 reaction centers in the adaxial surface of the leaf. Multiple palisade layers in the leaf mesophyll creates a low light environment, reduces the electron transport rate and the photochemical efficiency of photosynthesis, yet protects the apparatus from photoinhibition. Abaxial leaf surfaces reflect a greater percentage of incoming solar radiation than the adaxial leaf surface, also minimizing the deleterious effects of a high light/variable temperature habitat prevalent in Mediterranean regions.

Supercooling and winter pseudo-dormancy confer a slight to moderate level of cold hardening (i.e., down to -12°C); however, genotypic variation in cold hardiness does occur. Conversely, olive does have a cold requirement to insure adequate flowering and high qualitative and quantitative oil characteristics of the fruit.

ACKNOWLEDGMENTS

We are grateful to Prof. Enrico Baldini (University of Bologna, Italy) and Prof. Walter Larcher (Innsbruck University, Austria) for critical review. We should like to thank Judy Etherington for editorial assistance. Research was supported by National Research Council of Italy, Special Project RAISA, Sub-project N.2.

REFERENCES

1. Hoffmannsegg, J. C., Graf, V. and Link, G. F., Floure Portugaise, Hamerlang, C. F., Ed., Kronenstrasse No. 55, Berlin, 1809.

2. Piechura, J. E. and Fairbrothers, D. E., The use of protein-serological characters in the systematics of the family oleaceae, *Amer. J. Bot.,* 70, 780, 1983.
3. Mazzolani, G. and Altamura, B. M. M., Elementi per la revisione del genere *Olea* (Tourn.) Linn. I—Nota introduttiva, *Annali di Botanica,* 35–36, 463, 1976–77.
4. Mazzolani, G. and Altamura, B. M. M., Elementi per la revisione del genere *Olea* (Tourn.) Linn. II—Ciclo di *Olea chrysophylla* Lam., *Annali di Botanica,* 37, 127, 1978.
5. Lavee, S., Aims, methods, and advances in breeding of new olive (*Olea europea* L.) cultivars, *Acta Hort.,* 286, 23, 1990.
6. Sakai, A. and Larcher, W., *Frost survival of plants. Responses and adaptation to freezing stress,* Ecological studies, 62, Springer-Verlag, Berlin, 1987, 196.
7. Larcher, W., Moraes, J. A. P. V. and Bauer, H., Adaptive responses of leaf water potential, CO_2-gas exchange and water use efficiency of *Olea europea* during drying and rewatering, in *Components of productivity of mediterranean-climate regions. Basic and applied aspects,* Margaris, N. S. and Mooney, H. A., Eds., The Hague, Boston, London, 1981, 77.
8. Loukas, M. and Krimbas, C. B., History of olive cultivars based on their genetic distances, *J. Hort. Sci.,* 58, 121, 1983.
9. Scaramuzzi, F. and Roselli, G., Olive genetic improvement, *Olea,* 17, 11, 1986.
10. Bellini, E., Miglioramento genetico dell'olivo: osservazioni sul comportamento ereditario di alcuni caratteri, in *Proc. Giornate Scientifiche S.O.I.,* Grassi, F. and Pavia, R., Eds., Società Orticola Italiana, Firenze, Italy, 1992, 394.
11. Morettini, A., *Olivicoltura,* Ramo Editoriale degli Agricoltori, Eds., Roma, 1972.
12. Hartmann, H. T., Rootstock effects in the olive, *Proc. Am. Soc. Hort. Sci.,* 72, 242, 1958.
13. Hartmann, H. T., Schnathorst, W. C. and Whisler, J., 'Oblonga' a clonal olive rootstock resistant to *Verticillium* wilt, *Calif. Agric.,* 25, 12, 1971.
14. Ying, G., Shu-zhi, H., Zui-jun, S., Hui-chan, B., Hou-jun W. and Shan-an, H., Studies on variation and selection of olive seedlings for cold resistance, in *Olive acclimation and breeding,* Shanan, H. and Ying, G., Eds., PRC, Nanjing, 1984, 327.
15. Ju-zhen, L. and Shan-an, H., The prospect of olive cultivation in China, in *Olive acclimation and breeding,* Shanan, H. and Ying, G., Eds., PRC, Nanjing, 1984, 106.
16. Cimato, A., Cantini, C. and Sani, G., Climate-phenology relationship on olive cv. Frantoio, *Acta Hort.,* 286, 171, 1990.
17. Lavee, S., Olive, in *Handbook of fruit set and development,* Monselise, S. P., Ed., CRC Press, Inc., Boca Raton, Florida, 1986, 261.
18. Baldini, E. and Guccione, G., Osservazioni su di una razza di olivo con antere sterili, *Ann. Sperim. Agr.,* 6, 1, 1952.
19. Villemur, P., Musho, U. S., Delmas, J.-M., Maamar, M. and Ouksili, A., Contribution a l'etude de la biologie florale de l'olivier (*Olea europaea* L.): Sterilite male, flux pollinique et periode effective de pollinisation, *Fruits,* 39, 467, 1984.
20. Bellini, E., Behaviour of some genetical characters in olive seedlings obtained by cross-breeding, in *Proc. XXIII Int. Horticultural Congress,* Firenze, Italy, 1990, 3062.
21. Pannelli, G., Servili, M., Baldioli, M. and Montedoro, G. F., Osservazioni poliennali sulle variazioni quali-quantitative di oli ottenuti da cultivar di olivo con diverso modello di maturazione, in *Proc. Atti giornate scientifiche S.O.I.,* Pavia, R. and Grassi, F., Eds., Società Orticola Italiana, Firenze, Italy, 1992, 436.
22. Lavee, S. and Wodner, M., The specific nature of oil accumulation in fruit of different olive cultivars fact or artifact, in *Proc. XXIII Int. Horticultural Congress,* Firenze, Italy, 1990, 2389.
23. Tombesi, A. and Standardi, A., Effetti della illuminazione sulla fruttificazione dell'olivo, *Riv. Ortoflorofrutt. Ital.,* 6, 368, 1977.
24. Tombesi, A. and Cartechini, A., L'effetto dell'ombreggiamento della chioma sulla differenzazione delle gemme a fiore dell'olivo, *Riv. Ortoflorofrutt. Ital.,* 70, 277, 1986.
25. Hackett, Z. P. and Hartmann, H.T., Inflorescence in olive as influenced by low temperature, photoperiod and leaf area, *Bot. Gaz.,* 125, 65, 1964.
26. Walker, R. S., Automated measurement of leaf photosynthetic O_2 evolution as a function of photon flux density, *Phil. Trans. R. Soc. Lond.,* 323, 313, 1989.

27. Bongi, G. and Loreto, F., Broadleaved evergreens variations in Fv/Fm induced by photoinhibition are coupled to reductions in PSII rc unit size, in *Techniques and new developments in photosynthesis research,* Nato ASI series, Barber, J. and Malkin, R., Eds., Plenium Press New York, 1988, 168.
28. Bjorkman O., Low-temperature chlorophyll fluorescence in leaves and its relationship to photon yield of photosynthesis in photoinhibition, in *Photoinhibition,* Kyle, D. J., Osmond, C. B. and Arntzen C. J., Eds., Elsevier Science Publishers Biomedical Division, Amsterdam, The Netherlands, 1987, 123.
29. Groom, Q. J., Baker, N. R. and Long, S.P., Photoinhibition of holly (*Ilex aquifolium*) in the field during winter, *Physiol. Plant.,* 83, 585, 1991.
30. Demmig, B. and Bjorkman, O., Comparison of the effect of excessive light on chlorophyll fluorescence (77k) and photon yield of O_2 evolution in leaves of higher plants, *Planta,* 171, 171, 1987.
31. Bongi, G. and Lupattelli, M., Thermal damage of photosynthesis in broadleaved evergreens: photoinhibition at low temperatures, in *Plant and Temperature,* Soc. Expt. Biol., Symposium No. 42, Colchester, U.K., 1987.
32. Bongi, G. and Long, S.P., Light-dependent damage to photosynthesis in olive leaves during chilling and high temperature stress, *Plant Cell Environ.,* 10, 241, 1987.
33. Bongi, G., Luchetti, A. and Rocchi, P., Evergreen hardening against freezing produces long term photoinhibition, *Plant Physiol.,* 93, 957, 1990.
34. Bongi, B. and Loreto, F., Drought stress in high light produces fast photoinhibition of PS2 in broadleaved evergreens, *Giornale Botanico Italiano,* 122, 180, 1988.
35. Larcher, W., Low temperature effects on mediterranean sclerophylls: An unconventional viewpoint, in *Components of productivity of Mediterranean-climate regions. Basic and applied aspects,* Margaris, N. S. and Mooney H. A., Eds., The Hague, Boston, London, 1981, 259.
36. Larcher, W., Effects of low temperature stress and frost injury on plant productivity, in *Physiological processes limiting plant productivity,* Johnson C. B., Ed., Buttertcorths, London, 1981, 253.
37. Natali, S., Bignami, C. and Fusari, A., Water consumption, photosynthesis, transpiration and leaf water potential in *Olea europaea* L., cv. Frantoio, at different levels of available water, *Agr. Med.,* 121, 205, 1991.
38. Baldy, C., Lhotel, J. C. and Hanocq, J. F., Effetti della radiazione solare nella funzione fotosintetica dell'olivo (*Olea europea* L.), *Olivae,* 8, 18, 1985.
39. Bongi, G. and Loreto, F., Effect of diffuse light on stomatal opening, *Plant Physiol.,* 86, 4, (Suppl.), 118, 1988.
40. Hartmann, H. T. and Whisler, J. E., Flower production in olive as influenced by various chilling temperatures regimes, *J. Amer. Soc. Hort. Sci.,* 100, 670, 1975.
41. Denney, J. O. and MacEachern, G. R., An analysis of several climatic temperature variables dealing with olive reproduction, *J. Amer. Soc. Hort. Sci.,* 108, 578, 1983.
42. Hartmann, H. T., Effect of winter chilling on fruitfulness and vegetative growth in the olive, *Proc. Amer. Soc. Hort. Sci.,* 62, 184, 1953.
43. Hartmann, H. T. and Porlingis, I., Effect of different amount of winter chilling on fruitfulness of several olive varieties, *Bot. Gaz.,* 119, 102, 1957.
44. Lavee, S. and Harshemesh, H., Climatic effect on flower induction in semi-juvenile olive plants (*Olea europaea*), *Olea,* 17, 89, 1990.
45. Griggs, W. H., Hartmann, H. T., Bradley, M. V., Iwakini, B. T. and Whisler, J., Olive pollination in California, *Calif. Agric. Exp.,* Stn. Bull. 869, 1975.
46. Voyiatzis, D. G. and Porlingis, I. C., Temperature requirements for the germination of olive seeds (*Olea europea* L.), *J. Hort. Sci.,* 62, 405, 1987.
47. Mencuccini, M., Fontanazza, G. and Baldoni, L., Effect of basal temperature cycles on rooting of olive cuttings, *Acta Hort.,* 227, 263, 1987.
48. Bongi, G., Mencuccini, M. and Fontanazza, G., Photosynthesis of olive leaves: effect of light flux density, leaf age, temperature, peltates, and H_2O vapor pressure deficit on gas exchange, *J. Amer. Soc. Hort. Sci.,* 112, 143, 1987.
49. Drossopoulos, J. B. and Niavis, C. A., Seasonal changes of the metabolites in the leaves, bark and xylem tissues of olive tree (*Olea europaea* L.). II. Carbohydrates, *Ann. Bot.,* 62, 321, 1988.
50. Larcher, W. and Thomaser-Thin, W., Seasonal changes in energy content and storage patterns of mediterranean sclerophylls in a northernmost habitat, *Acta Oecol.,* 3, 271, 1988.

51. Martin, G. C., Olive flower and fruit population dynamics, *Acta Hort.*, 286, 141, 1990.
52. Barber, J. and Baker, N. R., Photosynthetic mechanisms and the environment, in *Topics in photosynthesis,* Vol. 6, Barber, J. and Baker, N. R., Eds., Elsevier Science Publishers Biomedical Division, Amsterdam, The Netherlands, 1985.
53. Schreiber, L. and Schönherr, J., Phase transitions and thermal expansion coefficients of plant cuticles. The effects of temperature on structure and function, *Planta,* 182, 186, 1990.
54. Valentini, R., Scarascia Mugnozza, G., Giordano, E. and Kuzminsky, E., Influence of cold hardening on water relations of three eucalyptus species, *Tree Physiol.,* 6, 1, 1990.
55. Scarascia Mugnozza, G., Valentini, R., Kuzminsky, E. and Giordano, E., Freezing mechanisms, acclimation processes and cold injury in eucalyptus species planted in the mediterranean region, *Forest Ecol. Management,* 29, 81, 1989.
56. Larcher, W., Kalteresistenz und ubertcinterungsvermogen medi-terraner holzpflanzen, *Oecol. Plant.,* 5, 267, 1970.
57. Larcher, W., Meindl, U., Ralser, E. and Ishikawa, M., Persistent supercooling and silica deposition in cell walls of palm leaves, *J. Plant Physiol.,* 139, 146, 1991.
58. Bongi, G., Palliotti, A. and Rocchi, P., Aspetti della fisiologia dell'olivo, *Frutticoltura,* 11, 83, 1992.
59. Palliotti, A., Bongi, G., and Rocchi, P., unpublished data, 1992.
60. Amiot, M-J., Fleuriet, A. and Macheix, J.-J., Importance and evolution of phenolic compounds in olive during growth and maturation, *J. Agr. Food Chem.,* 34, 823, 1986.
61. Callahan, A., Scorza, R., Morgens, P., Mante, S., Cordts, J. and Cohen, R., Breeding for cold hardiness: searching for genes to improve fruit quality in cold-hardy peach germplasm, *HortScience,* 26, 522, 1991.
62. Roselli, G., Benelli, G. and Morelli, D., Relationship between stomatal density and winter hardiness in olive (*Olea Europaea* L.), *J. Hort. Sci.,* 64, 199, 1989.
63. Loussert, R., Le aree ecologiche dell'olivo in Marocco, *Olivae,* 18, 32, 1987.
64. Fernandez, J. E., Moreno, F. and Martin-Aranda, J., Study of root dynamics of olive trees under drip irrigation and dry farming, *Acta Hort.,* 286, 263, 1990.
65. Palliotti, A. and Bongi, G., unpublished data, 1992.
66. Zhang, J. and Davies, W. J., Abscisic acid produced in dehydrating roots may enable the plant to measure the water status of the soil, *Plant Cell Environ.,* 12, 73, 1989.
67. Lavee, S., Nashef, M., Wodner, M. and Harshemesh, H., The effect of complementary irrigation added to old olive trees (*Olea europaea* L.) cv. Souri on fruit characteristics, yield and oil production, *Adv. Hort. Sci.,* 4, 135, 1990.
68. Proietti, P., Photosynthesis and respiration in olive fruit, *Acta Hort.,* 286, 211, 1990.
69. Proietti, P. and Tombesi, A., Changes in photosynthetic activity in olive fruits, in *Proc. Eighth Consultation of the FAO European cooperative research network on olives,* Bornova (Izmir) Turkey, 1991, in press.
70. Proietti, P., personal communication, 1993.
71. Leon, J. M. and Bukovac, M. J., Cuticle development and surface morphology of olive leaves with reference to penetration of foliar-applied chemicals, *J. Amer. Soc. Hort. Sci.,* 103, 465, 1978.
72. Blanke, M. M. and Lenz, F., Fruit photosynthesis, *Plant Cell Environ.,* 12, 31, 1989.
73. Doorenbos, J. and Pruitt, W. O., Les besoins en eau des cultures, *Bulletin d'irrigation et de drainage,* Roma, Italy, 1975.
74. Deidda, P., Dettori, S., Filigheddu, M. R. and Canu, A., Water stress and physiological parameters in young table-olive trees, *Acta Hort.,* 286, 255, 1990.
75. Natali, S., Indagine su i rapporti tra contenuto idrico del terreno, potenziale idrico fogliare e traspirazione, in *Proc. Incontro frutticolo su: L'irrigazione delle colture arboree,* Bologna, Italy, 1979.
76. Michelakis, N. and Vougioucalou, E., Water used, root and top growth of olive trees for different methods of irrigation and levels of soil water potential, *Olea,* 19, 17, 1988.
77. Agabbio, M., Dettori, S. and Azzena, M., Primi risultati sulle variazioni giornaliere e stagionali del potenziale idrico fogliare nella cultivar d'olivo "Ascolana tenera" sottoposta a differenti regimi idrici, *Riv. Ortoflorofrutt. It.,* 67, 317, 1983.
78. Milella, A. and Dettori, S., Confronto fra tre coefficienti colturali per l'irrigazione dell'olivo da mensa, *Riv. Ortoflorofrutt. It.,* 70, 231, 1986.

79. Monselise, S.P., Closing remarks, in *Handbook of fruit set and development,* Monselise, S.P., Ed., CRC Press, Inc., Boca Raton, Florida, 1986, 521.
80. Rhizopoulos, S., Meletiou-Christou, M. S. and Diamantagiou, S., Water relations for sun and shade leaves of four mediterranean evergreen sclerophylls, *J. Exp. Bot.,* 42, 627, 1991.
81. Moelwyn-Hughes, E. A., *Physical Chemistry,* Pergamon, Oxford, 1961, 803.
82. Wodner, M., Lavee, S. and Epstein, E., Identification and seasonal changes of glucose, fructose and mannitol in relation to oil accumulation during fruit development in *Olea europaea* (L.), *Sci. Hort.,* 36, 47, 1988.
83. Spiegel, P., The water requirement of the olive tree, critical periods of moisture stress and the effect of irrigation upon the oil content of its fruit, in *Proc. 14th Int. Cong. Hort. Sci.,* Wageningen, Netherlands, 1955, 1363.
84. Jones, H. G., Physiological aspects of the control of water status in horticultural crops, *HortScience,* 25, 19, 1990.
85. Thompson, R. G., Tyree, M. T., Lo Gullo, M. A. and Salleo, S., The water relations of young olive trees in a mediterranean winter: measurements of evaporation from leaves and water conduction in wood, *Ann. Bot.,* 52, 399, 1983.
86. Schulte, P. J., The units of currency for plant water status, *Plant Cell Environ.,* 15, 7, 1992.
87. Zimmermann, U., Physics of turgor- and osmoregulation, *Annu. Rev. Plant Physiol.,* 29, 121, 1978.
88. Bongi, G., unpublished data, 1988.
89. Nobel, P.S., *Physicochemical and environmental plant physiology,* Academic Press, Inc., New York, 1991.
90. Masle, J. and Farquhar, G. D., Effects of soil strength on the relation of water use efficiency and growth to carbon isotope discrimination in water seedlings, *Plant Physiol.,* 86, 32, 1988.
91. Larson, K. D., Schaffer, B., Davies, F. S. and Sanchez, C. A., Flooding, mineral nutrition and gas exchange of mango trees, *Sci. Hort.,* 52, 113, 1992.
92. Larson, K. D., Graetz, D. A. and Schaffer, B., Flood-induced chemical transformations in calcareous agricultural soils of south Florida, *Soil Sci.,* 152, 33, 1991.
93. Hassan, M. M. and Seif, S. A., Response of seven olive cultivars to waterlogging, *Gartenbauwissenschaft,* 55, 223, 1990.
94. Di Marco, L., Concimazione, in *L'Olivo,* Baldini, E. and Scaramuzzi, F., Eds., REDA, Roma, 1985, 80.
95. Ingestad, T., Relative addition rate and external concentration; driving variables used in plant nutrition research, *Plant Cell Environment,* 1, 443, 1982.
96. Jacoboni, N., Olivicoltura da reddito. Le cure culturali dell'oliveto, *L' Inf. Agr.,* 48, 81, 1987.
97. Bouat, A., Foliar analysis and the fertilizer requirements of the olive tree, *Modern olive-growing,* FAO, 85, 1977.
98. Priestley, C. A., The annual turnover of resources in young olive trees, *J. Hort. Sci.,* 52, 105, 1977.
99. Cartechini, A. and Tombesi, A., The effect of carbohydrate and lipid contents on flower bud differentiation in olive, *Riv. Ortoflorofrutt. It.,* 70, 287, 1986.
100. Hartmann, H. T., Some responses of the olive to nitrogen fertilizers, *Proc. Amer. Soc. Hort. Sci.,* 72, 257, 1958.
101. Hartmann, H. T., Uriu, K. and Lilleland, O., Olive nutrition, in *Fruit Nutrition,* Horticultural Publications, Rutgers University, New Brunswick, 1966, 252.
102. Therios, I. N. and Sakellariadis, S. D., Effect of nitrogen form on growth and mineral composition of olive plants (*Olea europea* L.), *Sci. Hort.,* 35, 167, 1988.
103. Tattini, M., Mariotti, P. and Fiorino, P., Fertirrigazione, crescita ed analisi fogliare di piante di olivo autoradicate (cv. Frangivento) allevate in contenitore, *Riv. Ortoflorofrutt. It.,* 70, 439, 1986.
104. Robinson, F. E., Growth potential of young olive with high chloride irrigation water, *HortScience,* 22, 509, 1987.
105. Therios, I. N. and Misopolinos, N. D., Genotypic response to sodium chloride salinity of four major olive cultivars (*Olea europea* L.), *Plant Soil,* 106, 105, 1988.
106. Bartolini, G., Mazuelos, C. and Troncoso, A., Influence of Na_2SO_4 and NaCl salts on survival, growth and mineral composition of young olive plants in inert sand culture, *Adv. Hort. Sci.,* 5, 73, 1991.
107. Bongi, G. and Loreto, F., Gas-exchange properties of salt-stressed olive (*Olea europea* L.) leaves, *Plant Physiol.,* 90, 1408, 1989.
108. Loreto, F. and Bongi, G., Control of photosynthesis under salt stress in the olive, in *Int. Conf. on Agrometeorology,* Prodi, F., Rossi, F. and Cristoferi, G., Eds, Fondazione Cesena Agricoltura, Cesena, 1987, 411.

109. Parkhurst, D. F., Wong, S. C., Farquhar, G. D. and Cowan, I. R., Gradients of intercellular CO_2 levels across the leaf mesophyll, *Plant Physiol.,* 86, 1032, 1988.
110. Wong, S. C., Covan, I. R. and Farquhar, G. D., Stomatal conductance correlates with photosynthetic capacity, *Nature,* 282, 424, 1979.
111. Loreto, F. and Sharkey, T. D., Low humidity can cause uneven photosynthesis in olive (*Olea europa* L.) leaves, *Tree Physiol.,* 6, 409, 1990.
112. Bongi, G., Soldatini, G. F. and Hubick, K. T., Mechanism of photosynthesis in olive tree (*Olea europaea* L.), *Photosynthetica,* 21, 572, 1987.
113. Leegood, R. C., Walker, D. A. and Foyer, C. H., Regulation of the Benson-Calvin cycle, in *Photosynthetic Mechanisms and the Environment,* Barber, J. and Baker, N. R., Eds., Elsevier Science Publishers Biomedical Division, Amsterdam, The Netherlands, 1985, 191.
114. Proietti, P. and Tombesi, A., Effect of girdling on photosynthetic activity in olive leaves, *Acta Hort.*, 286, 215, 1990.
115. Stuvier, M., The history of the recorded atmosphere as recorded by carbon isotopes, in *Atmospheric Chemistry,* Goldberg, E. D., Ed., Springer-Verlag, New York, 1982, 159.
116. Farquhar, G. D., O'Leary, M. H. and Berry, J. A., On the relationship between carbon isotope discrimination and the intercellular carbon dioxide concentration in leaves, *Austr. J. Plant Physiol.,* 9, 121, 1982.
117. Leavitt, S. W. and Long, A., Sampling strategy for stable carbon isotope analysis of tree rings in pine, *Nature,* 311, 145, 1984.
118. Bongi, G., Mencuccini, M. and Rocchi, P., Applicazioni della discriminazione isotopica naturale in orticoltura, in *Proc. Giornate Scientifiche S.O.I.,* Grassi, F. and Pavia, R., Eds., Società Orticola Italiana, Firenze, Italy, 1992, 160.
119. von Caemmerer, S. and Farquhar, G. D., Some relationships between the biochemistry of photosynthesis and the gas exchange of leaves, *Planta,* 153, 376, 1981.
120. Lloyd, J., Syvertsen, J. P., Kriedemann, P. E. and Farquhar, G. D., Low conductance for CO_2 diffusion from stomata to the sites of carboxylation in leaves of woody species, *Plant Cell Environ.,* 15, 873, 1992.
121. Loreto, F., Harley, P. C., Di Marco, G. and Sharkey, T. D., Estimation of mesophyll conductance to CO_2 flux by three different methods, *Plant Physiol.,* 98, 1437, 1992.
122. Evans, R. J., Sharkey, T. D., Berry, J. A. and Farquhar, G. D., Carbon isotope discrimination measured concurrently with gas exchange to investigate CO_2 diffusion in leaves of higher plants, *Aust. J. Plant Physiol.,* 13, 281, 1986.
123. Bongi, G. and Hubick, K., On-line carbon isotope discrimination of olive leaves, *Plant Physiol.,* 83, 34, 1987.
124. Carr, D. J., Oates, K. and Carr, S. G. M., Studies on intercellular pectic strands of leaf palisade parenchyma, *Ann. Bot.,* 45, 403, 1980.
125. Fahn, A., Structural and functional properties of trichomes of xeromorphic leaves, *Ann. Bot.,* 57, 631, 1986.

Chapter 7

Pear and Quince

Peter D. Mitchell, Ian Goodwin, and Peter H. Jerie

CONTENTS

I. INTRODUCTION

Both pear and quince belong to the Rosaceae family and along with the apple and some lesser known fruits, the subfamily Pomoideae. The pear belongs to the *Pyrus* genus of which there are at least 22 primary species. Present day pear varieties belong to those from the West (the European or French pears) or those from the East (the Oriental, Chinese or Asian pears). The European pears have evolved from *Pyrus communis* (Linnaeus), whereas of the 13 species native to China significant cultivars have been derived from *P. pyrifolia* (Nakai), *P. ussuriensis* (Maxim), and *P. bretschneideri* (Rehder).[1] Hybrids of *P. pyrifolia* × *P. ussuriensis* are also common. *P. pyrifolia* because of its fire blight tolerance has also been crossed with *P. communis,* the 2 common varieties of which are 'Keiffer' and 'LeConte'. Over time, thousands of varieties have been selected and bred, there being almost 3,500 in China alone.[1]

Among temperate fruits, pear is only surpassed by apple in economic importance and is widely distributed throughout the temperature zones. Commercial plantings are still expanding and production of European pears are a significant export industry in many countries, particularly in the Southern hemisphere. From the multitude of varieties several high quality broadly adapted varieties, Bartlett, Anjou, Comice, Bosc, and Packhams form the bulk of these commercial plantings.

It is only in the last 10 or so years that the Western world has shown interest in Oriental pears. This began with cultivars of *P. pyrifolia* often referred to as the Nashi or Japanese pear, Nashi being the word for pear in Japanese. It is now produced commercially in New Zealand, Australia, and the U.S.A. There is also interest in other species of Chinese pear.

0-8493-175-0/94/$0.00+$.50

The quince, *Cynodia oblonga* (Miller) though ancient and spread widely has not achieved great economic importance. Botanically the quince can be divided into 3 sub-species; *C. oblonga piriformis* (pear quince), *C. oblonga malformis* (apple quince), and *C. oblonga lusetanian* (Portugal quince). The Chinese quince *Chaenomelia sininesis* is also an edible fruit but not well known in Western culture. Within the above species and sub-species of quince there are a number of popular varieties and some areas of commercial production in Greece and Italy. Generally, however, quince is of little commercial value and there has been little research into its management and physiology. Notwithstanding the above, quince is very important horticulturally as a dwarfing rootstock for pear and has been so used for centuries, however, most quince rootstocks lack the adaptability generally associated with pear and are more prone to environmental stress.

Pear is well adapted to many of the environmental stresses experienced in its native range. The wide geographical range of commercial cultivation is evidence that species of pear evolving in nature or varieties of pear developed by breeding are adapted to a wide range of climatic and edaphic conditions. Nevertheless, the productivity of pear in all regions is limited by one or more often, several environmental stresses. An understanding of the environmental factors such as irradiance, temperature, and water on physiology can improve yield and fruit quality in current areas of production and may facilitate the further expansion of pear to new geographical regions. There is ample justification for further research in the area of stress physiology, because despite the economic importance of pear surprisingly little information concerning physiology appears in the literature.

II. IRRADIANCE

A. PHOTOSYNTHESIS

There have been very few experiments dealing with photosynthetic characteristics of *Pyrus* leaves; virtually no experiments have been performed for whole tree canopies. The light saturation curve has been estimated from intact leaves of *P. communis* (cv. Bartlett) in a laboratory experiment under varying light intensities.[2] The light saturation curve was hyperbolic with a light saturated net CO_2 assimilation (A) occurring at a photosynthetic photon flux (PPF) of approximately 1100 μmol $m^{-2}s^{-1}$. The light saturation curve, however, varies between pear species and varieties.[3] Measurements were made on detached leaves of *P. serotina* (cvs. Chojuro, Suishu, and Doitsu), *P. communis* (cv. Bartlett), and *P. betulaefolia.* All differed in A response to PPF. For example *P. communis* (cv. Bartlett) increased to near maximum A = 19.0 μmol $m^{-2}s^{-1}$ at PPF = 400 μmol $m^{-2}s^{-1}$ whereas *P. serotina* (cv. Suishu) increased slowly to A = 11.7 μmol $m^{-2}s^{-1}$ at PPF = 1200 μmol $m^{-2}s^{-1}$. Other workers have measured variation in response of photosynthesis to light between hybrids of *P. pyrifolia,*[4] varieties of *P. communis,*[5] and pears of different derivation.[6]

Under orchard conditions maximum photosynthesis (A = 18.9 μmol $m^{-2}s^{-1}$) of attached *P. communis* (cv. Bartlett) leaves occurred at 0830 hr with slow decline for the remainder of the day.[2] Light saturation (PPF = 830 μmol $m^{-2}s^{-1}$) was slightly below that measured in the laboratory (PPF = 1100 μmol $m^{-2}s^{-1}$). A similar decline in photosynthesis after the morning has been measured in apples.[7] Leaf water potentials, however, were much more negative at dawn (−1.2 MPa)[2] than the −0.46 MPa reported at dawn at the same location for regulated deficit irrigated trees receiving 25% of normal irrigation.[8]

When leaf gas exchange characteristics were measured on 11 species of field-grown fruit crops, *P. communis* cv. Flordahome had an intermediate A rate relative to the other species and similar to apple.[9] Net CO_2 assimilation was 11.8 μmol $m^{-2}s^{-1}$ at a PPF of 2049 μmol $m^{-2}s^{-1}$, an air and leaf temperature of 30.0 and 29.7°C, respectively, and a vapor pressure deficit of 3.5 kPa. Transpiration rate (E) was 5.7 mmol $m^{-2}s^{-1}$.

The single leaves of these experiments represent only a minute portion of the entire canopy and photosynthesis within the tree is highly complex. While it has been estimated that only 15 to 20% of the trees canopy may be light saturated,[10] the influence of alternating sunflecks, leaf orientation, a possible increase in apparent quantum yield of shade grown leaves and other dynamic factors must be considered in an assessment of whole tree photosynthesis.

B. SHADING

The adverse effect of shading on fruit size and quality of pear has been demonstrated. Individual branches of 35-year-old 'Bartlett' pears were covered with shade cloth to give 30, 50, 65, and 82.5% shade.[11] Fruit under all shade treatments at harvest were smaller and of lower quality than non-shade cloth-covered fruit. Size, total, and soluble solids were inversely related to shade levels in a linear manner and firmness was proportional to shade level. Shade also reduced specific leaf weight in a curvilinear manner with the biggest reduction (12.5%) between 0 and 30% shade with little further response past 50%. Shade-induced differences in fruit characteristics may have been largely due to differences in maturity. Although fruit may often ripen earlier in the top of the tree, the apparent influence of light on such responses could be influenced by positional effects within the tree, fruit maturing earlier as distance from tree base increases.[12] Shading also decreased specific leaf weight on young *P. pyrifolia* trees however dry matter production only decreased relative to a control treatment (full sun) when radiation on the shaded treatment decreased by 30 to 40% of the control treatment.[13]

Within well-trained pear trees shading is mainly intermittent and cloud and wind may also result in fluctuating irradiance. Net CO_2 assimilation rate of pear leaves exposed for 80 seconds to 100, 66, 33, and 10% full sun (100% PPF = 2049 ± 1.48 μmol $m^{-2}s^{-1}$) declined from 12.2 to 7.3 at 33% then to 1.8 μmol $m^{-2}s^{-1}$ at 10%.[14] Recovery was rapid and short term shading did not affect photosynthetic capacity. Transpiration rate declined from 5.9 to 4.8 mmol $m^{-2}s^{-1}$ between 100 and 10% full sun while stomatal conductance (g_s) was not affected by temporary shade (227 and 231 mmol $m^{-2}s^{-1}$ at 100 and 10% full sun, respectively). Water use efficiency (A/E) was thus low at 10% full sun. Pear leaves can therefore be considered as non-sun tracking having stomatal apertures independent of short term changes in irradiance.[15] The above lack of stomatal response to intermittent shade was consistent with slow changes in g_s measured by Yamamoto et al.[16] as light levels rose and fell, however, in this experiment stomatal movement was more sensitive to light when leaves reached full size than at an earlier or later stage of growth.

C. LIGHT AND FRUITING

The importance of maintaining an open canopy to intercept maximum sunlight has been demonstrated for pear trees;[17,18] photosynthetic activity was 15 to 33% higher on dwarf trees on quince rootstock due to better light interception than on larger trees.[18] However, in Italy, mechanical pruning of 'Abbe Fetel' and Passe Crassane pear trees over a five-year period did not reduce yield despite the development of a dense crown which resulted in poor light interception.[19] Nevertheless, yield could not be maintained on ultra-high density pear (4 × 0.5 m) presumably due to poor light penetration;[20] fruit set only at the crown and though fruit set was less at this spacing than at a wider spacing, fruit size was smaller.

Timing of adequate light penetration into the canopy is also important. Pears typically fruit on spurs and the formation of flower buds depends on light received by the spur leaves. The irradiance level 50 to 70 days from full bloom is generally considered critical for successful bud initiation. The level of irradiance required to promote successful bud initiation is not known for pear although 30% of full sun has been suggested for apple.[10] Maximum shoot growth of pear in a high density planting has occurred 40 to 60 days from full bloom[21] and the control of shoot growth at this time is important for fruit set and improved light penetration. For example a reduction in vegetative growth was probably responsible for a greater set response to regulated deficit irrigation (RDI) at 0.5 × 4 m compared to 1 × 4 m tree spacing.[20] RDI, an irrigation technique meant to limit shoot growth early in the season is reviewed in Section IV.A.

III. TEMPERATURE

A. HIGH TEMPERATURE

1. Summer Temperatures

High summer temperatures generally do not restrict the geographical range of pear and quince, however, they can have a negative impact on vegetative and reproductive growth.

Under conditions of high irradiance the temperature of plant tissue can be elevated considerably over air temperature. This is particularly true with darkly colored and or bulky tissue with low rates of transpiration. A threshold for the coagulation of protein for pear fruit has been reported to be 65°C, compared to 63°C for apple, and 50°C for peach.[22] Although leaf temperatures are seldom more than 5°C higher than air temperature, bark temperatures as high as 56°C have been recorded at air temperatures of 45.6°C.[23] Such air temperatures can occur in early spring in hot climates before the tree canopy is fully developed. Branch and trunk tissue may die and productivity may decline though the tree generally survives. Trunks can be painted white to minimize damage due to high or fluctuating temperature.

Though it is generally accepted that pears thrive in relatively hot climates, high temperatures can slow fruit growth, and temperatures above 24°C may delay ripening on some varieties.[24] Such varieties could only be suited to cooler climates. Nevertheless, greater growth and earlier ripening of 'Bartlett' pears have been reported where overhead sprinklers were used to hold air temperatures below 32°C.[25] After 50 minutes of continuous sprinkling air temperature was lowered 3.5 to 4.5°C, leaf temperature 7°C and internal fruit temperature 7 to 7.5°C. The fruit matured 7 days early and were 15% larger at harvest than a control treatment, though all treatments received ample water and were harvested at the same time. Under Australian conditions, pear fruit expand predominantly at night and high daytime temperatures would appear to delay the onset of the diurnal growth cycle.[26] Overhead sprinkling for temperature control has not been used commercially on pear probably due to high costs and management problems; wet leaves increase the risk of fungal disease while wet soil interferes with vital spray operations. Also, treatments which simply advance ripening are not of great advantage to the canning fruit industry. An optimum temperature for fruit development may exist for pear, however, even within varieties there is probably a wide range of temperatures over which pears will develop and mature well. Nevertheless, there are localities around the world renowned for their pears (e.g., the Goulburn Valley of Northern Victoria, Australia and Medford, Oregon). Such areas generally have warm to hot summer daytime temperatures. Localities with mean summer temperatures below 18°C are not suitable for Bartlett pear, and the firmest and best flavored pears are grown in areas much too hot for good apple flavor.[24]

Fruit development periods of pears and other fruits from 7 different localities in North America have been evaluated over a 9-year period.[27] Heat units were calculated as day degrees by averaging the maximum and minimum temperatures for each day and subtracting 10°C then summing these values over the period between blossoming and harvest. At the temperature extremes of this comparison were Kentville (772 ± 31 heat unit) in Nova Scotia and Vineland, Ontario (1150 ± 34 heat unit) in Canada. However, time to maturity was approximately the same 115.9 ± 1.34 and 117.2 ± 1.70 days, respectively, though both blossoming and harvest were 10 days earlier in the warmer regions. The pears from Nova Scotia were of good quality but required heavier thinning to produce fruit of acceptable size. Lower heat units apparently slowed growth but not maturity. Over 6 of the sites, fruit development period for 'Bartlett' pear ranged from 106.5 ± 2.00 to 117.2 ± 1.70 days. 'Bartlett' pears at Tatura in the forementioned Goulburn Valley of Southern Australia also have a similar fruit development period ranging from 110 to 120 days with heat units approximating 930.

2. Winter Dormancy Temperatures

The major factor limiting the expansion of pear production to warm regions is exposure to insufficient chilling temperature during the winter.

Deciduous fruit trees, including pear, in temperate climates enter a period of rest or dormancy in autumn and remain dormant. Rest is broken either at the accumulation of sufficient cold, by a lapse of time or both. Where dormancy is ended before the accumulation of sufficient cold fruit bud opening is delayed and leaf and flower buds open asynchronously, followed by flower abscission and poor fruit set.

In spite of research over many years the concept of "rest" in fruit crops is still far from being fully understood. The general approach to measuring accumulated cold is either to record hours of cold above 0°C and below some base temperature (e.g., 7.5 or 10°C) or by recording chilling units where temperature around 6°C are weighted higher.[29] Though the latter approach is widely favored, high correlations were obtained by both methods when 48 pear varieties and species were tested for chill requirement in Southern England.[30] Varieties were grouped into 5 dormancy classes with hours of accumulated cold (between 0 and 10°C) ranging from 850 to 2600 hours. The species *P. betulaefolia, P. calleryana,* and *P. serrulata* were ranked at the bottom of the 850 hour class, whereas *P. communis* cv. Fleurissant Tard required 3000 hours to break dormancy. In Table 1, abstracted from Speigel Roy and Alston data,[30] seven common varieties are grouped into their allotted dormancy class together with the mean accumulated hours of chilling recorded for each class.

The site specificity of a given chill hour accumulation can be illustrated by referring to conditions in the Goulburn Valley of Victoria, Australia. *P. communis* cvs. William Bon Chretien (syn. Bartlett) and Packham Triumph are the two major varieties. From the data in Table 1, only in the coldest winter would 'Packham Triumph' receive sufficient chilling whereas 'William Bon Chretien' would not be considered suitable for the area. For example, in 1990 both spur and lateral buds of 'Packham Triumph' growing in the Goulburn Valley, broke rest at approximately 60 to 70 days prior to full bloom after

Table 1 **Chilling requirements in Southern England of some common pear varieties**

Class	Variety	Chill units	Hours between 0–10°C	Hours below 7.2°C
1	Josephine	891	857	617
2	Doyenne du Comise Conference	1422	1412	1159
	Packham Triumph			
3	Buerre Hardy	1808	1912	1624
	Old Home			
	William Bon Chretien (syn. Bartlett)			

From Spiegel-Roy, P. and Alston, F. H., *J. Hort. Sci.*, 54, 115, 1979. With permission.

the accumulation of 464 hours below 7°C. This compared with 1159 hours below 7.2°C in Table 1. In addition, flower and leaf buds may vary greatly in chilling requirements. On 'William Bon Chretien' large differences seemed to exist between spur and lateral buds. Spur buds broke rest after 464 hours below 7°C at 70 to 80 days prior to full bloom, however, lateral buds required 623 hours and did not break rest till 40 to 50 days prior to full bloom indicating that reproductive and vegetative chilling requirements are quite different for 'William Bon Chretien'. The chilling requirements for the same variety in Southern England (Table 1) was 1624 hours below 7.2°C, while in Utah (USA) a statistical process of analyzing temperature and blossom dates over several years estimated 1100 to 1200 chill units for 'Bartlett' pear.[31]

The effect of post dormant temperatures on time of blossoming is also variety dependent. Spiegel-Roy and Alston[30] determined that varieties with a high chill requirement also had a high post dormant heat requirement (number of days at 15°C required for 50% bud break). Accumulated growing degree hours above 4.5°C (GDH°C) from end of dormancy to 80% full bloom were also recorded. Though for early and late flowering varieties GDH°C correlated well with chill and post dormant heat requirement, the correlation was poor for intermediate varieties. GDH°C from the completion of rest to full bloom has been estimated by a statistical method and validated in the field in Utah, USA for 'Bartlett' at 5644 GDH°C.[31]

Post dormant temperatures have been manipulated to delay blossoming by evaporative cooling[32] or to advance blossoming by plastic enclosures.[33] A potential may also exist to manipulate bloom date with chemical control of the cyanide-sensitive respiration pathway. Differences in chilling requirements and post dormant GDH°C between *Pyrus* species have been related to the mechanism controlling respiration of dormant buds. Respiration of *P. calleryana* buds, which blossom earlier, were more resistant to cyanide inhibition than buds of *P. communis*.[34] However, chemical control has not been investigated and the nature of bud respiration is not yet fully understood.

The concept of dormancy in deciduous fruit trees and its relationship to required cold is a highly pertinent topic given the forecast of imminent global warming. Improved methods of estimating effective chilling and post dormant requirements seem necessary. Optimum temperatures for chilling may vary between variety. Moreover, in the more temperate fruit growing areas where daytime winter temperatures rise to 10 to 15°C and fall to near or below 0°C at night the importance of day temperature outside the recognized chilling range needs to be understood. The latter seems highly important for it is in these regions that global warming will have the greatest impact on the production of pear and other temperate fruit crops.

B. LOW TEMPERATURES

1. Hardiness

As with other fruit crops, cold temperatures limit pear production to the temperate latitude. However, even within these zones, periodic freezes cause massive losses to the fruit industry. Larsen[35] in his introduction to a *Symposium on Cold Hardiness, Dormancy and Freeze Protection of Fruit Crops* claimed that "the economic losses from freeze damage to America's fruit industry exceeds the loss from insect, disease, rodents and weeds combined." Due to limited cold hardiness *P. communis* in North America is exclusively limited to the middle regions. Yet *P. ussuruensis* which is considered as being very cold hardy can exist in Siberian winters where temperatures are maintained at −50°C over long periods. The same species, however, lacked hardiness under much milder conditions in Oregon[36] indicative of the complex relationship of cold hardiness in the genus *Pyrus*.

Devoid of extraneous ice nucleators, water can be supercooled to −38°C without freezing. Pear along with other hardy fruits contain supercooled water in vegetative and reproductive tissue. Cold hardiness to −38°C was not reached in laboratory tests on a wide range of *Pyrus* species.[37] In this experiment the most hardy species was *P. caucasica;* visual stem damage being observed at −33°C and measured by electrical conductivity damage occurred at −37°C. Stem tissue of *P. ussuriensis,* however, was no more hardy than *P. communis.* This was not surprising given that the tested twigs were sampled from a collection of pear species in Oregon far from the Siberian habitat of *P. ussuriensis. P. ussuriensis* during dormancy has probably acclimated to supercool to close to −40°C in its native habitat. Its survival depends on low chilling requirement as this can only be achieved during a short spring season when temperatures rise above 0°C (i.e., temperatures below 0°C are not thought to contribute to chilling units). In the warmer regions of Oregon dormancy may be broken for varieties with low to moderate chilling requirements in late winter and be damaged during late winter or early spring freezes at temperatures no lower than −9 to 10°C.[36]

Similarly, freezes in late autumn before the tree is fully acclimated will result in damage. Though comprehensively reviewed, the process of acclimation is poorly understood.[38] Attempts to explain cold hardiness have generally considered the plasma membrane as the site of critical freezing injury. However, altered cell wall properties particularly changes in extra-cellular polysaccharides and deposition of callose in the cell wall have been associated with cold acclimation at 2°C of suspension cultured pear.[39]

Any condition which prolongs growth in late autumn such as high N levels, continued irrigation, vigorous rootstock, etc. will delay cold acclimation. Both trunk and lower branches are most susceptible to damage at these times.[23]

2. Blossoming Temperatures

As anthesis approaches buds of pear are at risk once temperatures drop to below 0°C. Damage can range from slight to drastically reduced fruit set and loss of the entire crop. Fruit can also be malformed, e.g., neck thickening or fruit elongation. As with wood damage at harsher temperatures, blossom damage rises when abnormally late winter warmth stimulates bud activity after chilling requirements are met.

The principal methods of frost protection, heaters, overhead irrigation, and wind machines, may increase average orchard temperatures by 1 to 2°C. The former is more reliable but expensive to operate. For instance, from estimates made by Gerber,[40] depending on conditions and placement, a heating system to be effective would need to produce 1.5 to 5 thousand kilowatts per hour per hectare to raise temperatures by 2 to 3°C. This could require up to 100 heaters per hectare within the orchard and individual heaters per tree on the orchard periphery, each burning up to 4.5 l of fuel per hour.

Critical temperatures have been defined as the lowest temperature that can be endured by buds, flowers, or fruit for 30 minutes without injury. Hardiness decreases with bud development. For example, critical temperatures for pear taken from the National Weather Service in the USA[41] in 1970 varied from −3.9°C as the blossom buds become exposed following scale separation to −2.2°C at full bloom and −1.1°C post bloom. Values were given for intermediate stages[42] but varied between locality.[43–46] Critical temperature are extremely variable at bud break though they became more constant at and after full bloom.[45] Also, the stage of bud development will vary within an orchard and within a tree, the exact stage is hard to define. Differences in critical temperature also occur between localities,[41,45,46] probably due to preconditioning of climate and microclimate. Another problem exists in relating tissue temperature to that of the surrounding air, given temperature variation within the tree associated with height and exposure.

The duration of and rate of freezing are important determinants of injury to fruit buds. For 'Bartlett' pear, freeze damage (i.e., internal tissue browning) for all stages of blossom development increased with up to 30 minutes exposure to critical temperature, but except for the small fruit stage, little further damage was measured beyond 60 minutes. Damage was greater at a rate of $-2.5°C\ hr^{-1}$ compared to $-1.9°C\ hr^{-1}$.[42] Evaporative cooling via overhead irrigation has also been used to delay bud break of pear by 14 days, however fruit size was decreased and maturity delayed.[32]

Lindow[47,48] reduced frost damage on 'Bartlett' pear from −3°C frost by spraying antagonistic bacteria to eliminate ice nucleation active bacteria (INA) that prevent supercooling beyond −2°C.[47,48] Results from further experiments indicated that although the absence of INA bacteria decreased frost injury of excised fruitlets, damage was not reduced on orchard trees.[49–51] Similarly, excised flower buds free from

INA bacteria supercooled to −5°C, a temperature 2 to 3°C lower than attached buds. The difference between intact and excised flower buds and fruitlets has been attributed to the overriding effect of ice formation initiated within the woody tissue spreading to the developing flowers and fruit.[52] An exception was a radiation frost when the developing flowers and buds cooled faster than adjacent woody tissue because of direct radiation to the sky.[52] Ice formation would be initiated in the blossom buds first, which could be prevented in the absence of INA bacteria. Proebsting and Gross[49] suggested that this occurred in Lindow's experiments during a mild radiation frost of −3°C. Similar frosts were not observed over 6 years in their experiment.[49]

Blossom damage will drastically reduce fruit set. High and low temperatures in the 10 days post bloom have also reduced fruit set of 'd'Anjou'[53] and 'Tsakoniki'[54] pear. During this period increased yield was positively related to temperature less than 19°C[53] and rapid pollen tube growth was recorded at 20°C.[54] However, at lower temperatures the pollination period is lengthened[54] and under favorable circumstances (low humidity) pollination could proceed at temperatures as low as 10°C.[55] Bee activity declines below 13°C and could be the limiting factor influencing set at low temperatures.[55]

3. Species Differences

As previously mentioned pear species are native to a wide range of latitudes from the sub Arctic to near tropical latitude. Thus it is not surprising that wide ranges of temperature tolerances exist across species.[56] *P. calleryana, P. pashia*, and *P. kawakami* are all classed as being highly tolerant to heat and susceptible to cold. *P. calleryana* is a popular vigorous stock in warm areas, while *P. pashia,* besides being a common stock in the Indian sub-continent has cultivars suited to hot climates. On the other hand *P. ussuriensis* is tolerant of winter cold but not of summer heat. Fortunately *P. communis* (common pear) is heat tolerant and moderately cold tolerant as is *P. betulaefolia,* a commonly used stock.

The prime economic importance of these different temperature traits relates to matching rootstock species to climatic zones to facilitate production of popular commercial varieties. The variation in intraspecific cold tolerance provide a rich gene pool for increasing temperature tolerance of both rootstock and scions. For example, hybrids of *P. ussuriensis* × *P. communis* have been bred for greater cold hardiness,[57] however, their use has been limited to the northern plains of America and Russia due to poor fruit quality.

Wide variation in intraspecific and interspecific chilling requirements also exists for germination of seed.[58] Trees grown from these seeds with a high chill requirement would also have a longer chilling requirement to break rest according to the general statement of Vegis.[59]

Different rootstock species may also influence the level of cold hardness of the scion. In Oregon, pear trees on *P. serotina* and *P. ussuriensis* stocks were damaged more in a cold winter than trees on *P. communis*, though no damage occurred below the union on all three stocks.[60] Observations of differential hardiness associated with rootstock have been observed elsewhere,[61] and is partly a consequence of the growth characteristic induced by different stocks (e.g., delayed defoliation reducing acclimation).

Nevertheless, there is good evidence for winter chilling characteristics being translocated across the union.[57] For example, 'Bartlett'/*P. calleryana* had lower chilling requirement to break rest than 'Bartlett'/ *P. communis*.[62] Westwood[61] discussed the possible movement of hardiness promoters and/or inhibitors across the graft union. Both the wide range of graft compatibility and the wide variation in cold hardiness led Westwood to conclude that *Pyrus* species could be used to test any general theory "of the origins, movement and fate of hardiness promoters and inhibitors".[58] Unfortunately, little progress has been made in this area.

C. TEMPERATURE EFFECTS ON QUINCE

As with other environmental factors there is little documented evidence on the effect of temperature on quince. As a rootstock for pear it is generally considered to have a low tolerance for cold and heat, however hardier strains such as the Causacian Quince and Fonteray Quince are graded as being moderately tolerant to cold.[56]

In spite of the above, quince generally grows well over a range of cool to hot climates. The buds have a low chilling requirement and flower early and evenly where winters are warm. Conversely, where winters are cold and long, flowering is delayed and only occurs on terminal buds after considerable growth has been made. Reduction in yield by frost is thus rare.[63]

IV. WATER STRESS

Water stress is a subjective term as it is difficult to quantify exactly when a plant water deficit (which terrestrial plants experience on a daily basis) may be sufficiently great to qualify as plant moisture stress. A slight water deficit is normal in most plants, and for pear there is no evidence that such deficits suppress short- or long-term productivity. Moreover, moderate deficits at specific times have been associated with increased yield.[20,64–66]

Pears are grown in a range of climates from cool humid areas where irrigation is either absent or supplementary, to arid zones where commercial yield is entirely dependent on irrigation. In humid areas frequent irrigation has often resulted in variable yield responses. This could relate to variations in environmental conditions (vapor pressure deficit and/or soil) influencing plant water status as in both humid and arid areas the water status at a specific phase of seasonal growth can be more important than actual deficit level. For instance, water deficit prior to harvest are likely to decrease fruit size and yield, whereas a similar deficit early in the season may reduce vegetative growth without decreasing yield or ultimate fruit size.

A. REGULATED DEFICIT IRRIGATION

RDI was developed in Australia in 1978 at Tatura (36° 26′S, 145° 16′E) on a medium to heavy duplex soil type to control vegetative growth in high density peach trees,[64] and was then extended to high density 'William Bon Chretien' (syn. 'Bartlett') pear trees.[20,65,66]

RDI involves limiting the supply of water to fruit trees during the period of rapid shoot growth, which for both peach and pear is from early to mid season while fruit growth is slow. In experiments at Tatura, Australia early irrigation was withheld to dry the soil after winter and spring rain. Cumulative net evaporation (class A pan evaporation (E) less rainfall) from full bloom was used to determine irrigation commencement (a cumulative deficit of 125 to 150 mm). This resulted in low soil matric potentials in the area of maximum root concentration in the tree line of −1.25 to −2.61 MPa at 1.0 m spacing and −1.80 to −1.94 at 0.5 m spacing (Table 2).[67]

When irrigation commenced, two RDI treatments based on replacing 20 and 40% of E were compared with a control treatment (100% E).

The influence of withholding irrigation followed by irrigation at low levels using trickle irrigation had two effects on soil moisture. First, as trees were irrigated at 1 or 2 day intervals the wetted volume of soils under the 20% treatment was reduced relative to the control (Figure 1)[67] and second, within this wetted volume soil matric potential (SMP) fluctuated between −0.4 and −0.1 MPa before and after irrigation. The control treatment wet a much larger volume of soil at a SMP between −0.04 and −0.01 MPa.

'Bartlett' on vigorous *P. callyerana* D6 stock in Australia required a considerable level of deficit irrigation at 40 days from full bloom to control shoot growth when the potential for rapid shoot growth was at its highest. Despite high soil moisture deficits, leaf water potentials (Ψ_l) did not approach those measured under drought conditions in Washington[68] (see Section IV.F).

RDI had major effects on shoot growth, fruit size, and number and these results are included in the following discussion on the effects of water deficit on these components of tree growth.

Table 2 **Soil matric potential (MPa) at the end of the withholding irrigation period from a grid of gypsum blocks located midway between two trees (i.e., Center) and at 30 and 60 cm each side of the tree line (i.e., East and West)**

Tree Spacing (m)	Depth (cm)	East			West	
		60 cm	30 cm	Center	30 cm	60 cm
0.5	10	−1.35	−1.72	−1.80	−0.58	−1.16
	25	−1.35	−1.20	−1.94	−0.47	−0.40
	40	−0.35	−0.37	−0.91	−0.41	−0.50
1.0	10	−0.61	−0.51	−1.25	−1.35	−0.57
	25	−0.81	−0.84	−2.61	−2.08	−0.63
	40	−0.51	−0.48	−0.61	−0.54	−0.48

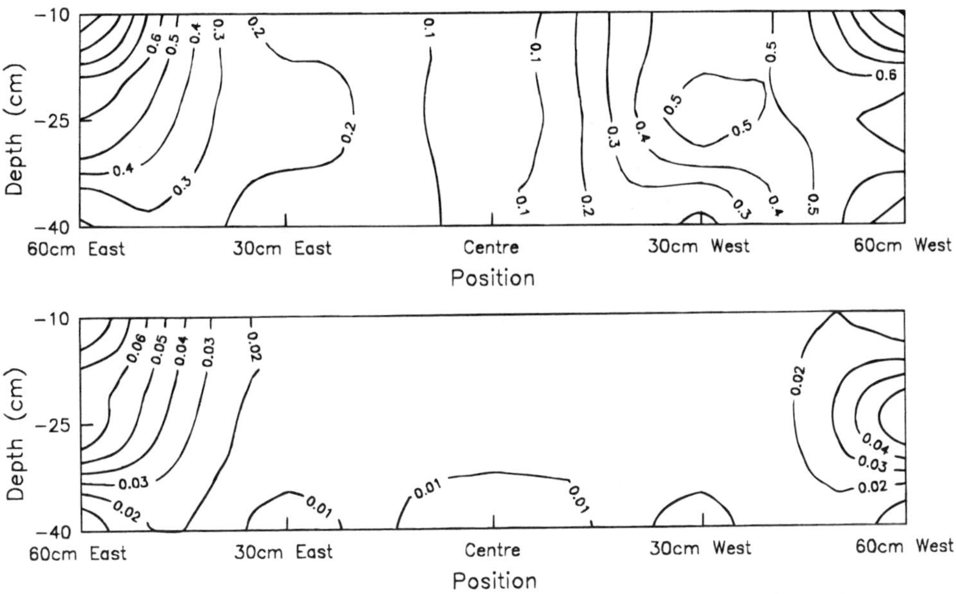

Figure 1 Contour diagram across the tree line of soil matric potential (SMP) distribution after irrigation joining points of equal SMP (MPa) from a gypsum block grid under trees irrigated at (top) 20 and (bottom) 80% replacement of evaporation during the RDI period.[67]

B. VEGETATIVE GROWTH

In general, shoot expansion of most cropping pears is at its maximum 40 to 60 days from full bloom and then declines as the fruit begin rapid growth. Cell expansion and cell division in the growing shoot and developing leaves are both linked to the essential components of growth; cell hydration and cell turgor. So within these distinct flush periods shoot growth and leaf expansion can be reduced by a water deficit. Given ample water availability and temperatures, terminal shoots can again grow strongly after harvest. The mechanism behind these distinct periods of growth is not fully understood. Presumably it is carbohydrate/hormone driven.

On young trees vegetative growth lays the foundation for future cropping, while on mature trees a large healthy canopy allows for maximum radiation to sustain maximum photosynthesis. The majority of pear varieties fruit on spurs and a large percentage of the current seasons growth is removed at pruning. Moreover, a dense canopy will reduce the irradiation level within the tree interior and when shoot growth is less vigorous more photoassimilate will be available for fruit growth.[8]

The threshold Ψ_l at which stomata closed was greater on terminal than on basal pear leaves.[16] For apples, it has been shown that young leaves due to a lower solute content lose turgor before mature leaves at a given Ψ_l thus slowing growth.[69] Osmoregulation, which is an adaptive mechanism to maintain turgor necessary for growth and stomatal opening for photosynthesis at a lower Ψ_l than would otherwise be possible, is more prevalent in mature rather than young leaves. For pear, both osmoregulation[70] and maintenance of photosynthesis at low Ψ_l has been shown to occur. Due to the greater sensitivity of young leaves to stress, significant reductions in cell elongation can occur at relatively high Ψ_l before photosynthesis in mature leaves is reduced. Goode[71] quotes numerous cases of shoot growth reductions on deciduous fruit trees at low moisture deficits. In RDI experiments, shoot extension of pear was linearly related to dawn and midday Ψ_l and growth was depressed at relatively high Ψ_l (Figure 2).[8]

Similarly shoot extension of *P. communis* (cv. Anjou) was less on a treatment receiving 100% replacement of E fortnightly (normal irrigation) compared with a wet treatment receiving 125% of E weekly.[72] In this experiment a comparatively small soil water deficit (−0.035 and −0.012 MPa for the normal and wet treatments, respectively) was measured.

Thus, pear vegetative growth and the structure of the tree framework can be manipulated by variations in soil moisture. In arid areas water can be applied by irrigation to achieve a desirable balance of vegetative and reproductive growth, although RDI is not successful in regions with abundant rainfall during the growing season.

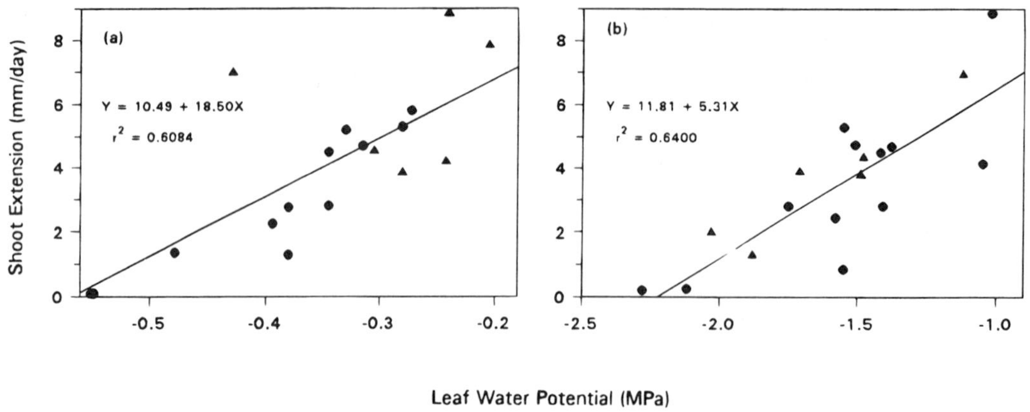

Figure 2 Influence of leaf water potential (Ψ_l) on shoot growth of 'Bartlett' pear trees at dawn (a) and midday (b). (▲) Ψ_l on 11 Nov. cf. shoot growth between 7 and 15 Nov. (•) Ψ_l on 20 Nov. cf. shoot growth between 14 and 21 Nov. From Chalmers, D. J., Burge, G., Jerie, P. H. and Mitchell, P. D., *J. Amer. Soc. Hort. Sci.*, 111, 904, 1986. With permission.

C. FRUIT GROWTH

There are three important concepts concerning the relationship of water deficits to fruit growth. First, fruit growth rate can decline at a relatively low moisture deficit.[72] Second, increasing soil drying may not further decrease fruit growth rate.[72] Third, a relatively large increase in growth rate is possible following the relief of water deficits.[64–66]

In a field experiment, Brun et al.[72] measured considerable loss in pear fruit size over 8 days on a dry treatment, however, in the next 13 days without irrigation growth rate did not decline further. In the first 8 days, Ψ_l at midday fell from −0.9 to −1.4 MPa (−0.9 to −1.0 MPa, respectively, on a wet control) and in the next 13 days continued to fall to −1.8 MPa. Leaf conductance of shaded leaves followed a similar pattern falling from 67 to 36 mmol $m^{-2}s^{-1}$ (76 to 54 mmol $m^{-2}s^{-1}$ on the wet treatment) and then fell to 18 mmol $m^{-2}s^{-1}$. These measurements were taken in mid season, however, similar fruit growth responses have been measured early in the season under RDI in 1984,[65] and in other years (unpublished). Likewise, Richards et al.[73] measured fruit diameter of *P. pyrifolia* (cv. Nijissike) fruit using a Linear Variable Differential Transformer (LVDT) and found that a water deficit lengthened the period of daily shrinkage and rehydration, and thus delayed the commencement of growth till late in the evening. This pattern was established early in the season (SMP of −0.1 MPa), and there was no marked tendency for the pattern to alter as the soil dried. The above three examples emphasize the importance of duration in addition to the level of water stress in determining total loss in fruit size. They also suggest some level of drought adaptation to water deficits that develop slowly which has been suggested for apples.[74]

Three possible mechanisms of plant adaptation to water deficits have been described by Turner.[75] Each is discussed in relation to pear:

1. The ability of a plant to complete its life cycle before serious soil and plant water deficits develop (i.e., drought avoidance). There is no evidence that pears can avoid water stress by early fruit maturation.
2. The ability of a plant to endure low tissue water potential as a result of a water deficit. Mechanisms in this category include desiccation tolerance and turgor maintenance by osmotic adjustment, increase in cell wall elasticity or decrease in cell size. As mentioned previously there is some evidence that mature leaves can osmotically adjust to maintain photosynthesis. There is also some evidence to support osmotic adjustment in the pear fruit to maintain turgor for maximum growth.[76] In most cases, however, fruit growth is slightly retarded from a water deficit during the slow fruit growth stage. Despite a decline in fruit growth rate during this stage, dry weight continues to accumulate[77] as has been reported in grapefruit[78] and apples.[79] The accumulation of the products of leaf photosynthesis by the fruit may also help to prevent a negative feedback on leaf photosynthesis. When water stress is relieved and Ψ_l and turgor increase, fruit growth accelerates which is facilitated by the increase in solute concentration that occurred during the water deficit. The increase in turgor allows the fruit to grow during the day whereas normally the fruit would shrink.[76] Such compensatory growth has been

measured in all RDI experiments as stress imposed by RDI is relieved.

3. The ability of a plant to endure water deficits while maintaining a high tissue water potential. The two mechanisms in this category are a reduction in water loss (by stomatal control, loss of leaf area or reduction in absorbed radiation) and/or maintenance of water uptake. Less and smaller leaves throughout the canopy could reduce leaf area and water loss. This would certainly seem to apply to pear in the RDI experiments where in the early stages of fruit growth, canopy growth is gradually reduced. An increase in root depth and density has also been suggested[9] which would help maintain water uptake during water deficits and could also enable root systems to extract more water when stress is relieved.

Adaptations to moderate levels of stress could explain the lack of fruit response to irrigation in humid climates where stress is moderate or intermittent,[80,81] and why in arid areas precisely scheduled irrigation may not dramatically increase yield. However, in the 3 to 4 weeks prior to harvest, the acceleration of fruit growth following improved water status is not sufficient for pear fruit growth to catch up. Indeed growth is so rapid during this stage that minor water deficits can permanently suppress fruit size. For example, similar sized fruit at day 116 from full bloom differed by approximately 15 g at day 144 on a treatment replacing 100% of E fortnightly, relative to a treatment replacing 120% of E weekly. At day 135 midday Ψ_l of shaded leaves was -0.5 MPa more negative on the 100% treatment and leaf conductance of shaded leaves, which had not previously differed, was 31.2 mmol $m^{-2}s^{-1}$ less.[72]

The timing of irrigation relative to the fruit growth cycle is thus probably more important than precise scheduling according to soil water deficits. In Australia, pear trees are usually irrigated more frequently in the period of high demand 3 to 4 week prior to harvest. Though harvest usually coincides with the hottest weather and a fully grown canopy, growth stage alone could impose a higher rate of water use (i.e., evapotranspiration per mm E) as in peach.[82,83] Chalmers et al.[84] found that transpiration rate of peach leaves correlated with assimilate demand by the fruit. In this regards pears, like peaches, accumulate assimilate and grow rapidly near harvest. A percentage decline in cell expansion during the RDI period has a small impact on final fruit size, relative to a similar percentage loss in fruit growth just prior to harvest.[82]

In general, temporary water deficits in humid areas on medium-to-heavy soil types may have minimal effect on final fruit size and yield of pear, except when deficits occur close to harvest. When grown on shallower, lighter soils a sustained water deficit will substantially decrease yield and tree health. In arid areas irrigation should be aimed at maintaining the maximum potential fruit growth rate, consistent with maintaining the maximum yield and fruit quality. Nevertheless, when tree vigor is high resulting in reduced irradiation within the canopy, less irrigation early in the season may increase yield and fruit size. This would be particularly so in high density planting on medium-to-heavy soil types.

D. CROP LOAD

Crop load has a major influence on final fruit size of pear and thinning may be necessary. On 'William Bon Chretien' (syn. 'Bartlett') pear in Australia, however, fruit number at harvest virtually determines canning yield, maximum canning size being restricted to diameters of <65 mm. In the absence of thinning, fruit number at harvest generally relates to bud initiation, bud development, and condition at blossoming and fruit set. Water deficit at these times will affect crop load.[85,86] Water deficits at 70 to 80 days[85] and at 75 to 90 days from full bloom[86] have increased fruit set the following year on apple and pear, respectively, presumable by increasing bud initiation. The major responses to RDI (up to 60 to 70 days from full bloom) have involved increased fruit set[20] and blossom density.[65] By contrast, fruit number on container-grown pear trees was increased by irrigation[87] while moderate levels of water stress (Ψ_l of -1.8 and -2.1 MPa at midday)[88] during rapid fruit growth (at 70 to 115 days from full bloom), and presumable bud initiation, coincided with decreased fruit set in the following year.[88]

A severe water stress during bloom is generally considered detrimental to fruit set[89] and critical periods prior to and during bloom for pear and other deciduous fruits have been proposed.[90] However, rapid shoot growth can also be antagonistic to fruit set and in humid areas, dry sunny conditions after blossoming could favor fruit set with cool moist conditions favoring shoot growth. In arid areas the necessity to irrigate provides the opportunity to manipulate both fruit set and shoot growth.

Data concerning the effect of climate and management on bud initiation and development is limited. It has been stated that stress may concentrate and localize the movement of carbohydrates and endogenous growth substances favoring flower bud formation whereas high transpirational rates may enable rapid movement and dispersal of the same growth substance.[91] On young expanding trees the above creates a dilemma. Long-term productivity relates to maintaining a certain level of vegetative growth and minimizing the water deficit. In the short term, however, on mature dense plantings the success of RDI emphasizes the value of moderate levels of stress at critical times.

E. QUALITY AND DISORDERS

Quality of pears is generally determined by sugar content and storage life, however, other factors such as texture, firmness, flavor and color are important. Soil water availability and climatic factors may influence the above parameters. However, as with other fruits, water may be an indirect cause of loss in quality. For instance, water supply will influence mineral composition and excess water in the vegetative stage limits light penetration.

Goode[92] cites various authors with conflicting data on the effect of water supply on fruit quality. For example, better keeping quality of pear[93,94] has been attributed to irrigation, however, the storage life of 'Bartlett' though not 'Buerre Bosc' was best when grown in dry compared to wet conditions.[95] Similarly, higher sugar contents have been attributed to favorable irrigation regimes[96] and poor flavor to dry conditions,[95] however, the flavor of 'Anjou' pear was improved by a water deficit near harvest.[97] Such conflicting reports probably relate to the poorly understood interactions of plant water status with stages of growth and other environmental variables.

The three common disorders associated with high water status (black-end, cork spot and alfalfa greening) are nutrition-mediated, and are greatly influenced by rootstock and scion. All reflect the balance of Ca and other minerals (e.g., K and N) in the fruit. Alfalfa greening and cork spot were positively correlated with increased irrigation and fruit K and N concentrations but negatively correlated to fruit Ca.[88]

F. DROUGHT TOLERANCE

Large water deficits can kill pear trees,[98] however, they can also survive extreme levels of water stress. Pear trees (cvs. Bartlett and Anjou) growing in 1.5 m deep fine sandy loam with 8% available water survived a whole season without irrigation (net E = 712 mm).[68] Fruit shrivelled and return bloom was delayed and reduced, as was fruit set. During the drought period, Ψ_l of shaded leaves fell to a minimum of -4.5 MPa. At the same sight the minimum Ψ_l of shaded leaves from droughted peach trees which later died was -3.9 MPa. Heavy winter pruning, lighter summer pruning and thinning improved productivity of the trees exposed to severe drought. In another experiment defruited peach trees used 50% of the water of fruited trees in the rapid stage of growth,[84] although similar results have not been reported for pear. When water stress is extreme heavy thinning and summer pruning of pear trees will ensure some level of commercial yield.

The drought tolerance of many pear varieties has been assessed in Russia.[98,99] The leaves of the drought-resistant varieties had a higher water retaining capacity during summer drought, and detached leaves lost water slower than drought susceptible varieties. This would suggest different levels of stomatal control between varieties and could reflect environmental pre-conditioning at their place of origin or genetic variability.

Similarly, the tolerance of rootstocks to dry conditions could relate to such pre-conditioning and/or variability in rooting characteristics. In Oregon, Lombard and Westigard[56] categorized the drought tolerance of pear and quince rootstocks from high to low as follows: *P. betulaefolia, P. calleryana* = *P. communis, P. ussuriensis* = *P. salicifolia, C. oblonga*. In Australia the *P. calleryana* D6 rootstock is recognized as being more drought tolerant than *P. communis*[101] and *P. salicifolia* outperformed *P. communis* in Russia and was claimed to be tolerant of drought, extreme temperature, and saline conditions.[101] Drought tolerance relates in part to genotype and rooting depth but may also be influenced by climate and edaphic pre-conditions.

G. FLOODING

"Pears grow in wet low land" (from Shi Jing a book written about 1000 B.C. in China).[102] Quince and pear are generally regarded as the most flood-tolerant fruit tree species.[103–107] Mature pear in Oregon flooded from April to August showed minimal damage, however, tree vigor declined after a second flooding two years later.[108] Despite generalization of extreme waterlogging tolerance, differences exist within and between pear and quince.[56,109–112] Observations have separated pear rootstock in order of decreasing tolerance: *P. betulaefolia, P. calleryana* = *P. communis, P. ussurienses, P. pyrifolia.*[56] Clones of 'Bartlett' and 'Anjou' and crosses within the *P. communis* species were also ranked. Furthermore Province Quince was rated equivalent to *P. betulaefolia* with Quince A less tolerant. Conversely quince stocks have been shown to be poorly tolerant.[112] This could relate to the known incompatibility of quince with pear and associated double stress. Quince may also be less tolerant to *Phytophora*.

The above observed difference has been confirmed by controlled pot experiments.[109–111] The flood tolerance of container-grown fruit tree species based on growth, leaf gas exchange, and survival was ranked as follows: *P. betulaefolia* > *P. calleryana* = *C. oblonga* > *P. communis* cv. Bartlett > *P. commumis* cv. OH × F97 = *P. pyrifolia* = *P. ussurienses* = *Malus domestica* > *Prunus persica*. *P. betulaefolia* was placed among the most flood tolerant mesophytic tree species having exhibited 100% survival after 20 months of flooding despite months of midday soil and air temperatures above 27 and 33°C, respectively. *P. communis* cv. OH × F97 but not cv. Bartlett defoliated after one month.[109] In the above experiment oxygen diffusion rate (ODR) stabilized after 30 days around 5×10^{-8} g $cm^{-2}min^{-1}$, a level similar to that measured after two months of flooding in an established orchard.[113] Rootstock selection is thus important for pears growing on heavy soil and or those prone to flooding.

One of the earliest physiological responses to flooding is a reduction in leaf conductance.[110,111] Measurements of leaf conductance have reflected differences in flooding tolerance between pear species and other fruits and was greater in spring than in autumn. The mechanism for flood induced stomatal closure is still unresolved[114] but it occurs independent of changes in Ψ_l.[115]

The decrease in leaf conductance on pear coincided with a simultaneous decline in root hydraulic conductivity.[111] Whether ABA is involved as a signal agent for stomatal closure of flooded pear trees is not known.[111] Nevertheless, in flooded soil, reduced stomatal aperture would seem to be an adaptive response to maintain leaf turgor under conditions of increased root resistance to water flow.[111]

The reasons for waterlogging tolerances of many *Pyrus* species is not clear. The movement of O_2 from shoot to root has been proposed as an adaptive mechanism to waterlogging.[116] Hypertrophied lenticels were more pronounced on pear and quince under flooded soil conditions[109] and movement of O_2 from shoot to root via lenticels has been proposed.[115–117] For many species, hypertrophied lenticels at or below the water line could also serve as exit points for toxic metabolites.[118] Rowe and Beardsell[103] dismissed O_2 transport from shoot to root as contributing to waterlogging tolerance of woody species including pear because of the limited shoot and root lengths through which O_2 may diffuse. Also Andersen et al.[110] found no evidence of aerenchyma on anaerobically-treated *P. betulaefolia* and *P. communis*. An alternative explanation for the flood tolerance of pear and quince could relate to the limitation of ethanol production by metabolic control over glycolysis (i.e., regulation of the Pasteur effect).[119,120]

Poor tree performance following waterlogging is generally associated with root death following oxygen deficiency or the build up of phytotoxins. However, sustained anaerobic conditions can markedly alter soil chemical and physical properties. For example, pear trees are susceptible to Mn toxicity[121] and sustained flooding can increase the availability of Mn to toxic levels by reduction of manganic ion to soluble manganous. Similarly Fe toxicity can also occur and S, Ca, Mn, N, Zn, Pb, and Co have been found more available after flooding.[122,123] Denitrification occurs in waterlogged soil and pH can change. These soil nutrient changes may affect tree vigor of waterlogging-tolerant fruits such as pear and quince even if root O_2 deficiency did not interfere with nutritional uptake. However, West and Nicholson,[124] found that after 20 days of flooding, container grown pears exhibited a decline in leaf concentration of K, Ca, Cu, and Zn. This decline was probably associated with interference to nutritional uptake, since on removal of flooding Ca and Cu rapidly equilibrated to the level of a non-flooded control with similar tendencies for K and Zn. Levels of leaf Mn and Fe did not rise. Uptake of Mn and Fe could be more related to greater soil concentrations of these two elements in soluble form under anaerobic conditions which may only occur under sustained flooding. Soil physical properties will also change as wet soil conditions combined with heavy orchard traffic compact soil and limit root growth. Though pear and quince may not die under waterlogged conditions numerous associated factors can combine to limit vigor and health. Fortunately pathogen invasion under wet conditions is not common on the above two species.

V. SALINITY

Though Bernstein[125] showed a yield reduction for pear of 10% at an electrical conductivity of a saturated soil extract (ECe) of 2.5 dS m^{-1} there is little available data for pear and quince on the effect of sustained high salinity levels in the soil profile. In Australia with mounting pressure to use moderately saline water for irrigation, 40 year old 'William Bon Chretien' (syn. Bartlett) pears were irrigated with five ranges of salinities from 0 to 2.7 dS m^{-1} over 10 years.[126] Despite soil ECe under the higher salinity levels stabilizing by year 4 at around 4.0 dS m^{-1},[127] there was no effect on yield until year 8. In year 7, leaf Cl and Na rose sharply after harvest while midday A rates and g_s were reduced with high salinity (2.7 dS m^{-1} treatment). Under this treatment in years 8 and 9 yields declined and 40% of the trees died. The yield reduction in year 9 coincided with a reduction in fruit number. However, in previous years, though fruit number was not measured, there were more undersized fruit at the high salinities with yields similar between treatments.[126] This infers a greater fruit number at the higher salinity levels. For canning pears, total tonnage depends largely on fruit set not fruit size. Salinity effects on productivity are therefore minimal, however, with increasing soil salinity, eventually yield and fruit number declines rapidly and death may follow.[126]

It is now well recognized that the effect of salinity increases under waterlogged conditions. Leaf chloride levels rose sharply on pear trees having access to a saline water table and waterlogged by successive irrigations.[128] Trees partially defoliated and yielded small fruit. Waterlogging in winter had no such effect.[108] In the pot experiment of West and Nicholson[124] where pear trees were flooded with saline water for 20 days, the combination of waterlogging and salinity also markedly depressed shoot growth rates.

These results pose problems of long-term management of pears in a saline environment (i.e., above a saline water table). While in the absence of spring and summer waterlogging, pears may produce profitability above a salty water table, long-term effects maybe more insidious. As toxic ion concentrations rise in both soil and plant, tree productivity will decline. In the heavy soils common to pear, leaching to remove salt build up poses the threat of waterlogging and rise in damage levels. Maintaining the water table below a level where toxic ions will not encroach significantly into the rootzone may be the only answers. Given the apparent slow nature of salt damage on pear this approach will require regular and careful monitoring of Cl and Na uptake and subsequent effects on tree performance.

VI. CONCLUSIONS

In compiling this chapter two things were obvious. First, the pear adapts well to a range of adverse conditions. Second, recent literature of environmental stress of pear is scarce. Perhaps the former predisposes the latter. Minimum winter temperatures are probably the major environmental limitation, a problem remaining complex and poorly understood. More importantly, freezing injury and chill and dormancy requirements require more study given present predictions of global warming. Large areas of pears, some comparatively new, may be at risk if mean temperatures rise 2 to 3°C. What seems needed is not detailed lists of chilling requirements for countless varieties but a better understanding of the relationship of temperature to dormancy.

As with other fruits most future pear plantings will be at higher density. This may increase environmental stress as temperature and humidity within the planting increase, light penetration to the lower canopy becomes more uneven, and water use of the dense canopy higher. None of these effects will be unique to pear. However, they could pose more of a problem on pear given the absence of suitable dwarfing rootstocks, the vigorous nature of many of the recognized rootstocks and the early flush period of shoot growth common to most European pears.

The success of the previously mentioned RDI experiment of pear at Tatura (a 20% yield increase over a 100 tonne/ha crop) shows what can be achieved by manipulations of the orchard environment. Even so it is still not clear as to what physiological plant responses increased yield. Similarly there are numerous accounts of increased yield from high density systems and pear plantings have been prominent in this regard. Claims based largely on yields are made for better and more efficient systems. Much less attention has been given to the more difficult problem of how and why yield and efficiency increased or in other words how man can manipulate the environment for his betterment.

ACKNOWLEDGMENTS

We wish to thank Miss J. Noonan and Miss S. Richardson for the arduous task of typing this chapter.

REFERENCES

1. Tsuin Shen., Pears in China, *HortSci.,* 15, 13, 1980.
2. Kriedemann, P. E. and Canterford, R. L., The photosynthetic activity of pear leaves (*Pyrus communis* L.), *Aust. J. Biol. Sci.,* 24, 197, 1971.
3. Honjo, H., Kotobuki, K., Asakura, T., and Kamota, F., Net photosynthesis and dark respiration of pear leaves infected by Japanese Pear Rust, *J. Japan. Soc. Hort. Sci.,* 99, 105, 1990.
4. Wang, B. P., Ding, X. C., Dai, W. S. and Xu, R. C., A study on the photosynthetic rate of sand pear (*Pyrus pyrifolia*) under field conditions, *Acta Hort.,* 14, 97, 1987.
5. Kovaleva, A. F. and Senin, V. I., The light regime and productivity of pears in the southern Ukraine, *Sadovodstvo i Vinogradarstvo,* 11, 14, 1988.
6. Orsyannikov, A. S., Photosynthesis of pear cultivars of different derivations, Sbornik Nauchnykk Rabot Vsesoyuznyi Nauchno—Issledovatel'skii Institut Saclovodstvo imeni I. V. Michurina, 25, 61, 1977.
7. Lakso, A. N. and Sealey, E. J., Environmentally induced responses of apple tree photosynthesis, *HortSci.,* 13, 646, 1978.
8. Chalmers, D.J., Burge, G., Jerie, P.H. and Mitchell, P.D., The mechanism of regulation of 'Bartlett' pear fruit and vegetative growth by irrigation withholding and regulated deficit irrigating, *J. Amer. Soc. Hort. Sci.,* 111, 904, 1986.
9. Anderson, P. C., Leaf gas exchange characteristics of eleven species of fruit crops in North Florida, *Proc. Fla. State Hort. Soc.,* 102, 229, 1989.
10. Cain, J. C., Foliage canopy development of McIntosh apple hedgerows in relation to mechanical pruning. The interception of solar radiation and fruiting, *J. Amer. Soc. Hort. Sci.,* 98, 357, 1973.
11. Kappel, F., Artificial shade reduces 'Bartlett' pear fruit size and influences fruit quality, *HortSci.,* 24, 595, 1989.
12. Dann, I. R., Mitchell, P. D. and Jerie, P. H., The influence of branch angle on gradients of growth and cropping within peach trees, *Scientia Hort.,* 43, 37, 1990.
13. Honjo, H., Asakura, T., Kamota, F. and Nakagawa, Y., Responses in growth and development of fruit trees to controlled environment. 1. Measurement of the physical properties of the phytotron and analysis of the influence of light intensity on growth of young Japanese pear trees, *Bull. Fruit Tree Res. Station Jap.,* A. (Yatabe), 10, 91, 1983.
14. Anderson, P. C., Leaf gas exchange of species of 11 fruit crops with reference to sun tracking/non-sun tracking responses, *Can. J. Plant. Sci.,* 71, 1183, 1991.
15. Knapp, A. K. and Smith, W. K., Stomatal and photosynthetic responses to variable sunlight, *Physiol. Plant.,* 78, 160, 1990.
16. Yamamoto, T., Watanabe, S. and Harada, H., Studies of leaf burn of pear trees. Part X. A slowing down in the closing and opening of stomata in the leaves, *J. Jap. Soc. Hort. Sci.,* 48, 267, 1979.
17. Wagenmakers, P., Make provision for light in the tree, *Fruitteeit,* 78, 18, 1988.
18. Kovaleva, A. F. and Senin, V. I., The light regime and productivity of pears in the southern Ukraine, *Sadovodstvo i Vinogradarstva,* 11, 14, 1988.
19. Baldino, E., Rossi, F., Baraldi, R. and Marangoni, B., The availability of radiant energy in mechanically pruned pear trees, *Acta Hort.,* 161, 201, 1984.
20. Mitchell, P. D., Van den Ende, B., Jerie, P. H. and Chalmers, D. J., Responses of 'Bartlett' pear to withholding irrigation, regulated deficit irrigation and tree spacing, *J. Amer. Soc. Hort. Sci.,* 114, 15, 1989.
21. Chalmers, D. J., Mitchell, P. D. and Jerie, P. H., Break the rules and boost orchard dollars, *Austr. Country,* 4, 54, 1984.
22. Zulavshaja, M. N., Maksimenko, K.I. and Medvedeva, T.N., Heat resistance in fruit plants, Ixv. Akad. Nauk mold ssr, ser. biol. him. Nauk. No. 3, 1970, cited *Hort. Abst.,* 41, 8076.
23. Chandler, W. H., Killing Temperatures, In *Deciduous Orchards,* Chandler W. H., Ed., Lea & Febiger, Philadelphia, 1947, chap. 8.
24. Chandler, W. H., Pome Fruits, In *Deciduous Orchards,* Chandler W. H., Ed., Lea & Febiger, Philadelphia, 1947, chap. 16.
25. Lombard, P. B., Westigard, P. H. and Carpenter, D., Overhead sprinkler system for environmental control and pesticide application in pear orchards, *HortSci.,* 1, 94, 1966.
26. Richards, S. M., personal communication, 1990.

27. Fischer, D. V., Heat units and number of days required to mature some pome and stone fruits in various areas of North America, *Proc. Amer. Soc. Hort. Sci.,* 80, 114, 1962.
28. Ryall, A. I., Smith, E. and Pentzer, W. T., The elapsed period of full bloom as an index of harvest maturity in pears, *Proc. Amer. Soc. Hort. Sci.,* 38, 273, 1941.
29. Richardson, E. A., Seeley, S.D. and Walker, D.R., A model for estimating the completion of rest for Redhaven and Elberta peach trees, *HortSci.,* 9, 331, 1974.
30. Spiegel-Roy, P. and Alston, F. H., Chilling and post dormant heat requirements as selection criteria for late-flowering pears, *J. Hort. Sci.,* 54, 115, 1979.
31. Ashcroft, G. L., Richardson, E. A. and Seeley, S. D., A statistical method of determining chill units and growing degree hour requirements for deciduous fruit trees, *HortSci.,* 12, 347, 1977.
32. Collins, M. D., Lombard, P. D. and Wolfe, J. W., The effects of evaporative cooling for bloom delay on 'Bartlett' and 'Bosc' pear fruit maturity and quality, *J. Amer. Soc. Hort. Sci.,* 103, 187, 1978.
33. Seeley, S. D., Seeley, E. T. and Richardson, E. A., Use of phenological model to time acceleration of pollinizer bloom, *HortSci.,* 22, 51, 1987.
34. Cole, M. E., Solomos, T. and Faust, M., Growth and respiration of dormant flower buds of *Pyrus communis* and *Pyrus calleryana, J. Amer. Soc. Hort. Sci.,* 107, 231, 1982.
35. Larsen R. P., Introduction to symposium on cold hardiness, dormancy and freeze protection of fruit crops, Wash. State Univ., Washington, *Hort. Sci.,* 5, 402, 1970.
36. Brown, G. G., Mellenthin, W. M. and Childs, L., Observation of winter injury to apple and pear trees in the Hood River Valley, *Ore. Agr. Exp. Sta. Bul.,* 595, 1964.
37. Rajashekar, C., Westwood, M. N. and Burke, M. J., Deep supercooling and cold hardiness in genus *Pyrus, J. Amer. Soc. Hort. Sci.,* 107, 968, 1982.
38. Weiser, C. J., Cold resistance and acclimation in woody plants, Scientific J. Series Articles No. 7185 of the Minnesota Agricultural Experiment Station; condensed review cited *HortSci.,* 5, 403, 1970.
39. Wallner, S. J., Wu, M. T., and Anderson-Krengel, S. J. Changes in intracellular polysaccharides during cold acclimation of cultured pear cells, *J. Amer. Soc. Hort. Sci.,* 111, 769, 1986.
40. Gerber, J. F., Crop protection by heating, wind machines and overhead irrigation, Florida Agric. Exp. Stat. J. Series No. 3452, cited *HortSci.,* 5, 428, 1970.
41. Rogers, W. J. and Swift, H. L., Frost and the prevention of frost damage, U.S. Dept. of Commerce, NOAA, Silver Spring, MD, 1970.
42. Strange, J. C., Lombard, P. B., Westwood, M. N. and Weiser, C. J., Effect of duration and rate of freezing and tissue hydration on 'Bartlett' pear buds, flowers and small fruit, *J. Amer. Soc. Hort. Sci.,* 105, 102, 1980.
43. Lombard, P. B., Bauman, B. and Strang, J., Susceptibility of several pear varieties to frost injury, *Proc. Ore. Hort. Soc.,* 68, 85, 1977.
44. Peters, N. L. and Rackhams, R., Guide to orchard heating of pears, Ore. State Univ. Ext. Serv. Bul., 1974.
45. Proebsting, Jr., E. L. and Mills, H. H., Low temperature resistance of developing flower buds of six deciduous fruit species, *J. Amer. Soc. Hort. Sci.,* 103, 192, 1978.
46. Ballard, J. K., Proebsting, E. L. and Tukey, R. B., Critical temperature for blossom buds of pears, *Wash. State Univ. Ext. Cir.,* 370, 1971.
47. Lindow, S. E., Population dynamics of epiphytic ice nucleation active bacteria on frost sensitive plants and frost control by means of antagonistic bacteria, in *Plant Cold Hardiness and Freezing Stress: Mechanisms and Crop Implications,* Vol. 2, Li, P. H. and Sakai, A., Eds., Academic, New York, 1982.
48. Lindow, S. E., Methods of preventing frost injury caused by epiphytic ice-nucleation-active bacteria, *Plant Dis.,* 67, 327, 1983.
49. Proebsting, E. L. Jr. and Gross, D. C., Field evaluation of frost injury to deciduous fruit trees as influenced by ice nucleation—active Pseudomonas syringae, *J. Amer. Soc. Hort. Sci.,* 113, 498, 1988.
50. Ashworth, E. N., Anderson, J. A., Davis, G. A. and Lightner, G. W., Ice formation in *Prunus persica* under field conditions, *J. Amer. Soc. Hort. Sci.,* 110, 287, 1985.
51. Ashworth, E. N. and Davis, G. A., Ice nucleation within peach trees, *J. Amer. Soc. Hort. Sci.,* 64, 552, 1984.
52. Gross, D. C., Proebsting, E. L. Jr., and Andrews, P. K., The effects of ice nucleation-active bacteria on temperatures of ice nucleation and freeze injury of *Prunus* flower buds at various stages of development, *J. Amer. Soc. Hort. Sci.,* 109, 375, 1984.

53. Mellenthin, W. M., Wang, C. Y. and Wang, S. Y., Influence of temperature on pollen tube growth and initial fruit development in 'd'Anjou' pear, *HortSci.,* 7, 557, 192.
54. Vasilakakis, M. and Porlinis, I. C., Effect of temperature on pollen germination, pollen tube growth, effective pollination period and fruit set of pear, *HortSci.,* 20, 733, 1985.
55. Langridge, D. F. and Jenkins, P. T., A study of pollination of Packham's Triumph pears, *Aust. J. Exp. Agric. Animal Husb.,* 12, 328, 1971.
56. Lombard, P. B. and Westigard, M. N., Pear rootstocks, in *Rootstocks for Fruit Crops,* Rom, R. C. and Carlson, R. F., Eds., Wiley Interscience, New York, 1987, chap. 5.
57. Layne, R. E. C. and Quamme, H. A., Pears, in *Advances in Fruit Breeding,* Janich, J. and Moore, J. N., Eds., Purdue University Press, Indiana, 38, 1975.
58. Westwood, M. N. and Bjornstad, H. O., Chilling requirements of dormant seeds of 14 pear species as related to their climatic adaption, *Proc. Amer. Soc. Hort. Sci.,* 92, 141, 1968.
59. Vegis, A., Climatic control of germination, bud break and dormancy, in *Environmental Control of Plant Growth,* Evans, L. T., Ed., Academic Press, New York, 1963.
60. Brown, G. G., Mellenthin, W. M. and Childs, L., Observation on winter injury to apple and pear trees in the Hood River Valley, *Ore. Agr. Exp. Sta. Bul.,* 595, 1964.
61. Westwood, M. N., Rootstock—scion relationship in hardiness of deciduous fruit trees, *HortSci.,* 5, 418, 1970.
62. Westwood, M. N. and Chestnut, N. E., Rest period chilling requirement of Bartlett pear as related to *Pyrus calleryana* and *P. communis* rootstocks, *Proc. Amer. Soc. Hort. Sci.,* 84, 82, 1964.
63. Chandler, W. H. Pome Fruits, in *Deciduous Orchards,* Chandler, W. H., Ed., Lea and Febiger, Philadelphia, 1947, chap. 16.
64. Chalmers, D. J., Mitchell, P. D. and van Heek, L., Control of peach tree growth and productivity by regulated water supply tree density and summer pruning, *J. Amer. Soc. Hort. Sci.,* 106, 307, 1981.
65. Mitchell, P. D., Jerie, P. H. and Chalmers, D. J., The effects of regulated water deficits on pear tree growth, flowering, fruit growth and yield, *J. Amer. Soc. Hort. Sci.,* 109, 604, 1984.
66. Mitchell, P.D., Chalmers, D.J., Jerie, P.H. and Burge, G., The use of initial withholding of irrigation and tree spacing to enhance the effect of regulated deficit irrigation on pear trees, *J. Amer. Soc. Hort. Sci.,* 111, 858, 1986.
67. Goodwin, I., Mitchell, P. D. and Jerie, P. H., Measurements of soil matric potential under 'William Bon Chretien' pear comparing regulated deficit with normal irrigation, *Aust. J. Exp. Agric.,* 32, 487, 1992.
68. Proebsting, E. L., Jr. and Middleton, J. E., The behaviour of peach and pear trees under extreme drought stress, *J. Amer. Soc. HortSci.,* 105, 380, 1980.
69. Syvertsen, J. P. Integration of water stress in fruit trees, *Hort. Sci.,* 20, 1039, 1985.
70. Fereres, E. and Goldhamer, D. A., Deciduous fruit and nut trees, in *Irrigation of Agricultural Crops,* Stewart, B. A. and Nielson, D. R., Eds., Amer. Soc. Agronomy, 1990, chap. 33.
71. Goode, J. E., Fruit and other tree and bush crops—vegetative growth, in *Crop Responses to Water at Different Stages of Growth,* Salter, P. J. and Goode, J. E., Eds., Commonwealth Agricultural Bureaux, 1967, chap. 10.
72. Brun, C. A., Raese, J. T. and Stakly, E. A., Seasonal response of Anjou pear trees to different irrigation regimes. I. Soil moisture, water relations, tree and fruit growth, *J. Amer. Soc. Hort. Sci.,* 110, 830, 1985.
73. Richards, S. M., Jerie, P. H. and Mitchell, P. D., The effect of withholding irrigation on shoot and fruit growth of Asian pear (Nashi) Res. Rep. 1989–90 Instit. for Sustainable Agr. Tatura, Kyabram, 54, 1989.
74. Powell, D. B. B., Some effects of water stress on the growth and development of apple trees, *J. Hort. Sci.,* 51, 75, 1976.
75. Turner, N. C., Adaption to water deficits: a changing perspective, in *Plant Growth, Drought and Salinity,* Turner, N. C. and Passioura, J. B., Eds., CSIRO, Melbourne, 1988.
76. Jerie, P. H., Mitchell, P. D. and Goodwin, I., Growth of William Bon Chretien pear fruit under regulated deficit irrigation (RDI), *Acta Hort.,* 240, 271, 1989.
77. Goodwin, I., unpublished data, 1988.
78. Cohen, A. and Goell, A., Fruit growth and dry matter accumulation in grapefruit during periods of water with holding and after reirrigation, *Aust. J. Plant Physiol.,* 15, 633, 1988.
79. Failla, O., Zocchi, G., Treccani, C. and Cocucci, S., Growth, development and mineral content of apple fruit in different water status conditions, *HortSci.,* 67, 265, 1962.

80. Gayner, F. C. H., Studies in the non-setting of pears. IV. The effect of irrigation and injection on the June drop of Conferance pear, *Ann. Rep. E. Malling Res. Stat.,* 36, 1941.
81. Taerum, R. Effects of moisture stress and climatic conditions on stomatal behaviour and growth in Rome Beauty apple trees, *Proc. Amer. Soc. Hort. Sci.,* 85, 20, 1964.
82. Mitchell, P. D., Pear fruit growth and the use of diameter to estimate fruit volume and weight, *HortSci.,* 21, 1004, 1986.
83. Boland, A.M., Mitchell, P.D., Jerie, P.H. and Goodwin, I., The effect of Regulated Deficit Irrigation on tree water use and growth of peach, *J. Hort. Sci.,* 68, 261, 1993.
84. Chalmers, D. J., Olsson, K. A. and Jones, T. R., Water relations of peach trees and orchards, in *Water Deficits and Plant Growth,* Vol. V11, T. T. Kozlowski, Ed., Academic Press, New York, 1983.
85. Magness, J. R., Soil moisture in relation to fruit tree functioning, Rep. 13th Int. Hort. Congr. London, 1, 230, 1952.
86. Aldrich, W. W. and Work, R. A., Effect of leaf-fruit ratio and available soil moisture in heavy clay soil upon amount of bloom of pear trees, *Proc. Amer. Soc. Hort. Sci.,* 31, 57, 1934.
87. Barss, A. F., Effect of moisture supply on development of *Pyrus communis, Bot. Gaz.,* 90, 151, 1930.
88. Brun, C. A., Raese, J. T. and Stahly, E. A., Seasonal response of "Anjou" pear trees to different irrigation regimes. II. Mineral composition of fruit and leaves. Fruit disorders and fruit set, *J. Amer. Soc. Hort. Sci.,* 110, 835, 1985.
89. Skepper, A. H., Irrigation needs of fruit trees, *Agric. Gaz. NSW,* 75, 1138, 1964.
90. Goode, J. E., Fruit and other tree and bush crops—fruit set and development, in *Crop Responses to Water at Different Stages of Growth,* Salter, P. J. and Goode, J. E., Eds., Commonwealth Agricultural Bureaux, 1967, chap. 15.
91. Goode, J. E., Fruit and other tree and bush crops—flower-bud formation and development, in *Crop Responses to Water at Different Stages of Growth,* Salter, P. J. and Goode, J. E., Eds., Commonwealth Agricultural Bureaux, 1967, chap. 15.
92. Goode, J. E., Fruit and other tree and bush crops—disorders and diseases, in *Crop Responses to Water at Different Stages of Growth,* Salter, P. J. and Goode, J. E., Eds., Commonwealth Agricultural Bureaux, 1967, chap. 15.
93. Candiola, P., Alcune notizie su un impianto di irragaziona a pioggia del Ravenate, *Agric. Venezie,* 7, 141, 1953.
94. Piolanti, G., Prime osservazioni sull' irrigazione dei frutteti nel comprensario del Canole Emiliano Romagnola, *Frutticolturca,* 26, 1964.
95. Ryall, A. L. and Aldrick, W. W., The effect of water supply to the tree upon water content, pressure test, and quality of Bosc pears, *J. Agric. Res.,* 68, 121, 1944.
96. Barss, A. F., The pear as affected by moisture supply, *Bien. Crop Pest. Rep. Ove. 1913–1914,* 28, 1915.
97. Aldrick, W. W., Lewis, M. R., Work, R. A., Ryall, A. L. and Reiner, F. C., Anjou pear responses to irrigation in a clay abode soil, *Bull. Ore. Agric. Exp. Stat.,* 370, 1940.
98. Nosonenko, N. A., Yakimov, V. A. and Babina, R. D., The problem of drought resistance and productivity of pear cultivars in the Crimea, Skornik Nauchnykh Trudov Vsesoyuznyl Nauchno Issledovatiloku. Institut. Sadovodstva Imeni E.V. Michurina 186, 46, 55, cited *Hort. Abst.,* 6875, 1987.
99. Maksimova, I. N., Degree of drought resistance and productivity of pears under Zaporzke conditions, Skornik Nauchnykk Trudov Vsesoyuznyi Nauchno Issledovatel' skill Institut. Sudovodstva imeni, E. V. Michurino, 46, 52, 1986.
100. Cole, C. E., The fruit industry of Australia and New Zealand, *Proc. 17 Int. Hort. Congr.,* 3, 321, 1966.
101. Kuznetzov, P. V., The role of Pyrus salicifolia Pall. in the development of fruitgrowing in arid regions, Sovetsh Bot. No. 1–2, 103, 1941, cited *Plant Brdg. Abst.,* 13, 359.
102. Anon, Shi Jing, (A comprehensive compilation of ancient poems and songs), 1,000 B.C.
103. Rowe, R. N. and Beardsell, D., Waterlogging of fruit trees, *Hort. Abst.,* 43, 533, 1973.
104. Bini, G., Studies of peach and pear trees to root asphyxia, *Riv. Ortoflorofruthic* ital., 47, 22, 1963.
105. Jawanda, J. S., The effects of waterlogging on fruit trees, *Punjab Hort. J.,* 150, 1961.
106. Morita, V., Studies of orchard soils. II. Soil atmosphere and plant growth, *Bull. Nat. Inst. Agric. Sci. Hiratsuka,* Ser. E. No. 4, 88, 1955.
107. Remy, P. and Bidabe, B., Root asphyxia and collar rot in pome fruit trees, the influence of the rootstock, *Congr. Pomol. 92nd Sess. Proc.,* 17, 1962.
108. Kiëntolz, J. R., Performance of pear orchard with flooded soil, *Proc. Amer. Soc. Hort. Sci.,* 47, 10, 1946.

109. Anderson, P. C., Lombard, P. B. and Westwood, M. W., Leaf conductance, growth, and survival of willow and deciduous fruit tree species under flooded soil conditions, *J. Amer. Soc. Hort. Sci.,* 109, 132, 1984.
110. Andersen, P. C., Montana, J. M. and Lombard, P. B., Root anaerobiosis, root respiration and leaf conductance of peach, willow, quince and several pear species, *HortSci.,* 20, 248, 1985.
111. Andersen, P. C., Lombard, P. B. and Westwood, M. N., Effect of root anaerobiosis on the water relations of several *Pyrus* species, *Physiol. Plant.,* 62, 245, 1984.
112. Westwood, M. N. and Lombard, P. B., Pear rootstocks, *Proc. Ore. Hort. Soc.,* 58, 61, 1966.
113. Andersen, P. C., unpublished data.
114. Schaffer, B., Andersen, P. C. and Ploetz, R. C., Responses of fruit crops to flooding, *Hort. Rev.,* 13, 257, 1992.
115. Sena Gomez, A. R. and Kozlowski, T. T. Growth responses and adaption of *Fraxinus pennsylvanica* seedlings to flooding, *Plant Physiol.,* 66, 267, 1980.
116. Kawase, M., Anatomical and morphological adaption of plants to waterlogging, *HortSci.,* 1, 30, 1981.
117. Leyton, L. and Rousseau, L. Z., Root growth of tree seedling in relation to aeration, in *The Physiology of Forest Trees,* Thimann, K. V., Ed., Ronable Press, New York, 1958, 467.
118. Chirkova, T. V. and Gutman, T. S., Physiological role of branek lenticels in willow and popular under conditions of root anaerobiosis, *Sov. Plant Physiol.* 19, 289, 1972.
119. Crawford, R. M. M. and Baines, M. A., Tolerance of anoxia and metabolism of the ethanol in tree roots, *New Phytol.,* 79, 519, 1977.
120. Rowe, R. N., Anaerobic metabolism and cyanogenic glycoside hydrolysis in differential sensitivity of peach, plum and pear roots to water saturated conditions, Ph. D. Thesis Univ. Calif. Davis, 1963.
121. Keatley, J. and Selimi, A., Lime trial on Josephine pears at Tatura, *Victorian Hort. Digest,* Dept. of Agric., 55, 63, 1962.
122. Ng, S. K. and Bloomfield, C., The effect of flooding and aeration on the mobility of certain trace elements in the soils, *Plant Soil,* 16, 108, 1962.
123. Hodgson, J. F., Chemistry of the micro-nutrient elements in the soil, *Adv. Agrom.,* 15, 119, 1963.
124. West, D.W. and Nicholson, unpublished data.
125. Bernstein, L., Crop growth and salinity, in *Drainage for Agriculture,* Van Schilfgaarde, J., Ed., Am. Soc. Agron., 1965, chap. 3.
126. Myers, B. A. and West, D. W., Effects of saline irrigation on mature pear trees, *Acta Hort.,* 240, 279, 1989.
127. Myers, B. A. and West, D. W., unpublished data, 1991.
128. West, D. W. and Black, J. D. F., Irrigation timing—its influence on the effects of salinity and waterlogging stresses in tobacco plants, *Soil Sci.,* 125, 367, 1978.

Chapter 8

Persimmon

Alistar D. Mowat and Alan P. George

CONTENTS

I. INTRODUCTION

The cultivation and climatic requirements of persimmon have been extensively covered by previous authors.[1–7] These reviews have shown that persimmon can be cultivated over a wide range of climatic environments, from temperate through tropical zones. This review describes the ecophysiology of persimmon and discusses the environmental limitations to production.

A. ORIGIN AND DISTRIBUTION

Persimmon is a member of the genus *Diospyros* containing over 400 species[8] which occur predominantly in tropical and subtropical environments. Species tend to be evergreen, but several species are deciduous and adapted to warm temperate climates. *Diospyros kaki* L. is the major commercial species. Other species, *D. virginiana* L., *D. lotus* L., *D. oleifera* Cheng, *D. digyna* Jacq., and *D. blancoi* A. DC., are of relatively minor economic importance.

The mountains of central China are the center of origin and primary center of diversity for *D. kaki,* and Japan is a secondary center.[9] The wild form of persimmon, *D. kaki* var. *sylvestris* Makino grows in mountain areas in central and western China to an altitude of 1150 m.[10] *D. kaki* is hexaploid (2n = 90)[11] and the wide diversity in fruit shape and form suggest probable alloploid origins.[12] Morphological, geographical, and cytological evidence suggests that *D. oleifera* (syn. *D. roxbergia* Carriere) is a parent of *D. kaki.*[13] The origin of the other parental species is not known but it could be *D. hexameria* Wu, a species closely related to *D. oleifera* from southern China that has a ribbed fruit.

Persimmon has been cultivated in China for several thousand years, where over 2000 cultivars have been selected.[14] *D. kaki* was introduced to Japan about 1300 years ago.[15] In Japan two distinct forms have been selected, astringent and non-astringent, the latter having a much lower concentration of water-

0-8493-175-0/94/$0.00+$.50

Table 1 **Climatic characteristics of persimmon-growing locations**

Sites	Latitude	Mean monthly temperature Minimum (°C)	Mean monthly temperature Maximum (°C)	Mean annual rainfall (mm)
Batumi, Georgia	41.37 N	2.0	25.0	100
Fukushima, Japan	41.31 N	−3.0	30.5	1455
Pulsan, South Korea	35.06 N	−3.0	30.5	1380
Nara, Japan	34.41 N	−0.5	32.0	1375
Xian, China	34.16 N	2.5	31.5	100
Los Angeles, U.S.	34.00 N	8.0	28.0	380
Haifa, Israel	32.49 N	9.5	32.0	670
Cairo, Egypt	30.03 N	5.0	35.5	25
Chang sha, China	28.10 N	6.0	30.0	140
Homestead, U.S.	25.29 N	12.0	32.5	1570
Canton, China	23.08 N	8.5	32.5	1580
Mackay, Australia	21.10 S	16.8	27.4	1694
Limeria, Brazil	22.34 S	13.0	30.0	150
Nambour, Australia	26.40 S	16.0	29.0	1800
Santiago, Chile	33.16 S	3.0	29.0	35
Loxton, Australia	34.38 S	4.0	31.0	274
Kerikeri, New Zealand	35.12 S	10.0	20.0	1650
Hastings, New Zealand	39.39 S	7.5	19.5	770

soluble tannins.[4,10] Persimmon can be further subdivided into pollination constant and pollination variant types.[10] The pollination constant astringent (PCA) form is the oldest and comprises almost all the Chinese cultivars and several Japanese cultivars. Pollination variant astringent (PVA) and non-astringent (PVNA) forms originated in Japan about 1200 years ago and exhibit flesh darkening when pollinated. The flesh darkening reaction is associated with a low ascorbic acid content in the flesh,[16] high polyphenol oxidase activity,[17] and ethanol production in seeds.[18] The pollination constant non-astringent (PCNA) persimmon is the most recent form, originating approximately 500 years ago in Japan.[19] It was derived from PCA cultivars, but has a different tannin composition.[20] The PCNA form is generally free of astringency at harvest; post-harvest treatments to remove astringency are not required.

The distribution of persimmon to countries outside Asia has occurred within the last 150 years. Commercial cultivation occurs between latitudes 45° north and south of the equator (Table 1). Over one million tonnes of persimmon are produced annually from an area of approximately 235,000 hectares.[21] The major producers are China, Japan, Brazil, Korea, and Italy. Minor producers include Israel, U.S.A., New Zealand, Australia, Spain, Georgia, Egypt, and Chile.

II. PHENOLOGY

The growth cycle of persimmon is described in Figure 1. The phenological sequence of development is; (1) bud burst; (2) shoot elongation; (3) cessation of shoot growth; (4) full bloom; (5) fruit growth; (6) floral bud development for the subsequent season; (7) resting of floral bud development; (8) fruit maturation; and (9) leaf-fall.[22] This sequence of development is specific to cultivar, environment, tree condition, and management practices.

A. VEGETATIVE DEVELOPMENT

Trunk and branch cambial activity begins at bud-burst and continues for 24 weeks.[23] Shoots, emerging from buds that developed the previous season, cease growing before full-bloom. Conditions that stimulate vigor (i.e., juvenility, low crop load, hard pruning, excess nitrogen, and soil moisture) can induce a second flush of the terminal buds of current season growth. Shoots emerging from older buds, a result of hard winter pruning, limb breakage, or frost damage, are generally vigorous and cease growth late in the season. In Japan, canopy development reaches a maximum approximately 75 days from budbreak[24] and the leaves are present on the tree for 170–200 days.

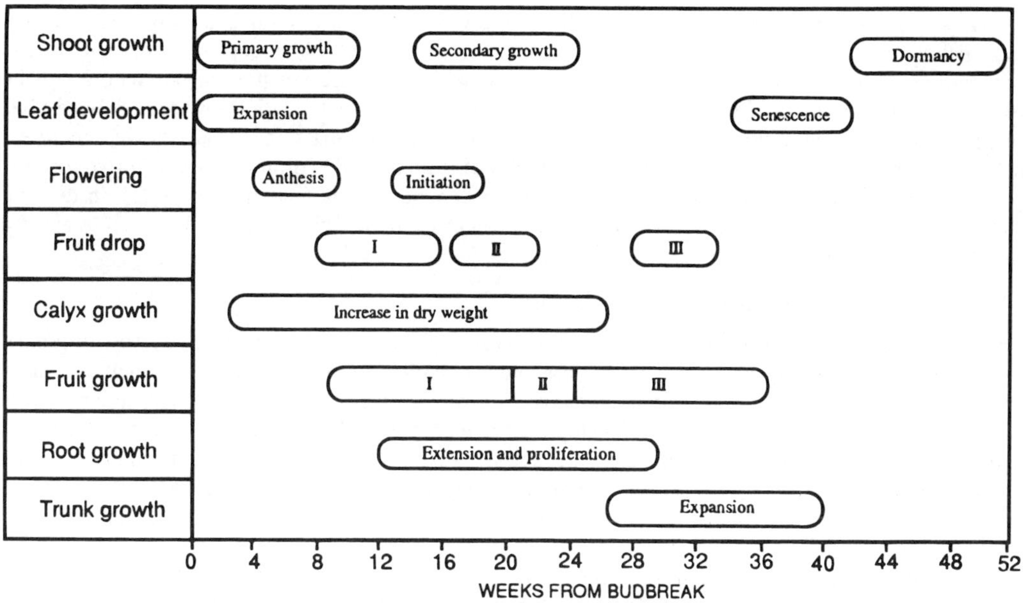

Figure 1 The phenology of persimmon.

Trunk and branch expansion occur over the latter part of the growing season. Trunk expansion is sensitive to competition from the fruit. In Queensland, Australia, the relative increase in butt cross-sectional area for a range of cultivars was found to decline as the annual yield increased.[25] Since fruit harvest generally occurs near or at leaf fall there is no opportunity for post harvest accumulation of reserves in heavy cropping years.

In bearing trees the biomass is distributed between the trunk and branches, leaves, fruit, and roots. A Japanese study, on 9-year-old bearing persimmon, found the proportion of biomass (DM) contained by specific organs was as follows: leaves 11.4%, fruit 14.6%, one-year-old shoots 3.3%, trunk and branches 38.4%, fine roots 1.7%, and other roots 30.5%.[26] The biomass associated with new tissue was 52.2% of the total biomass. In 25-year-old bearing 'Fuyu' persimmon the distribution of new tissue biomass (DM) between specific organs was as follows: leaves 23.2%, fruit 29.2%, trunk, branches, and shoots 28%, and roots 19.6%.[27] The ratio of shoot biomass (including trunk and branches) to the root biomass may vary with tree maturity and bearing. The shoot to root ratio of 3-year-old persimmon trees, on a dry weight basis, was 0.73[28] compared with a ratio of 1.29 in 9-year-old bearing trees.[26] Root growth may be more affected by the partitioning of resources to fruit than shoot growth. Bearing may affect the total amount of biomass contained in vegetative organs. In water culture studies, the new tissue biomass of vegetative organs was 40–50% lower in bearing trees in comparison to non-bearing trees.[29]

Root growth occurs in one or two flushes over the growing season. In Japan root flushes occurred from mid-summer to autumn in bearing trees.[30] In Queensland, Australia, the timing of the root flush occurred in late spring/early summer and appears to be influenced by shoot and fruit growth cycles.[31] Peak root growth tends to occur between the spring vegetative flush and fruit growth. The effect of soil temperature on persimmon root activity, although not studied, may be an important factor in explaining differences in the timing of root growth between temperate and sub-tropical environments.

B. REPRODUCTIVE DEVELOPMENT

Flower evocation begins after a terminal bud has formed on the current seasons growth and the production of new leaf primordia within the bud has ceased.[32] Shoots terminating early in the growing season have a greater ability to differentiate flower buds than those terminating in the mid or late part of the growing season.[33] Only the sepals and petals differentiate prior to the winter rest period and all other development occurs in the spring.[34]

The tree has a tip bearing habit with flowers occurring in the more basal nodes of current season shoots. The flowering habit is complex, ranging from dioecious to monoecious.[35] Most commercial cultivars are pistillate or pistillate-sporadically monoecious, though several androdioecious and tri-monoecious cultivars exist.[36] Fruit can be set parthenocarpically, but this is dependent on cultivar and environment.[37] If suitable cultivar and environment combinations are not present pollinators will be required to ensure reliable fruit set. The flower is pollinated by insects, with the European honey bee being the predominant pollinator.[38,39] The transfer of pollen between male and female flowers by insect activity can be poor[40] and flowers may require hand pollination.[41] Fruit drop can occur in two to three phases after fruit set.[42,43] Competition between fruit and shoots causes seedless fruit to fall in the first two phases of drop while fruit and root competition can cause both seeded and seedless fruit to abscise in a third phase of drop.

C. FRUIT DEVELOPMENT

The fruit of persimmon is botanically a berry[44] consisting of a parenchymatous mesocarp surrounded by an epicarp, covered by a cuticle.[45] Fruit mass and size curves follow a sigmoid[39] or double sigmiod growth pattern,[5,36,46] consisting of two active stages of growth, stage I and stage III separated by a less active stage (stage II) (Figure 2). Fruit development ranges from 120–190 days, depending on cultivar and environment; and the duration of stages I, II and III is 60–100, 20–40 and 40–50 days, respectively. Growth stage I is thought to be associated with cell division/differentiation and growth stage III with cell expansion/maturation. The significance of growth stage II is not clear, but it does not appear to be related to seed development as the growth curves of seed bearing and seedless cultivars are similar.[36,47–49] Growth stage II may reflect a change point between cell division/differentiation and cell expansion/ maturation. Therefore the duration of stage II may simply reflect the influence of genetic, physiological, or environmental factors on the duration of stage I and the start point of stage III. Seed growth is completed by the end of stage I, after which physiological changes associated with seed coat coloring and endosperm hardening occurs.[47] Calyx growth commences before flowering and is completed by the end of stage II. At full bloom the calyx may comprise more than 50% of the flower weight,[5] and though the calyx can assimilate carbon,[50] it is not known if the calyx is a significant source of photosynthates for the developing fruit. The calyx is an important gas exchange organ for the fruit,[50] since fruit lack stomata or lenticels and is covered by a layer of wax.[36] The calyx has a major influence on fruit growth and development. Removal of the lobes of the calyx or damage by agrochemicals and disease can enhance fruit drop and reduce fruit size and soluble solids.[51,52]

Soluble sugars accumulate in the fruit by translocation[53] of sucrose where it is hydrolyzed to fructose and glucose.[54] The rate of sugar unloading in the fruit is controlled by fructose invertase activity and varies with fruit age.[55] The sugar composition of mature ripe fruit is ca. glucose (48%), fructose (49%) and sucrose (3%).[56] Fruit color increases with maturation as a result of chlorophyll degradation and carotenoid synthesis.[57]

Water-soluble tannin is contained in specialized tannin cells dispersed within the parenchyma of the fruit mesocarp.[36] In astringent cultivars, tannin cells enlarge from flowering until stage III while in PCNA cultivars they cease growth at the end of stage I.[58] Tannin accumulation stops when cell growth ceases because of pore closure in the cell wall.[49] Loss of astringency in PCNA cultivars is a combination of two processes, first a dilution of soluble tannins, and second the coagulation (i.e., polymerization) of soluble tannins into an insoluble form. Astringent cultivars require treatment to remove astringency.

D. PRODUCTIVITY

The productivity of mature orchards is moderate compared to other fruit crops, ranging from 8.6–25 t ha^{-1}.[36,59] The factors limiting the productive potential of persimmon have not been adequately defined. Productivity could have a genetic component. Productivity differences have been reported between cultivars under a range of environmental conditions.[60–64] Cultivars can differ in their precocity and parthenocarpic setting ability.[25,37] Bud mutations of cultivars can also differ in productivity.[65] 'Supahiratanenashi', a dwarf bud mutation of 'Hiratanenashi', has been found to have lower vigor, earlier bearing, and higher yields than 'Hiratanenashi'.[66]

Detailed studies on the physiological differences between a dwarf bud mutation and the original cultivar could provide a better understanding of the factors influencing persimmon productivity. In Japan, a number of studies between a dwarf bud mutation of 'Nishimura wase' and 'Nishimura wase' have been carried out on 20-year-old trees planted at the same density. The dwarf strain produced twice the yield on a canopy volume basis but only the same yield as 'Nishimura wase' on a yield per unit of land area basis.[67] As the canopy volume of the dwarf strain was 83% smaller than 'Nishimura wase'

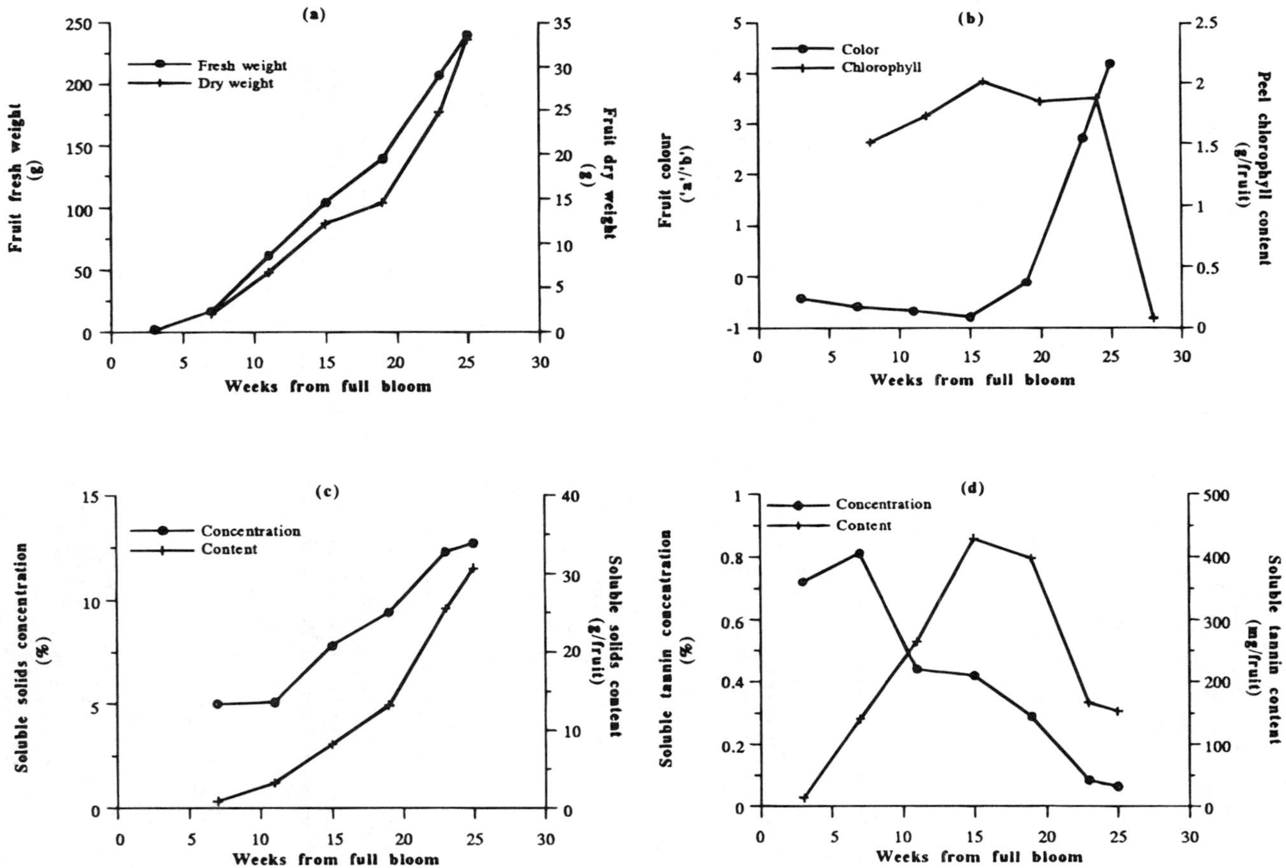

Figure 2 Seasonal changes in fruit composition (cv Fuyu): (a) fresh and dry weight; (b) color and peel chlorohpyll content; (c) soluble solids concentration and content; and (d) soluble tannin concentration and content.

then the yield per unit of land area could be increased by planting the trees at a higher density. The improvement in productivity of the dwarf strain could be attributed to a number of factors. Improved partitioning between vegetative and reproductive growth could be another factor influencing productivity. The dwarf form had 80% less wood than 'Nishimura wase'. The percentage of current season growth to total wood weight was 6.3% for the dwarf form and 1% for 'Nishimura wase'. As the net CO_2 assimilation (A) rate of a tree is strongly affected by the ratio of the non-photosynthetic system to the photosynthetic system[68] then the dwarf strain could have a higher A rate through a reduction in total respiration of non-photosynthetic organs. The leaves of the dwarf strain also had a higher set CO_2 assimilation rate, smaller leaf size, thicker leaves, and contained higher levels of chlorophyll and nitrogen than 'Nishimura wase'.[69] Shoots on the dwarf strain ceased growth earlier in the season than those on 'Nishimura wase'.[70] The dwarf strain had a higher flower density fruit set than 'Nishimura wase'.[71]

Rootstocks and interstocks can also influence persimmon productivity.[72,73] Persimmon is generally cultivated on seedling rootstocks from a range of species. In California, 'Haichiya' was found to be lower yielding on *D. lotus* than other rootstocks.[72–74] The main cause of low yields was an excessive fruit drop associated with high tree vigor rather than low flower density. The development of clonal rootstocks in persimmon has been limited by suitable propagation techniques.[75] In Japan, fruit yield of trees on dwarfing rootstocks was higher than on standard rootstocks on a canopy volume basis but not on a canopy area basis.[76] Shoot growth was shorter on the dwarf trees in comparison to standard trees, although the graft union of the dwarf trees showed abnormal thickening and stock over growth.

Persimmon is prone to fluctuations in bearing. Bearing can range from bienniel to much longer cycles. In Japan, persimmon was found to have an alternate bearing cycle of 4.2 years.[77] This contrasts with a biennial bearing habit of 'Pomelo IAC 6-22' in Brazil.[78] Fluctuation in bearing could be caused by over cropping or through environmental influences. Overcropping during the 'on' year results in competition for assimilates between fruit and shoots.[78–80] The low carbohydrate status of a tree in an 'on' year can reduce flower evocation[81] and may also cause pre-bloom flower bud abscission in the following season. In New Zealand, persimmon often shows a longer fluctuation in bearing that could be associated with a shortage of reserves. Where cropload is used to control tree vigor, a high cropload in one season can stimulate flower initiation by causing shoots to cease growth early and subsequently cause heavy flowering and cropping in the second season. A heavy crop carried in the second season may then deplete tree reserves leaving limited resources for shoots, trunk and roots. This may then reduce flower evocation or cause pre-bloom flower abscission. Longer bearing cycles may also occur as a result of a slow depletion of reserves by a succession of heavy crops.

Environmental factors can influence bearing by affecting assimilation or through physical damage of reproductive organs. Flower evocation and fruit drop are particularly sensitive to environmental stress.[42,81] Flower buds and fruitlets can be prone to damage from cold, hail, and wind.[82–84]

Management factors influencing productivity include plant density, training system, pruning and crop thinning. In Japan, the yield per unit of land area of 'Matsumoto wase Fuyu' was found to increase with increasing planting density until full canopy closure was reached.[85] Training systems such as the 'Y' and palmette have been developed to improve the productivity of persimmon (Photo 1). In Japan, a 'Y' training system was found to increase the yield per tree and per trunk cross sectional area, and

Photo 1 Persimmon trained on a Y trellis to improve light distribution within the canopy.

reduce fluctuation in annual yields in comparison with the traditional open vase training system.[86] Renewal pruning can increase persimmon productivity by stimulating new replacement growth from old wood.[87,88] In comparison, a lack of pruning may cause a decline in productivity. This could be due to a reduction in flower evocation. In unpruned trees, flower evocation occurs mainly on the tips of bearing shoots where competition from fruit for assimilates can be high. Productivity of persimmon can increase with fruitload.[89] However, as fruit weight is negatively correlated with fruit load then the marketable yield per tree tends to be lower than the maximum potential yield per tree.[89,90] Biennial bearing can be reduced by thinning fruit to an optimum cropload that is based on shoot type and leaf number.[91] Phytotoxicity from agrochemical sprays can reduce productivity by stimulating fruit drop.[42,92] Fruit drop may occur as result of a loss of leaf surface through spray damage or as a stress response.

Nutrition can improve persimmon productivity by increasing tree size or reducing fruit drop.[42,93,94] The response of persimmon to nutrition is dependent on the initial nutritional status of the tree and applications of additional nutrients may have no effect on productivity.[95] In New Zealand, leaf nitrogen levels can be high (>2g kg^{-1} DM) when persimmon is cultivated on high fertility soils. Additional applications of nitrogen under these conditions can reduce productivity by stimulating vigor and enhancing fruit drop or reducing flower evocation by stimulating secondary growth flushes. In Japan, annual yields can fluctuate on soil types that stimulate tree growth.[96]

Pest and disease problems can reduce persimmon productivity via damage to the root system,[97,98] a reduction in leaf area,[99,100] a loss of flowers/fruit,[101] or a reduction in fruit size. Pest damage may have an adverse effect on physiological processes in the plant. In Georgia, an infestation of citrus white fly on persimmon was found to reduce photosynthesis, leaf carbohydrate and protein content and enhance respiration.[102]

III. IRRADIANCE

Irradiance affects photosynthetic activity, productivity, and fruit quality of persimmon. Climate, shelter height, proximity of adjacent trees, and the canopy density all influence the radiation environment.

A. PHOTOSYNTHESIS

Photosynthetic activity increases with leaf age. In Japan photosynthetic activity reached a maximum ca. 85 days after bud-break, then declined slowly through the growing season until ca. 50 days before leaf fall when it declined rapidly.[102,103] Initially, the photosynthetic rate of the leaf is related to the development of palisade and spongy mesophyle tissue.[102] Leaf thickness differs between cultivars, and photosynthetic activity is highest in cultivars with thick leaves.[104,105] Leaves in the lower part of the canopy have a higher photosynthetic activity than those in the upper part during leaf expansion.[103] This could be due to the high nitrogen content of lower canopy leaves increasing the rate of leaf development,[106] or to photoinhibition caused by high irradiance levels in the upper canopy. The light environment of individual leaves changes with canopy development which results in higher photosynthetic activity in the upper part of the canopy compared with lower parts.[103]

Net CO_2 assimilation decreases as irradiance falls from light saturation. In a Japanese study, photosynthetic activity decreased as irradiance declined from light saturation at 720 $\mu mol\ m^{-2}s^{-1}$ (A = 7.9 $\mu mol\ CO_2\ m^{-2}s^{-1}$) to light compensation of 90 $\mu mol\ m^{-2}s^{-1}$.[107] Fluctuating irradiance on exposed persimmon leaves can cause a rapid decline in A. Net CO_2 assimilation of 'Fuyu' decreased from 7.8 $\mu mol\ CO_2\ m^{-2}s^{-1}$ in full sun [photosynthetic photon flux (PPF) = 2106 $\mu mol\ m^{-2}s^{-1}$] to 0.2 $\mu mol\ CO_2\ m^{-2}s^{-1}$ in heavy shade (PPF = 211 $\mu mol\ m^{-2}s^{-1}$), followed by a rapid recovery in full sun to 7.4 $\mu mol\ m^{-2}s^{-1}$.[108] The irradiance efficiency (IE = total radiation at leaf surfaces/total radiation at soil surface of area occupied by leaf surface $\times$ 100%) of the canopy reaches a maximum (IE = 90%) at approximately the same time as the cessation of canopy leaf development (leaf area index = 1.9).[24] At full canopy development, differences in leaf area index and shoot composition within the canopy (Figure 3) can expose leaves to lower than optimum levels of irradiance.[104] Leaves in shaded parts of the tree are characterized by low specific leaf weights, low nitrogen contents and low dry weights. The decline in late season photosynthetic activity could be due, in part, to low irradiance, falling temperatures, leaf aging, senescence, or a reduction in leaf nitrogen content. In the subtropical environment of southeast Queensland, Australia, irradiance and temperature conditions after harvest are sufficient to maintain high photosynthetic activity (8–12 $\mu mol\ CO_2\ m^{-2}s^{-1}$).[110] Declining leaf nitrogen could also lower photosynthetic rates, since a large portion of organic nitrogen is in the form of ribulose bisphosphate

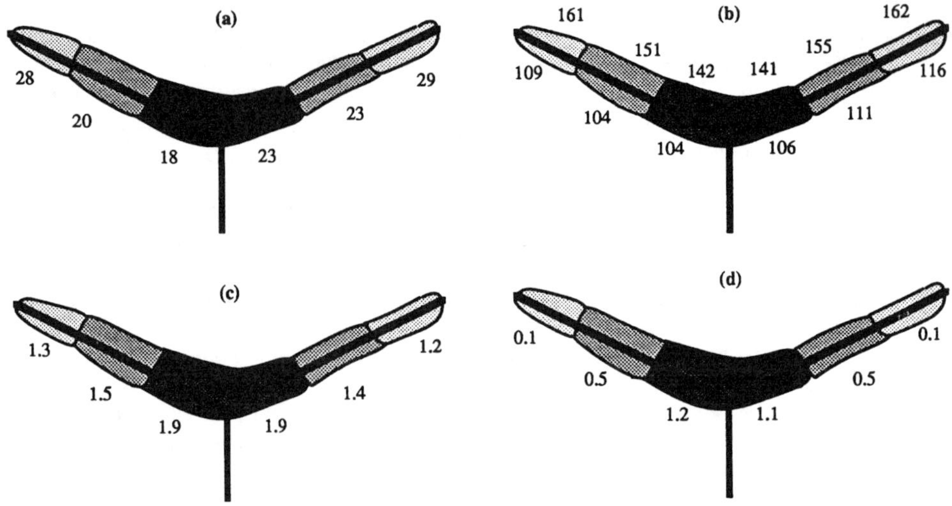

Figure 3 The influence of canopy location (shaded areas denote upper, middle, and lower parts of the canopy) on: (a) PAR (percent of full sun); (b) specific leaf weight (μg mm^{-2}) (above and below canopy surface); (c) leaf area index; and (d) ratio of nonfruiting to fruiting shoots.

carboxylase/oxygenase. Also, late season nitrogen applications can improve assimilate translocation to the fruit and delay leaf-abcission.[111]

B. EVOCATION AND FRUIT SET

Irradiance can influence the productivity of persimmon by affecting flower evocation and fruit set. Evocation and fruit set have been positively correlated with the total hours of irradiance over late summer and spring, respectively.[112] The reduction in the shoot carbohydrate content caused by lower irradiance[113] can adversely affect evocation.[81] Low irradiance from cloud cover and artificial shade stimulated fruit drop.[39] In Queensland, a low irradiance of 29.4 mol m^{-2} day^{-1} or less, can significantly increase fruit drop over the post-flowering period and less than 12.6 mol m^{-2} day^{-1} can prevent fruit set.[114] Unpollinated fruit are more susceptible to fruit-drop than pollinated fruit,[37,115,116] as the presence of seed increased fruit sink strength.[117] Fruit set of cultivars with a high parthenocarpic ability may be adequate without pollination when grown in regions with high irradiance hours or days following flowering.[37] However, in regions with low levels of irradiance after flowering, pollination is needed to control fruit drop. Fruit drop during and after flowering is due to severe assimilate competition between shoots and fruit as the sink strength shifts from vegetative to reproductive organs.[118,119] A reduction in photosynthetic activity, due to low irradiance, reduces the assimilate pool for competing sinks. The effect of low irradiance can also be localized within a tree, as fruit on shoots growing in shaded areas of the canopy are more prone to drop than fruit on exposed shoots.[114]

C. FRUIT QUALITY

The quality of fruit from a given tree can be location dependent. Fruit in exposed parts of the canopy are of larger size, better colored, higher in soluble solids, and lower in astringency than fruit from lower or interior parts of the canopy.[106,109,117] Pruning has been effective in improving light distribution within the tree and has increased fruit quality.[106,120,121] It is not clear if irradiance affects fruit quality by influencing the A rate of adjacent leaves and the calyx and/or through the thermal effects of direct radiation. Fruit weight and soluble solids could be influenced by irradiance through changes in the assimilate supply of adjacent leaves. Fruit size can be reduced by shading the calyx or entire fruit over the first 40 days of fruit development.[114] Irradiance influences fruit temperature. In Japan, fruit exposed to full sun were 6.5 to 11°C higher than shaded fruit.[122] Fruit color, soluble solids, and astringency were also affected by temperature,[123,124] (and by the thermal effects of irradiance). An irradiance level of 25 to 30% of full sun was adequate for inducing the red peel color of mature fruit associated with lycopene development.[122]

Alternatively, fruit can be injured by high irradiance levels. Sunburn severly reduces fruit quality and can be a major quality defect in Australia.[25,125] Sunburn affects exposed parts of the fruit by bleaching the surface and causing localized damage to fruit tissue. Sunburn occurs through the final stages of fruit development and damage appears to be less in seasons of low irradiance.[31] The exposure of fruit to ultraviolet radiation can cause black spots to develop on immature fruit.[126]

IV. TEMPERATURE

A. GROWTH AND DEVELOPMENT

Persimmon has a relatively low chilling requirement for dormancy release compared to other deciduous fruit crops, and has been successfully cultivated in areas with low winter chilling,[127–129] including the tropics.[130,131] Some reports however, suggest that the tree does not thrive in tropical lowland environments,[132,133] though it may not be due to a lack of chilling. Chilling requirements differ between cultivars and those with higher chilling requirements, are grown at higher elevations in the tropics than those with lower requirements.[134]

The physiological basis of dormancy and the degree of chilling needed to satisfy dormancy in persimmon is not clearly defined. Shanks[135] noted that a rest period of 800–1000 hours below 7°C was adequate to permit normal bud-break and subsequent vegetative and reproductive growth. In Queensland, Australia, persimmon can be grown satisfactorily in regions which receive 100 hours of chilling below 7°C. However, more uniform and rapid bud-break occurs in regions which receive greater than 300 hours of chilling. The use of 7°C as a base temperature for persimmon may be inappropriate, as dormancy can be satisfied by a mean minimum temperature of 14°C.[136] Root temperature rather than bud temperature may be the more important factor in defining the chilling requirements, as detached shoots break bud with minimal exposure to chilling.[137] Root temperature could influence the initiation of growth through the conversion of starch to sugars in roots and then subsequently affecting the rate of water uptake. In a controlled environment study, root temperature was found to influence the time of bud-break. A moderate root temperature (13°C) promoted bud-break, which was delayed by high (23°C) and low (7°C) temperatures.[138]

Once chilling requirements have been satisfied, the cumulative daily mean air temperature, above 10°C, required for bud-break and anthesis is 90 and 300 degrees, respectively.[139] Raising soil temperatures by clear polyethene mulches can advance bud-break, and root and shoot growth.[140] Temperature influences the amount of time between bud-break and anthesis. In a controlled environment, simulated low spring temperatures (17/12°C day/night) delayed this period by 28 days and increase flower size at anthesis by 60% in comparison with a higher temperature (27/22°C day/night); suggesting that cool temperatures can increase cell division cycles.[137]

Temperature over the flowering period can influence the duration of flowering. In southeast Queensland, flowering is concentrated over a 7–10 day period and temperatures average 21°C. In contrast, cooler spring temperatures (14°C) in New Zealand extended the length of flowering to 14–28 days. Temperature also influences the length of the fruit development period. Fruit development of the non-astringent cultivar 'Fuyu' often takes 24 to 27 weeks under ambient conditions in Australia. However, when the tree is grown at a constant temperature of 25°C, fruit development can be as short as 20 weeks.[139] Fruit quality and composition are also influenced by temperature. Fruit quality, based on fruit weight, peel color, soluble solids and soluble tannins, was highest when the day/night temperature during stage II and stage III respectively was 25/25 and 25/20°C, compared to higher (30/30°C) and lower temperatures (15/15°C).[139]

B. LOW TEMPERATURE STRESS

D. kaki is cold tolerant and may survive temperatures, down to −18 to −20°C.[141,142] However, in regions where winters are mild, the tree can be injured by a 3 h exposure to −7°C.[143] In Korea, frost injury of vegetative and floral buds increased with lowering temperature, with 10% injury at −10°C to 75% at −17°C, and 100% at −20°C.[82] The duration of cold can also influence the severity of frost injury. Dormant trees can be killed by exposure to −7°C for more than 30 days, but injury to vegetative buds occurred after 15 days at −7°C.[144] Tree and wood age can influence the sensitivity of the tree to low temperatures. Mature trees are more resistant to freezing temperatures than younger trees[142] as older wood is hardier than younger wood.[82] Severely frost-injured trees recover rapidly over the following season by producing vigorous shoots from adventitious and dormant buds in the trunk and major branches.[135,142]

A history of heavy cropping or poor growing conditions over previous seasons increases the risk of severe frost injury.[129] Freeze resistance of the woody parts of the tree could be associated with carbohydrate reserves, as fruiting suppresses starch accumulation of shoots and branches.[80]

Cold tolerance in persimmon also has a genetic basis, with cultivars differing in cold tolerance.[135,145] The introduction of cold tolerance genes into *D. kaki* has been achieved by intra-specific hybridization with *D. virginiana,* in combination with backcrossing to high quality cultivars.[146–148]

Cold tolerance is also influenced by rootstock. In temperate regions, persimmon is cultivated on *D. lotus* and *D. virginiana* rootstock,[14,132,149] while the less hardy *D. kaki* and *D. oleifera* rootstocks are generally restricted to warm temperate and subtropical regions. In regions subject to freezing winter conditions, persimmon is grafted well above the ground on *D. lotus* and *D. virginiana.*[129,149] Persimmon on *D. virginiana* was found to be more frost resistant than those on *D. lotus.*[142] The cold tolerance of *D. virginiana* could be due to this stock being better adapted to alternating warm/cold cycles and not initiating cambial activity as readily as *D. kaki.*[129] The physiological basis of cold tolerance within the deciduous species of *Diospyros* has not been adequately investigated and warrants further investigation. It is not clear if different rootstocks impart cold tolerance by causing early cessation of shoot growth and subsequent shoot hardening during the growing season or through other mechanisms that induce and maintain deep dormancy within the top growth during the winter months. Temperatures over the dormant season could be a significant factor for inducing dormancy, as trees exposed to temperatures below 0°C over dormancy have a delayed bud-break in comparison to trees exposed to temperatures above 0°C.[150–152]

Newly emerging and expanding vegetative and reproductive buds are most sensitive to cold (Photo 2). In Japan severe damage to vegetative and floral buds of 'Fuyu' occurred after exposure to −3°C for 1 hour.[153] Low spring temperatures influence fruit set by affecting flower development and pollination. In Japan abnormal embryo development occurs when flower differentiation is exposed to a mean temperature of 15°C or lower during the time between bud-break and anthesis.[154] Fruit set is reduced because abnormal embryos are not capable of producing viable seeds and pollen tube growth is inhibited by temperatures of 15°C or lower.[155] As persimmon flowers are naturally pollinated by European honey bees and other bee species, low temperatures during flowering may also reduce pollinator activity and pollination effectiveness.

Temperatures over the latter stage of the growing season affect maturation and senescence. When fruit are exposed to lower temperatures (14°C) during fruit growth stage III, rapid chlorophyll degradation occurs in the peel,[156] unmasking the carotenoids that produce the typical orange/red color of a mature fruit. In temperate regions, frost (<−3°C) in autumn can cause premature leaf fall and frost damage to fruit. The symptoms of frost damage to fruit include surface pitting, rupturing of flesh cells, and

Photo 2 The effect of prebud burst frost damage on persimmon buds (right: undamaged buds; left: damaged buds).

fruit softening. Ethylene production from damaged leaf tissue and defoliation can also cause fruit softening and can lead to abscission of undamaged fruit within the canopy.[157]

Mature fruit can be stored at low temperatures (−1 to 1°C) for 3 months under controlled atmosphere (CA) or modified atmosphere (MA) storage.[36] However, chilling injury can occur during low temperature storage, reducing the effective storage life.[158–162] Chilling injury of mature fruit has also been reported to occur when fruit are held at temperatures below 13°C for extended periods.[163] Symptoms of chilling injury include changes in odor, flavor, texture, and flesh color.[160] The sensitivity of the fruit to chilling injury is influenced by maturity, seasonality, and storage environment.[163,164]

C. HIGH TEMPERATURE STRESS

Persimmon can be successfully cultivated in desert regions[165] where air temperatures reach 48°C.[166] However, under hot dry conditions unprotected young trees can be killed by sunburn.[1] Persimmon trees are more resistant to summer heat than apple, pear, and grape, tolerating temperatures up to 50°C, and have a high optimum night temperature for vegetative growth (32°C).[167] However, high temperatures may reduce A and the optimum leaf temperature for maximum A was reported at 20°C.[107] High day/night temperatures (32/27°C) have induced excessive vegetative growth and increased flower and fruitlet abscission.[137] The influence of temperature on flower and fruitlet abscission may be due to high temperatures causing an increase in shoot to fruit competition and a reduction in carbon assimilation.

Fruit quality and composition are affected by high temperature stress. Fruit growth is sensitive to high temperatures during growth stage II. A temperature of 30°C or greater extended the duration of stage II and delayed maturation in comparison to fruit temperatures of 20°C.[139,156] Persimmon growing in hot, dry climates can have a flatter fruit shape when compared to fruit from cool, damp climates.[168] Temperature can have a negative effect on fruit soluble solids, as fruit exposed to high temperatures (30°C) during stage III have a lower soluble solids than fruit exposed to lower temperatures (15°C).[124]

High temperatures during growth stages II and III are important for causing natural astringency loss in PCNA cultivars.[123] However, the mechanism for natural loss of astringency has not been clearly defined. It is known that soluble tannins coagulate in the presence of acetaldehyde to form insoluble tannins[169] and that acetaldehyde accumulates in the tannin cells of fruit exposed to high temperatures (>25°C) during growth stage II.[139] PCNA cultivars would require exposure to anaerobic conditions during fruit development since acetaldehyde accumulation occurs during anaerobic respiration.[170] A study on seasonal changes in the internal gas composition of persimmon fruit has shown that internal oxygen levels drop from approximately 5 to 1% v/v during stage II for a short period.[171] The cause of this phenomena is unknown but may be associated with an increase in resistance to oxygen diffusion into the fruit by cuticular wax formation. When changes in cuticular resistance to oxygen are coupled with an increase in oxygen demand through the effect of high temperature on fruit respiration,[51] then it is possible that persimmon fruit increase the proportion of anaerobic/aerobic respiration. Excess exposure to anaerobic conditions may reduce internal fruit quality, as fruit exposed to high temperatures (30°C) during growth stage III can contain brown specks in the flesh.[172]

Thus, the natural loss of astringency in PCNA cultivars, by soluble tannin coagulation, could be associated with two factors: (1) early cessation of tannin cell development coinciding with changes in cuticular resistance to oxygen diffusion; (2) mid summer temperatures inducing anaerobic conditions within the fruit.

V. WATER

Persimmon is adaptable to regions with a wide range of soil moisture regimes, including drought prone environments,[165,166] the humid tropics[173] and high rainfall areas in warm temperate zones (Table 1). The response of persimmon to water is dependent on the organ or tissue and the phenological stage of development.

A. DROUGHT STRESS

Persimmon seedlings show greater sensitivity to water stress than apple, peach, or grape seedlings.[174,175] The rate of water loss from the tree is influenced by soil moisture, leaf area, and environmental conditions.[176] Growth of seedling persimmon can be poor in soils with low moisture.[177] The concentration of roots in persimmon orchards is greatest within 200–300 mm of the surface,[178,179] and changes in soil moisture in the upper soil layers (100–150 mm in depth) can have a major influence on persimmon tree development and fruit yield.[180] Water stress influences the productivity of persimmon by adversely affecting vegetative growth, fruit set, and fruit development. Water stress affects shoot development,[1] reducing the potential bearing surface and leaf area for the present and following seasons crop. Severe water stress during flowering can stimulate fruit drop,[181,182] and cause a reduction in yield. During fruit growth, water stress can reduce fruit weight and increase soluble solids,[183] by either concentrating soluble solids and/or by retarding of shoot growth and increasing assimilate partitioning to the fruit.

Rootstock can influence the tolerance of persimmon to low soil moisture, as the drought tolerance of *D. kaki* and *D. virginiana* is higher than *D. lotus.*[184] In southern China, persimmon can be successfully cultivated in drought prone regions when it is grown on a *D. oleifera* rootstock. Biological characteristics differ between rootstocks,[185] but the physiological basis of drought tolerance between species is presently not well understood.

Mineral nutrition can be influenced by moisture stress. In extremely dry summers, leaf symptoms associated with magnesium deficiency can occur on fruiting trees.[186] A moderate moisture stress was found to improve the leaf nitrogen content of persimmon in comparison with a higher and lower moisture stress.[175] Manganese toxicity causes 'green blotch', a physiological disorder of the fruit peel.[187] A high calcium to manganese ratio can ameliorate the severity of manganese toxicity.[188] In soils prone to 'green blotch' the severity of the disorder is increased by low transpiration rates reducing the calcium supply to the fruit. Because the major period of calcium uptake coincides with fruit growth stage I,[47] the calcium content of fruit may be reduced by water stress occurring over this growth stage.

Plant water relations within the persimmon tree have received limited attention. The stomatal conductance and transpiration ratio of persimmon is low to moderate (207 mmol m^{-2} s^{-1} and 1.4 mmol CO_2 mol H_2O^{-1}, respectively) in comparison with other fruit crops.[108] Persimmon appears to be very sensitive to water stress with stomatal conductance falling rapidly with only slight increases in leaf water potential from 0.2–0.6 MPa.[189]

B. WATERLOGGING STRESS

Persimmon has been reported to grow in wet sites on heavy soils.[1] However, the drainage of the soil is more important than soil type,[190] because the growth of persimmon seedlings is retarded when soils are maintained at a high moisture content throughout the year.[174] The rootzone is predominantly found in the upper soil layer,[178,179] and growth can be retarded when the water table rises to <350 mm below the soil surface for extended periods.[191] Tree growth and fruit yield can be poor when grown under waterlogged conditions (soil moisture tension < 2 kPa) for 80% of the year. In contrast, moist soils (soil moisture tension > 2 kPa for 50% of the year) enhance tree growth and yield.[180] Waterlogged soils may cause damage to root systems by oxygen deficiency as a result of changes in the redox potential.[192] Persimmon seedlings were found to exhibit normal root growth down to a soil oxygen concentration of 5% v/v,[193] and have been reported to tolerate a lower soil oxygen concentration (1% O_2) than apple, pear, and peach seedlings.[194] Persimmon may have a greater resistance to Fe^{++} and H_2S injury than fig, peach, and pear.[192]

Waterlogging can reduce the yield of persimmon by increasing fruit drop[181] and by reducing fruit size. Techniques such as forced soil aeration by compressed air have been found to be effective in increasing the soil oxygen concentration and improving shoot growth and the soluble solids content of fruit.[195]

C. HUMIDITY STRESS

Fruit staining, a physiological disorder, significantly reduces the external quality of persimmon. This disorder is characterized by dark staining of the surface of ripe fruit and may also have a cloudy, fine crazing, or smudged appearance. Staining develops over the final stages of maturation and is associated with a high relative humidity (>90% RH) over the 2 months prior to harvest.[196] During fruit enlargement, cuticular cracks can form which extend into sub-epidermal cells.[197] Cuticular cracks occurring in stage I of fruit growth form a protective suberized cell layer, but cracks occurring in stage III form no protective layer.[198] The severity of cuticular cracking can differ between cultivars and is related to differences in fruit growth patterns and cuticle thickness.[199] The incidence of cuticular cracking increases with irrigation and other factors stimulating excessive fruit growth during stage III.

The presence of unprotected cuticular cracks can provide entry to fruit tissue by pathogens, pesticide sprays and water under conditions of high relative humidity or rainfall. Pesticides and rainwater cause black stains by the oxidation of polyphenolic compounds in exposed epidermal and sub-epidermal

Photo 3 Plastic covers to reduce the effect of water on the fruit surface.

cells.[126,200] Anthracnose (*Glomerella cingulata* Spaulding et Schrenk) and alterneria (*Alternaria* sp.) can invade the fruit through cuticle cracks under conditions of high humidity. Water can also cause physical damage to the fruit surface by osmosis drawing water into exposed sub-epidermal tissue, resulting in localized splitting of the cuticle and subsequent fruit softening. Cultivars with a high soluble sugar concentration (i.e., >16% soluble solids) are more prone to this disorder than those with lower soluble solids.[201] This is consistent with the disorder having an osmotic component. The change in turgor pressures in localized regions of water absorption may exceed the strength of the cuticle or of cell walls in sub-epidermal regions.

Staining can be reduced by lowering within-canopy humidity through summer pruning, the application of protective calcium and plastic polymer sprays to cuticular cracks, and the exclusion of water from the fruit surface by bagging or growing trees under plastic covers (Photo 3).[201,202]

VI. WIND

Wind can modify the surrounding thermal environment of a tree and temperatures of specific plant organs. In temperate zones, the amount of heat units over the growing season can be marginal for growth and development. Wind during the period between bud-break and anthesis can delay flowering and reduce leaf size by lowering orchard temperatures. Fruit development can also be delayed by wind cooling exposed fruit. In areas prone to wind, natural and artificial wind breaks are used to raise orchard temperatures[203] and reduce the effects of wind damage.[204]

Wind can cause physical damage to plant organs. Persimmon leaves are large (70–100 cm^2) and prone to damage. Physical damage of emerging shoots and leaves, by spring winds, reduces the potential bearing surface and leaf area in an orchard. Bacterial blast (*Pseudomonas syringae* van Hall) and grey mold (*Botrytis cinerea* Persoon) can enter leaf tissue through wind damaged leaves and shoots causing further loss in leaf area. Persimmon is also prone to branch and trunk breakage in strong winds. The trees require staking when young and the development of a strong branch framework to reduce tree damage by wind in future years. Salt laden wind can cause injury and leaf fall in persimmon[205] when the salt content on leaves reaches 0.3 mg/cm^2.[206]

The loss of leaf area by wind damage in the early stages of fruit growth can cause fruit drop, while damage in later stages reduces the soluble solids content of fruit.[84] A premature loss of leaf area lowers the carbohydrate status of a tree and reduces flower evocation and fruiting in the following season. Wind rub can be major cause of fruit blemish. Trellising of trees and the use of wind breaks have been effective in reducing the incidence of wind rub in New Zealand orchards (Photo 4).

Persimmon pollen can be carried great distances by wind,[207] but it is not effective in transferring pollen to female flowers.[208] Some air movement around persimmon is beneficial. Wind can lower relative humidity in and around the tree, and subsequently reduce humidity related problems. Low wind flows of 2 l^{-2} dm^{-2} min^{-1} can enhance photosynthetic activity in persimmon.[107]

VII. CONCLUSIONS

Persimmon will grow under a wide range of environmental conditions, but when grown as a fruit crop, production and market needs tend to restrict the suitable environmental range. The length of the growing

Photo 4 An artificial windbreak with natural shelter in the background.

season required to mature fruit, rather than winter temperatures, may be the major limitation to persimmon cultivation in cool temperate climates. The recent selection of early ripening cultivars[148] may expand the range of persimmon cultivation to climates with a short growing season. Frost damage during the growing season can adversely affect plant growth and fruiting in temperate regions. Areas prone to frost damage over the growing season may require frost protection by using overhead irrigation, wind machines, or careful site selection.

The low dormancy requirement of many persimmon cultivars allows the tree to be grown in the tropics. However, careful selection of site would be necessary to minimize the affect of environmental stress on productivity, fruit quality, and tree health.

The productivity of persimmon is very sensitive to assimilate supply during the period from flower evocation to fruit set.[81,209] Environmental factors that reduce photosynthetic activity including low irradiance, high temperatures, drought stress, and defoliation by wind may all have a deleterious effect on fruit set and fruit development. Techniques to minimize stress include the use of pruning and training systems to improve light distribution in the canopy, pollination to increase fruit sink strength, irrigation, and the use of windbreaks. The quality and composition of fruit can also be affected by factors influencing photosynthetic activity. Temperature can affect flower and fruit growth, color development, and fruit astringency.[139] High temperature stress during flower development reduces flower size and over growth stage II, delays maturation, and reduces fruit size. The cultivation of non-astringent cultivars is unsuitable in climatic regions with low summer temperatures, as the fruit can remain astringent at harvest. Fruit temperature can be increased by exposing fruit to full sun, or by raising ambient temperatures through the use of shelter and plastic covers.[123,202,210] However, excess exposure of fruit should be avoided in areas with high irradiances as this can cause blemishes through sunburn and fruit spotting[59,127] (see Section III.C, Irradiance). Wind, humidity, rainfall, and hail over fruit development can also cause blemish through physical damage and cuticular cracking.[31,83,127,196,200]

Opportunities exist for reducing some of the present environmental limitations to persimmon cultivation by better understanding the whole plant effects of a particular environmental stress. Carbon assimilation and allocation within a persimmon tree has received little attention, though this process appears to be very sensitive to irradiance, temperature, and drought stress. The recent development of models describing seasonal changes in the tree and canopy structure of mature persimmon trees, and the radiation flux at leaf surfaces within the canopy[23,211,212] will assist in better understanding whole plant interception of light. The use of phenological approaches to describe and quantify whole plant responses to stress could also assist in better defining the effect of environment on the growth and development of persimmon.[22,213,214] Plant improvement could be used to select cultivars with high A rates and better partitioning. A dwarf bud mutation of the cultivar Nishimura wase has been shown to have thicker leaves and higher photosynthetic activity on a leaf area basis than the original form of Nishimura wase.[67] The selection of dwarf cultivars with lower wood to leaf biomass ratio may increase carbon gain through a reduction in whole plant respiration. The selection of rootstocks that improve plant water relations may increase tolerence to conditions of drought or flooding and may serve to maximize carbon gain under a range of climatic and edaphic conditions. The selection of cultivars and rootstocks with improved partitioning could improve productivity and fruit quality, reduce fluctuations in bearing and enhance fruit set under adverse conditions.

Expansion of the area of persimmon under cultivation is occurring outside the major production areas. This has been due to consumer demand for "out of season" fruit in traditional markets and an increasing demand for exotic fruit in non-traditional markets. Knowledge and management of environmental stress, particularly in areas of marginal adaptation, will be critical in meeting these market demands and for minimizing production failure.

REFERENCES

1. Condit, I., The kaki or oriental persimmon, *California Agricultural Experimental Station Bulletin,* 316, 231, 1919.
2. Popenone, W., *Manual of tropical and subtropical fruits,* 1920.
3. Rynerson, K., Culture of the oriental persimmon in California, *California Agricultural Experimental Station Bulletin,* 416, 63, 1927.
4. Ito, S., Persimmon, in *Tropical and subtropical fruits—composition and uses,* Nagy, S., and Shaw, P., Eds., AVI Publishing Inc., Westport, Connecticut, 1980, 442.
5. Kitagawa, H. and Glucina, P., *Persimmon culture in New Zealand,* Science Information Publishing Centre, Wellington , New Zealand, 1984, 74.
6. Bellini, E., Panoramica varietale del kaki nel mondo e orientamenti produttivi per l'Italia, *L'informatore Agrario,* 47, 47, 1988.
7. George, A. and Nissan, R., Persimmon, in *Fruits: tropical and subtropical,* Bose, T. and Mitra, S., Eds., 1990.
8. Sponberg, S. A., Notes on persimmons, kakis, date plums, and chapotes, *Arnoldia,* 39(5), 290, 1979.
9. Zeven, A. C. and Zhukovsky, P. M., Dictionary of cultivated plants and their centres of diversity, Wageningen, 1975.
10. Rehder, A. and Wilson, E. H., Ebenaceae, in *Plantae Wilsonianae,* Sargent, C. S., *Arnoldia Arboretum,* No. 4, The University Press, Cambridge, 1916, 587.
11. Zhuang, D. H., Kitajima, A., Ishida, M., and Sobajima, Y., Chromosome numbers of *Diospyros kaki* cultivars, *Journal of the Japanese Society of Horticultural Science,* 59(2), 289, 1990.
12. Hume, H. H., A kaki classification, *Journal of Heredity,* 5, 400, 1914.
13. Ng, F. S. P., Diospyros roxburghii and the origin of *Diospyros kaki, Malaysian Forester,* 41, 43, 1978.
14. Pienazek, S. A., Fruit production in China, *Proceedings of the XVII International Horticultural Congress, IV,* 427, 1967.
15. Ikegami, T., Morphological studies on the origin of *Diospyros kaki* in Japan, *Mem. Osaka Kyoiku University,* 16(2), 55, 1967.
16. Tsukamoto, Y., On the vitamin C and oxidizing enzymes in kaki fruit, *Journal of the Horticultural Association of Japan,* 11(3), 287, 1940.
17. Taira, S., Sugiura, A., and Tomana, T., Relationships between natural flesh darkening and polyphenoloxidase activity in Japanese persimmon (*Diospyros kaki* Thunb.) fruits, *Journal of Japanese Society of Food Science and Technology,* 34(9), 612, 1987.
18. Sugiura, A., and Tomana, T., Relationships of ethanol production by seeds of different types of Japanese persimmons and their tannin content, *Horticultural Science,* 18(3), 319, 1983.
19. Yamada, M., personal communication, 1988.
20. Yonimori, K., Matsushima, J., and Sugiura, A., Differences in tannins between nonastringent and astringent type fruits of Japanese persimmon, *Journal of the Japanese Society of Horticultural Science,* 52(2), 135, 1983.
21. Mowat, A., The World Scene—Production and Marketing, in *Charting the Future,* Collins, R. J., The University of Queensland, Gatton College, Proceedings of the First National Non-astringent Persimmon Industry Workshop, 37, 1990.
22. Takano, T., and Nakijima, M., Phenological pattern of development in Japanese persimmon trees, *Scientific Reports of the Faculty of Agriculture, Meijo University,* 9, 16, 1973.
23. Archer, C. J., and Cameron, S. H., Cambial activity in the Hachiya persimmon, *Journal of the American Society of Horticultural Science,* 37, 127, 1939.

24. Yamamoto, T., and Hata, R., An analyzing system of seasonal changes of tree form and structure of leaf canopy of fruit trees, *Bulletin of the Yamagata University of Agricultural Science,* 11(2), 343, 1991.
25. George, A. P., and Nissan, R. J., Yield, growth and fruit quality of the persimmon (*Diospyros kaki* L.) in south-east Queensland, *Queensland Journal of Agriculture and Animal Sciences,* 39(2), 149, 1982.
26. Sato, K., and Ishihara, M., The amounts of the nutrient elements absorbed by Japanese persimmon trees: I, *Journal of the Horticultural Association of Japan,* 22(1), 1, 1953.
27. Sato, K., Ishihara, M., and Harada, R., The amounts of the nutrient elements absorbed by Japanese persimmon trees: II, *Journal of the Horticultural Association of Japan,* 24(4), 1, 1956.
28. Hirano, S., On the specific differences of the growth of fruit trees planted in the different amounts of soil, *Journal of the Horticultural Association of Japan,* 33(4), 23, 1964.
29. Tanaka, K., Aoki, M., Kinbara, T., Tsuda, K., and Kawabuchi, A., Studies on the absorption of nutrient elements by kaki (*Diospyros kaki*) in water culture, *Research Bulletin of the Aichi-Ken Agricultural Research Center: Series B (Horticulture),* 8, 74, 1976.
30. Nakamura, S., On the root activities of some fruit trees, *Journal of the Horticultural Association of Japan,* 6(2), 305, 1935.
31. George, A. P., Collins, R. J., and Nissan, R. J., Growth, yield and fruit quality of non-astringent persimmon in subtropical Australia, *Australian Journal of Experimental Agriculture*, 34, in press, 1994.
32. Harada, H., The relationship between shoot growth, auxillary bud development and flower initiation in Japanese persimmon, *Journal of the Japanese Society of Horticultural Science,* 53(3), 271, 1984.
33. Glucina, P. G., and Toye, J. D., Flower bud and fruit development of persimmons, in Persimmon, *Proceedings of the Ruakura Horticultural Conference, Hamilton, New Zealand, Ministry of Agriculture and Fisheries,* 11, 1985.
34. Moncur, M. W., *Floral development of tropical and subtropical fruit and nut species,* CSIRO, 1988.
35. Hodgson, R. W., Floral situation, sex condition and parthenocarpy in the oriental persimmon, *Journal of the American Society of Horticultural Science,* 37, 250, 1939.
36. Itoo, S., *CRC handbook of fruit set and development,* Monselise, S. P., Ed., CRC Press, Florida, 1986, 355.
37. Yamada, M., Kurihara, A., and Sumi, T., Varietal differences in fruit bearing in Japanese persimmon (*Diospyros kaki* Thunb.) and their yearly fluctuations, *Journal of the Japanese Society of Horticultural Science,* 56(3), 293, 1987.
38. Yokozawa, Y., Insect visitors on the flowers of Japanese persimmon, *Journal of the Horticultural Association of Japan,* 20(1), 58, 1951.
39. McGregor, S. E., Persimmon (oriental or kaki), in *Insect pollination of cultivated crop plants,* United States Department of Agriculture Handbook, 1976, 296.
40. Yokozawa, Y., Insect visitors on the flowers of Japanese persimmon, *Journal of the Horticultural Association of Japan,* 21(1), 25, 1952.
41. Mori, H., and Hamaguchi, K., Some experiments on the commercial hand pollination of Kaki (Japanese persimmon), *Journal of the Horticultural Association of Japan,* 19(2), 98, 1950.
42. Kadiura, M., The physiological dropping of fruits in the Japanese persimmon, *Journal of the Horticultural Association of Japan,* 14(1), 5, 1943.
43. Bargioni, G., Pisani, P. L., Ramina, A., and Castelli, F., Aspetti fisiologici dell'allegagione, cascola ed accrescimento di frutti partenocarpici e derivati da fecondazione di Diospyros kaki (L.), *Rvista Ortofloro Frutticoltura Italaina,* 63, 81, 1979.
44. Schroeder, C. A., Fruit growth in the oriental persimmon, *Yearbook of the California Avocado Society,* 44, 130, 1960.
45. Ida, G., Anatomical structure of fruit in Japanese persimmon or Kaki (*Diospyros kaki* Linn. fil.), *Journal of Okitsu Horticultural Society,* 28, 126, 1932.
46. Clark, C. J., and Smith, G. S., Seasonal changes in the composition, distribution and accumulation of mineral nutrients in persimmon fruit, *Scientia Horticulturae,* 42, 99, 1990.
47. Otani, K., An observation of the relationship between physiological drop and growth of persimmon (fruit, seeds and embryos), *Journal of Agricultural Science (Setagaya),* 6, 438, 1961.
48. Sugiura, A., and Tomana, T., Relationships of ethanol production by seeds of different types of Japanese persimmons and their tannin content, *Horticultural Science,* 18(3), 319, 1983.
49. Yonimori, K., and Matsushima, J., Changes in tannin cell morphology with growth and development of Japanese persimmon fruit, *Journal of the American Society of Horticultural Science,* 112(5), 817, 1987.

50. Maeda, S., Histological and physiological studies on the calyx of persimmon fruit, *Research Bulletin of the Tokushima Fruit Experimental Station,* 2, 50, 1968.
51. Atsumi, K., and Nakamura, M., Physiological and ecological studies on the calyx of Japanese persimmon fruit. I. Development of fruit affected by removal of the calyx lobes, *Journal of the Horticultural Association of Japan,* 28(3), 28, 1959.
52. Yonimori, K., Hirano, K., and Sugiura, A., Effect of calyx lobe removal on fruit development and sugar metabolism in Japanese persimmon cv. Hiratanenashi, *Abstracts of the XXIII International Horticultural Congress,* 612, 1990.
53. Sobajima, Y., Ishida, M., Inaba, A., and Masui, K., Studies on the fruit development of Japanese persimmon (*Diospyros kaki* L.). III. Translocation and accumulation of photosynthates, *Reports of Kyoto Prefecture of Agriculture,* 28, 18, 1976.
54. Matsui, T., and Kitagawa, H., Sucrose synthetase, sucrose phosphate synthetase and invertase in persimmon, *Abstracts of the XXIII International Horticultural Congress,* 25, 1990.
55. Zheng, G. H., and Suguira, A., Changes in sugar composition in relation to invertase activity in the growth and ripening of persimmon (*Diospyros kaki*) fruits, *Journal of the Japanese Society of Horticultural Science,* 59(2), 281, 1990.
56. Ito, S., The Persimmon, in *The Biochemistry of Fruits and Their Products,* Hulme, A. C., Academic Press, London and New York, 1971, 281.
57. Ebert, G., and Gross, J., Carotenoid changes in the peel of ripening persimmon (*Diospyros kaki*) cv. Triumph, *Phytochemistry,* 24, 29, 1985.
58. Yonimori, K., and Matsushima, J., Morphological characteristics of tannin cells in Japanese persimmon fruit, *Journal of the American Society of Horticultural Science,* 112(5), 812, 1987.
59. Monselise, S. P., Closing remarks, in *CRC handbook of fruit set and development,* Monselise S. P., ed., CRC Press, Boca Raton, Florida, 1986, 521.
60. Iikubo, S., Sato, T., and Nishida, T., New Japanese persimmon variety 'Suruga', *Bulletin of the Natinal Tokai-Kinki Agricultural Experiment Station: Okitsu Horticultural Station,* 6, 33, 1961.
61. Watanabe, Y., Yamamoto, M., Sakuma, F., Kasumi, M., Adachi, M., Karashima, N., Doi, A., Iijima, K., Hoshino, M., Hiyama, H., Saotome, T., and Ichimura, T., Studies on the ecological characteristics of fruit tree cultivars in Ibaraki (2) on the chestnut, grape, Japanese persimmon and mume cultivars, *Bulletin of the the Ibaraki-ken Horticultural Experiment Station,* 13, 31, 1987.
62. Pinheiro, R. V. R., Costa, A. N., da Souza, A. C. G., and de Conde, A. R., Producao e qualidade dos frutos de diferentes varietntes de caqui, visando a industrializacao, *Revista Ceres,* 32, 453, 1985.
63. Miller, E., P., Oriental persimmon in Florida, *Proceedings of the Florida State Horticulture Society,* 79, 374, 1984.
64. Yamane, H., Kurihara, A., Nagata, K., Yamada, M., Kishi, T., Yoshinaga, K., Matsumoto, R., Kanto, K., Sumi, T., Hirabayashi, T., Ozawa, T., Hirose, K., Yamamoto, M., and Kakutani, M., New Japanese persimmon cultivar 'Youhou', *Bulletin of the Fruit Tree Research Station,* 20, 49, 1991.
65. Isoda, R., Strains of Saijo persimmon in the Chugoki and Shikoku regions, *Bulletin of the Hiroshima Agricultural College,* 165, 1983.
66. Murakami, K., Kurokami, K., and Maeda, S., A new strain of persimmon, 'Supahiratanenashi', *Agriculture and Horticulture,* 323, 1975.
67. Fumuro, M., and Murata, R., Physiological and ecological characteristics of a dwarf strain of Japanese persimmon cv. 'Nishimurawase'. 1. Growth characteristics and fruit productivity, *Bulletin of the Shiga Prefecture Agricultural Experiment Station,* 30, 45, 1989.
68. Kishimoto, O., Evaluation of pruning methods and training systems in Japanese pears, in *The pear,* Zwet T., and Childers N., Eds., Hort Publications, Florida, 1982, 354.
69. Murata, R., Oishi, R., and Okishima, H., Characteristics and cultural value of dwarf and normal 'Nishimurawase' persimmon. 4. Production efficiency of dwarf and normal trees, *Bulletin of the Shiga Prefecture Agricultural Experiment Station,* 26, 48, 1985.
70. Fumuro, M., and Murata, R., Physiological and ecological characteristics of a dwarf strain of Japanese persimmon cv. 'Nishimurawase'. 2. Relationship between growth characteristics and endogenous growth substances, *Bulletin of the Shiga Prefecture Agricultural Experiment Station,* 30, 57, 1989.
71. Fumuro, M., and Murata, R., Physiological and ecological characteristics of a dwarf strain of Japanese persimmon cv. "Nishimurawase". 3. Flower setting and fruiting, *Bulletin of the Shiga Prefecture Agricultural Experiment Station,* 30, 45, 1989.

72. Hodgson, R., Rootstocks for the oriental persimmon, *Proceedings of the American Society for Horticultural Science,* 37, 338, 1937.
73. Aoki, M., Takase, S., Kimura, N., and Kawabuchi, A., Studies on method of top grafting to 'Nishimura-wase' kaki (*Diospyros kaki* Linn, f.): Effects of difference of top grafting and intermediate stock on the growth of tree and yield, *Research Bulletin of the Aichi-Ken Agricultural Research Center,* 15, 268, 1983.
74. Schroeder C., Rootstock influence on fruit-set in the Hachiya persimmon, *Proceedings of the American Society for Horticultural Science,* 50, 149, 1950.
75. Yamada, M., Sumi, T., Kurihara, A., and Yamane, H., Effects of various factors on the regeneration of Japanese persimmon (*Diospyros kaki* Thunb.) from root cuttings, *Bulletin of the Fruit Tree Research Station Series E (Akitsu),* 7, 9, 1988.
76. Kimura, N., Kawabuchi, A., Aoki, M., Okada, N., Manago, N., and Suzari, S., On the search for the dwarfing stocks for kaki and their utilization. 1. Growth characteristics and productivity of dwarf trees, *Research Bulletin of the Aichi-Ken Agricultural Research Center,* 17, 273, 1985.
77. Asakura, T., Kamota, F., and Honjo, H., Yield variability of fruit crops in Japan, *Bulletin of the Fruit Tree Research Station, Series A (Yatabe),* 15, 69, 1988.
78. Ojima, M., Dall'orto, F. A. C., Barbosa, W., Yombolato, A. F. C., and Rigitano, O., Frutificacao alternada em caqui cultivar pomelo (IAC 6-22), *Bragantia Campinas,* 44(1), 481, 1985.
79. Hodgson, R. W., and Schroeder, C. A., On the bearing behavior of the kaki persimmon (*Diospyros kaki*), *American Society for Horticultural Science,* 50, 145, 1947.
80. Archer, C. J., The starch cycle in the Hachiya persimmon, *Journal of the American Society of Horticultural Science,* 38, 187, 1941.
81. Nagasawa, K., and Hirai, T., Flower bud formation in Fuyu kaki under different ecological conditions, *Technical Bulletin of the Faculty of Horticulture, Chiba,* 7, 27, 1959.
82. Kim, Y. S., Joung, S. B., Son, D. S., Lee, K. K., Park, J. S., and Lee, U. J., Studies on the establishment of the safety-cultivating region of non-astringent persimmon, *Research Report of the Rural Development Administration,* 30(3), 56, 1988.
83. Fumuro, M., Murata, R., and Kawai, F., Effects of hailstone on fruiting of Japanese persimmon and grapevines, *Bulletin of the Shiga Prefecture Agricultural Experiment Station,* 32, 53, 1991.
84. Chujo, T., Effect of defoliation on the growth, sugar content and coloration of kaki fruit, *Technical Bulletin Faculty of Agric Kagawa,* 21, 14, 1970.
85. Iimuro, S., Fukunaga, S., Matsumoto, Y., Iwamoto, K., and Kuroda, K., Increasing Japanese persimmon yields using trees in high density plantings dwarfed by pruning, *Bulletin of the Nara Agricultural Experiment Station,* 12, 22, 1981.
86. Himeno, S., Yoshinaga, F., Tsuru, T., Shoda, K., Moriata, A., and Tsuneto, M., Influence of the Y-shaped training on yield and fruit quality of persimmon, *Bulletin of the Fukuoka Agricultural Research Centre: Series B Horticulture,* 11, 89, 1991.
87. Omarov, M. D., Effect of different pruning methods on the growth and cropping of persimmon trees, *Subtropicheskie Kul'tury,* 1, 71, 1980.
88. Omarov, M. D., Efficary of rejuvenating pruning in *Diospyros kaki*, *Sadovodstvo,* 6, 23, 1986.
89. Kishimoto, O., Estimations of optimum range of degree of fruit thinning and desirable fruit weight in Japanese pear and persimmon tree, *Journal of the Japanese Society of Horticultural Science,* 48(4), 368, 1975.
90. Kishimoto, O., Factors affecting fruit weight of Japanese persimmons (*Diospyros kaki* Linn. f.), *Journal of the Japanese Society of Horticultural Science,* 33(4), 31, 1964.
91. Kishimoto, O., Studies on the standard of fruit thinning in Japanese persimmon (*Diospyros kaki* Linn. f.), *Journal of the Japanese Society of Horticultural Science,* 32(3), 20, 1963.
92. Stevens, P., Mowat, A., Rohitha, H., and Wilcox, B., Pesticide spray formulation, *Report for the New Zealand Persimmon Export Council,* 38pp, 1992.
93. Kobayashi, M., Effect of organic matters on higher plants, *Japan Research Quarterly,* 10(1), 7, 1976.
94. Gasanov, Z. M., Nitrogen nutrition and productivity of persimmon, *Sadovodstov,* 12, 37, 1984.
95. Aoki, M., Tanaka, K., and Okada, N., Effects of nitrogen fertilization and ringing treatment on intial yield in 'Wase Jiro' kaki (*Diospyros kaki* Linn. f.), *Research Bulletin of the Aichi-Ken Agricultural Research Center: Series B (Horticulture),* 9, 119, 1977.

96. Di Maio, F., Il nematode *Tylenchulus semipenetrans* Cobb., su radici di loto, *Informatore Fitopatologico,* 13, 1979.

97. Gonzalez, H., El kaki (*Diospyros kaki* L.), un nuevo huesped de tylenchulus semipenetrans Cobb, en Chile, *Agricultura Tecnica,* 56, 1988.

98. Herbas, A., R., El mosaico del caqui (*Diospyros kaki*) y algunas propiedades fisicas de su agente causal, *Turrialba,* 19, 480, 1969.

99. Sikharulidze, A., M., Tetrapod mites, pests of sub-tropical persimmon, *Subtropicheskie Kul'tury,* 6, 136, 1970.

100. Rossetto, C. J., Ojila, M., Rigitano, O., and Igue, E. T., Queda dos frutos do caquizeiro, associada a infestacao de *Aceria diospyri* K. (Acarina, Eriophyidae), *Fitotecnia Latinoamericana,* 9(1), 22, 1973.

101. Demetradze, T. Ya., and Sikharulidze, A. M., The effect of citrus white fly damage on some physiological and biochemical processes in the leaves of sub-tropical persimmon, *Subtropicheskie Kul'tury,* 4, 129, 1971.

102. Hino, A., Amano, S., Sawamura, Y., Sasaki, S., and Kuraoka, T., Studies on the photosynthetic activity in several kinds of fruit trees. II. Seasonal changes in the rate of photosynthesis, *Journal of the Japanese Society of Horticultural Science,* 34(3), 209, 1974.

103. Mu, Y., and Li, X., Studies on the photosynthetic characteristics of some deciduous fruit trees, *Acta Horticulturae Sinica,* 13(3), 157, 1986.

104. Naito, R., Yamamura, H., and Mikamori, T., On spraying time, concentration and other factors involved in using NAA as a fruit-thinning agent for Saijo persimmon, *Bulletin of the Faculty of Agriculture,* 1, 1973.

105. Murata, R., Oishi, R., and Okishima, H., Characteristics and cultural value of dwarf and normal Nishimurawase persimmon. 4. Production efficiency of dwarf and normal trees, *Bulletin of the Shiga Prefecture Agricultural Experiment Station,* 26, 48, 1985.

106. Mowat, A. D., and Ah Chee, A., A comparative study on Fuyu fruit quality and maturation in different growing districts of New Zealand, *Report for the New Zealand Persimmon Export Council,* 1991, 107.

107. Amano, S., Hino, A., Daito, H., and Kuraoka, T., Studies on photosynthetic activity in several kinds of fruit tree. I. Effect of some environmental factors on the rate of photosynthesis, *Journal of the Japanese Society of Horticultural Science,* 41, 144, 1972.

108. Andersen, P. C., Leaf gas exchange of 11 species of fruit crops with reference to sun-tracking/non-sun-tracking responses, *Canadian Journal of Plant Science,* 71, 1183, 1991.

109. Mowat, A. D., and Ah Chee, A., Studies on improving fruit quality of Fuyu by canopy microclimate manipulation, *Report for the New Zealand Persimmon Export Council,* 1990, 57.

110. George, A., and Mowat, A. D., unpublished data.

111. Iimuro, S., Okamura, T., Sawamura, Y., Matsumoto, Y., and Fukunaga, S., Studies on nutrition of persimmon trees and the sugar content of the fruit. I. Relationship between leaf nitrogen and fruit sugar content in the autumn, *Bulletin of the Nara Agricultural Experiment Station,* 6, 9, 1974.

112. Kaneko, M., Relationships between yields of Japanese persimmon and hours of sunshine, *Research Bulletin of the Aichi-Ken Agricultural Research Centre,* 9, 131, 1977.

113. Hisada, H., Distribution of carbohydrate in persimmon tree (*Diospyros kaki* L. cv. Jiro) and decrease of it by shading, *Bulletin of the Sizuoka Experimental Station,* 19, 65, 1983.

114. George, A. P., Nissan, R. J., Morley Bunker, M. J., and Collins, R. J., Effects of pollination and irradiance on fruiting of persimmon in subtropical Australia, *Journal of Horticultural Science,* 68, 447, 1993.

115. Yamamura, H., Effects of pollination and fertilization of fruit set in 'Saijo' Japanese persimmon, *Bulletin of the Faculty of Agriculture, Shimane University,* 16, 8, 1982.

116. Horie, Y., Hirasima, K., and Turu, T., Influence of fruit-thinning and artificial-pollination on physiological fruit drop of Japanese persimmon 'Izu', *Bulletin of the Fukuoka Agricultural Research Center,* 8, 15, 1988.

117. Hasegawa, K., and Nakajima, Y., Effects of seediness on fruit quality of Japanese persimmon cv. Maekawa-Jiro, *Journal of the Japanese Society for Horticultural Science,* 59(2), 255, 1990.

118. Kitajima, A., Fujiwara, T., Kukizaki, T., Ishida, M., and Sobajima, Y., Relationships between early fruit drop and dry matter accumulation on bearing shoots in Japanese persimmon (*Diospyros kaki* Thunb.), *Scientific Reports of the Kyoto Prefectural University, Agriculture,* 39, 1, 1987.

119. Kitajima, A., Matsumoto, T., Ishida, M., and Sobajima, Y., Relationship between dry matter prosuction of bearing shoots and physiological fruit drop of Japanese persimmon, by shading treatments, *Journal of the Japanese Society for Horticultural Science,* 59(1), 75, 1990.
120. Okishima, H., Oishi, R., and Murata, R., On remodelling of young trees of Fuyu persimmon, *Bulletin of the Shiga Prefecture Agricultural Experiment Station,* 25, 77, 1983.
121. Takano, S., Nishino, S., and Kuroda, K., The establishment of a technique for producing high-quality fruits of Japanese cultivar Fuyu with two low branches. 1. The relationship between the type of branch or shoot, degree of thinning and fruit quality with regard to exposure to sunlight, *Bulletin of the Nara Agricultural Experiment Station,* 22, 29, 1991.
122. Chujo, T., Studies on the coloration in the fruits of Fuyu kaki. II. Effect of light intensity on the development of reddish color of the peel, *Kagawa Daigaku Ngakubu Gakuzyutu Hokoku,* 23, 35, 1971.
123. Chujo, T., Hashimoto, T., and Ashizawa, M., The effect of temperature on the growth and quality of kaki fruit. II. Day and night temperature treatments during the stage of fruit enlargement, *Kagawa Daigaku Ngakubu Gakuzyutu Hokoku,* 25(1), 25, 1973.
124. Chujo, T., and Ashizawa, M., Studies on the coloration in the fruits of *Fuyu kaki.* IV. On the seasonal difference in favorable temperature for the development of reddish color in the maturity of detached kaki fruits, *Kagawa Daigaku Magakuba Gazukyatu Hokoku,* 24, 137, 1973.
125. George, A. P., The Australian scene—Production, in Charting the Future, Collins, R. J., The University of Queensland, Gatton College, *Proceedings of the First National Non-astringent Persimmon Industry Workshop,* 13, 1990.
126. Isoda, R., Cause of a certain physiological disorder 'Kokuhen-Ka' of 'Saijo' persimmon (*Diospyros kaki* L.), *Studies from the Institute of Horticulture, Kyoto University,* 9, 36, 1979.
127. Oppenheimer, C., *Fruit growing,* Ed. Margalit, Hassadeh, Tel Aviv, Israel, 1961.
128. Opitz, K. W., and LaRue, J. H., Growing persimmons, *Leaflet of the Cooperative Extension, California University,* 1975, 8 pp.
129. Miller, E. P., Oriental persimmons (*Diospyros kaki* L.) in Florida, *Proceedings of the Florida State Horticultural Society,* 97, 340, 1984.
130. Terra, G. J. A., Diospyros kaki in Garut, Java, *Landbouw,* 11 326, 1936.
131. Ng, F. S. P., Diospyros kaki L.f., in *Plant Resources of South-East Asia,* Vol. 2, *Edible Fruits and Nuts,* Verheij, E. W. M., and Coronel, R. E., Eds., Pudoc Wageningen, the Netherlands, 1991, 154.
132. Nuttonson, M. Y., Ecological crop geography and field practices of Japan. Japan's natural vegetation, and agro-climatic analogues in North America, American Institute of Crop Ecology, Washington, D.C., 1951.
133. Morton, J. F., Ebenaceae, in *Fruits of warm climates,* Dowling, C. F., 1988, 411.
134. Rigitano, O., Instrcoes para a cultura do caquizeiro, *Bol. Institute of Agronomic Secret. Agric. Est. S. Paulo. Campinas,* 30, 2, 1956.
135. Shanks, J. B., Cold hardiness of oriental persimmons (*Diospyros kaki*) in Maryland, *68th Annual Report Northern Nut Growers Association,* 111, 1977.
136. Rozanov, B. S., and Vorob'eva, L. T., Winter dormancy in subtropical and nut crops, *Temat. Sbornik Nauch. Tr. Tadzh. Nii Zemiedeliya,* 8, 28, 1976.
137. George, A. P., Nissan, R. J., and Collins, R. J., Effects of temperature and pollination on growth, flowering and fruit set of non-astringent persimmon cultivar 'Fuyu' under controlled environment conditions, *Journal of Horticultural Science*, 69, in press, 1994.
138. Mowat, A.D., The effect of root temperature on the bud dormancy of persimmon, Unpublished.
139. Chujo, T., Studies on the effects of thermal conditions on the growth and quality of fruits of Fuyu kaki, *Mem Faculty of Agriculture, Kagawa University,* 37, 1, 1982.
140. Son, D. S., Hong, K. H., Kim, Y. S., Kim, K. Y., and Lee, U. J., Studies on reducing transplanting shock in persimmon, *Research Reports of the Rural Development Administration, Horticulture, Korea Republic,* 30(1), 83, 1988.
141. Zareckii, A. J., The Japanese persimmon in central Asia, *Orchard and Garden,* 1, 42, 1950.
142. Zivotinskaja, S. M., and Kul'Kov, O. P., On the frost resistance of oriental persimmon in Uzbekistan, *Subtrop. Kul'tury,* 5, 87, 1968.
143. Sharpe, R. H., Persimmon variety and rootstock observations, *Proceedings of the Florida State Horticultural Society,* 79, 374, 1966.

144. Yoshimura, F., Influence of winter temperature on the dormancy of some deciduous fruit trees. VI. Chilling requirements of peaches, Japanese pears, grapes and Japanese persimmons to break the dormancy of buds, *Journal of the Japanese Society of Horticultural Science,* 30, 351, 1961.
145. Ebel, M., The kaki, *Rev. Hort. Agric. Afr. N.,* 42, 232, 1938.
146. Massover, B. L., The problem of breeding winter-hardy forms of persimmon, *Doki. Akad. Fanhoi RSS Tocikiston,* 56, 1972.
147. Pasenkov, A. K., New varieties and forms of oriental persimmon obtained by intraspecific hybridization, *Doki. Sov. Uchenykh k XIX Mezhdunarod. Kongr. po sadovodstvu Varshava,* 203, 1975.
148. Kazas, A. N., Use of the interspecific hybrid Rossiyanka in breeding persimmon, *Bulletin of Gosudarstvennogo Nikitskogo Botanicheskogo Sada,* 51, 1986.
149. Manaresi, A., Observations on grafting and cold sensitivity in persimmons, *Frutticoltura,* 28, 231, 1966.
150. Yoshimura, F., Influences of winter temperature on the dormancy of some deciduous fruit trees. I. Studies on bud-opening, shoot development and root growth in young trees of Japanese persimmon, peach and Japanese pear after cold, mild or warm winters, *Journal of the Horticultural Association of Japan,* 25, 265, 1957.
151. Yoshimura, F., and Kamohara, T., Influence of winter temperature on the dormancy of some deciduous fruit trees. III. Effect of time and length of chilling treatment on bud-opening and shoot and root-growth of young trees of peach and Japanese persimmon, *Journal of the Horticultural Association of Japan,* 28, 177, 1959.
152. Yoshimura, F., and Kawamura, Y., Influence of winter temperature on the dormancy of some deciduous fruit trees. V. Effect of day temperature during winter on the spring shoot growth of peach and Japanese persimmon trees, *Journal of the Horticultural Association of Japan,* 29, 47, 1960.
153. Nakagawa, Y., and Sumita, A., Studies on favourable climatic environments for fruit culture. VII. The critical temperatures for frost damage to deciduous fruit trees, *Bulletin of the Horticultural Research Station of Hiratsuka,* 8, 95, 1969.
154. Fukui, H., Wakayama, Y., and Nakamura, M., Effect of night temperature on the development of abnormal embryosacs in Japanese persimmon 'Nishimurawase', *Journal of the Japanese Society of Horticultural Science,* 59(1), 59, 1990.
155. Fukui, H., Demachi, M., Yamada, M., and Nakamura, M., Effect of temperature and cross- and self-pollination of pollen tube growth in styles of Japanese persimmon 'Nishimurawase', *Journal of the Japanese Society of Horticultural Science,* 59(2), 275, 1990.
156. Sugiura, A., Zheng, G. H., and Yonimori, K., Growth and ripening of persimmon fruit at controlled temperatures during growth stage III, *Horticultural Science,* 26(5), 574, 1991.
157. Itamura, H., Yamabori, S., Kitamura, T., and Fukushima, T., Effect of defoliation on ethylene evolution of immature Japanese persimmon fruits, *Bulletin of the Faculty of Agriculture, Shimane University,* 23, 11, 1989.
158. Guelfat-Reich, S., and Ben-Arie, R., The effect of maturity at harvest and storage temperatures on keeping quality of Fuyu fruits, *Alon HaNotea,* 28, 3, 1974.
159. El-Wahab, F. K. A., Aziz, A. B. A., El Latif, F. A., and Maksoud, M. A., Behaviour of persimmon fruits under cold storage, *Annals of Agricultural Science,* 28(1), 287, 1983.
160. MacRae, E., Storage and shelf life of Fuyu and Flat Fuyu persimmon in New Zealand, 1984–1986, *DSIR Postharvest Bulletin,* 335, 1987.
161. Taira, S. et al., Comparative studies of postharvest fruit quality and storage quality in Japanese persimmon (*Diospyros kaki* L. cv. Hiratanenashi) in relation to different methods for removal of astringency, *Journal of the Japanese Society of Horticultural Science,* 215, 1987.
162. Toye, J. D., Glucina, P. G., and Minamide, T., Removal of astringency and storage of Hiratanenashi persimmon fruit, *New Zealand Journal of Experimental Agriculture,* 15, 351, 1987.
163. MacRae, E., Development of chilling injury in New Zealand grown Fuyu persimmon during storage, *New Zealand Journal of Experimental Agriculture,* 15, 333, 1987.
164. Mason, K. A., Glucina, P. G., and MacRae, E. A., Maturation and chilling sensitivity of Fuyu persimmon fruit in New Zealand, *New Zealand Journal of Crop Horticultural Science,* 17, 251, 1989.
165. McFarlane, S., and Winwright, G. L., Desert Agriculture, *Circular of the California Extension Service,* 176, 56, 1951.
166. Guseva, E. P., Orchard in the desert, *Sadovodstvo,* 5, 13, 1960.

167. Kobayashi, A., Studies on thermal conditions for grapes. 1. Effects of night temperatures on the growth and respiration of several fruit trees including grapes, *Bulletin of the Research Institute of Food Science, Kyoto University,* 24, 20, 1960.
168. Hodgson, R. W., and Schroeder, C. A., Effect of climate on fruit form in the kaki persimmon, *Proceedings of the American Society of Horticultural Science,* 48, 71, 1946.
169. Matsuo, T., and Ito, S., A model experiment for de-astringency of persimmon fruit with high carbon dioxide treatment: in vitro gelation of kaki-tannin by reacting with acetaldehyde, *Agriculture Biological Chemistry,* 46(3), 683, 1982.
170. Matsuo, T., and Ito, S., On mechanisms of removing astringency in persimmon fruits by carbon dioxide treatment. I. Some properties of the two processes in the de-astringency, *Plant and Cell Physiology,* 18, 17, 1977.
171. Taira, S., Suguira, A., and Tomana, T., Seasonal changes of internal gas composition in fruits of Japanese persimmon (*Diospyros kaki* Thunb.) and effects of different gas environments on production of ethanol by their seeds, *Journal of the Japanese Society for Horticultural Science,* 55(2), 228, 1986.
172. Chujo, T., Kataoka, M., Yamauchi, S., and Ashizawa, M., The effect of temperature on the growth and quality of persimmon fruit. I. Temperature treatments during the stage of fruit enlargement, *Journal of the Japanese Society of Horticultural Science,* 41(4), 339, 1972.
173. Khan, M. A., Japanese fruit or persimmon (*Diospyros kaki*), *Punjab Fruit Journal,* 4, 846, 1940.
174. Morita, Y., and Yoneyama, K., Studies on physical properties of soil in relation to fruit tree growth. III. Soil moisture and tree growth. (1) Effects of soil moisture on the growth of peach, pear and persimmon seedlings, *Journal of the Horticultural Association of Japan,* 18(3,4), 156, 1949.
175. Morita, Y., and Yoneyama, K., Studies on physical properties of soils in relation to fruit tree growth. III. Soil moisture and tree growth. (2) Effects of soil moisture on the growth of apple, chestnut, and persimmon (*D. lotus* Linn.) seedlings and grape cuttings, *Journal of the Horticultural Association of Japan,* 19, 185, 1950.
176. Ashizawa, M., and Nakagawa, M., Studies on fruit physiology in relation to water. I. On seasonal variations in the water relations of fruit trees, *Technical Bulletin of the Kagawa Agricultural College,* 6, 260, 1955.
177. Morita, Y., Nishida, T., and Oguro, E., Studies on physical properties of soils in relation to fruit tree growth. III. Soil moisture and growth. (5) Relation between soil moisture and oxygen decrease in the soil atmosphere where peach and persimmon (*D. lotus* Linn.) seedlings are growing, *Journal of the Horticultural Association of Japan,* 20, 158, 1952.
178. Gasanov, Z. M., The effect of summer pruning on root growth in subtropical persimmon during the growth period, *Nauchnye Doklady Vysshoi Shkoly Biologicheskikh Nauk,* 5, 82, 1972.
179. Kamio, A., Characteristics of root development in deciduous fruit tree orchards surrounded by paddy fields, *Bulletin of the Yamagata University (Agricultural Science),* 10(3), 523, 1988.
180. Hase, Y., Machida, Y., and Maotani, T., Influence of subsoil moisture on the growth of Japanese persimmon and its fruit yield, *Bulletin of the Fruit Tree Research Station, Akitsu, Japan,* 7, 31, 1988.
181. Suzuki, A., Murakami, Y., and Maotami, T., Physiological studies on physiological fruit drop of persimmon, Diospyros kaki Thunb. IV. Effect of fruit growth on physiological fruit drop of persimmon, *Bulletin of the Fruit Tree Research Station, A (Yatabe), Japan,* 15, 41, 1988.
182. Hodgson, R. W., Girdling to reduce fruit-drop in the Hachiya persimmon, *American Society for Horticultural Science,* 36, 405, 1938.
183. Tanaka, K., and Aoki, M., Effects of irrigation and nitrogen fertilizer application in summer on the fruiting of 'Fuyu' kaki (*Diospyros kaki* Linn. f.), *Bulletin of the Aichi-Ken Agricultural Research Center,* B(3), 9, 1971.
184. Tanaka, Y., Experiments on the rootstocks for the kaki or Japanese persimmon, *Bulletin of the Horticultural Experiment Station Okitsu, Japan,* 14, 30pp, 1930.
185. Omarov, M. D., Choosing rootstocks for persimmon, *Subtropicheskie Kul'tury,* (4), 94, 1984.
186. Tanaka, K., Aoki, M., Hirota, K., Kawabuchi, A., and Sinoda, H., Magnesium deficiency in Kaki (*Diospyros kaki* Linn f.). I. Influences of foliar sprays of Magnesium salts and the mineral composition of the leaves, fruits and shoots, *Bulletin of the Aichi Horticultural Experiment Station,* 5, 19, 1966.
187. Aoba, K., Excess manganese disorder in fruit trees, *Japanese Agricultural Research Quarterly,* 20(1), 38, 1986.
188. Iimuro, S., Ono, Y., Sugimoto, Y., Fukunaga, S., and Kuroda, K., Studies on the green-spot disorder of Japanese persimmon, *Bulletin of the Nara Agricultural Experiment Station,* 11, 1, 1980.

189. Morley-Bunker, M., and George A., unpublished data.
190. Camp, A. F., and Mowry, H., The cultivated persimmon in Florida, *Florida Agriculture Experimental Station Bulletin,* 124, 31pp, 1945.
191. Susa, T., Aoba, T., Ishizuka, S., and Abe, S., Studies on root development of fruit trees influenced by ground water conditions, *Journal of the Horticultural Association of Japan,* 21(2), 49, 1952.
192. Hayashi, S., and Wakisaka, I., Studies on the water logging injury of fruit trees from the viewpoint of soil chemistry, *Journal of the Horticultural Association of Japan,* 25(1), 59, 1956.
193. Morita, Y., and Nishida, T., Studies on physical properties of soils in relation to fruit growth. II. Soil atmosphere and tree growth, (4). Growth of peach and persimmon (*D. lotus* L.) seedlings as influenced by various concentrations of oxygen in the soil atmosphere, *Journal of the Horticultural Association of Japan,* 20(3,4), 1, 1952.
194. Morita, Y., and Nishida, T., Studies on physical properties of soils in relation to fruit growth. II. Soil atmosphere and tree growth, (5). Growth of peach, apple, pear and persimmon (*D. lotus* L.) seedlings as influenced by various concentrations of oxygen in the soil atmosphere, *Journal of the Horticultural Association of Japan,* 20(3,4), 8, 1952.
195. Ishii, T., Iwasaki, K., Mizutani, F., Hino, A., and Amano, S., Shoot growth, mineral nutrient absorption, fruit size, quality and cold hardiness of Japanese persimmon (*Diospyros kaki* L.) and citrus (*C. unshiu Marc*., and *C. iyo hort*. ex. Tanaka) trees as influenced by forced soil aeration, *Memoirs of the College of Agriculture, Ehime University,* 25, 63, 1981.
196. Sugimoto, Y., and Yasui, A., Numerical analyses on the effects of environmental factors (especially orchard humidity and Bordeaux mixture) on black spotting of Japanese persimmons, *Bulletin of the Nara Agricultural Experiment Station,* 13, 1, 1982.
197. Chujo, T., Kataoka, M., Yamauchi, S., and Ashizawa, M., The effect of temperature on the growth and quality of persimmon fruit. I. Temperature treatments during the stage of fruit enlargement, *Journal of the Japanese Society of Horticultural Science,* 41(4), 339, 1972.
198. Tarutani, T., Kitagawa, H., and Fukuda, K., Studies on the utilization of persimmons (*Diospyros kaki* Linn. F.) VIII. Effect of injuries to the skin on keepability of the fruit, *Bulletin of the Faculty of Agriculture, Kagawa,* 21, 34, 1970.
199. Yamamura, H., Bessho, H., and Naito, R., Occurrence of black stains on fruit skin (black spots) in relation to growth and development of pericarp tissues in Japanese persimmons, *Journal of the Japanese Society of Horticultural Science,* 53(2), 115, 1984.
200. Yamamura, H., and Yamane, N., Occurrence of black stain on fruit skin (black spots) in relation to ascorbic acid contents in pericarp tissues of Japanese persimmon, *Bulletin of the Faculty of Agriculture, Shimane University,* 21, 18, 1987.
201. Yamane, H., Kurihara, A.. Nagata, K., Yamada, M., Kishi, T., Yoshinaga, K., Matsumoto, R., Ozawa, T., Sumi, T., Hirabayashi, T., and Kakutani, M., New Japanese persimmon cultivar 'Shinsyuu', *Bulletin of the Fruit Tree Research Station,* 19, 13, 1991.
202. Fukuyo, K., Causes and control of black stain of persimmon fruits in Japan, *Bulletin of the Citrus Experimental Station, Shizuoka,* 16, 29, 1980.
203. Uehara, M., Studies on micrometeorology in sloping orchards. II. *Journal of the Horticultural Association of Japan,* 25, 17, 1956.
204. Nakahara, M., and Aso, T., Studies on fruit trees in situations exposed to wind, with special reference to the Matsudo city area, *Technical Bulletin of the Faculty of Horticulture, Chiba,* 16, 83, 1968.
205. Kuraoka, O., Investigations on damage caused by salt-laden air brought by No. 13 typhoon (1953) to the fruit trees in the Hamamatsu region, *Bulletin of the Citrus Experimental Station, Shizuoka,* 2, 3, 1956.
206. Ikubo, S., and Nishda, T., Studies on briny wind damage to fruit trees. I. *Bulletin of the Tokai-Kinki. Agricultural Experimental Station Horticultural Station,* 5, 77, 1959.
207. Fletcher, W. F., The native persimmon, *U. S. Dept. Agriculture Farmers' Bulletin,* 685, 22, 1942.
208. Asami, Y., and Chow, C. T., Is the pollen of Japanese persimmons carried by wind?, *Journal of the Horticultural Association of Japan,* 7(2), 246, 1941.
209. Watanabe, S., Morphological studies of flower initiation in deciduous fruit trees by scanning electric microscope. Japanese persimmon, apple, pear and sweet cherry, *Bulletin of the Yamagata University, Agriculture Science,* 9(4), 515, 1985.

210. Mason, K. A., Glucina, P. G., and MacRae, E. A., Effects of polyethylene film cover on the maturation and quality of Fuyu persimmon fruit in New Zealand, *New Zealand Journal of Crop and Horticultural Science,* 19, 37, 1991.
211. Yamamoto, T., and Yamaguchi, E., A quick analyzing system of tree form and structure of leaf canopy of fruit trees—Development of a new system with two side-photographs, *Bulletin of the Yamagata University of Agriculture Science,* 10(4), 869, 1989.
212. Yamamoto, T., Ueta, J., Koizumi, H., and Nishizawa, T., Several extinction coefficients of radiations and their estimating parameters in several deciduous fruit trees, *Bulletin of the Yamagata University of Agriculture Science,* 11(1), 155, 1990.
213. Nii, N., Seasonal changes in fruit growth of Japanese persimmon *Diospyros kaki* cv. Fuyu, in relation to development of vascular tissue in the fruit stalk, *Journal of the Japanese Society of Horticultural Science,* 49(2), 160, 1980.
214. Nii, N., Current shoot and leaf growth in Japanese persimmon, *Diospyros kaki* cv. Fuyu, in relation to the development of the tissue system in the leaf, *Journal of the Japanese Society of Horticultural Science,* 49(2), 149, 1980.

Chapter 9

Stone Fruit

James A. Flore

CONTENTS

0-8493-175-0/94/$0.00+$.50

I. INTRODUCTION

When does the environment limit yield, fruit quality, and/or longevity of stone fruit, and which environmental conditions are most limiting? Clearly, these questions cannot be resolved unless one has a thorough understanding of the crop, coordination of growth and development in that crop, and the impact of the environment on the physiology of the plant. Herein, the effect of light, water, and temperature and other environmental inputs are reviewed in relation to carbon assimilation and the longevity and productivity of *Prunus*. Emphasis is on cherry and peach, the principles, however, can be applied to most stone fruit crops in different locations.

The reader is referred to several reviews and books on *Prunus* growth and physiology, or on environmental physiology for more information.[1–5]

A. BOTANICAL DESCRIPTION, ORIGIN, AND HABITAT

Stone fruit are members of the Rosaceae, subfamily Prunoidea.[1] The genus *Prunus* includes peach, nectarine, plum, cherry, apricot, and many species used as rootstocks. The basic chromosome number varies from 16 to 176. Stone fruit may be deciduous or evergreen. The fruit is classified as a drupe, and is usually one seeded. Fruit growth is described as double sigmoidal, with rapid growth during cell division, followed by a period of slow growth during pit hardening, and then a period of rapid cell enlargement.[6]

Commercially important species of *Prunus* originated between eastern Europe and western China. They have a broad range of adaptability to environmental stress and wide distribution between the latitudes of 30–40° where light levels are high, skies are clear, and seasons are long and dry. In humid environments *Prunus* spp. are sensitive to a broad range of diseases. Compared to apple, stone fruit generally are less winter hardy and bloom earlier then apple, thus limiting their range of commercial importance. They are considered to be more sensitive than apple to high water tables, and require sites that have good water and air drainage, or must be grafted (e.g., plum) to rootstocks that allow them to tolerate anoxic conditions.

II. GROWTH AND DEVELOPMENT

In relation to many other fruit trees, *Prunus*, particularly cherry, peach, and nectarine, have a shorter life cycle. *Prunus* generally bear fruit at an early age, and the fruiting period lasts 5–35 years. Each stage of plant growth and development is strongly limited by the environment which can have a major effect on survival and development (Table 1).

Time to establish full canopy varies with species (peach is generally faster than cherry), and location. For example, peach canopy development in Michigan may require 4–6 years and only 3–4 years in California. Most stone fruit have the potential to fruit the year after planting. However, the primary objective during this period is to maximize vegetative growth to fill the allotted space in the orchard. For many crops (i.e., apple, pear), it is recognized that cropping is essential for growth control. In cherry, for example, cropping may not be desirable until trees are 6 or more years of age when the trunk diameter is large enough to allow for mechanical harvesting. The strategy during this phase of

Table 1 **Life cycle phases for *Prunus* and major environmental constraints during each phase**

Species	Propagation	Canopy development	Maturity cropping	Senescence
		Phase of development		
		Duration in years		
Peach	1–2	3–6	3–25	>25
Plum	1–2	5–10	5–35	>35
Apricot	1–2	3–6	3–35	>35
Cherry	1–2	5–10	5–35	>35
		Management strategy		
	Promote vegetative growth	**Promote vegetative growth**	**Reproductive growth**	**Environmental or biological stress**
		Major environmental factors		
	Nutrition H	Water M	Irradiance H	Temperature H
	Water H	Nutrition M	Temperature H	Water H
	Soil pests H	Pest control H	Water M	Pest H
	Foliage pests H	Site selection H	Nutrition M	
	Weed control H	Temperature H	Pest control M	

Note: H = high; M = moderate; L = low.

development is to fill canopy space and promote vegetative growth followed by reproductive bud development.

Cold temperatures during dormancy and in the spring are the most limiting factor for fruit trees grown in regions near the latitude range for peach and cherry. Nearer the equator the opposite is true and inadequate cold temperature for chilling is a problem. A plant architecture that maximizes light interception and minimizes shading is particularly important to conserve moisture during the fruiting stage unless irrigation is provided as in many arid regions of the world. Soil moisture during the growing season may also be a major limitation. When grown in the northern U.S., temperature is the main problem affecting tree longevity, while biological stress is more important in areas with milder temperatures.

A. ANNUAL LIFE CYCLE OF MATURE TREES

Several physiological, developmental, and growth processes are initiated at different times during the annual cycle for stone fruit. Periods of initiation and maximum activity are highly dependent upon the species and the cultivar, the location, crop load, and the environment. Since it is difficult to develop a common generic model that describes each cultivar at each location, this section will emphasize sour cherry in Michigan conditions as an example (Figure 1). Similar models could or have been developed for other *Prunus* spp.[7,8]

1. Vegetative Growth

Root growth development occurs before shoot development. A bimodal periodicity of root growth occurs for many fruits crops, characterized by a flush of growth in the spring prior to vegetative bud break and again in the fall when vegetative growth has stopped.[9] The common explanation for this observation is that the above and below ground parts of the tree are in competition with one another for resources, and that the above ground parts have priority. The few studies that have been conducted on the seasonal root growth periodicity in *Prunus* spp. do not support the observation that root growth stops during periods of high sink demand by the above ground part of the plant. Atkinson and Wilson[10] reported a single peak of growth for 'Merton Glory' sweet cherry that extended from May until mid-July or mid-August for F12/1 and Colt rootstocks, respectively, in southeast England. Beckman[11] found that *Prunus mahaleb* L. and *P. avium* L. cv. Mazzard rootstocks displayed a marked reduction in root

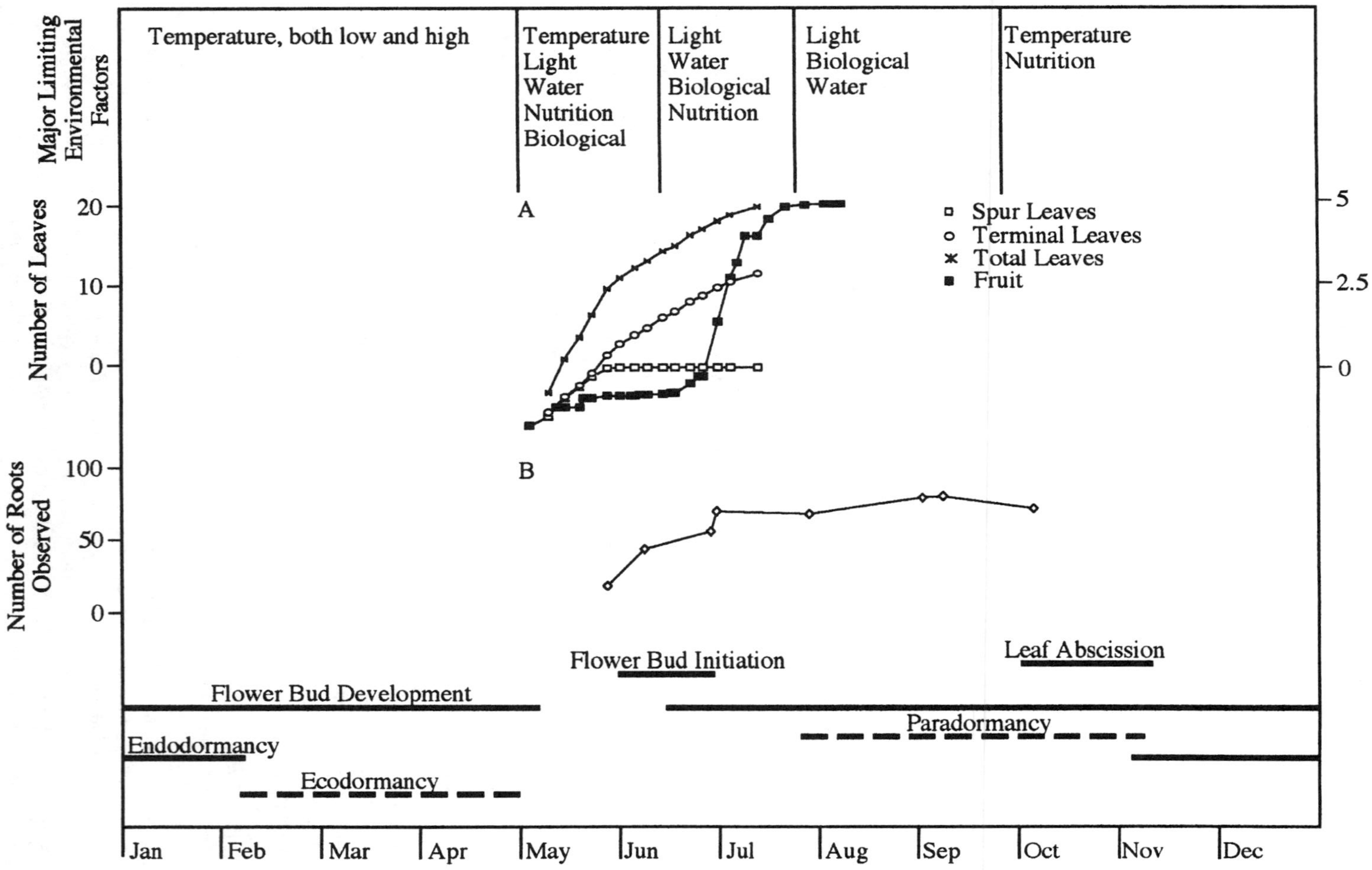

Figure 1 Vegetative and reproductive growth and the important environmental limitation during the annual growth cycle of mature 'Montmorency' sour cherry under Michigan conditions. (A, from Eisensmith et al.,[14] B, from Flore et al.[13])

growth potential during first leaf expansion but recovered later in the season. Williamson and Coston[12] demonstrated that maximum root growth occurred for peach during the summer, after the cessation of shoot growth but prior to abscission. Recently, using a minirhizotron camera to observe root growth of sweet cherry ('Emperor Francis' on Giessen 148/1 rootstock), Flore et al.[13] observed a single peak of root activity that extended throughout the summer period. Most of these studies have been conducted on young non-bearing trees, thus there is a need for further study to determine if there is preferential partitioning of carbon to roots or arboreal plant parts during different phases of growth, or in response to different types of environmental stress.

After sufficient chilling temperatures have occurred to release dormancy, the initiation of bud growth in the spring is a temperature dependent process. Accumulation of heat is generally calculated in terms of the number of growing degree days (GDD) or of growth degree hours (GDH). Eisensmith et al.[14,15] found that sour cherry leaf emergence was highly correlated with degree-day accumulation at a base of 4°C, and that shoot extension ceased about 350 (20–23 days) and 850 (55–65 days) degree-days after first leaf emergence for spur and terminal shoots, respectively. Leaf size increased linearly with degree-day accumulation until full expansion. This approach has also been used for peach.[16]

Vegetative and reproductive buds may have different chilling requirements or different temperature thresholds for growth. Vegetative and reproductive buds do not always break at the same time. At locations nearer the equator reproductive growth usually occurs first, followed by vegetative growth, but may coincide with vegetative development at distances farther from the equator.[17,18] Temperature is locally perceived by buds; and has been shown not to be affected by soil temperature. In an attempt to delay bud break, Hammond and Seeley[19] modified soil temperatures around roots of peach and cherry prior to anthesis, but there was no advance or delay of bud development from any of the treatments except when the soil was frozen.

Vegetative bud development depends upon the temperature to which each bud is exposed. Buds on apple[20] or cherry[21] can be locally affected by mist cooling, which may delay bud break up to 3 weeks. On the other hand, localized heat can accelerate bud break and growers are placing plastic bags over dormant buds to advance their development.[22] Terminal buds outside the bag were unaffected.

Foliage development of *Prunus* can be classified as either terminal shoots or as lateral shoots, and/or spurs. The proportion of each depends on the species, the cultivar, the degree of fruiting, and tree vigor. Peach and cherry have simple buds, either vegetative or reproductive. Flower buds are found in the lateral position on shoots and spurs; the proportion between shoots and spurs is variety and age dependent.[23] As vigor decreases the number of nodes that have flower buds increases and the number of spurs decreases.[24] A similar response occurs in peach under drought.[25]

Foliage and fruit development may be in competition for the same carbon resources early in the season[8,26,27] In sour cherry, vegetative growth is greatest during stage I and early stage II of fruit growth, and is completed by fruit harvest. In contrast to cherry, terminal growth on shoots and laterals of peach progress throughout the summer, often setting terminal bud during periods of environmental stress at or just before harvest, and resuming shortly after the stress has been eliminated. Once the terminal bud is set, shoot growth does not stop. Shoot and trunk growth of *Prunus* will continue to grow in girth until leaf fall. This initial phase of vegetative development is primarily driven by increasing temperature, but sustained growth is contingent upon internal factors (e.g., degree of competition and environmental factors such as soil moisture, vapor pressure deficit, light, temperature, and nutrition).

2. Reproductive Growth

The double-sigmoidal growth curve of stone fruit has been described as having three stages: stage I, the period of cell division; stage II, the slow growth phase that coincides with the period of pit hardening; and stage III, the period of cell enlargement.[6,28] It has been estimated that 50–80% of peach and cherry fruit growth may occur during stage III. More recently, DeJong and Goudriaan[8] have reevaluated the double-sigmoidal growth pattern for peach and have related the relative growth rates of the fruit to degree-days after flowering (Figure 2). Their results indicated that there were actually two physiologically distinct phases of sink activity instead of the three stages that are traditionally recognized, and that the traditional stage II of fruit growth is apparently a function of the timing of the shift between these two phases.

They also showed that each of these phases were linearly related to degree day accumulation using an upper and lower threshold of 35 and 7°C, respectively. The slopes of the relationships are similar between growth and degree days between varieties, but the transition points occurred at different points

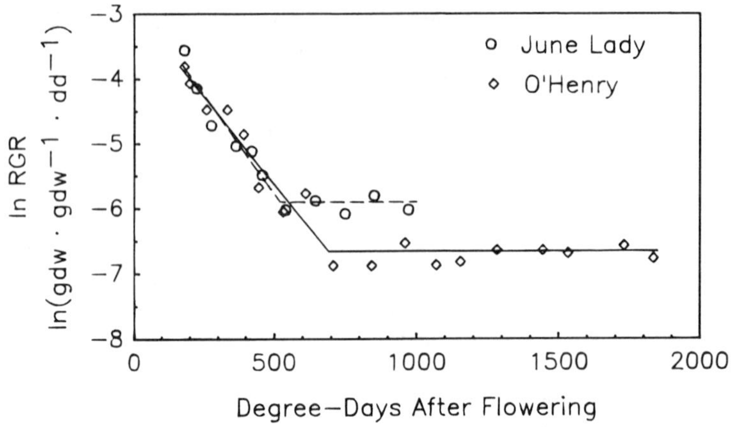

Figure 2 Relative fruit growth rates of 'Spring Lady' under California conditions. Reprinted with permission from DeJong, T. M. and Goudriaan, J., *J. Am. Soc. Hortic. Sci.*, 114, 800, 1989.

after flowering. Stages of fruit development for 'Montmorency' sour cherry can also be predicted by degree day accumulations.[7]

Fruit of sour cherry 'Montmorency' accumulated 164 mg (7.5 mg/day), 28.9 mg (2.4 mg/day), and 320 mg (15.2 mg/day) of carbohydrate during stages I, II, and III, for periods of 22, 12, and 21 days of growth, respectively.[26,27] Haun and Coston[29] predicted daily growth and development of peach leaves and fruit to different environmental factors. They found that maximum and minimum temperature, soil moisture, and age were among the most important variables affecting leaf and fruit growth.

Under Michigan conditions flower bud initiation for sour cherry and peach occurs from 3–4 weeks after bloom through terminal bud set and does not seem to be influenced by photoperiod.[1,30] For other stone fruit in locations around the world, flower bud initiation can range from 3–4 weeks after bloom until just after terminal bud set, or fruit harvest.[1] Location, cultivar, and cultural practices can have a profound effect on the time of initiation within a given year. The initial change from vegetative to the reproductive stage in sour cherry under Michigan conditions is evident 4 weeks after anthesis, sepal primordia form by July 30 and petal primordia are evident by August 15.[30]

3. Dormancy

Prunus vegetative and reproductive buds go through three distinct stages of dormancy, and Lang[31] has defined those stages as follows. During paradormancy, buds are dormant from the inhibitory influences of other plant parts. Endodormancy is related to factors inside the affected structure, and during ecodormancy, the dormant structure has the capability to resume growth once favorable environmental conditions are met. Endodormancy cannot be broken until a certain chilling requirement has been met. During dormancy, chilling is an important limiting factor at latitudes near the equator. At distances farther from the equator, cold temperature injury to the buds and tree is more important (see next section for a discussion of cold hardiness).

4. Storage Carbohydrates

Storage carbohydrates are necessary to sustain growth under periods of stress, during dormancy, and are important during initiation of growth in the spring. Nonstructural carbohydrates of stone fruit change quantitatively and qualitatively during the season.[17,32] Nonstructural carbohydrates in sweet cherry consist mainly of starch, sorbitol, sucrose, fructose, glucose, and raffinose. Keller and Loescher[17] reported that total nonstructural carbohydrates (TNC) were the highest at leaf abscission. Shortly before budbreak, TNC decreased in all perennial tissues except the spurs where they increased. TNC then increased slowly once provided with current year's assimilate until fruit harvest, after which they accumulated at a higher rate. Starch was the most common storage material, sucrose the most predominant soluble carbohydrate during dormancy and sorbitol the most common soluble carbohydrate during spring bud break. Keller and Loescher[17] speculated that if flower buds developed earlier than vegetative buds as they do in the northwest U.S., then flower development must rely upon TNC reserves until enough photosynthetic leaf area was present to provide sufficient photosynthate.

Kappes[27] found that for sour cherry, net export from a terminal shoot did not occur until 17 days after leaf emergence when the shoot was approximately one quarter (27%) of total dry matter accumulation. Under simulated conditions, Johnson and Lakso[33] found that spurs and shoots on apple did not begin to export carbon to the rest of the plant until 15–25 days after budbreak, and that short shoots began earlier than long shoots.

The amount of storage TNC is dependent upon the duration of the foliage in the fall between harvest and leaf fall. Respiration rate during the winter months has also been negatively related to yield the next season.[34] Both of these processes are temperature dependent. In the spring the plant must rely on carbohydrate storage until it becomes a net assimilator. During this period there are considerable demands on storage carbohydrate from roots, leaves, and developing fruits. Therefore environmental conditions at and immediately after bud break are important in determining current carbohydrate supply. Light and temperature are the most critical. During stage II of fruit development, demand for carbohydrate by the fruit decreases[8,27] and most of the carbon produced is partitioned into vegetative growth. Environmental constraints on carbon assimilation and supply could be limiting at the end of stage II and the beginning of stage III of fruit growth if sink activity is greater than the daily supply of carbon. (See Section VI.C.1 for a discussion on deficit irrigation.) Moreover, light, temperature, and evaporative demand are at their peak this period. Demands for carbohydrate at this time include: fruit enlargement and flower bud development; if carbon is limiting, it could affect flower bud initiation or development.

After harvest, carbohydrate demands are reduced considerably especially if new vegetative growth has ceased. Biological stress, such as leaf damage due to disease or insects, remain important because damage to leaves, or loss of leaves, can reduce the amount of storage carbohydrate available for next season's growth. There is a positive relationship for the length of time between harvest and leaf abscission and fruit production the following year.

III. PHOTOSYNTHESIS

A. INTRODUCTION

The process of photosynthesis is fundamental to plant productivity, and any factor or stress that limits this process may potentially reduce yield. However, only under certain circumstances does photosynthetic rate limit yield.[3,35] It was proposed[35] that photosynthetic potential in fruit crops is under two forms of control: (1) environmental, which influences physical and biochemical reactions directly and/or by affecting the development of the leaf and the manufacture of photosynthetic machinery available to convert CO_2 to carbohydrate; and (2) through sink demand which controls photosynthetic rate through some type of feedback signal from the sink itself. (See Reference 3 for an in-depth discussion.)

Scurlock et al.[36] presented an equation with four factors that define the quantitative relationship between photosynthesis and plant productivity, where biomass gain (Pn) is affected by the quantity of incident light (Q), the proportion of the light that is intercepted by the plant (β), the efficiency of photosynthetic conversion of the intercepted light into biomass (ϵ), minus respiratory losses of biomass (R).

$$Pn = Q \cdot \beta \cdot \epsilon - R \tag{1}$$

The quantity of the incident light is determined by the location, day of the year (day length), and length of the season, and cannot be easily manipulated unless the plants are grown in a controlled environment facility supplemented with artificial light. The efficiency of light interception β is directly related to the development, magnitude, and duration of foliage, as well as the light absorption characteristics of the individual leaves. Light absorption characteristics of individual leaves can be optimized by changing canopy structure, tree architecture, and planting system, and likely has a greater effect on biomass productivity than any other factor.

Efficiency of energy conversion (ϵ) refers directly to the photosynthetic process, and is most usually reported as a rate of CO_2 fixed per unit area, per unit time. Flore and Lakso[3] have suggested units for expression that shall be used throughout the text (Table 2).

The ϵ is fairly constant and not of great concern unless the leaf is below light saturation, which may occur either in the interior of the canopy or under environmental stress.

Carbon loss due to dark respiration: can range from 10–20% of maximum net photosynthesis; is temperature dependent, has Q_{10} that range from 1.8–2.1;[37] and can account for a substantial loss in

Table 2 **Suggested items, units, and abbreviations for reporting gas exchange parameters**

Term	Abbreviation	Units
Net carbon dioxide exchange (i.e., photosynthesis or assimilation rate)	A	μmol $m^{-2}s^{-1}$
Gross carbon exchange	A_g	μmol $m^{-2}s^{-2}$
Long term dry weight (24 h; net assimilation rate or crop growth)	C	kg dry matter m^{-2} day^{-1}
Transpiration (water loss by the leaf)	E	mmol $m^{-2}s^{-1}$
Conductance		
H_2O	g	mmol $m^{-2}s^{-1}$
CO_2	g′	mmol $m^{-2}s^{-1}$
Leaf	g_l	mmol $m^{-2}s^{-1}$
Stomata	g_s	mmol $m^{-2}s^{-1}$
Temperature		
Air	T_a	C
Leaf or other plant surface	T_l	C
Partial pressure of water vapor	e	kPa
Vapor pressure gradient (leaf to air)	VPG	kPa
Leaf area (single surface for photosynthesis)	L	m^2
Respiration		
Dark	R_d	μmol $m^{-2}s^{-1}$
Photorespiration	R_l	μmol $m^{-2}s^{-1}$
Concentration of CO_2		
Air	C_a	μmol mol^{-1}
Internal	C_i	μmol mol^{-1}
Photosynthetic photon flux (incident)	PPF	μmol $m^{-2}s^{-1}$
Quantum flux of photons (absorbed)	Q	μmol $m^{-2}s^{-1}$

Reprinted from Flore, J. A. and Lakso, A. N., *Hortic. Rev.*, 11, 111, 1989. With permission.

carbon. Other than leaves and fruit, few studies have characterized respiration on woody plants (see Chapter 2 for references) or on the effect of temperature on respiration rates of woody plants. It has been estimated that 40–60% of gross photosynthesis of cool temperate forests is lost due to respiration.[38] Grossman and DeJong[39] have reported Q_{10} of approximately two for leaves and stems of peach between 20° and 30°C, with lower values for leaves above 30°C. They reported that the specific respiration rate at 20°C of leaves and stems declined rapidly from the spring to a constant rate by summer, and that although trunk biomass accounted for 80–90% of above-ground vegetative biomass, trunk respiration accounted for only 8–12% of above-ground diurnal respiration. They speculated that the large demand for carbohydrate imposed by the respiratory need of vegetative and reproductive organs, rather than a reduction in photosynthesis supply lead to source limited fruit growth.

It has been suggested that reducing dark respiration will increase biomass accumulation. Small adjustments in rate or Q_{10} probably would have little effect on carbon balance over the year, however, length and temperature of the night period and/or the dormant period would likely have a greater influence.

Re-examination of Equation 1 then would lead us to conclude that for *Prunus* the greatest gains in productivity would be achieved by increasing light interception, while optimizing environmental factors that affect the development of photosynthetic machinery rather than affecting the photosynthetic process directly. Current environmental and preconditioning effects of the environment will be discussed in the following sections.

B. PHOTOSYNTHETIC CHARACTERISTICS OF STONE FRUIT

The net photosynthetic (A) characteristics of *Prunus* have been described in several studies (Table 3) and are typical of C_3 plants. It is interesting to note that four of the five *Prunus* species, except apricot, have similar values. Maximum values of A (A_{max}) ranged from a low of 13.3 μmol CO_2 m^{-2} s^{-1} in peach to a high of 21.5 μmol CO_2 m^{-2} s^{-1} for sour cherry.[3]

DeJong[40] found that A, expressed on a leaf area basis, mesophyll conductance (g_m), and leaf conductance (g_l) to water vapor, were all linearly related to leaf nitrogen content. Leaf intercellular CO_2 (C_i)

Table 3 **Reported and calculated gas exchange characteristics for peach, sweet cherry, sour cherry, plum, and apricot**

Parameter	Peach	Sour cherry	Sweet cherry	Plum	Apricot
Maximum A at 350 μl $l^{-1}CO_2$ (μmol CO_2 m^{-2} s^{-1})	13.3 ± 3.8	21.5 ± 6.2	17.9 ± 5.3	20.6	7.0 ± 0.1
Carboxylation efficiency (mol CO_2 m^{-2} s^{-1})	0.1–0.08	0.1–0.09	0.1–0.08	0.1–0.08	—
Apparent quantum yield (μmol CO_2 fixed/mol PPF)	0.04–0.03	0.06–0.05	0.05	0.04–0.06	0.02–0.04
Light compensation point (μmol quanta s^{-1} m^{-2})	20–30	15–30	25–30	15–30	25–40
CO_2 compensation point (μl l^{-1} CO_2)	66	40–60	56	61	69
Net carbon exchange per unit of N (A_N nmol CO_2 mg N^{-1} s^{-1})	6.39	—	6.75	6.17	3.76

Data from References 3 and 40.

concentrations decrease slightly with increasing A rate and leaf N content, indicating that A rate was not being restricted by low C_i or g_l with lower assimilation capacity. Maximum A per unit of N was 6–7 nmol CO_2 mg N^{-1} s^{-1}. DeJong[40] and DeJong and Doyle[41] argue that photosynthetic rates expressed in terms of CO_2 fixed per unit of leaf N are just as valid as those expressed on a leaf area basis.

C. SPECIES AND CULTIVAR DIFFERENCES

There is little genetic variation of A in *Prunus* reviewed in this chapter. Among the deciduous temperate fruits, A_{max} of sour cherry and sweet cherry rank near the highest, followed by peach which is similar to apple. DeJong[40] reported that the mean A_{max} rates of peach and cherry, as well as other *Prunus* except apricot, were very similar. Early reports indicated that peach had low photosynthetic rates. However, recently reported values were comparable to or higher than apple (18.0 vs. 15.7 μmol CO_2 m^{-2} s^{-1}). Flore and Sams[42] reported differences between sweet and sour cherry and between different clones of sour cherry, and noted that spur type selections often had higher rates of A than the standard variety 'Montmorency'. Differences of this magnitude have also been shown for apple by Looney.[43] Respiration (R) could influence the overall carbon balance of the plant, especially if temperatures during the dormant period are high. Genetic variation for R exists in other plants but we know of no studies of this type for *Prunus*. It is a question that needs resolution. In comparison to other factors, genetic variation in ϵ and R likely has little practical effect on fruit production in *Prunus*.

D. LEAF AGE

Photosynthesis increases with leaf expansion and reaches a peak just before full development, then remains steady for a period of time before declining.[3,44] Roper and Kennedy[45] found that chlorophyll, stomatal conductance (g_s), and quantum efficiency increased in sweet cherry as leaves expanded to a leaf plastachron index of 10, then remained constant until senescence. Andersen and Brodbeck[46] noted that g_l to water vapor approached maximum values in peach before A_{max} was reached. Net CO_2 assimilation and water use efficiency were greatest for recently expanded leaves and gradually declined with leaf age after full expansion. Leaf water potential (Ψ_l) was similar for all leaf ages under field conditions. Leaf dry weight and chlorophyll per leaf increased with leaf age after expansion. Diurnal patterns of g_l, transpiration (E), and Ψ_l were similar for expanded spring- and summer-flush leaves. Minimum Ψ_l during midday (i.e., −2.4 MPa) were not associated with reduced g_l.

Normal decline in A after full leaf expansion can be ameliorated by the presence of fruit,[47–49] by decapitation or debudding,[50] by defoliation in sour cherry,[51] or by summer pruning.[52,53]

E. SEASONAL INFLUENCE

Several whole season studies have been conducted on stone fruit. Usually single leaf or branch measurements are made at predetermined intervals, then are used to calculate an estimated seasonal profile.

Crop load and seasonal differences can affect this pattern. Sams and Flore[44] showed two very different patterns of A for sour cherry on the same trees in 1978 and 1979. They concluded that the differences were due in part to different crop loads (one heavy and the other light), and rate of leaf emergence and canopy development due to very warm vs. cool temperatures during the spring. There was a strong positive correlation between A and leaf emergence rate when vegetative growth was rapid under warm conditions, but not when conditions were cool which resulted in slow leaf emergence rates. Chalmers et al.[54] found that daily A rate for peach was closely related to demand, time of year, and location in the canopy.

F. SINK ACTIVITY

Increased sink activity, either reproductive or vegetative or reduction in source supply, has been associated with an increase in rate of photosynthesis (see review by Flore and Lakso[3]). Direct positive effects of fruit sink strength for stone fruit have been shown for peach[55–57] and sour cherry,[48] for defruited cherry and plum,[49,58] and for defoliation.[51,59] DeJong[56] found that A and g_l were 11–15 and 30% higher, respectively, for fruiting compared to defruited trees. Minor changes in g_m and leaf nitrogen content were observed. It was concluded that the fruit effect on A was primarily related to stomatal behavior. The results of this study clearly indicate that the presence of fruit influenced stomatal regulation relative to plant water status in mature peach trees. Fruited trees maintained higher g_l for a given Ψ_l than in the de-blossomed trees. These data indicate that fruit may have specific influences on stomatal regulation in bearing trees rather than a more generalized "feedback inhibition" involving CO_2 metabolism. Gucci et al.[49] investigated a possible feedback regulation of A by starch accumulation in defruited field grown sweet cherry. Defruiting decreased A by 43% and increased leaf starch content by 59% within 24 hours. Similar results were obtained with sour cherry[48] and by calculation from A/C_i curves, the stomatal contribution to the decrease of A from defruiting ranged from 32–40% in a controlled environment and from 25–45% under field conditions. There was a larger difference in A in the afternoon compared to the morning for both plum and sour cherry[48,49] (Figure 3). The large non-stomatal contribution to the

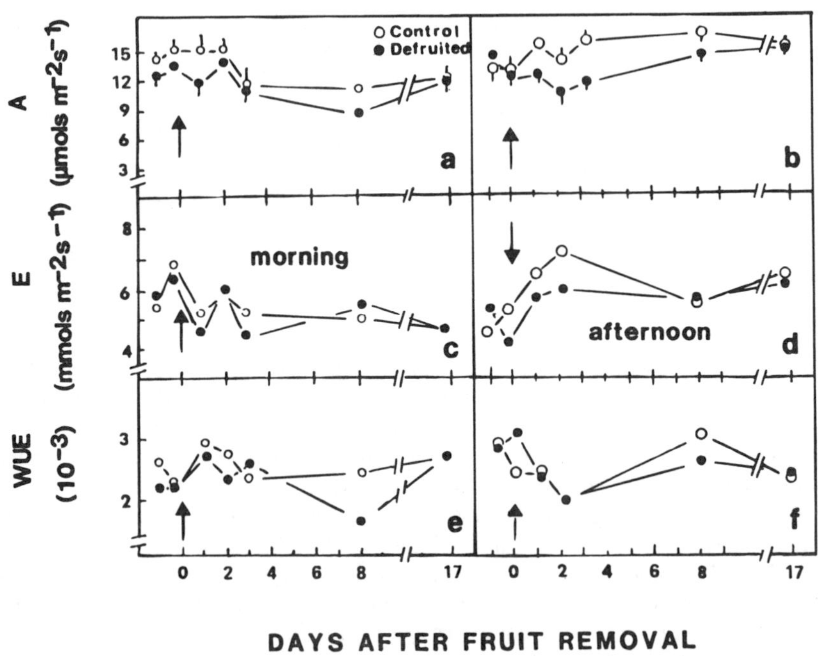

Figure 3 Changes of leaf photosynthetic rate (A), transpiration (E) and water use efficiency (WUE) of fruiting and defruited plum trees during stage II of fruit development in the morning and in the afternoon in 1987. Symbols are means ± SE of 18–24 replicates (SE bars not shown if smaller than the symbol). Arrows indicate the date of fruit removal (16 June). Reprinted with permission, Gucci, R., Xiloyannis, C., and Flore, J. A., *Physiol. Plant*, 83, 497, 1991.

decrease of A after harvest and the changes in starch levels supported the hypothesis that defruiting results in end-product inhibition of photosynthesis.

Layne[59] altered assimilate supply in sour cherry trees by whole-plant partial defoliation or by continuous illumination.[60] Partial defoliation resulted in a photosynthetic enhancement within one day that was attributed to higher g_l, higher estimated photochemical and carboxylation efficiencies and higher ribulose-1,5-bisphosphate (RuBP) regeneration rates. Stomatal limitation to A increased from 20–35% within 24 hours, but fell back to predefoliation levels by day 3. The increase in A was associated with reduced carbon partitioning to starch and increased partitioning to sucrose and sorbitol. Photosynthetic enhancement was maintained over a 32 day period and senescence was delayed. Continuous illumination resulted in an inhibition of A within 24 hours that was attributed to lower g_l, lower estimated photochemical and carboxylation efficiencies and lower RuBP regeneration rates. Another study[51] demonstrated that photosynthetic compensation occurred in response to a partial reduction in leaf area in sour cherry. The threshold level of leaf area removal before A was significantly reduced by 20% defoliation. Carboxylation efficiency increased at 10 or 20% defoliation, and there was no significant increase in g_l or change in stomatal limitation to A. These studies are consistent with a type of feedback inhibition controlling A.

Maximum potential for efficiency of conversion can be affected by the environment, either by affecting the process directly as is the case for temperature, CO_2 concentration, light, and humidity, or longer term by affecting the development of the leaf and the amount of photosynthetic machinery present. Since A is influenced by many factors, such as season, age of tissue, fruit load, and environment, it seems dangerous to try to draw any conclusions concerning overall efficiency of conversion of energy without extensive studies.[42]

IV. IRRADIANCE

The importance of light to stone fruit crop productivity has been the subject of several recent studies. Irradiance influences the energy balance of the plant, the photosynthetic capacity of the leaves and whole tree, and may influence responses that are either photomorphogenic (non-time-dependent) or photoperiodic (time-dependent). The amount of irradiation received by a plant is dependent upon the intensity and the duration of photosynthetic active radiation (PAR). The factors are predetermined by latitude, elevation, and slope of the site. In addition, light interception and shading within and between trees depends upon the degree of foliage development, tree shape and size, foliage density, and the geometry of the orchard. For many crops light interception has been positively related to dry matter production and yield.[61] Few studies have been conducted on stone fruit but in a preliminary report, Flore and Layne[62] found a linear relationship between yield and light interception for 8-year-old 'Montmorency' sour cherries when fruit set was high, with a strong tree shape interaction ($r = 0.81$ for triangular shaped trees, and $r = 0.4$ for rectangular shaped trees). Additionally, earlier high yields in terms of tree age have been found to be associated with high tree density, the implication being that light interception is maximized early in the development of the orchard.

A. PHOTOSYNTHETIC RESPONSE

Light is the main driving force for photosynthesis, while temperature, water, CO_2, and humidity play regulatory roles. Typical asymptotic photosynthetic response curves to increasing photosynthetic photon flux (PPF) have been reported for stone fruit with saturation occurring between 400–800 μmol m^{-2} s^{-1}.[55,63–65] Apparent quantum efficiencies (Q) vary from 0.04–0.06 μmol CO_2 fixed/mol PPF (Table 3) for sun grown leaves. There seems to be little genetic variation for this trait when trees are grown under similar nonstress conditions. DeJong[40] compared photosynthetic responses in light for five *Prunus* spp., and although not reported, it appeared from his data that Q were about the same among species except apricot which was considerably lower. Light compensation points were similar (15–30 μmol m^{-2} s^{-1}), however saturation points differed, and ranged from 20% for sweet cherry to 40–50% for peach and other species (Figure 4). Sams and Flore,[63] and Kappel and Flore[64] have reported similar values for 'Montmorency' sour cherry, and 'Redhaven' peach, respectively. It is interesting to note that a small change on the lower part of the curve can result in a major change in A, while above saturation changes in PPF have little effect. Light response curves for whole canopies or branches often do not show a point of saturation (Figure 5). We did not find a point of saturation when using a whole plant chamber on mature sour cherry grown in a hedge row situation. Lack of saturation in a hedge row is not surprising

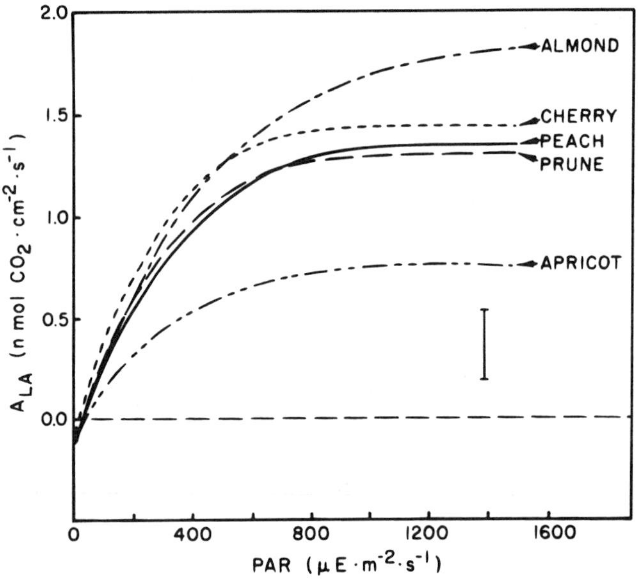

Figure 4 Relationship between net CO_2 assimilation (A_{LA}) and level of photosynthetically active radiation (PAR) for 5 *Prunus* species. Each curve represents the mean response of 3 individual leaves for a given species. Reprinted with permission, DeJong, T. M., *J. Am. Soc. Hortic. Sci.*, 108, 303, 1983.

because of inter- and intra-tree shading. Leaves in the interior receive irradiance levels less than saturation, and may be exposed to levels near the compensation point.

Photoinhibition[66] may occur after several days of cloudy cool conditions followed by high light. In a preliminary experiment on sweet cherry grown under high light, then transferred to the dark at 5°C for 4 hours, followed by normal temperatures (20°C) for the rest of the dark period (10-hours total) and high light intensity, A dropped significantly for a period of several days,[67] and was associated with damage to PSII. This situation may occur under field conditions in the spring when temperatures are low, growth has started, and is followed by clear sunny days. This needs further study.

B. ADAPTATION TO SHADE

Adaptation to shade or sun during or after leaf development is well documented for several species (see Flore and Lakso[3] for review). In general leaves developed under shade are thinner, less dense, have fewer layers of palisade cells, more intracellular spaces, have lower A_{max}, lower compensation

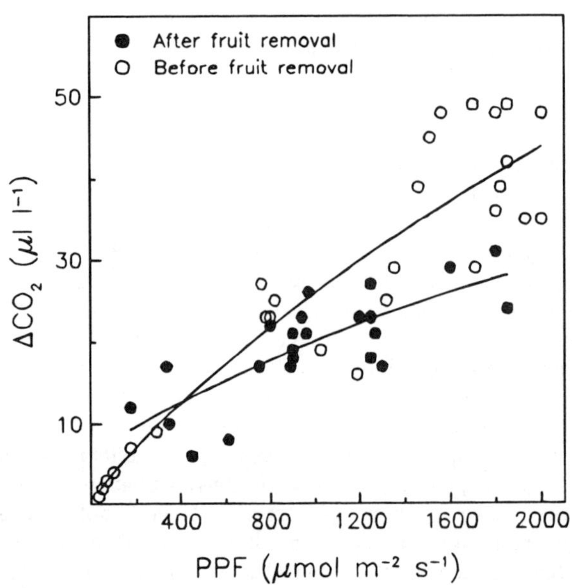

Figure 5 Whole cherry tree photosynthetic response to increasing PAR grown in a hedge row, before and after fruit harvest. From Flore, J. A., unpublished data, 1993.

Table 4 **Available percent of full sunlight in cherry tree canopies after full foliage development at different locations within the tree, under sunny conditions**

Distance from apex of canopy (m)	Available light (% full sun) distance from tree center (m)						
	1.5	1.0	0.5	0.0	0.5	1.0	1.5
	North						**South**
1.0	36	10	5	5	6	6	58
2.0	39	5	4	4	4	6	32
3.0	4	2	3	4	3	5	5
	West						**East**
1.0	35	18	15	5	10	10	43
2.0	21	7	8	4	4	6	14
3.0	13	5	3	4	3	3	8

Note: Data are the means of three tree replications, determined between 1200 and 1400 hrs, August 25, 1978, standard 'Montmorency' 11-year-old, planted 24′ × 20′. Reprinted from Flore, J. A., *Annu. Rep. Mich. St. Hortic. Soc.*, 111, 161, 1981. With permission.

points, and lower light saturation points. For fully expanded leaves, A rate was correlated to previous exposure, and shade induced loss of A was not readily reversible by re-exposure of the leaves through summer pruning or canopy manipulation. Peach[64] and sour cherry[63,68] responded in a similar manner in that A_{max}, light saturation point and light compensation points were reduced with no change in quantum efficiency when leaves developed in shade. Leaf thickness, leaf density, and chlorophyll a/b were decreased while leaf area and chlorophyll (on a dry weight basis) increased. In both species, leaves continued to thicken after full expansion, and as a consequence, shading had a similar but greater effect on leaves that were in the bud stage at the time of shading. Shading also increased shoot length and decreased shoot diameter.

C. LIGHT DISTRIBUTION AND LEVELS IN TREE CANOPIES

Light interception and distribution depend on the leaf area index (LAI) or density of the foliage, the spacial distribution of leaves, and the shape and color of individual leaves. The penetration of light into a canopy decreases logarithmically with LAI from the top of the canopy to the bottom, and can be defined using Beers law with the following equation reported by Jackson,[69]

$$I_L/I_O = e^{-kL} \tag{2}$$

where I_O = incident light above the canopy, I_L = light under the canopy, L = leaf area index, and k = the extinction coefficient for visible radiation. Reported k values range from 0.33 to 0.77 for apples, with means of 0.43,[70] 0.56,[71] and 0.60.[72] Recalculating the data in Kappel et al.,[73] for different training systems of peach, k ranged from 0.27 to 0.34 for peach. For sour cherry we have calculated a k of 0.33 for non-hedged trees (LAI of 6.2).[74]

The characteristic growth habit of peach and sour cherry may result in a mantle of dense foliage on the perimeter of the tree. This is more pronounced on sour cherry than peach, depends on the variety, the degree of spurs vs. long shoots, and is enhanced by summer hedging or topping. We have characterized light distribution in four different types of sour cherry orchard designs of different spacing, different tree shape, plus and minus summer hedging.[62,75–77] Regardless of system, light levels decreased rapidly from the exterior to the interior of the tree (see Table 4). Summer hedging increased the density of the foliage on the perimeter of the tree and caused a greater decrease in available light 0.5 m from the exterior of the canopy (Table 5). Triangle-shaped trees had higher light levels in the top of the tree (Table 6). Regardless of system, the available percent of full sunlight was often less than 20%. Leaf area index of non-hedged control trees and summer hedged trees for a period of 3 years were 6.2 and 7.5, respectively. It was concluded that: (1) foliage, and therefore shade, was fully developed within 60–65 days from bloom (see Section II.A.1, on foliage development); (2) triangular shaped trees in a N/S orientation had better light distribution in the top and center of the canopy when compared with standard non-hedged trees; and, (3) repeated hedging increased leaf density on the exterior of the tree

Table 5 **Influence of summer hedging on light intensity in hedge row 'Montmorency' cherry tree canopies**

		Sunlight (% sky)		
		Date		
Pruning[a] system	**Canopy position**	**5/5**	**7/1[b]**	**8/29**
Control	N	52	69	52
	W	82	60	6
	S	50	56	4
	E	41	52	67
	3m	89	94	46
	2m	63	78	12
	1m	61	47	9
P-1[c]	N	43	22	4
	W	82	60	6
	S	50	39	27
	E	41	52	67
	3m	68	71	10
	2m	64	29	5
	1m	14	59	5
P-3[d]	N	84	16	7
	W	68	14	5
	S	69	18	6
	E	27	15	8
	3m	55	37	3
	2m	57	24	9
	1m	55	20	3

[a]Trees summer-pruned three 3 weeks prior to harvest. [b]Readings taken under cloudy conditions (320 μE). [c]Summer pruned in 1976, 1977. [d]Summer pruned in 1974, 1975, 1976, 1977.
Reprinted from Flore, J. A., *Annu. Rep. Mich. St. Hort. Soc.*, 111, 161, 1981. With permission.

causing increased internal shading. Since currently there are no commercial plantings of sour cherry on dwarfing rootstocks, summer hedging is the only practical way to control tree size in a dense planting. Therefore it is necessary to supplement repeated hedging with dormant pruning to open "windows" of light into the interior of the tree.[77]

Kappel et al.[73] studied light levels in four different peach hedgerow canopies: oblique fan (OF), canted oblique fan (COF), modified central leader (MCL), and open center (OC). They found little

Table 6 **Available percent of full sunlight in tapered cherry tree canopies after full foliage development at different locations within the tree, under sunny conditions**

Distance from apex of canopy (m)	Available light (% full sun) distance from tree center (m)				
	1.0	**0.5**	**0.0**	**0.5**	**1.0**
	West				**East**
1.00		40	19	76	
2.00		44	5	12	
3.00	17	4	10	8	10
	North				**South**
1.00	70	83	19	100	97
2.00	47	5	5	50	53
3.00	8	6		15	4

Note: Data were determined between 1200 and 1400 hrs, September 11, 1980, 'Montmorency' spaced 20′ × 12′.
Reprinted from Flore, J. A., *Annu. Rep. Mich. St. Hortic. Soc.*, 111, 161, 1981. With permission.

difference in light level in the canopy due to canopy training system except at 1 m above the ground. Hedging increased light in the interior of the canopy, but this advantage was lost by the end of the season. Mean LAI for these canopies were 4.57 for OF and 2.92, 3.30, and 2.96, respectively, for COF, MCL, and OC. Values below 30% full sun were rare, even in the center of the tree. As canopy depth increased, absorption of radiation of all wavelengths decreased, but not equally. Light wavelengths of 400 nm penetrated peach canopies the least. Thus interior leaves were exposed to a higher ratio of near infrared to red light than occurs in direct sunlight. We have found similar results for cherry,[74] and these results agree with those reported for apple.[79]

Light distribution and levels in peach canopies seem then to be highly dependent on tree vigor, crop load, duration of season, degree of pruning, and orchard system. When peach was grown in a non-hedgerow system, under a traditional open center training system, light levels were greatest at the tree perimeter, lowest midway between the periphery and the trunk, and intermediate at the center.[80] Light levels were very low at the tree center in mid-July, prior to pruning and topping, regardless of treatment (4–15% full sun). Summer pruning increased light penetration in the center of both 'Loring' and 'Cresthaven' trees while summer topping only improved it in 'Cresthaven'. The authors concluded that this was probably because topping stimulated lateral shoot growth in 'Cresthaven' more than in 'Loring'. This indicates that there may be strong cultivar differences in response to summer pruning and topping, and that regrowth is likely to be highly dependent upon timing. Also climate, variety, and vigor can affect light distribution.

D. MORPHOLOGICAL RESPONSE

Leaf adaptation to shade has been reviewed. In general, as shade increases, vegetative morphology of the tree changes to facilitate light harvesting. In shade, leaves are larger, flatter, thinner, and have more chlorophyll per unit weight or area. In addition, shoots are longer and thinner, leaf area and the number of lateral shoots and internodes are shorter, and whole tree dry weight is decreased.[65,68,73,75,77]

E. REPRODUCTIVE RESPONSE

Shade affects reproductive development of the current crop for peach and cherry (reduced color, reduced soluble solids, smaller fruit, delayed maturity, delayed abscission, reduced fruit set) and affects flower bud formation and fruit set the following year.[65,68,73,75,76,81,82] Timing of shade can influence the response. Shading had a profound effect on fruit maturation in sour cherry[62,76] as indicated by color and fruit retention force of the fruit. Fruit in the interior of the tree were often greener and harder to remove than those on the exterior. The degree of response depends on variety and on crop load (leaf to fruit ratio). Similar results have been reported for peach.[65,73,80,81,83] As load increases, shade delays maturity to a greater extent than for trees with a low crop. In a differential shading study, Marini et al.[84] found that time of shading was important for 'Biscoe' peach and that shading during the final swell of fruit development had the greatest effect on fruit weight and quality, while shading during stage II had little effect.

Light levels have also been associated with flower bud initiation in cherry. Through shading and correlations studies in the orchard, we have found that 15–20% of full sun is needed for flower bud formation in sour cherry.[68,75] Fruit set in large trees is adversely affected by levels of light below 20%.

Wood and bud hardiness are also influenced by shade. Cold hardiness of wood and buds during the acclimation period in the fall for 'Redhaven' and 'Montmorency' trees grown in pots was reduced as much as 9°C due to shading. Hardiness decreased as shading increased from 37% of full sun to 9%, and was associated with a decline in storage carbohydrate.[85] Shading whole peach trees also reduced hardiness during the acclimation period by 1–2°C.[86]

Light thresholds for optimum growth and production have been developed for apple (see review by Jackson[69]). Herein we present our estimate of threshold values for sour cherry.[75,76] Note that these are based on Michigan conditions and may need to be adjusted for areas that receive greater or lesser amounts of incident solar radiation or have longer or shorter growing seasons (Table 7).

V. TEMPERATURE

A. VEGETATIVE AND REPRODUCTIVE GROWTH

Temperature above freezing is the main regulator of metabolic and therefore growth processes in the plant. It is the driving force for bud break, and is closely correlated with vegetative and reproductive growth, as well as fruit maturity. Young and Werner[87] found that chilling temperatures of the roots were

Table 7 **Summary of the effect of shading on the morphology and fruiting of sour cherry**

Parameter	Response to shading	Light threshold[a] (% full sun)
Leaf size	+	36
Leaf thickness	−	36
Leaf chlorophyll (mg/gm)	+	36
Stem length	+	21
Stem diameter	−	21
Flower bud formation	−	20–25
Fruit set	−	10
Fruit retention force	+	36
Fruit color	−	36
Hardiness	−	21
Maximum photosynthetic rate		36

[a]Highest percent of sky value at which a significant effect was observed. Reprinted from Flore, J. A., *Annu. Rep. Mich. St. Hortic., Soc.*, 111, 161, 1981. With permission.

not required for the onset of growth for peach, although chilling the rootstock delayed budbreak. By contrast, chilling temperatures are required to terminate dormancy of shoot buds (followed by a period of warming). The reader is referred to several models that have been proposed to predict the chilling requirement necessary to end rest for peach and cherry for an in-depth discussion.[88–91] The range of chilling hours (total number of hours below 45°F) requirement for peach, sour and sweet cherry are 100–1100, 600–1500, and 500–1300, respectively.[2] Once endodormancy is completed, budbreak is regulated by temperature. The amount of heat required varies with the species, the physiological condition of the plant, the stage of development, and geographic location of the trees.

B. GAS EXCHANGE AND RESPONSE TO TEMPERATURE AND HUMIDITY

Prunus spp. exhibit a typical parabolic relationship between photosynthesis and temperature, with a flat response between 17° and 30°C.[63,92] Cherry had higher A at low temperatures than peach, while the converse is true at high temperatures. A true high temperature threshold has not been well established for two reasons. First, under field conditions plants can adapt to higher temperatures.[93–95] Lange et al.[95] demonstrated that under desert conditions, apricot exhibited characteristic seasonal shifts in temperature optimum for A. Second, unless vapor pressure of water is controlled in assimilation cuvettes, vapor pressure deficits (VPD) between the leaf and air increase with increased temperature. Vapor pressure deficits greater than 1.5 kPa have been shown to cause stomatal closure and decreased photosynthesis in sour cherry.[96,97] The decrease in A in response to temperature could in part be due to the direct effect of low humidity on the stomata, and in turn, affect A. A typical response to VPD when temperature is held constant was demonstrated for control and flooded cherry trees as shown in Figure 6.

With the advent of portable photosynthetic equipment, it is now possible to measure A in the field throughout the growing season. Often under field conditions, A of *Prunus* is still 50–70% of maximum when temperatures are between 35° and 38°C.[98,99] It is doubtful that an upper threshold for *Prunus* can be fully characterized without taking into consideration preconditioning. The genotypic potential for preconditioning and adaptation are not known, but warrants further study.

C. LOW TEMPERATURE RESPONSE

Parker[100] has indicated that cold temperature is the single most important factor limiting the range and distribution of plant species. Peach and cherry are no exception and seem especially vulnerable. Examples include the late November freeze of 1950 in the Midwest which killed most of the peach trees in the region, and the November 1955 and December 1964 freezes in the Pacific Northwest which killed hundreds of thousands of fruit trees. Fluctuation in yield due to frost are also common and are probably the single most important marketing problem for the sour cherry industry in Michigan.

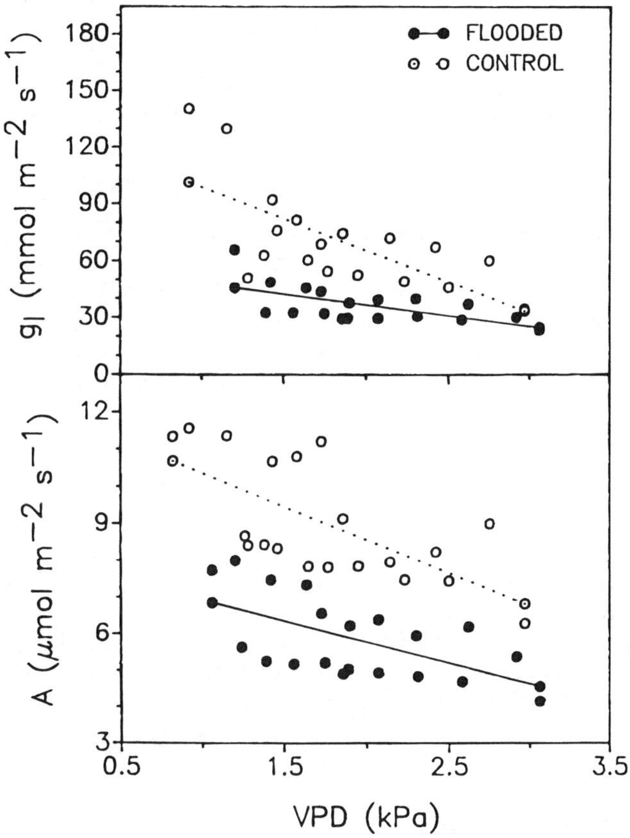

Figure 6 Stomatal conductance to water (g_l) and net CO_2 assimilation rate (A) as functions of vapor pressure deficit (VPD) in sour cherry trees after 5 days of flooding (all regressions significant at $p < 0.01$). Fitted lines—control: $g_l = 131.9 - 33.3 \cdot VPD$, $r^2 = 0.47$; flooded: $g_l = 59.4 - 11.4 \cdot VPD$, $r_2 = 0.37$; control: $A = 12.2 - 1.8 \cdot VPD$, $r^2 = 0.45$; flooded: $A = 8.1 - 1.1 \cdot VPD$, $r^2 = 0.35$. Reprinted with permission from Beckman, T. G., Perry, R. L., and Flore, J. A., *Hort Science*, 27, 1297, 1992.

D. TYPES OF COLD DAMAGE

Cold damage can be manifested in several ways: either as damage to buds, flowers or fruits, or as damage to the regenerative tissue of the tree itself. Tree damage has been termed crown injury, crotch injury, sunscald, bark splitting, southwest injury, or black heart. Several reviews have been written on the mechanisms of cold damage, we will summarize them briefly, but refer the reader to the following for an in-depth discussion.[101-106] It is generally thought that death to stone fruit trees occurs either by extracellular or intracellular freezing. In extracellular freezing, ice forms on the surface of cells or between the protoplast and the cell wall. Because of differences in osmotic potentials (Ψ_π), water moves through the plasma membrane to the ice crystal, causing dehydration, and if severe plasmolysis follows, this can cause injury and permanent damage. When intracellular ice forms within the cell, the cells are killed due to the destruction of membranes.

The ability to resist cold below the normal freezing point of water is called supercooling.[107] Ice does not form until a small volume of liquid crystallizes (see Ashworth[107] for a thorough discussion). Ashworth[107] has reviewed many factors which affect the ability of a plant tissue to supercool and form ice; among them are: specimen size, cooling rate, surface moisture, presence of ice nucleation active bacteria, intrinsic ice nucleating agents, tissue moisture content, tissue structure, and the cell wall. It is generally now accepted that ice crystallization is a prerequisite for injury, and is particularly important in plants which lack tolerance to ice formation.[107] For these plants, the temperature of ice formation is associated with lethal damage,[103] however, the temperature at which ice crystallization occurs is not constant, nor is it necessarily just at or below 0°C. The importance of supercooling and ice formation to survival of wood and stems seems to be different from that of buds. In many deciduous woody plants, plant death and ice nucleation occurred at the same temperature in living xylem tissues that supercooled.[107] It has been hypothesized that intracellular ice formation was responsible for cell death. However, when ice formed in overwintering buds, a single nucleation event resulted in freezing of an entire flower resulting in death. Tissue structure has been implicated as being responsible for differential ice formation and spread of ice within plant tissues. It is clear that there is segregation of ice when overwintering buds are frozen. Theories to explain the compartmentalization include: the presence of bud scales[108]; that the scales contain intrinsic ice nucleating agents[108]; tissue water potentials are different; and there are differences in bud morphology and vascular development.[109] Quamme and Gusta[108] presented compelling evidence that the supercooling of flower primordia within dormant peach buds is dependent on water migration from the base of the flower primordium to preferential sites of freezing in the flower bud scales and pith during the initial stages of freezing.

Trunk splitting is manifested by a vertical split in the tree trunk extending through the bark and cambium into the sapwood, and is a serious problem for many trees species in temperate regions. It has a serious debilitating or even lethal effect on the tree. This problem has been studied extensively, and results from measurements indicate 17–20°C differences of temperature within the trunk tissue on clear sunny days.[110] Norris et al.[110] concluded that this type of injury was mainly due to differences in thermal expansion, and noted significant differences among the bark, sapwood, and heartwood, as well as thermal coefficients of expansion in both tangential and radial directions.

Winter injury that causes death of the bark (cambium, phloem, or cortex) or sap wood (xylem parenchyma and pith) at or near ground level is less well explained, and is thought to be caused by freezing death of the tissues. Some trees may survive damage to the phloem and xylem during the winter, but seldom survive if the cambium around the tree is substantially damaged. Layne and Flore[111] showed that if the trunk was injured to the cambium in a band 10 cm in length, 75% around the circumference of the tree, it could survive but not if damage was more extensive. White paint applied to the trunk and scaffolds can minimize cold damage to stone fruit in the colder temperate region.

E. COLD ACCLIMATION, DEEP WINTER HARDINESS, AND DEACCLIMATION

The hardiness cycle for stone fruit is usually divided into 3 stages: acclimation, deep winter hardiness, and deacclimation. Depending on the location and the cultivar, each of these phases can have more or less importance in the survival of the buds and the tree. Acclimation or hardening coincides with cessation of growth in late summer or early fall.[106,112] Leaves provide substrate, receive signals, and promote hardiness. It has been postulated that a hardiness promoting factor is produced in the leaves which is responsible for the first phase in the acclimation process. This process is thought to be induced by photoperiod in some plants, is reversible, and the stimulus is translocatable.[113] Although short days have been implicated as the trigger, they are not effective during spring growth, and some plants will acclimate in the fall regardless of photoperiod.[106,113] Once induction has occurred, high temperature will not reverse it. The hardiness factor does not seem to be genotypically specific. In peach, the youngest leaves seem to be the most active in promoting hardiness development.[114]

A second phase of acclimation is usually induced by frost or low temperature, which causes the plant to go through a further hardening process. As temperature drops, plants become more hardy, the process is reversible and seems to be quantitative. There may be a third transition to deeper hardiness induced by very low temperature (−30° to −50°C) in some species.

There are differences among species in developing bud hardiness in the fall. Once hardening occurs it does not remain constant. Apple will deharden when exposed to warm temperatures and reharden when exposed to low temperatures. The degree of rehardening seems to be dependent upon the stage of dormancy, and the length of time that they are exposed to warm temperatures. As dormancy proceeds toward spring, the capacity to reharden decreases, and after a certain point they are permanently dehardened.[115] Andrews and Proebsting[116] found that sweet cherries developed the capacity to deep supercool 1–2 weeks earlier than peach. Increase and decrease in hardening follows temperature fluctuations. Cherry lost 6.1°C in hardiness when exposed to 24°C for 4 hours, while cherry and peach reharden at a rate of 1.9°C, and 2.2°C per day, respectively, if temperatures are below −1.1°C.[117] The dehardening-rehardening cycle is not completely reversible, and dehardening occurs faster than hardening. Peach loses the capacity to reharden after rest is complete and some after bud development has occurred.[112]

Synchronization of hardiness does not occur equally in all parts of the plant in the fall. It coincides with cessation of growth and begins in the periphery of the tree, first and last in the trunk.[112,118] In Michigan it has been reported[119] that the apical part of peach shoots are hardier than the base early in the acclimation period, however, this reverses later in the season. Not only do different organs and tissues acclimate at different rates, but their overall sensitivity to cold is different. In general, roots are more sensitive than shoots, fruit buds are more sensitive than shoots, and shoots are the most hardy. The cambium is generally the most sensitive tissue in the shoot.

Dehardening is strictly related to temperature. As buds develop, they are more vulnerable to frost, which can be identified by stage of phenological development. Killing temperatures (LD_{50}) for 'Montmorency' sour cherry buds rise rapidly as buds swell in the spring, reaching $-5°C$ just before separation of bud scales and $-3°C$ at full bloom.[120] They seem to lose hardiness in the water stage, then regain some protection in the next stage when air insulates the pistle from the rest of the flower. Similar responses are found for the other stone fruit as they develop.

F. SPECIES, VARIETY, AND ROOTSTOCK

Plant genotype is the most important factor in determining the potential for cold tolerance.[112] Cultural and environmental factors can only modify this potential. It has long been recognized that there is a wide tolerance to cold within the genus *Prunus*, and between cultivars within a species. Scion hardiness is also influenced by rootstock.[121–127] It is generally accepted that within *Prunus*, tolerance to cold from most to least occurs as follows: *P. domestica* > *P. cerasus* > *P. avium* > *P. armeniaca* > *P. persica.*[128–130] Within a species rootstock can play a large role. Layne[130] classified the following rootstocks as: very hardy (Siberian C, Tzim Pee Tao, Chui Lum Tao); medium hardy (Bailey, Harrow Blood), medium tender (Yeh Hsiemtung Tao, Sinung Chui Mi, Rutgers Red Leaf); and tender (Gold Drop, Lemon Free, and Elberta).

Resistance to cold stress is a complex phenomenon involving several genes and is expressed through a multiplicity of physiological processes. Among them are carbohydrate production and acclimation, cessation of growth, acclimation, deep winter hardiness, super cooling, length of endodormancy, deacclimation, and bloom date. As a result, it is difficult to predict how a certain scion/rootstock combination will perform in a new combination or environment. Traditionally, hardiness has been evaluated during test winters or to determine which cultivars survive after being subjected to very cold temperatures sometime during the dormant season, and then by ranking the cultivars according to their survival rate in the spring. This technique has several deficiencies, among those are that no two seasons are ever exactly alike either in regard to cultural and environmental preconditioning, or in terms of the time and degree of the cold event.[118] In recent years researchers have been using artificially controlled events to evaluate cold hardiness in an attempt to decrease that component of variability that is associated with the cold event. Some of these tests have been conducted at the peak of winter dormancy in an attempt to characterize maximum winter tolerance to cold, but Howell[131] would argue that in order to fully characterize tolerance to cold, one must assess tolerance to cold at 2 to 4 week intervals throughout the acclimation, deep winter hardiness, and deacclimation stages.

The effect of season, variety, and rootstock can be illustrated by an experiment where we compared the hardiness of 'Redhaven', 'Redskin', and 'Cresthaven' on 'Halford' and 'Siberian C' rootstock in Michigan throughout the dormant season (Figure 7). The scions, rootstocks, and their combinations differ in response depending on the season.[132] 'Redhaven' on 'Halford' was hardier than 'Redhaven' on 'Siberian C' on September 10, 1985, but this relationship was reversed by November 19, 1985. In addition 'Siberian C' was several degrees hardier than 'Halford' by the end of December.

Thus the rootstock scion combination should be matched for the particular climatic situation. If acclimation in the fall is important then a combination which assures early development of hardiness is probably most desirable and may be more important than midwinter hardiness. On the other hand, in more northern climates, midwinter hardiness might be the most important factor in relation to survival of the tree. The length of endodormancy is also important. If the chilling requirement is too low, then active growth may resume too early in the spring and a warm period during the winter could trigger active growth, making the tree more vulnerable to cold. This illustrates two points: (1) that hardiness ranking depends upon the time and stage of development, and (2) that a test winter, or controlled freezing test at one time during the winter may not adequately describe tolerance to cold in a scion/rootstock combination for a certain climate.

G. CULTURAL AND ENVIRONMENTAL FACTORS

A healthy tree will better resist cold damage than a tree that is in low vigor, or one that has been under environmental or biological stress. Cessation of growth which coincides with the trigger for the first stage of the hardening process is desirable, but should not result in growers inducing early bud set by early defoliation, by water stress or poor nutrition. These are all detrimental to cold tolerance.

Circumstantial evidence indicates that a certain minimum level of carbon assimilation is necessary for maximum winter hardiness. Levels of carbohydrate in peach buds and shoots have been positively

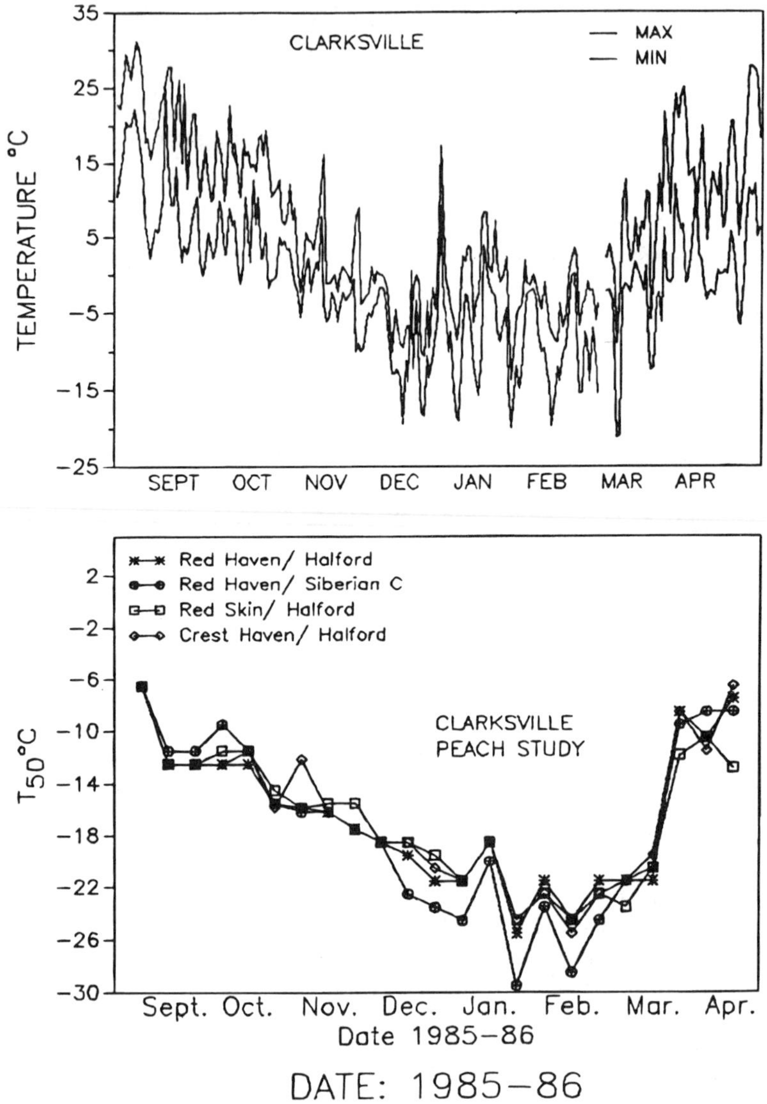

Figure 7 Maximum and minimum temperatures at the Clarksville Horticultural Research Station, Clarksville, MI (top) and T_{50} values for Redhaven/Halford, Redhaven/Siberian C, Red Skin/Halford and Crest Haven/Halford during the dormant period, 1985–86. Reprinted with permission from Flore, J. A., Howell, G. S. Jr., Gucci, R., and Perry, R. L., *Compact Fruit Trees*, 20, 60, 1987.

correlated with cold hardiness.[126,133] Cold acclimation is an active process and requires energy to occur. However, this does not necessarily mean that an increase in carbohydrate will increase hardiness, or that sugar or polyols are the hardiness promoting factor. It is likely that some minimum level of carbohydrate is required to attain maximum hardiness, beyond which no additional benefit can be derived. The same cannot be said for the hardiness factor produced by the leaf and translocated to the rest of the tree as more research is needed in this area. Important factors related to carbohydrate accumulation and cold hardiness are outlined in the following sections.

1. Defoliation and Duration of Foliage

Early cold acclimation is often associated with early defoliation and low vigor. However, duration of foliage on the tree should be maximized and early senescence avoided. Howell and Stackhouse[134] showed that early defoliation in sour cherry due to pathogens or by artificial means decreased hardiness the following spring. Regression analysis of Cooper's data on peach[135] revealed a positive relationship between later leaf abscission and percent bud mortality the next spring, while Flore et al.[132] did not find a correlation between hardiness and percent leaf abscission in the fall for peach in a rootstock trial.

Early defoliation has been considered as an indicator of early acclimation to cold damage, and there is very good evidence that actively growing shoots are less hardy than those that have stopped growing. However, we should not equate early defoliation with adequate vigor and health of the tree. Clearly active growth in the fall does not allow for sufficient acclimation in time to resist cold temperatures. By contrast premature leaf abscission should not be encouraged since this might result in a carbohydrate limitation, and once abscission occurs, the acclimation factor present in the leaves is no longer available to the rest of the plant.

2. Light

Light is important in cold acclimation and winter resistance process. Acclimation in many plants is triggered by day length. It is also necessary for adequate carbohydrate assimilation and translocation in trees. Cultural practices which reduce exposure of leaves to light presumably affect hardiness by their influence on carbon assimilation. Shading studies on potted trees of peach and cherry[85] and on field grown trees[132] indicated that there was a decrease in hardiness with increasing shade, which was also associated with a decrease in shoot carbohydrate content.

3. Nutrition

Any nutritional factor that is limiting enough to show deficiency symptoms will generally decrease winter hardiness. A review of the effect of mineral supply on cold hardiness has been published by Levitt,[136] and here we will only place emphasis on nitrogen because of its influence on tree vigor, acclimation, and hardiness.

Late applications of N have generally been discouraged because of their possible stimulation of growth. However, later applications after active growth has stopped, or which have prolonged foliage health have been associated with higher yields in stone fruit. Savage[137] showed that N applied after harvest in the fall on peaches in Georgia prolonged tree life over applications in the Spring. Fall N seems to delay leaf fall and may help increase the duration of foliage. Savage[137] speculated that this might also delay the emergence from rest the next spring.

4. Fruiting

Crop load decreases cold tolerance.[118,138] The fruit on the tree are the primary sink for carbon produced by photosynthesis. Overcropping reduces vegetative growth, vigor, and decreases storage carbohydrate. In grape vines, excess cropping reduced cold resistance in cane and bud tissues.[139] We have found that non-thinned peach trees are similar in hardiness to shaded trees, which are less hardy than thinned fully exposed 'Newhaven' peach. Heavy crop load is generally not a problem on cultivars that mature their fruit early in the season, but may be a problem with the later maturing varieties of peach and plum in extreme climates where seasons are short. Once the optimum crop load has been determined for a given cultivar, excessive cropping should be prevented.

5. Pruning

Pruning may affect cold hardiness positively by increasing the level of light in certain areas of the tree, or negatively by reducing carbon production, by removing leaf area, by reducing uptake of water and nutrients (root pruning), or by stimulation of shoot and cambium activity. For the purposes of this review, pruning terms will be defined as follows: summer pruning = selective removal of shoots or branches during the growing season; dormant pruning = the same as summer pruning, except that shoots are removed during the dormant season before active new growth; and summer or dormant hedging or topping = the indiscriminate removal of all branches within a plane either in the summer or dormant period.

Reports relating cold tolerance to the time and degree of pruning have been inconsistent. In general the earlier that pruning occurs in the dormant period, the more negative the effect on cold hardiness. Dormant pruned trees often defoliate slightly earlier than summer pruned trees, but the onset and duration of dormancy seem to be just about the same. Pruning just prior to the occurrence of low temperatures tends to increase injury.[1,140] Therefore in extreme climates it is often recommended that growers prune their hardiest varieties first on their best sites followed by the less hardy varieties.

When compared to dormant, summer pruning reduces total shoot growth, decreases shoot diameter, increases lateral shoot growth, and flower bud density is decreased while lateral branching is often increased.[52] The earlier in the season that summer pruning or hedging is conducted the greater the reduction in current season growth. Marini[141] found that summer pruning in late July slightly delayed defoliation of mature 'Sunqueen', while Hayden and Emerson[142] found that summer hedging hastened defoliation. Summer pruning may induce late fall growth which does not have an adequate period of time to acclimate before the first cold temperatures. Early dormant pruning can influence the cold resistance of cambial tissue close to the cut surface and can affect the resistance of shoots and buds if inadequate healing does not occur.[105] In general, summer pruning or hedging may reduce cold hardiness of peach.[141,143]

When considering if summer pruning will have an effect on cold hardiness, timing is especially crucial. If done too early in the season, it will reduce leaf area for carbon production and have a negative effect; if done too late it will also have a negative effect by stimulating shoot growth and delaying dormancy; and if done very late in the season, the tissue may not have an adequate time to heal. In Michigan, we recommend that summer pruning coincide with final swell of fruit growth. During this period, most of the carbon will be partitioned to fruit growth and shoot growth will usually only be slightly stimulated. Unless the variety is particularly late in maturing, this should give adequate time for the foliage to produce carbohydrate for the next year. Clearly in our studies[52,105] during the acclimation period shoots from summer, pruned peach were 1–3 degrees less hardy than dormant-pruned trees. In many years this difference may not be significant, but in Michigan, where cold periods can occur in the late fall and early winter, this is a very critical period of time. The benefits of summer pruning in terms of increased fruit quality and reduced pruning costs the following spring must be considered in relation to the possible negative effect on winter hardiness.

6. Water Content, Irrigation, and Ground Cover

Water content of the tissue is often negatively correlated with hardiness.[144,145] Irrigation and rain may influence cambial activity in the fall, and the heat holding capacity of the soil. Adequate water should be supplied to promote good vigor since reduced vigor often results in lower hardiness. Layne and Tan[146] found that irrigation reduced peach tree mortality on sandy soils, presumably due to decreased winter injury. A similar response was observed in several cherry orchards in northern Michigan in 1988 when conditions were much hotter and drier than normal.[147] In areas where rainfall is high in the late summer or fall, cover crops might be sown to compete with the trees and hasten acclimation. However, in experiments at Clarksville Horticultural Research Station, Clarksville, MT in 'Newhaven' with sorghum sown on July 1 or August 1, we found no benefit and in some cases found that the cover crop reduced hardiness.[148]

In areas where irrigation is used, water should be reduced in late summer and early fall to stop tree growth, then increased near the time of defoliation to assure that the soil is wet to slow the rate of drop in temperature of the soil to prevent cold injury to the roots.[112] Ground cover from best to worst for spring frost protection would be wet soil > mulch > herbicide strips > sod cover.

VI. WATER

After irradiation, water is the second most important factor limiting growth of *Prunus* during the summer period. Several excellent reviews have been recently published concerning the effect of drought[2,4,149] and flooding[150,151] on trees. Since we have covered the effects of humidity under Temperature, examples of drought and anoxia on stone fruit will be stressed in this section. In this review emphasis will be on: vegetative, reproductive, and physiological response to drought and anoxia; current theory concerning irrigation scheduling and strategy, and the importance of root restriction and root zone wetting on management decisions.

A. CULTIVAR AND SPECIES AND ROOTSTOCK DIFFERENCES

Water use and sensitivity to drought differs within and between species in the genus *Prunus*. Faust[2] has classified fruit trees according to their relative requirement for water and Kenworthy[152] has ranked them according to drought tolerance, while Rieger and Dummel[153] have made a comparison of morphological and physiological characteristics for different species within the genus *Prunus* (Table 8).

Table 8 **Relative water use and sensitivity to drought within the genus *Prunus***

Relative response	Water requirements[a] most to least	Tolerance[b] to drought most to least	Habitat[c] dry to wet
Greatest	Quince	Pear	Nevada desert almond
	Pear	Prune	Western sand cherry
	Plum	Apple	Beach plum
	Peach	Grape	Peach
	Apple	Walnut	Sierra or Pacific plum
	Sweet cherry	Plum	Nanking cherry
	Sour cherry	Almond	
	Apricot	Citrus	
		Peach	
		Cherry	
		Olive	
Least		Apricot	

[a]According to Faust.[2] [b]According to Kenworthy.[152] [c]According to Rieger and Dummel.[153]

Drought tolerance in *Prunus* from most to least is as follows: plum > almond > peach > sweet cherry > sour cherry > apricot. However, there is a high degree of diversity in tolerance to drought and anoxia within and among *Prunus* spp.[150,153] Many *Prunus* spp. are graft compatible. For example, a plum rootstock could be grafted with a peach or apricot cultivar to improve it's tolerance to drought or flooding. There seems to be a greater diversity to anoxia than drought and the reader is referred to the recent book edited by Rom and Carlson[154] for a comprehensive discussion of rootstocks for each species.

1. Peach Rootstocks

Peach, interspecific peach hybrids, plums, plum hybrids, and other *Prunus* species have been used as rootstocks for peach. There are few examples of wide tolerances to drought for peach. Couvillon et al.[155] reported that peach trees on their own roots were able to withstand greater levels of water stress than trees propagated on 'Lovell', 'Nemaguard', or 'Halford' seedling rootstocks. In a rootstock trial for peach in Illinois, Kjelgren and Taylor[156] concluded that there were no major differences for cv. Redhaven own rooted, or on cvs. Halford, Amandier, Damas, or Lovell, but that St. Julien had lower g_l and greater hydraulic resistance than the others. They also noted considerable differences in plant size, so it is difficult to know if differences in drought tolerance were due to the rootstock or to size of the plant.

Peach rootstock response to anoxia is diverse. Layne[130] classified resistance to waterlogging from most to least as follows: **very good**, 'Damas GF 1869'; **good**, cvs. St. Julien No 1 and 2, St. Julien GF 655.2; **moderate**, wild peach, cvs. Nemaguard, GF 305, Brompton, Prunier GF 43, GF 677; and **low**, peach, apricot and almond seedlings. However, Chaplin et al.[157] reported tolerance of peach rootstocks from most to least as follows: 'Ruters Red Leaf', 'New', 'Siberian C', 'Harrow Blood', and 'Lovell'. Rom and Brown[158] reported that 'Halford' and 'Lovell' were more sensitive to waterlogging than 'Nemaguard'. It is agreed that almond and apricot are among the most sensitive of all species in the Rosaceae family.[150] Through the correct choice of rootstock, the grower can utilize a wide range of sites with different soil moisture characteristics.

2. Cherry Rootstocks

There is little reported genotypic diversity among cherry rootstocks for drought tolerance, however, trees on mahaleb are usually more vigorous than mazzard under dry conditions.[128] The opposite seems to be true in wet soils where mazzard survives longer. Tolerance to anoxia has been ranked from most to least as: *P. cerasus* L. > *P. avium* > *P. mahaleb* by Norton et al.,[159] while Beckman[160] has ranked cherry rootstocks as follows on the basis of A and shoot extension: M × M 2 least sensitive; mahaleb, 148/1, colt, mazzard, Montmorency, M × M 60, 148/9, 195/1, 195/2 as sensitive, and 196/4, M × M 39 as most sensitive. It must be pointed out that in comparison to rootstocks available for peach and plum, all cherry rootstocks are relatively sensitive to anoxia.

3. Apricot and Plum Rootstocks

There seems to be little diversity in response to drought or flooding for apricot and the important aspects concerning plum have already been discussed. We refer the reader to Okie[161] and Crossa-Raynaud and Audergon[129] for an in-depth discussion. In conclusion, under flooding conditions, plum and prune are the most tolerant, followed by Japanese plum, sour cherry, sweet cherry, with peach and apricot being the most sensitive.[150]

B. GENERAL CONSEQUENCES OF DROUGHT

The physiological control of water status in temperate fruit crops has been recently reviewed by Jones, Lakso, and Syvertsen,[4] and the influence of water and anoxia on photosynthesis and stomatal conductance has been covered by Flore and Lakso.[3] Herein, the consequences of drought and anoxia on vegetative and reproductive growth and physiological process in *Prunus* are reviewed with emphasis placed on sensitivity, and how water can be managed to improve yield.

1. Diurnal Changes

Diurnal changes in physiology and growth in *Prunus* are well documented. Stomata open in response to light (maximum g_l levels are similar to those for A or approximately at PPF of 500–800 μmol m^{-2} s^{-1}). As a consequence, the negative Ψ_l that develop within the leaf and stem are thought to follow g_l and are not the cause of stomatal closure unless the trees are under moderate to extreme stress.[3,4,46] Diurnal shrinkage in fruit, trunks, and scaffolds can be noted in *Prunus* in response to diurnal changes in water potential, and some have suggested that the degree of diurnal shrinkage can be used as a measure of plant water status and thus could be used to schedule irrigation.[162–164]

2. Shoot and Leaf Expansion

Shoot and leaf expansion are among the most sensitive plant parameters to Ψ_l. Andersen and Brodbeck[46] found that as water deficits developed, shoot and leaf expansion were inhibited prior to reductions in A or g_l. They concluded that a moderate level of water stress could reduce the rate of vegetative growth of peach trees without a reduction in carbon assimilation. This would be consistent with the results found in deficit irrigation experiments (see next section).

3. Leaf Gas Exchange

The influence of water status on g_l and A have been reviewed by Flore and Lakso.[3] Values of g_s for *Prunus* are between 150 to 400 mmol m^{-2} s^{-1} and are typical of most fruit crops. For different species, reported values are: apricot, 108–440 mmol m^{-2} s^{-1}; peach, from 60 to 280 mmol m^{-2} s^{-1}; and plum and cherry, 280 mmol m^{-2} s^4, converted from mmol according to Nobel.[165] Previously it was thought that g_l, and therefore A, were almost entirely dependent upon Ψ_l and loss of turgor,[3] however, in recent research, it has been shown recently that *Prunus* undergo adaptation to drought, one method being osmotic adjustment.[46,166,167] Thus, keeping turgor potential high and stomata open even at very negative Ψ_l.

Stomatal conductance is strongly coupled to A in *Prunus*[96] and is responsive to vapor pressure gradients above 1.5 kPa for cherry, however peach was not affected until Ψ_l greater than -2.0 MPa were reached. Photosynthesis was not affected adversely until these levels of stress were reached. Therefore under field conditions, it seems that g_l and A are not affected until more severe stress conditions are imposed.

Recently it has been suggested that ABA is produced in the root in response to drought or anoxia and that it is transported to the leaves where it plays a regulatory role in the control of g_l and A (see review by Flore and Lakso[3]). Neri and Flore[168] have shown in split root studies with peach that ABA increased with soil drying, but g_l or A did not decrease until several days after leaf and shoot growth were affected. There was an increase in hydraulic conductivity of the dry roots.

In non-irrigated peach trees, ABA concentration increased sharply after mid-summer; this was associated with a decrease in g_l and Ψ_l. ABA concentrations in leaves from irrigated and non-irrigated trees increased as g_l decreased, however in stressed trees, high levels of ABA in the morning were not associated with closed stomata.[169] Stomatal conductance and A were inversely related to Ψ_l, when drought studies were conducted on peach grown in pots.[170] However, there was no 'threshold value' for g_s, as it decreased gradually as Ψ_l became more negative.

Stomatal conductance and A are both affected by flooding in fruit crops.[150] Recently in a very sensitive crop, sour cherry, Beckman et al.[97] showed that A was inhibited 24 hours after flooding was imposed, and that nonstomatal factors contributed greatly to the inhibition (Figure 8).

The effect of a mild water stress was investigated for container grown sour cherry. Under optimum conditions, there was a strong relationship between A and g_l. Stomata of plants subjected to a low level of stress were more responsive to VPD and seemed less responsive to CO_2 gradient than on highly stressed plants. Stomatal aperture decreased in response to increased external CO_2 concentration, and the degree of response varied with stress. Water use efficiency of sour cherry increased under mild stress conditions.[96]

Leaf water potential of container grown peach was also correlated with g_l, A, and VPD. Maintenance of soil moisture slightly above field capacity (FC) for even short periods may decrease A as compared to maintaining soil moisture slightly below FC. Decreasing the soil water potential from −0.05 MPa to the permanent wilting point resulted in decreased photosynthetic rates.[171]

4. Water Use and Dry Matter Production and Partitioning

There is a direct relationship between water use and dry weight increase in peach. A functional equilibrium may exist between water usage and growth increment, which is mediated through the capacity of the root system to increase its water uptake as the demand/supply ratio increases.[172,173] The use of deficit irrigation, and root restriction as a means to control growth without a negative effect on fruiting are based in part on this premise.

Dry matter partitioning between different plant parts is influenced by drought, and is dependent on season, competition between different organs, and the presence or absence of fruit. Personal observations[18] would indicate that for stone fruit, the order of dry matter partitioning in relation to drought seems to follow the following pattern. Early in the season roots have preference, followed by vegetative shoot growth until the end of pit hardening. During final swell of fruit growth, fruit have preference. In

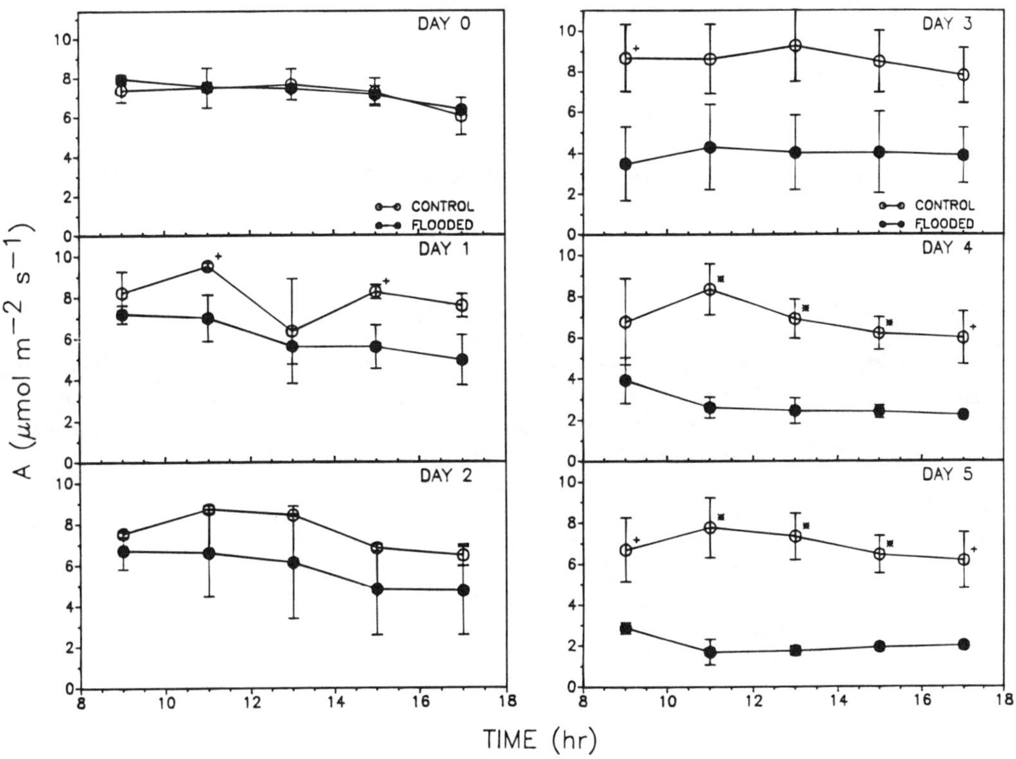

Figure 8 Effects of 1 to 5 days of flooding on net CO_2 assimilation (A) of sour cherry trees. Data points are means of two plants/time ± SD [+I,* indicate significance at $p = 0.10$ and 0.05, respectively, otherwise nonsignificant, F test]. Reprinted with permission from Beckman, T. G., Perry, R. L., and Flore, J. A., *Hort. Sci.*, 27, 1297, 1992.

general, as soils dry, partitioning to roots increases, and roots develop at greater depths in the soil profile.[13,174] Root growth can also be locally stimulated by water, and often there will be large masses of fine roots near the source of water. Under areas of high rainfall, roots will proliferate in the soil where moisture is favorable. In orchard systems sod cover may restrict the growth of roots, presumably due to competition for water and fertilizer. Layne et al.,[174] in studies on irrigation in a mature peach orchard found that the total number, and the number of small diameter roots were highest in nonirrigated plots, and that they decreased with increasing levels of irrigation.

5. Sensitivity of Vegetative Growth to Drought

Shoot growth, leaf growth, leaf emergence, shoot diameter, and trunk diameter are all adversely affected by drought or anoxia. Generally in order of sensitivity tc drought, leaf growth is the most sensitive followed by shoot growth, leaf emergence and shoot or trunk diameter. This pattern was confirmed in experiments on peach under a rapid (zero water vs. control) or a slow water stress situation (watering at 50 or 25% of control).[164] Growth characteristics were observed several times per week and the occurrence of a statistical difference between treatments was used to determine sensitivity to stress. Processes that were most to least affected by drought were: leaf growth > shoot extension > leaf emergence > trunk diameter > Ψ_l > g_s. Similar results have been observed with sweet and sour cherry.[13] In recent experiments on nonfruiting 2-year-old 'Cavalier' on 148/1 rootstock grown in a rain exclusion shelter, it was observed that the first response to a slow stress drought was an increase in root numbers, followed by a decrease in shoot growth, trunk growth, A, and g_l.[13]

Changes in trunk, scaffold, and shoot growth occur daily in response to changes in diurnal Ψ_l, but are somewhat dependent upon the level of drought, daily evaporative demand, the location on the tree, and the presence or absence of fruit. Diurnal shrinkage of trunks, shoots, and fruits have been reported for peach,[162] and 'Montmorency' sour cherry.[163] Daily shrinkage is proportional to the period of evaporation following irrigation and the magnitude and duration of daily shrinkage in peach is increased with fruit load.[175] Seasonal increase in trunk or scaffold diameter may be used as a tool to determine the amount of water stress the tree has undergone during the growing season.

A positive turgor potential Ψ_p is necessary for cell expansion and is often correlated with g_l and A. The turgor potential at a given Ψ_l is a function of Ψ_π in the cell. Increasing Ψ_π in response to lower water potential is a mechanism that has been observed in other plants. Osmotic adjustment has been confirmed for leaves and roots of cherry[167,176] and leaves of peach.[46,166] Jones et al.[4] have hypothesized that sorbitol may be an important osmolyte contributing to osmotic adjustment in fruit trees. Ranney et al.[167] suggests that the possibility exists to select superior drought resistance cultivars on the basis of their ability to osmotically adjust.

6. Response to Extreme Stress

Under arid conditions, water is supplied by irrigation during the growing season. Proebsting[177,178] has studied the effect of extreme water stress on the survival of peach, cherry, and plum. Peach was grown without irrigation and received only 86-mm rainfall during the growing season under Washington State conditions.[177] Trees died after experiencing Ψ_l below -3.0 MPa in July and August. Defoliation began in July, fruit growth was decreased, and fruit flavor was astringent. Flower buds did not differentiate. Heavy pruning ("dehorning") delayed the appearance of drought symptoms until very late in the season and resulted in 100% tree survival. Heavy thinning in early June did not affect current season's symptoms but reduced dieback and death of trees.

Under similar conditions Proebsting et al.[178] subjected mature bearing cherry and prune trees to very low rates of irrigation and observed their response in the year of application and the following year. Trees were trickle irrigated daily or weekly at 100, 50, or 15% of E_c (evaporation adjusted to the area of the tree canopy). The Ψ_l reached as low as -2.8 MPa for prunes in late July and August. Growth of fruit and vegetative parts were reduced by severe stress, but the trees survived on 15% E_c. Prunes recovered to normal yield and growth by the second year after treatment. Peripheral branches of cherries died back during the year of treatment and in the following year. Cherries grew and fruited normally by the third year, but had a reduced bearing surface area.

C. MANAGEMENT OF WATER

1. Deficit Irrigation

Chalmers and his co-workers[179,180] have proposed that timing amount of irrigation can control vegetative growth in peach without a decrease in fruit yield or quality. They proposed that under Australian arid conditions irrigation is needed early in the life of the tree to fill in allotted space. Once this has been achieved, vegetative growth during the current season can be controlled by timing of irrigation. Adequate irrigation is needed during the early part of the season when canopy and fruit cell division are occurring, but once stage II of fruit growth begins, most of the carbon and water are needed for vegetative development. Water is again needed during final swell, and at that point the fruit will compete strongly for water and photosynthate. Water may be withheld after harvest to reduce vegetative growth and pruning costs. Increased yields resulted from increased fruit growth (relative to the control) after the level of irrigation was raised to the control level following a period of reduced irrigation.[179,181] There was an interaction between water, tree density, and tree age. They hypothesize that spacing and water supply interact synergistically to reduce vegetative growth and increase fruit growth.

There is a strong physiological relationship between size of the root system and vegetative framework above ground,[182] and the root to shoot rate remains relatively constant over a broad range of conditions.[172,173] Since root growth can be controlled by both drip irrigation and row spacing, the result is control of vegetative growth without inhibiting fruit growth. Is this concept economically feasible? Girona et al.[183] pointed out that when growers receive a premium for large fruit and water is relatively cheap, the risk/reward ratio does not favor deficit irrigation on late maturing tree fruit.

Since stone fruit may mature a crop early in the season (55–65 days for cherry), or in the first third to half of the season under California conditions, the question of after harvest requirements for irrigation is an appropriate one for growers. This is especially important in arid areas with long growing seasons where the interval between harvest and leaf fall may be a period of months. Early reports on stone fruit have indicated that postharvest water stress can have a detrimental effect on flower bud development and subsequent fruiting in apricot,[184,185] but little effect on vegetative tree growth.[186]

Johnson et al.[25] determined the long-term response of early maturing peach to postharvest water deficits under California conditions. They found that flower, fruit density, and the occurrence of double fruit were greater in the dry treatment than the control. After normal hand thinning, yields and fruit size were not affected over a 4-year period. Vegetative growth was reduced, but there was no indication of a decline in vigor. A well-timed irrigation before carpel differentiation reduced the occurrence of double fruit. This management strategy could reduce water use while preserving marketable yields.

Three levels of postharvest irrigation on peach production under California conditions were investigated by Larson et al.[187] Control trees were irrigated at 3-week intervals after harvest, the medium-treatment received one, and dry-treatment trees received no irrigation. The seasonal increase in trunk radius of the dry treatment trees was reduced by 33% relative to either wet or medium treatments. The amount of daily trunk radial shrinkage was inversely proportional to irrigation level. Dormant pruning weights were less in dry treatments, return bloom was greater, and fruit set was increased in the dry treatments. After fruit thinning, there were no significant differences among treatments for fruit yield or fruit size, but fruit maturity was slightly delayed in the dry treatment.[187]

2. Methods to Determine Water Stress and Schedule Irrigation

The quantification of plant stress or the scheduling of irrigation can be based on physical (soil moisture, VPD, evapotranspiration), morphological (leaf growth, fruit growth, shoot growth, stem growth), or physiological (water potential, stem water potential, stomatal conductance, photosynthesis, chlorophyll fluorescence) factors.

Stem Ψ has been found to be a sensitive and reliable indicator of water stress in prune trees[188] under California conditions. However, Garnier and Berger[189] demonstrated that Ψ_l in peach trees was only slightly affected during a drying period and concluded Ψ_l was not very sensitive to irrigation, and that stomatal closure acted as a regulator of Ψ_l while soil water status was a strong regulator of stem Ψ. Xiloyannis et al.[169] found that in peach trees irrigated at 50% ET, Ψ_e and g_l were slightly affected compared with trees irrigated at 100% ET, and strong differences in Ψ_l were only found with non-irrigated trees.

In a study comparing drought resistance among *Prunus* species from divergent habitats, Rieger and Dummel[153] demonstrated that photosynthesis decreased linearly with stem Ψ in all species tested, however, the stem Ψ at which A reached zero was not constant.

Shoot and leaf elongation rates were exponentially dependent on Ψ_p and were reduced drastically below 1.0 and 0.7 MPa, respectively.[46] As leaf water deficits developed, shoot and leaf expansion were inhibited prior to a reduction in g_l or A. Thus, a moderate level of water stress can reduce the rate of vegetative growth of peach trees without concomitant reductions in carbon assimilation. It seems, therefore, that Ψ_l is not the most sensitive parameter to stress, and that stem water potential or pre-dawn Ψ_l (which is ca. equal to stem water Ψ), would be the matter of choice.

Irrigation based on soil water content or using estimates of evapotranspiration have been used extensively for stone fruit. However, one major limitation in the use of both is that they are not directly related to the occurrence of plant water stress. Worthington et al.[190] have shown that the crop coefficient (amount of water loss in relationship to Epan) for 5-year-old mature peach may vary (0.31 to 0.98) during the season, and that it is very dependent upon crop load and leaf age. Soil moisture values used to indicate irrigation requirements may not be critical values for the plants under all atmospheric conditions, nor do we know where and at what depth to measure soil moisture, especially in glacial till soils where soils can vary greatly within a short distance.

Infrared thermometry has been used to estimate the water use of peach trees[191] by measuring the temperature of the leaves. Leaves not under water stress with open stomata should be significantly cooler than leaves with closed stomata due to high heat of vaporization of water, and the leaves will be cooler than stressed trees. Monitoring foliage temperature minus air temperature may be used to determine water use, and schedule irrigation in peach orchard systems, however modification may need to be made, especially under high VPD. At VPD greater than -2 kPa, stomatal response to VPD introduces a curvilinear response which must be taken into consideration.

Estimating sap flow using the heat balance technique on woody stems is a recent technique that is gaining in interest. In this method, the sap velocity is estimated from the time required for a heat pulse, injected into the xylem to travel a finite distance downstream. Several variations of this method have been tried.[192–195] Good agreement between estimates made with this method and E determined by lysimeters has been reported for trees,[194–196] but Shackel et al.[197] have reported substantial errors on peach under field conditions. However, depending on the accuracy needed and the ability to substantiate results, this seems to be a promising technique for the future.

3. Root Zone Wetting

Several split root studies have shown that a proportion of the root system was able to supply the water needs of the tree, and restriction to greater than 50% of the root system was required before there were major effects on shoot system. Tan and Buttery[170] found that watering 50% of the root system could meet the entire need of the plant. That suggested that drip irrigation could be an effective means of applying water to peach. Proebsting et al.[198] found that the effect of restricted irrigated soil volume was similar to that of deficit irrigation. Increasing root:shoot ratios by adjusting the soil volume or by pruning the shoot always increased g_l. Williamson and Coston[12] restricted peach root growth using several different techniques and evaluated the effect of two levels of irrigation on growth. They concluded that the effects of irrigation rate on vegetative growth were small compared to differences among planting treatments which restricted root volume.

D. FLOODING

Flooding has negative effects on most growth and physiological process of woody plants.[149–151,199,200] The response is clearly due to a lack of oxygen, early symptoms are similar to drought, and death often occurs without diseases if the species is flood sensitive.

1. Species and Rootstock

Prunus are generally considered to be quite sensitive to anoxia. Rowe and Beardsell[199] ranked waterlogging tolerance of fruit trees and from that ranking *Prunus* sp. would be classified as follows: moderately tolerant, plums (*P. domestica* L. and *P. cerasifera* L.); sensitive, plum (*P. salicina* Lindl.); very sensitive, cherry, apricot, and peach. It is generally accepted that tolerance to waterlogging is mainly determined by rootstock and not the scion.[150,199,201,202] According to Schaffer et al.[150] almond and apricot are the most sensitive, followed by peach, cherry, plum, and prune.

Although not considered the most tolerant of fruit species to flooding, *Prunus* spp. exhibit a wide range of diversity, and due to graft compatibility allow for this trait to be transmitted via rootstock to the more sensitive species.[150,158] Schaffer et al.[150] have rated the relative tolerance of *Prunus* species to flooded soil conditions and we refer the reader to their review article for a more thorough discussion of species and cultivar differences. Tolerance may range from several months for plum and prune rootstocks[203] to less than 5 days for peach[157] and cherry.[160,204]

2. General Responses

Soil flooding results in low oxygen levels, and soils may be oxygen depleted within 1 day of waterlogging.[205] Oxygen diffusion rates are inhibited and rates below 0.2 $\mu g\ O_2\ cm^{-2}\ hr^{-1}\ min^{-1}$ have been correlated with reduced hydraulic conductivity and/or growth in peach.[201] Inhibition of shoot, leaf, and/or root growth have been reported in response to flooding for peach[158] and cherry.[160,204] Plant metabolism can also be affected (see Schaffer et al.[150] for a thorough review).

Growth and survival is dependent upon the time of flooding; with flooding during the growing season having a greater effect than during the dormant period.[158,160,204] Cyanogenic glycosides are common in tissues of *Prunus* spp. and Rowe[206] observed that under an aerobic stress, detached roots of these species evolved phytotoxic amounts of hydrogen cyanide. Additionally, Rowe and Catlin[200] have demonstrated that differential sensitivity of peach, apricot, and plum was correlated with cyanogenic glycoside content of the root tissue, and has been related to their ability to metabolize prunasen.[203]

3. Gas Exchange

The mechanisms by which anoxia inhibits A and g_s in fruit crops is not fully understood. Short-term effects include a decrease in A and g_l and an increase in root resistance to water flow. Typically, hydraulic conductivity of the root system is reduced[150,202,207] followed by typical water stress symptoms (more negative Ψ_l, lower g_l, and reduced shoot growth). Using A/C_i curves to partition out stomatal vs. biochemical limitations, and using a bioassay system, Beckman et al.[97,160] presents evidence that toxins produced by the root under anaerobic conditions contributed to the nonstomatal inhibition of photosynthesis in sour cherry. The identity of the toxin or inhibitory compound is not known, although ABA is known to increase in concentration in roots of pea in response to flooding.[208] This raises the possibility that g_s was not the major factor in reducing A.

The hydraulic conductivity of the root system is disrupted soon after flooding is imposed. This could result in more negative Ψ_l and an indirect effect on A through inhibition of g_l. Although this occurred in cherry, Beckman et al.[209] concluded that if inhibition of A occurred prior to stomatal closure it was in part due to an unknown signal from the root.

Our current understanding of anoxia on the mechanism of photosynthetic inhibition in *Prunus* is not complete. There is likely more than one cause, and the appearance of a toxin or lack of growth regulator produced in the root, followed by a disruption of hydraulic conductivity, is consistent with observed responses.

VII. OTHER ENVIRONMENTAL FACTORS

A. SALINITY

Prunus spp. are generally considered to be sensitive to salinity compared to most fruit crops.[210–212] Salt can be a problem in certain arid regions (Middle East, North Africa, Arizona) where rainfall levels are low. Berstein et al.[210] demonstrated that six different varieties ('Golden Blush' peach, 'Texas' and 'Nonpareil' almonds, 'Santa Rosa' plum, 'French Improved' plum, and 'Royal' apricot) of stone fruit on Lovell rootstock were sensitive to moderate levels of salinity (EC_i = 0.9 mmhos), and determined that 50% of the response was due to chloride injury and 50% due to the increase in Ψ_π of the soil. Salt tolerance of all six varieties was similar except that the apricot was more sensitive than the others. When compared to 'Lovell' rootstock, 'Shalill' had higher chloride accumulation and less growth in peach and almonds; the Yunnan root increased chloride accumulation and toxicity in apricots, while the 'Mariana' reduced chloride accumulation and improved growth in plum and prune.

B. AIR POLLUTION

Ozone inhibited growth and photosynthesis of almond, plum, apricot, and pear linearly with increasing concentration[213] while peach and cherry are unaffected. Apricot cultivars differed in their sensitivity, 'Nonpareil' was most sensitive, 'Butte', 'Carmel', and 'Sonora' were intermediate, and 'Mission' was unaffected.[214] Exposure of 'Casselman' plum to increased atmospheric ozone partial pressure (−0.44 to 0.111 $\mu Pa \cdot Pa^{-1}$) over two growing seasons reduced A, caused premature leaf fall, decreased trunk growth, and reduced yield. Symptoms appear as either flick or stipple on the upper surface of the leaves,[215] and chlorosis and abscission may occur.

Fluoride causes injury to leaves and fruit of many stone fruit[215] species. Sensitive species show leaf necrosis at concentrations of 0.5 ppm or less. A disorder in peach known as soft suture is associated with fluoride. Heggestad[215] considered *Prunus* as among the most sensitive species to fluoride injury, and ranks apricot as highly sensitive, prune (*P. domestica*), peach, and mahaleb cherry as sensitive, while sweet cherry, oriental flowering cherry, and myrabalam plum (*P. cerasiters*) are moderately tolerant.

Sulfur dioxide and acid rain are major air pollutants near industrial areas that affect vegetation. Zimmerman and Hitchcock[216] demonstrated that when plants were exposed to SO_2 plants of 0.2 to 1.1 ppm for 2 to 8 hours, apricot was only slightly affected, while Italian prune showed moderate phytotoxicity. Acid rain does not seem to have a direct effect of foliate, but may be a problem indirectly by lowering soil pH to 3 or less.[217]

VIII. CONCLUSIONS

The questions raised in this chapter indicate definite needs for future research. In the author's opinion the following questions need to be resolved.

1. Since we are dealing with a perennial plant, any environmental or biological limitation that occurs in one season can have a profound effect on future seasons growth and yield. The effect may be obvious as would be the case with winter injury to the trunk of a tree, or it might be quite subtle, as is the case with premature defoliation. At present we do not have a "currency" within the plant that we can measure that will indicate that future effect. Research is needed in this area. What variable should be measured when should it be measured, and how can the variable be related to plant performance?
2. Quantitative date on the interaction between multiple environmental stresses and between environmental and biological stress are needed. For example how much damage can a tree sustain from an insect before economic damage results, and what effect would water stress have on this threshold.
3. The development of good predictive models that can help us anticipate the effect of an environmental constraint so that changes in management might be initiated to maximize future production have just begun to be developed. This area needs considerable emphasis in the future.
4. Are some stresses desirable? Is it possible to precondition a plant with an initial stress, or simulated stress, to increase the plants resistance to a future stress? How universal is this response?

ACKNOWLEDGMENTS

The author greatly acknowledges the assistance of Lynne Sage and Gloria Blake in the manuscript preparation.

REFERENCES

1. Westwood, M. N., *Temperate zone pomology*, 3, Timber Press, Portland, OR, 1993.
2. Faust, M., *Physiology of temperate zone fruit trees*, John Wiley & Sons, New York, 1989.
3. Flore, J. A., and Lakso, A. N., Environmental and physiological regulation of photosynthesis in fruit crops, *Hort. Rev.*, 11, 111, 1989.
4. Jones, H. G., Lakso, A. N., and Syvertsen, J. P., Physiological control of water status in temperate and subtropical fruit trees, *Hort. Rev.*, 7, 301, 1985.
5. Pearcy, R. W., Ehleringer, J., Mooney, H. A., and Rundel, P. W., *Plant physiological ecology, field methods and instrumentation*, Chapman and Hall, London, 1989.
6. Tukey, H. B., Growth of the peach embryo in relation to growth of fruit and season of ripening, *Proc. Amer. Soc. Hort. Sci.*, 30, 209, 1933.
7. Kappes, E. M., and Flore, J. A., Carbohydrate balance models for 'Montmorency' sour cherry leaves, shoots and fruits during development, *Acta Hort.*, 184, 123, 1986.
8. DeJong, T. M., and Goudriaan, J., Modeling peach fruit growth and carbohydrate requirements: reevaluation of the double-sigmoid pattern, *J. Amer. Soc. Hort. Sci.*, 114, 800, 1989.
9. Atkinson, D., The distribution and effectiveness of the roots of tree crops, *Hort. Rev.*, 2, 424, 1980.
10. Atkinson, K., and Wilson, S. A., The growth and distribution of fruit tree roots: some consequences for nutrient uptake, in *The mineral nutrition of fruit trees*, D. Atkinson, J. E. Jackson, R. O. Sharples and W. M. Waller, Eds., Butterworths, Borough Green, U. K., 1980, 259.

11. Beckman, T. B., Seasonal patterns of root growth potential (RGP) of 2 containerized cherry rootstocks. *P. mahaleb* L. and *P. avium* L. cv. Mazzard, Thesis for MS degree, Michigan State University, East Lansing, 1984.
12. Williamson, J. G., and Coston, D. C., The relationship among root growth, shoot growth, and fruit growth of peach, *J. Amer. Soc. Hort. Sci.*, 114, 180, 1989.
13. Flore, J. A., Longstroth, M., and McLean, M., unpublished data, 1993.
14. Eisensmith, S. P., Jones, A. L., and Flore, J. A., Predicting leaf emergence of sour cherry (*Prunus cerasus* L. 'Montmorency') from degree-day accumulations, *J. Amer. Soc. Hort. Sci.*, 104, 75, 1980.
15. Eisensmith, S. P., Jones, A. L., Goodman, E. D., and Flore, J. A., Predicting leaf expansion of 'Montmorency' sour cherry from degree-day accumulations, *J. Amer. Soc. Hort. Sci.*, 107, 717, 1982.
16. Grossman, Y. L., and DeJong, T. M., Carbohydrate requirements for dark respiration by peach vegetative organs, *Tree Physiol.*, in press, 1994.
17. Keller, J. D., and Loescher, W. H., Nonstructural carbohydrate partitioning in perennial parts of sweet cherry, *J. Amer. Soc. Hort. Sci.*, 114, 969, 1989.
18. Flore, J. A., unpublished observation, 1992.
19. Hammond, M. W., and Seeley, S. D., Spring bud development of *Malus* and *Prunus* species in relation to soil temperature, *J. Amer. Soc. Hort. Sci.*, 103, 655, 1978.
20. Anderson, J. L., Ashcroft, G. L., Richardson, E. A., Alfaro, J. F., Griffin, R. E., Hanson, G. R., and Keller, J., Effects of evaporative cooling on temperature and development of apple buds, *J. Amer. Soc. Hort. Sci.*, 100, 229, 1975.
21. Dennis, F. G., Can sprinkling save cherries?, *The Great Lakes Fruit Grower News*, 20, 5, 30, 1981.
22. Anon., Branches in the bag in Michigan orchards, *The Great Lakes Fruit Grower News*, 32, 6, 1, 1993.
23. Chang, L. S., Iezzoni, A. F., and Flore, J. A., Variation in yield components between 'Montmorency' and 'Meteor' sour cherry, *J. Amer. Soc. Hort. Sci.*, 112, 247, 1987.
24. Kenworthy, A. L., Sour cherry tree vigor as related to higher yields and better fruit quality, *Mich. Agr. Expt. Sta. Rpt.*, 223, 1974.
25. Johnson, R. S., Handley, D. F., and DeJong, T. M., Long-term response of early maturing peach trees to postharvest water deficits, *J. Amer. Soc. Hort. Sci.*, 117, 881, 1992.
26. Kappes, E. M., and Flore, J. A., The influence of phyllotoxy and stage of leaf and fruit development on the initiation and direction of gross carbohydrate export from sour cherry (*Prunus cerasus* L. 'Montmorency') leaves, *J. Amer. Soc. Hort. Sci.*, 114, 642, 1989.
27. Kappes, E. M., Carbohydrate production, balance and translocation in leaves, shoots and fruits of 'Montmorency' sour cherry, Ph. D. dissertation, Michigan State University, East Lansing, 1985.
28. Tukey, H. B., Embryo abortion in early-ripening varieties of *Prunus avium, Bot. Gaz.*, 44, 433, 1933.
29. Haun, F. R., and Coston, D. C., Relationship of daily growth and development of peach leaves and fruit to environmental factors, *J. Amer. Soc. Hort. Sci.*, 108, 666, 1983.
30. Diaz, D. H., Rasmussen, H. P., and Dennis, F. G., Jr., Scanning electron microscope examination of flower bud differentiation in sour cherry, *J. Amer. Soc. Hort. Sci.*, 106, 513, 1981.
31. Lang, G. A., Dormancy: a new universal terminology, *HortScience*, 22, 817, 1987.
32. Dowler, W. M., and King, F. D., Seasonal changes in starch and soluble sugar content of dormant peach tissues, *Proc. Amer. Soc. Hort. Sci.*, 89, 80, 1966.
33. Johnson, R. S., and Lakso, A. N., Carbon balance model of a growing apple shoot II. Simulated effects of light and temperature on long and short shoots, *J. Amer. Soc. Hort. Sci.*, 111, 164, 1986.
34. Jackson, J. E., and Hamer, P. M. C., The causes of year-to-year variation in the average yield of Cox's Orange Pippin apple in England, *J. Hort. Sci.*, 55, 149, 1980.
35. Flore, J. A., and Sams, C.E., Does photosynthesis limit yield of sour cherry (*Prunus cerasus*)?, in *Regulation of photosynthesis in fruit trees*, A. Lakso and F. Lenz, Eds., NY State Agr. Exp. Sta. Sp. Bull., 1986, 105.
36. Scurlock, J. M. O., Long, S. P., Hall, D. O., and Coombs, J., Introduction, in *Techniques in bioproductivity and photosynthesis*, J. Coombs, D. O. Hall, S. P. Long and J. M. O. Scurlock, Eds., Pergamon Press, New York, 1985, xxi.
37. Flore, J. A., unpublished results, 1993.
38. Sprugel, D. G., and Benecke, U., Measuring woody-tissue respiration and photosynthesis, in *Techniques and approaches in forest tree ecophysiology*, J. P. Lassoie and T. M. Hinckley, Eds., CRC Press, Boca Raton, FL, 1991, 329.

39. Grossman, Y. L., and DeJong, T. M., Peach: a simulation model of reproductive and vegetative growth in peach trees, *Tree Physiol.*, 14, 329, 1994.
40. DeJong, T. M., CO_2 assimilation characteristics of five Prunus tree fruit species, *J. Amer. Soc. Hort. Sci.*, 108, 303, 1983.
41. DeJong, T. M., and Doyle, J. F., Seasonal relationships between leaf nitrogen content photosynthetic capacity and leaf canopy light exposure in peach *(Prunus persica), Plant Cell Environ.*, 8, 701, 1985.
42. Flore, J. A., and Sams, C. E., Does sink strength control photosynthetic rate?, in *Regulation of photosynthesis in fruit crops*, T. M. DeJong, Ed., Symp. Proc. Publ. Univ. of California, Davis, 1986, 30.
43. Looney, N. E., Comparison of photosynthetic efficiency of two apple cultivars with their compact mutants, *Proc. Amer. Soc. Hort. Sci.*, 92, 34, 1968.
44. Sams, C. E., and Flore, J. A., Factors affecting net photosynthetic rate of sour cherry (*Prunus cerasus* L. 'Montmorency') during the growing season, *Photosynthetic Res.*, 4, 307, 1986.
45. Roper, T. R., and Kennedy, R. A., Photosynthetic characteristics during leaf development in 'Bing' sweet cherry, *J. Amer. Soc. Hort. Sci.*, 111, 938, 1986.
46. Andersen, P. C., and Brodbeck, B. V., Water relations and net CO_2 assimilation of peach leaves of different ages, *J. Amer. Soc. Hort. Sci.*, 113, 242, 1988.
47. Downtown, W. J. S., Grant, W. J. R., and Loveys, B. R., Diurnal changes in the photosynthesis of field-grown grape vines, *New Phytol.*, 105, 71, 1987.
48. Gucci, R., The effect of fruit removal on leaf photosynthesis, water relations, and carbohydrate partitioning in sour cherry and plum, Ph.D. dissertation, Michigan State University, East Lansing, 1986.
49. Gucci, R., Xiloyannis, C., and Flore, J. A., Diurnal and seasonal changes in leaf net photosynthesis following fruit removal in plum, *Physiol. Plant.*, 83, 497, 1991.
50. Satoh, M., Kriedemann, P. E., and Loveys, B. R., Changes in photosynthetic activity and related processes following decapitation in mulberry trees, *Physiol. Plant*, 41, 203, 1977.
51. Layne, D. R., and Flore, J. A., Photosynthetic compensation to partial leaf area reduction in sour cherry, *J. Amer. Soc. Hort. Sci.*, 117, 279, 1992.
52. Flore, J. A, The influence of summer pruning on the physiology and morphology of stone fruit trees, *Acta Hort.*, 322, 257, 1992.
53. Rom, C. R., and Ferree, D. C., Time and severity of summer pruning influences on young peach tree net photosynthesis, transpiration, and dry weight distribution, *J. Amer. Soc. Hort. Sci.*, 110, 455, 1985.
54. Chalmers, D. J., Canterford, R. L., Jerrie, P. H., Jones, T. R., and Ugalde, T. D., Photosynthesis in relation to growth and distribution of fruit in peach trees, *Austral. J. Plant Physiol.*, 2, 634, 1975.
55. Crews, C. E., Williams, S. L., and Vines, H. M., Characteristics of photosynthesis in peach leaves, *Planta*, 126, 97, 1975.
56. DeJong, T. M., Effects of reproductive and vegetative sink activity on leaf conductance and water potential in *Prunus persica* (L.) Batsch, *Sci. Hort.*, 29, 131, 1986.
57. DeJong, T. M., Fruit effects on photosynthesis in *Prunus* persica, *Physiol. Plant.*, 66, 149, 1986.
58. Gucci, R., Petracek, P. D., and Flore, J. A., The effect of fruit harvest on photosynthetic rate, starch content, and chloroplast ultrastructure in leaves of *Prunus avium, Adv. Hort. Sci.*, 5, 19, 1991.
59. Layne, D. R., Physiological responses of *Prunus cerasus* to whole plant source manipulation, Ph.D. dissertation, Michigan State University, East Lansing, 1992.
60. Layne, D. R., and Flore, J. A., Physiological responses of *Prunus cerasus* to whole-plant source manipulation. Leaf gas exchange, chlorophyll fluorescence, water relations and carbohydrate concentrations, *Physiol. Plant*, 88, 44, 1993.
61. Palmer, J. W., Annual dry matter production and partitioning over the first 5 years of a bed system of Crispin-M.27 apple trees at four spacings, *J. Appl. Ecol.*, 25, 569, 1988.
62. Flore, J. A., and Layne, D. R., The influence of tree shape and spacing on light interception and yield in sour cherry (*Prunus cerasus* cv. 'Montmorency'), *Acta Hort.*, 285, 91, 1990.
63. Sams, C. E. and Flore, J. A., The influence of age, position and environmental variables on net photosynthetic rate of sour cherry leaves, *J. Amer. Soc. Hort. Sci.*, 107, 339, 1982.
64. Kappel, F., and Flore, J. A., Effect of shade on photosynthesis, specific leaf weight, leaf chlorophyll content, and morphology of young peach tree, *J. Amer. Soc. Hort. Sci.*, 108, 541, 1983.
65. Kappel, F., Characterization of the light microclimate in four peach tree canopies and the effect of shading on the growth and leaf photosynthesis of peach trees, M. S. Thesis, Michigan State Univ., East Lansing, 1980.

66. Powles, S. B., Photoinhibition of photosynthesis induced by visible light, *Annu. Rev. Plant Physiol.*, 35, 15, 1984.
67. Catania, M., and Flore, J. A., unpublished data, 1992.
68. Sams, C. E., Factors affecting the leaf and shoot morphology and photosynthetic rate of sour cherry (*Prunus cerasus* L. 'Montmorency'), Ph.D. dissertation, Michigan State University, East Lansing, 1980.
69. Jackson, J. E., Light interception and utilization by orchard systems, *Hort. Rev.*, 2, 208, 1980.
70. Proctor, J. T. A., Apple photosynthesis: microclimate of the tree and orchard, *HortScience*, 13, 641, 1975.
71. Cain, J. C., Foliage canopy development of McIntosh apple hedgerows in relation to mechanical pruning, the interception of solar radiation and fruiting, *J. Amer. Soc. Hort. Sci.*, 98, 357, 1973.
72. Jackson, J. E., Utilization of light resources by high density planting systems, *Acta. Hort.*, 65, 61, 1978.
73. Kappel, F., Flore, J. A., and Layne, R. E. C., Characterization of the light microclimate in four hedgerow canopies, *J. Amer. Soc. Hort. Sci.*, 108, 102, 1983.
74. Flore, J. A., and Sage, L. E., unpublished results, 1983.
75. Flore, J. A., The effect of light on cherry trees, *Annu. Rpt. Mich. State Hort. Soc.*, 110, 119, 1980.
76. Flore, J. A., Influence of light interception on cherry production and orchard design, *Annu. Rpt. Mich. St. Hort. Soc.*, 111, 161, 1981.
77. Flore, J. A., and Kesner, C., Orchard design for stone fruit based on light interception, *Compact Fruit Tree*, 25, 159, 1982.
78. Flore, J. A., and Sams, C. E., unpublished data, 1980.
79. Proctor, J. T. A., Kyle, W. I., and Davies, J. A., The penetration of global solar radiation into apple trees, *J. Amer. Soc. Hort. Sci.*, 100, 40, 1975.
80. Marini, R. P., Vegetative growth, yield, and fruit quality of peach as influenced by dormant pruning, summer pruning, and summer topping, *J. Amer. Soc. Hort. Sci.*, 110, 133, 1985.
81. Erez, A. and Flore, J. A., The quantitative effect of solar radiation on anthocyanin production in 'Redhaven' peach fruit skin color, *HortScience*, 21, 1424, 1986.
82. Langord, L. R., Seasonal influences upon the effect of shading in regard to setting of sour cherry fruits, *Proc. Amer. Soc. Hort. Sci.*, 33, 234, 1935.
83. Grappadelli, L. C., Effects of flowering and fruit quality in peach [*Prunus persica* (L.) Batsch], MS Thesis, Clemson Univ., Clemson, SC, 1985.
84. Marini, R. P., Sowers, D., and Marini, M. C., Peach fruit quality is affected by shade during final swell of fruit growth, *J. Amer. Soc. Hort. Sci.*, 116, 383, 1991.
85. Flore, J. A., Howell, G. S., and Sams, C. E., The effect of artificial shading on cold hardiness of peach and sour cherry, *HortScience*, 18, 321, 1983.
86. Flore, J. A., Howell, G. S., Jr., and Perry, R. L., High density peach production related to cold hardiness, *Compact Fruit Tree*, 20, 60, 1987.
87. Young, E., and Werner, D. J., Effects of rootstock and scion chilling during rest on resumption of growth in apple and peach, *J. Amer. Soc. Hort. Sci.*, 109, 548, 1984.
88. Erez, A., and Couvillon, G. A., Characterization of the influence of moderate temperatures on rest completion in peach, *J. Amer. Soc. Hort. Sci.*, 112, 677, 1987.
89. Felker, C. F., and Robitaille, H., Chilling accumulation and rest of sour cherry flower buds, *J. Amer. Soc. Hort. Sci*, 110, 227, 1985.
90. Gilreath, P. R., and Buchannan, D. W., Rest prediction model for low-chilling 'Sungold' nectarine, *J. Amer. Soc. Hort. Sci*, 106, 426, 1981.
91. Richardson, W. A., Seeley, S. D., and Walker, D. R., A model for estimating the completion of rest for 'Redhaven' and 'Elberta' peach trees, *HortScience*, 9, 331, 1974.
92. Flore, J. A., and Houle, M. E., unpublished data, 1983.
93. Berry, J. A., and Bjorkmann, O., Photosynthetic response and adaptation to temperature in higher plants, *Annu. Rev. Plant Physiol.*, 31, 491, 1980.
94. Caldwell, J., Hancock, J. F., and Flore, J. A., Strawberry leaf photosynthesis acclimation to temperature (abstr.), *HortScience*, 25, 1166, 1990.
95. Lange, O., Schulze, E.-D., Evenari, M., Kappen, L., and Buschbom, U., The temperature-related photosynthetic capacity of plants under desert conditions. I. Seasonal changes in the photosynthetic response to temperature, *Oecologia* (Berl.), 17, 97, 1978.

96. Flore, J. A., Moon, J. W., and Lakso, A. N., The effect of water stress and vapor pressure gradient on stomatal conductance, water use efficiency, and photosynthesis of fruit crops, *Acta Hort.*, 171, 209, 1985.
97. Beckman, T. G., Perry, R. L., and Flore, J. A., Short-term flooding affects gas exchange characteristics of containerized sour cherry trees, *HortScience*, 27, 1297, 1992.
98. DeJong, T. M., personal communication, 1993.
99. Gucci, R., personal communication, 1993.
100. Parker, J., Seasonal changes in cold resistance and free sugars in some hardwood tree barks, *For. Sci.*, 8, 225, 1963.
101. Ashworth, E. N., Formation and spread of ice in plant tissues, *Hort. Rev.*, 13, 215, 1992.
102. Ashworth, E. N., and Wisniewski, M. E., Response of fruit tree tissues to freezing temperatures, *HortScience*, 26, 501, 1991.
103. Burke, M. J., Gusta, L. V., Quamme, H. A., Weiser, C. J., and Li, P. H., Freezing and injury in plants, *Annu. Rev. Plant Physiol.*, 27, 507, 1976.
104. Burke, M. J., and Stushnoff, C., Frost hardiness: a discussion of possible molecular causes of injury with particular reference to deep supercooling of water, in *Stress physiology in crop plants*, H. Mussel and R. C. Staples, Eds., Wiley, NY, 1979.
105. Flore, J. A., and Howell, G. S., Environmental and physiological factors that influence cold hardiness and frost resistance in perennial crops, in *Int. conf. on agrometeorology*, I. Prodi, F. Rossi and G. Cristoferi, Eds., Fondazione Cesene Agr. Publ., Cesana, Italy, 1987, 139.
106. Weiser, C. J., Cold resistance and acclimation in woody plants, *HortScience*, 5, 403, 1970.
107. Ashworth, E. N., Deep supercooling in woody plant tissues, in *Advances in plant cold hardiness*, P. H. Li, and L. Christersson, Eds., CRC Press, Boca Raton, FL, 1993, 203.
108. Quamme, H. A., and Gusta, L. V., Relationships of ice nucleation and water status to freezing patterns in dormant peach flower buds, *HortScience*, 22, 465, 1987.
109. Ashworth, E. N., Rowse, D. J., and Billmeyer, L. A., The freezing of water in woody tissues of apricot and peach and the relationship to freezing injury, *Plant Cell Environ.*, 12, 521, 1989.
110. Norris, E. R., Stout, B. A., and Mecklenburg, R. A., Thermal studies of fruit trees in relation to low temperature trunk splitting, *Amer. Soc. Ag. Eng.*, 71, 344, 1971.
111. Layne, D. R., and Flore, J. A., Response of young, fruiting sour cherry trees to one-time trunk injury at harvest date, *J. Amer. Soc. Hort. Sci.*, 116, 851, 1991.
112. Proebsting, E. L., Adapting cold hardiness concepts to deciduous fruit culture, in *Plant cold hardiness and freezing stress. Mechanisms and crop implications*, P. H. Li and A. Sakai, Eds., Academic Press, New York, 1978, 267.
113. Howell, G. S., Jr., and Weiser, C. J., The environmental control of cold acclimation in apple, *Plant Physiol.*, 45, 390, 1970.
114. Holubowicz, T., The effect of defoliation and the age of the leaves on development of frost resistance of fruit shoots, *XXI Intl. Hort. Cong.*, 1, 1008 (abstr.), 1982.
115. Howell, G. S., and Weiser, C. J., Fluctuations in the cold resistance of apple twigs during spring dehardening, *J. Amer. Soc. Hort. Sci.*, 95, 190, 1970.
116. Andrews, P. K., and Proebsting, E. L., Jr., Effects of temperature on the deep supercooling characteristics of dormant and deacclimating sweet cherry flower buds, *J. Amer. Soc. Hort. Sci.*, 112, 334, 1987.
117. Proebsting, E. L., and Mills, H. H., A comparison of hardiness responses in fruit buds of 'Bing' cherry and 'Elberta' peach, *J. Amer. Soc. Hort. Sci*, 97, 802, 1972.
118. Bradford, F. C., and Cardinell, H. A., Eighty winters in Michigan orchards, *Michigan AES Bull.*, 149, 1926.
119. Howell, G. S., Flore, J. A., and Dittmer, T., Factors affecting cold hardiness of peach, *Annu. Rpt. Mich. State Hort. Soc.*, 118, 137, 1988.
120. Dennis, F. G., Jr., Carpenter, W. S., and MacLean, N. J., Cold hardiness of 'Montmorency' sour cherry flower buds during spring development, *HortScience*, 10, 529, 1975.
121. Chandler, W. H., The killing of plant tissue by low temperature, *Univ. Missouri Res. Bull.*, 8, 1913.
122. Blake, M. A., Relative hardiness of 157 varieties of peaches and nectarines in 1933, and of 14 varieties in 1934 at New Brunswick, N. J., *N. J. Agr. Expt. Sta. Cir.*, 303, 1934.
123. Cain, D. W., and Andersen, R. L., Sampling procedures for minimizing non-generic wood hardiness variation in peach, *J. Amer. Soc. Hort. Sci*, 101, 668, 1976.

124. Gucci, R., Flore, J. A., and Perry, R. L., A two-year study on cold hardiness of 'Redhaven' peach shoots as influenced by eight different rootstocks, *Compact Fruit Tree*, 21, 113, 1988.
125. Layne, R. E. C., Jackson, H. O., and Stroud, F. D., Influence of peach seedling rootstocks on defoliation and cold hardiness of peach cultivars, *J. Amer. Soc. Hort. Sci.*, 102, 89, 1977.
126. Layne, R. E. C., and Ward, G. M., Rootstock and seasonal influences on carbohydrate levels and cold hardiness of 'Redhaven' peach, *J. Amer. Soc. Hort. Sci.*, 103, 408, 1978.
127. Scorza, R., Lightner, G. W., Gilreath, L. I., and Wolf, S. J., Reduced-stature peach tree growth types: pruning and light penetration, *Acta Hort.*, 146, 159, 1991.
128. Perry, R. L., Cherry Rootstocks, in *Rootstocks for fruit crops*, R. C. Rom and R. F. Carlson, Eds., John Wiley & Sons, Inc., New York, 1987, 217.
129. Crossa-Raynaud, P., and Audergon, J. M., Apricot rootstocks, in *Rootstocks for fruit crops*, R. C. Rom and R. F. Carlson, Eds., John Wiley & Sons, Inc., New York, 1987, 295.
130. Layne, R. E. C., Peach rootstocks, in *Rootstocks for fruit crops*, R. C. Rom and R. F. Carlson, Eds., John Wiley & Sons, Inc., New York, 1987, 18.
131. Howell, G. S., personal communication, 1993.
132. Flore, J. A., Howell, G. S., Jr., Gucci, R., and Perry, R. L., High density peach production related to cold hardiness, *Compact Fruit Tree*, 20, 60, 1987.
133. Lasheen, A. M., Choplin, C. E., and Harmon, R. N., Biochemical comparison of fruit buds of five peach cultivars of varying degrees of cold hardiness, *J. Amer. Soc. Hort. Sci.*, 95, 177, 1970.
134. Howell, G. S., Jr., and Stackhouse, S. S., The effect of defoliation time on acclimation and dehardening in tart cherry, *J. Amer. Soc. Hort. Sci.*, 98, 132, 1986.
135. Cooper, J. R., Factors affecting winter injury to peach trees, *Univ. of Arkansas Agr. Exp. Sta. Bull.*, 536, 1953.
136. Levitt, J., *Frost killing and hardiness of plants. A critical review*, Burgess, Minneapolis, MN, 1941.
137. Savage, E. F., Cold injury as related to cultural management and possible protective devices for dormant peach trees, *HortScience*, 5, 425, 1970.
138. Howell, G. S., Jr., and Dennis, F. G., Jr., Cultural management of perennial plants to maximize resistance to cold stress, in *Analysis and improvement of plant cold hardiness*, C. R. Olien and M. N. Smith, Eds., CRC Press, Boca Raton, FL, 1981, 175.
139. Stergios, B. G., and Howell, G. S., Jr., Effects of defoliation, trellis, height, and cropping stress on the cold hardiness of Concord grape, *Amer. J. Enol. Vitic.*, 28, 35, 1977.
140. Edgerton, L. J., and Shaulis, N. J., The effect of time of pruning on cold hardiness of Concord grape vines, *Proc. Amer. Soc. Hort. Sci.*, 62, 209, 1953.
141. Marini, R. P., Defoliation, flower bud cold hardiness, and bloom date of peach as influenced by pruning treatment, *J. Amer. Soc. Hort. Sci.*, 111, 391, 1986.
142. Hayden, R. A., and Emerson, F. H., Pruning high density peach hedge plantings, *Compact Fruit Tree*, 12, 76, 1979.
143. Walsh, C. S., and Thompson, A. H., Influence of the time of summer shearing on yield and fruit quality in hedgerow peach plantings, *HortScience*, 19, 565 (abstr.), 1984.
144. Bittenbender, H. C., and Howell, G. S., Jr., Interaction of temperature and moisture content on spring deacclimation of flower buds of highbush blueberry, *Can. J. Plant Sci.*, 55, 447, 1975.
145. Johnston, E. S., Moisture relations of peach buds during winter and spring, *Maryland Agric. Exp. Sta. Bull.*, 225, 59, 1923.
146. Layne, R. E. C., and Tan, C. S., Long term influence of irrigation and tree density on growth, survival, and production of peach, *J. Amer. Soc. Hort. Sci.*, 109, 795, 1984.
147. Flore, J. A., unpublished data, 1988.
148. Gucci, R., and Flore, J. A., La resistenza al freddo nel pesco in relazione a fattori ambientali e pratiche colturali (The resistance of peach to cold in relation to environmental and cultural conditions), *Frutticoltura*, 51, 13, 1989.
149. Kozlowski, T. T., Kramer, P. J., and Pallardy, S. G., *The physiological ecology of woody plants*, Academic Press, San Diego, 1991.
150. Schaffer, B., Andersen, P. C., and Ploetz, R. C., Responses of fruit crops to flooding, *Hort. Rev.*, 13, 257, 1992.
151. Kozlowski, T. T., Responses of woody plants to flooding, in *Flooding and plant growth*, T. Kozlowski, Ed., Academic Press, Orlando, FL, 1984, 129.

152. Kenworthy, A. L., *Tree fruit and landscape plants*, Dept. of Hort., Michigan State Univ., East Lansing, MI, 1978.
153. Rieger, M., and Dummel, M. J., Comparison of drought resistance among *Prunus* species from divergent habitats, *Tree Physiol.*, 11, 369, 1992.
154. Rom, R. C., and Carlson, R. F., Eds., *Rootstocks for fruit crops*, John Wiley & Sons, Inc., New York, 1987.
155. Couvillon, G. A., Rieger, M., Harrison, R., and Daniell, J., Stress-mediated responses of own rooted peach cultivars, *Acta Hort.*, 243, 1989.
156. Kjelgren, R. K., and Taylor, B. H., Rootstock effect on the water relations of 'Redhaven' peach, *XXIII Intl. Hort. Cong.*, Firenze, 1633, 1990.
157. Chaplin, E. E., Schneider, G. W., and Martin D. C., Rootstock effect on peach tree survival on a poorly drained soil, *HortScience*, 9, 28, 1974.
158. Rom, R. C., and Brown, S. A., Water tolerance of apples on clonol rootstocks and peaches on seedling rootstocks, *Compact Fruit Tree*, 12, 30, 1979.
159. Norton, R. A., Hansen, C. J., O'Reilly, H. J., and Hart, W. H., Rootstocks for sweet cherries in California, *Calif. Agr. Exp. Sta. Ext. Serv. Lft.*, 159, 1963.
160. Beckman, T. G., Flooding tolerance of sour cherries, Ph.D. dissertation, Michigan State University, East Lansing, 1988.
161. Okie, W. R., Plum rootstocks, in *Rootstocks for Fruit Crops*, R. C. Rom and R. F. Carlson, Eds., John Wiley & Sons, New York, 1987.
162. Chalmers, D. J., and Wilson, I. B., Productivity of peach trees: tree growth and water stress in relation to fruit growth and assimilate demand, *Ann. Bot.*, 42, 285, 1978.
163. Kozlowski, T. T., Diurnal changes in diameters of fruits and tree stems of Montmorency cherry, *J. Hort. Sci.*, 43, 1, 1968.
164. Olien, M. E., and Flore, J. A., Effect of a rapid water stress and a slow water stress on the growth of 'Redhaven' peach trees, *Fruit Var. J.*, 44, 4, 1990.
165. Nobel, P. S., *Physicochemical and environmental plant physiology*, Academic Press, San Diego, 1991.
166. Young, E., Hand, J. M., and West, S. C., Osmotic adjustment and stomatal conductance in peach seedlings under severe water stress, *HortScience*, 17, 791, 1982.
167. Ranney, T. G., Bassuk, N. L., and Whitlow, T. H., Osmotic adjustment and solute constituents in leaves and roots of water-stressed cherry (*Prunus*) trees, *J. Amer. Soc. Hort. Sci.*, 116, 684, 1991.
168. Neri, D., and Flore, J. A., Effect of soil moisture stress and root pruning on ABA content, photosynthesis and root hydraulic conductivity of "Cresthaven" peach trees with split root systems, *XIII ISHS Congress*, 1, 1612, 1990.
169. Xiloyannis, C., Uriu, K., and Martin, G. C., Seasonal and diurnal variations in abscisic acid, water potential and diffusive resistance in leaves from irrigated and non-irrigated peach trees, *J. Amer. Soc. Hort. Sci.*, 105, 412, 1980.
170. Tan, C. S., and Buttery, B. R., The effect of soil moisture stress to various fractions of the root system on transpiration, photosynthesis, and internal water relations of peach seedlings, *J. Amer. Soc. Hort. Sci.*, 107, 845, 1982.
171. Harrison, R. D., Daniell, J. W., and Cheshir, J. M., Jr., Net photosynthesis and stomatal conductance of peach seedlings and cuttings in response to changes in soil water potential, *J. Amer. Soc. Hort. Sci.*, 114, 986, 1989.
172. Richards, D., Root-shoot interactions: a functional equilibrium for water uptake in peach (*Prunus persica* L. Batsch.), *Ann. Bot.*, 41, 279, 1976.
173. Richards, D., and Rowe, R. N., Effects of root restriction, root pruning, and 6-benzylaminopurine on the growth of peach seedlings, *Ann. Bot.*, 41, 729, 1977.
174. Layne, R. E. C., Tan, C. S., and Perry, R. L., Characterization of peach roots in fox sand as influenced by sprinkler irrigation and tree density, *J. Amer. Soc. Hort. Sci.*, 111, 670, 1986.
175. Chalmers, D. J., Olsson, K. A., and Jones, T. R., Water relations of peach trees and orchards, in *Water deficits and plant growth*, T. T. Kozlowski, Ed., Academic Press, New York, 7, 1983, 197.
176. Ranney, T. G., Bassuk, N. L., and Whitlow, T. H., Turgor maintenance in leaves and roots of 'Colt' cherry trees (*Prunus avium* × *pseudocerasus*) in response to water stress, *J. Hort. Sci.*, 66, 381, 1991.
177. Proebsting, E. L., Jr., and Middleton, J. E., The behavior of peach and pear trees under extreme drought stress, *J. Amer. Hort. Sci.*, 105, 380, 1980.

178. Proebsting, E. L., Jr., Middleton, J. E., and Mahan, M. O., Performance of bearing cherry and prune trees under very low irrigation rates, *J. Amer. Soc. Hort. Sci.*, 16, 243, 1981.
179. Mitchell, P. P., and Chalmers, D. J., The effect of reduced water supply on peach tree growth and yields, *J. Amer. Soc. Hort. Sci.*, 107, 853, 1982.
180. Chalmers, D. J., Mitchell, P. D., and van Heek, L., Control of peach tree growth and productivity by regulated water supply, tree density, and summer pruning, *J. Amer. Soc. Hort. Sci.*, 106, 307, 1981.
181. Chalmers, D. J., Mitchell, P. D., and Jerie, P. H., The physiology of growth control of peach and pear trees using reduced irrigation, *Acta Hort.*, 146, 143, 1984.
182. Chalmers, D. J., and van den Ende, B., Productivity of peach trees: factors affecting dry-weight distribution during tree growth, *Ann. Bot.*, 39, 423, 1974.
183. Girona, J., Ruiz-Sanchez, M. C., Goldhamer, D., Johnson, S., and DeJong, T., Late maturing peach response to controlled deficit irrigation: seasonal and diurnal patterns of fruit growth, plant and soil water status, CO_2 uptake, and yield, *XXIII ISHS Congress*, 1, 1632, 1990.
184. Brown, D. S., The effects of irrigation on flower bud development and fruiting in the apricot, *Proc. Amer. Soc. Hort. Sci.*, 61, 119, 1953.
185. Uriu, K., Effect of postharvest soil moisture depletion on subsequent yield of apricots, *Proc. Amer. Soc. Hort. Sci.*, 84, 93, 1964.
186. Veihmeyer, F. J., The growth of fruit trees in response to different soil moisture conditions measured by widths of annual rings and other means, *Soil Sci.*, 119, 448, 1975.
187. Larson, K. D., DeJong, T. M., and Johnson, R. S., Physiological and growth responses of mature peach trees to postharvest water stress, *J. Amer. Soc. Hort. Sci.*, 113, 296, 1988.
188. McCutchan, H., and Shackel, K. A., Stem-water potential as a sensitive indicator of water stress in prune trees (*Prunus domestica* L. cv. French), *J. Amer. Soc. Hort. Sci.*, 117, 607, 1992.
189. Garnier, E., and Berger, A., Testing water potential in peach trees as an indicator of water stress, *J. Hort. Sci.*, 60, 47, 1985.
190. Worthington, J. W., McFarland, M. J., and Rodgigue, P., Water requirements of peach as recorded by weighing lysimeters, *HortScience*, 19, 90, 1984.
191. Glenn, D. M., Worthington, J. W., Welker, W. V., and McFarland, M. J., Estimation of peach tree water use using infrared thermometry, *J. Amer. Soc. Hort. Sci.*, 114, 737, 1989.
192. Baker, J. M., and van Bavel, C. H. M., Measurement of mass flow of water in stems of herbaceous plants, *Plant Cell Environ.*, 10, 777, 1987.
193. Sakuratani, T., A heat balance method for measuring water flux in the stem of intact plants, *J. Agr. Meterol.*, 37, 9, 1981.
194. Steinberg, S. L., van Bavel, C. H. M., and McFarland, M. J., Improved sap flow gauge for woody and herbaceous plants, *Agron. J.*, 82, 851, 1990.
195. Steinberg, S. L., van Bavel, C. H. M., and McFarland, M. J., Comparison of trunk and branch sap flow with canopy transpiration in pecan, *J. Expt. Bot.*, 41, 653, 1990.
196. Steinberg, S. L., van Bavel, C. H. M., and McFarland, M. J., A gauge to measure mass flow rate of sap in stems and trunks of woody plants, *J. Amer. Soc. Hort. Sci.*, 114, 466, 1989.
197. Shackel, K. A., Johnson, R. S., Medawar, C. K., and Phene, C. J., Substantial errors in estimates of sap flow using the heat balance technique on woody stems under field conditions, *J. Amer. Soc. Hort. Sci.*, 117, 351, 1992.
198. Proebsting, E. L., Jerie, P. H., and Irvine, J., Water deficits and rooting volume modify peach tree growth and water relations, *J. Amer. Soc. Hort. Sci.*, 114, 368, 1989.
199. Rowe, R. N., and Beardsell, P. V., Waterlogging of fruit trees, *Hort. Abstr.*, 43, 533, 1973.
200. Rowe, R. N., and Catlin, P. B., Differential sensitivity to waterlogging and cyanogenesis by peach, apricot, and plum roots, *J. Amer. Soc. Hort. Sci.*, 96, 305, 1971.
201. Andersen, P. C., Lombard, P. B., and Westwood, M. N., Leaf conductance, growth and survival of willow and deciduous fruit tree species under flooded soil conditions, *J. Amer. Soc. Hort. Sci.*, 109, 132, 1984.
202. Andersen, P. C., Lombard, P. B., and Westwood, M. N., Effect of root anaerobiosis on the water relations of several *Pyrus* species, *Physiol. Plant*, 62, 245, 1984.
203. Salesses, G., and Juste, C., Recherches sur l'asphyxie radiculaire des arbres fruitieres a noyau 11. Comportement des porte-greffes de types pecher et prunier: Etude de leur teneru en amygdaline et des facteures intervenaut dans l'hydrolyse de celle-ci, *Ann. Amelior. Plantes*, 21, 265, 1971.

204. Beckman, T. G., Flore, J. A., and Perry, R. L., Flooding tolerance in sour cherry, *Compact Fruit Tree*, 19, 136, 1986.
205. Patrick, W. H., Jr., and Mahapatra, I. C., Transformation and availability to rice of nitrogen and phosphorus in waterlogged soils, *Adv. Agron.*, 20, 323, 1968.
206. Rowe, R. N., Anaerobic metabolism and cyanogenic glycoside hydrolysis in differential sensitivity of peach, plum and pear roots under water-saturated conditions, Ph.D. dissertation, Univ. Calif., Davis, 1966.
207. Syvertsen, J. P., Zablotowicz, R. M., and Smith, M. L., Jr., Soil temperature and flooding effects on two species of citrus. I. Plant growth and hydraulic conductivity, *Plant Soil*, 72, 3, 1983.
208. Zhang, J., and Davies, W. J., ABA in roots and leaves of flooded pea plants, *J. Exp. Bot.*, 38, 649, 1983.
209. Beckman, T. G., Flore, J. A., and Perry, R. L., Sensitivity of various growth indices and the production of a translocatable photosynthesis inhibitor by containerized one-year-old tart cherry trees during soil flooding, *HortScience*, 22, 1028 (abstr.), 1987.
210. Berstein, L., Brown, J. V., and Hayward, H. E., The influence of rootstocks on growth and salt accumulation in stone fruit trees and almonds, *Proc. Amer. Soc. Hort. Sci.*, 68, 86, 1956.
211. Wadleigh, C. H., Hayward, H. E., and Ayers, A. D., First year growth of stone fruit trees on saline substrates, *Proc. Amer. Soc. Hort. Sci.*, 57, 31, 1951.
212. Hayward, H. E., and Long, E. M., Vegetative responses of the 'Elberta' peach on 'Lovell' and 'Shalil' rootstocks to high chloride and sulfate solutions, *Proc. Amer. Soc. Hort. Sci*, 41, 149, 1942.
213. Retzlaff, W. A., Williams, L. E., and DeJong, T. M., The effect of different atmospheric ozone partial pressures on photosynthesis and growth of nine fruit and nut tree species, *Tree Physiol.*, 8, 93, 1991.
214. Retzlaff, W. A., Williams, L. E., and DeJong, T. M., Photosynthesis, growth, and yield response of 'Casselman' plum to various ozone partial pressures during orchard establishment, *J. Amer. Soc. Hort. Sci.*, 117, 703, 1992.
215. Heggestad, H. E., Air pollution injuries to stone fruit trees, *Virus diseases and noninfectious Disorders of Stone Fruits in North America*, U.S.D.A. Agric. Handbook, 437, 355, 1976.
216. Zimmerman, P. W., and Hitchcock, A. E., Susceptibility of plants to hydrofluoric acid and sulfur dioxide gasses, *Boyce Thompson Inst. Contrib.*, 18, 263, 1956.
217. Rinallo, C., Effects of acidity of simulated rain on the fruiting of Summered apple trees, *J. Environ. Qual.*, 21, 61, 1992.

Chapter 10

Strawberry

Kirk D. Larson

CONTENTS

I. INTRODUCTION

The modern cultivated strawberry (*Fragaria x ananassa* Duch.) is the most widely distributed fruit crop world wide due to genotypic diversity and a broad range of environmental adaptation.[1-3] World production of strawberries is increasing. The major strawberry producing countries are the United States, Poland, Japan, Italy, Spain, the former USSR, Republic of South Korea, and Mexico,[4] with Europe accounting for nearly half of the world's production. In the United States, California produces over 75% of the commercial crop, followed by Florida, Oregon, Michigan, Louisiana, and Washington. Many cultivars have specific regional adaptation due to critical photoperiod and temperature requirements,[5] and strawberry cultural systems are highly variable.[2,3,6]

A. ORIGIN AND GENETICS

Fragaria x ananassa is a monoecious octoploid hybrid of two largely dioecious octoploid species, *F. chiloensis* (L.) Duch. and *F. virginiana* Duch.[2,6] Some strawberry cultivars also derive from a subspecies of *F. virginiana, F. virginiana glauca* Staudt.[2,3,6] Hybridization of *F. chiloensis* and *F. virginiana* occurred spontaneously in Europe in the 1700s when female plants of *F. chiloensis* of Chilean origin were grown in proximity to male *F. virginiana* plants of North American origin.[2,3,6] Since that time, extensive

0-8493-175-0/94/$0.00+$.50

hybridization between the parent species and their descendents has occurred, making *F. x ananassa* a highly variable, highly heterozygous species.

There is a wide range of morphological and physiological characteristics among individual strawberry cultivars, with considerable variation in environmental responses.[2,3,7,8] The progenitor species of *F. x ananassa* exhibit considerable ecological variability, due in part to dioecy.[9,10] *F. chiloensis*, which is indigenous to Chile, Hawaii, and the Pacific coast of North America, typically has thicker leaves, more leaf mesophyll tissue and higher net CO_2 assimilation rates (A) (on a leaf area basis) than *F. virginiana*, a native of eastern and central North America.[2] *F. chiloensis* generally is considered more tolerant of drought and salinity, but less tolerant of heat and cold, than *F. virginiana*.[2]

Strawberry cultivars in the eastern and northern United States may have a larger proportion of *F. virginiana* than *F. chiloensis* characteristics, probably as a result of selection in an environment best suited to that species.[2] Modern-day California cultivars may have a relatively larger percentage of germplasm derived from *F. chiloensis* than cultivars developed in other parts of the United States.[11] Present-day cultivars can be grouped according to geographic origin, with the greatest genetic differences occurring between those cultivars developed in California and those developed outside of California.[12] Based on genetic differences, cultivars developed outside of California can be grouped into two geographical regions, a western or northern group, and a southern or eastern group. Thus, physiological differences among cultivars are partly due to the extent to which individual parent species or clones have contributed to a particular genotype.[2]

In addition to *F. x ananassa*, the genus *Fragaria* includes at least eleven other species, including diploids, tetraploids, octoploids, and a hexaploid.[2,3,6] Many of these species have been cultivated at one time or another, and some are still grown on a limited basis.[6] Environmental physiology studies of *Fragaria* have often focused on species other than *F. x ananassa*, and thus do not necessarily describe the physiological responses of the modern cultivated strawberry.

B. BOTANY

The anatomy, morphology, and growth habit of the strawberry plant have been described in detail in numerous publications.[2,7,13-20] In brief, the plant is a determinate perennial in which the stem is compressed into a rosetted crown,[6] with internodes about 2 mm in length.[7] Axillary buds in the leaf nodes of the crown either remain dormant, or develop into branch crowns, or stolons (runners), depending on prevailing environmental conditions. Inflorescences form terminally, and vegetative growth is continued by the uppermost axillary bud of the crown, resulting in a sympodial growth habit. Although the strawberry is often referred to as an herbaceous perennial,[7,15,21,22] aging results in lignification of the crown, producing a hard, woody tissue.[6] With time, extensive branch crown development produces a structure resembling a highly compacted tree scaffold,[6,8] with certain physiological responses to the environment similar to those of deciduous fruit trees.[6] However, vegetative and reproductive growth of strawberry are more sensitive to environmental conditions, particularly photoperiod and temperature, than most other fruit crops.[3,6-8] Studies concerning the environmental physiology of strawberries must be interpreted carefully due to the genetic diversity in strawberry, and to the variability in environmental response among cultivars.

This chapter discusses the effects of light, temperature, water, and other environmental factors on the growth and development of the modern, cultivated strawberry, *F. x ananassa*. The discussion also includes other *Fragaria* species, particularly when pertinent citations are unavailable for *F. x ananassa*.

II. IRRADIANCE

A. IRRADIANCE LEVEL

In their natural habitats, the progenitor species of *F. x ananassa* are found in divergent light environments. *F. virginiana* is generally considered an understory species, although it occurs infrequently in mature forests,[23-25] whereas *F. chiloensis* is native to high light environments typical of beaches or alpine regions.[26,27]

1. Leaf Gas Exchange

For *F. x ananassa* cultivars, A ranges from about 10 to 23 μmol CO_2 m^{-2} s^{-1},[28-31] which is intermediate between A of *F. virginiana* (7–15 μmol CO_2 m^{-2} s^{-1}) and that of *F. chiloensis* (15–30 μmol CO_2 m^{-2} s^{-1}).[24,25,29-33] There is a positive correlation between A and the percentage of *F. chiloensis* germplasm

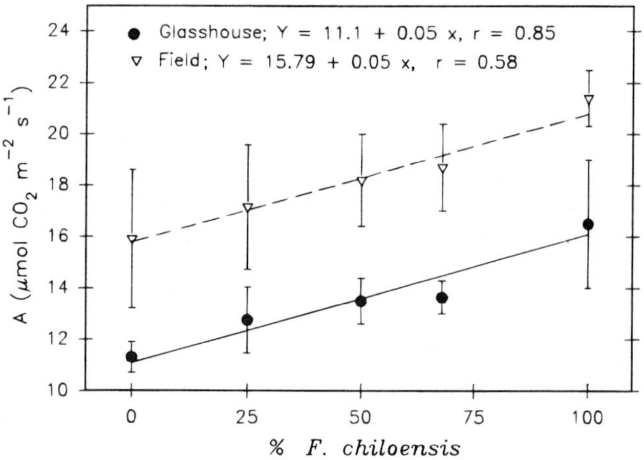

Figure 1 Influence of percentage of *Fragaria chiloensis* genetic contribution on net CO_2 assimilation (A) in breeding populations of *F. chiloensis* and *F. x ananassa*. Symbols represent mean A ± SE. Redrawn from Hancock, J. F., Flore, J. A., and Galletta, G. J., *Scientia Hort.*, 40, 139, 1989.

in *F. x ananassa* cultivars (Figure 1).[30] The greater A of *F. chiloensis* compared to *F. virginiana* or *F. x ananassa* may be due to thicker leaves and greater amount of leaf mesophyll tissue for *F. chiloensis*.[2,29]

Several studies have been conducted on the effects of irradiance level on A in strawberry, often with differing results. Campbell and Young[34] reported that light saturation for A of 'Quinault' strawberry plants occurred at a photosynthetic photon flux (PPF) between 600 and 800 μmol quanta m^{-2} s^{-1}. Similarly, light saturation for A for glasshouse-grown 'Elsanta' plants occurred at about 700 μmol quanta m^{-2} s^{-1}, with a light compensation point for A at about 50 μmol quanta m^{-2} s^{-1},[35] and A of 'Earliglow' strawberry plants increased with increasing light intensity, reaching a maximum rate at a PPF between 600 and 800 μmol quanta m^{-2} s^{-1}.[36] However, Van Elsacker et al.[37] reported that light-saturated A of leaves of glasshouse-grown 'Primella' strawberry plants occurred at a PPF between 400 and 600 μmol quanta m^{-2} s^{-1}, whereas Cameron and Hartley[29] reported that A of four cultivars in a glasshouse increased with increases in PPF up to about 1000 μmol quanta m^{-2} s^{-1}, or greater (Figure 2). In a growth chamber, light-saturated A of leaves of 'Bogota' plants occurred at about 50 klx.[38] Although kilolux is a photometric rather than radiometric measurement, 50 klx represents about one half of full sunlight, or approximately 1000 μmol quanta m^{-2} s^{-1}.

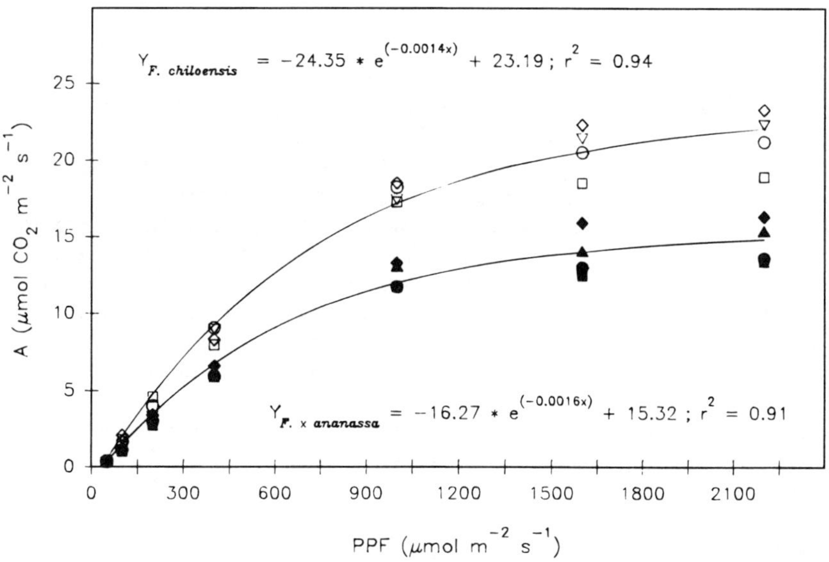

Figure 2 Influence of photosynthetic photon flux (PPF) on net CO_2 assimilation (A) for *Fragaria chiloensis* (open symbols) and *F. x ananassa* (closed symbols) genotypes. Different symbols within a species represent different genotypes. Redrawn from Cameron, J. S. and Hartley, C. A., *HortScience*, 25, 327, 1990.

Differences in the reported values for photosynthetic light saturation of strawberry may be cultivar, season, or leaf age dependent.[24] Measurement conditions, or environmental preconditioning, particularly in regard to light, temperature, nutrients, and ambient CO_2 concentration, may also account for some of the observed variability in photosynthetic light saturation. For example, light saturation studies by Campbell and Young[34] used plants preconditioned at 21°C and a PFF of 300 μmol quanta m^{-2} s^{-1}, whereas Cameron and Hartley[29] used plants maintained at 25.6°C and a PFF of 1100 μmol quanta m^{-2} s^{-1}.

Thus, photosynthetic light saturation in strawberry appears to occur at about one third to one half of full sunlight. Although cultivated strawberries are typically grown in well-illuminated sites where radiant energy is non-limiting, interplant competition for light, especially in mature, matted row production systems or in nurseries, could have adverse effects on whole plant A and productivity.[33]

2. Growth and Development

Light is a major environmental factor regulating plant growth and development of *Fragaria* species. Increased irradiance level resulted in increased leaf, root, and crown dry weights,[36,39] stamen development,[40,41] fruit set,[40–43] yield,[36] fruit size,[42] stolon formation,[39,44–46] and stolon dry weight.[39] Although constant shading at 40% of ambient sunlight reduced fruit yield and leaf and crown dry weights, fruit yields and root dry weights increased when shading was applied during the period of stolon formation.[36] Constant shading, or shading during stolon development, either had no effect on stolons or resulted in reduced stolon formation, whereas petiole length and leaf size increased while leaf dry weight decreased. Shading during the spring growth and fruiting period delayed fruit maturity 5 to 7 days. However, shade-induced reductions in irradiance level resulted in decreased air, plant, and soil temperatures, which may also have affected plant growth and development.[36] Thus, shading during a relatively cool year resulted in decreased plant growth and dry weights, but had no effect on fruit yields the following spring.[39]

Ferree and Stang[36] acknowledged the difficulty in defining the specific cause of the shading responses due to the close relationship between light and temperature. Intrepreting the effects of irradiance level on strawberry physiology and morphology is often difficult, particularly under field conditions, due to irradiance level, temperature, and photoperiod interactions. Dennis et al.[44] considered that the effects of irradiance on stolon formation were mainly quantitative, and that photoperiod was the most important factor affecting plant response.

Maximum A (A_{max}), on a leaf area basis, leaf thickness, specific leaf weight, and mesophyll cell volume of *F. virginiana* all increased with increasing total daily PPF,[47–50] and total daily PPF had a greater effect on these variables than peak PPF.[49] Jurik et al.,[24] observed differences in leaf expansion rates and leaf longevity for *F. virginiana* plants maintained under different PPFs. Leaves from plants maintained in high (678 μmol quanta m^{-2} s^{-1}) and low (64 μmol quanta m^{-2} s^{-1}) light environments completed leaf expansion in 14 and 19 days, respectively, and had median life spans of 51 and 79 days, respectively. For all plants, A_{max} was achieved within 3 days of cessation of leaf growth, with a gradual reduction observed starting 4 to 7 days later. Although A_{max} on a leaf area basis was similar for leaves that developed in high and in low light, A_{max} was reached earlier, and declined more rapidly for leaves that developed in high-light. High-light plants had higher light compensation points, greater specific leaf weight (SLW), and thicker leaves with more mesophyll cell area and less air space than plants grown in low light. Changes in leaf anatomy and A were observed for developing leaves when plants were transferred from one light environment to another, indicating a potential for adaptation to changing light environments. This adaptive potential decreased with leaf expansion and increasing leaf age. Light-saturated A in *F. vesca* occurred at PPFs between 400 and 600 μmol quanta m^{-2} s^{-1}; A, specific leaf weight, leaf thickness, flowering, stolon production, and biomass all increased with increasing irradiance level.[47]

Strawberry seed germination is promoted by exposure to light, and germinating seed should be left uncovered.[2,6,51,52] The physiological mechanism underlying light promotion of strawberry seed germination has not been elucidated.

B. PHOTOPERIOD

Strawberry cultivars are often categorized on the basis of photoperiodic responses, particularly in regard to floral induction and flower bud initiation. Leaf size, petiole length, and stolon development are also highly sensitive to photoperiod and temperature. In many cultivars there are strong photoperiod × temperature interactions in which temperature modifies the photoperiodic response.[8,26,44,46,53–58] Although vegetative and reproductive responses of strawberry plants to photoperiod and temperature are well documented, the physiological processes that regulate these responses have not been elucidated.

Most strawberry cultivars are classified as 'short-day' (SD) (also known as 'single cropping', 'Junebearing', or 'noneverbearing') types, in which floral induction occurs with photoperiods of less than about 14 hrs. However, most SD cultivars exhibit a *facultative* SD response, in which floral induction will occur more or less continuously regardless of daylength, provided that temperatures are less than about 16°C.[5,8,26,59] Thus, although SD cultivars may have different critical photoperiods,[26] and a few SD cultivars are obligate single-croppers that have only one short fruiting season per year,[2] most will bloom independently of daylength with moderately cool temperatures. It is uncertain whether an obligate single-crop bearing habit is an inherent varietal characteristic, or is the result of excessive vegetative vigor due to winter chilling[26,60–63] (see Section III).

A second photoperiodic group of strawberries is comprised of the everbearing (EB) cultivars, also referred to as day-neutrals (DN), that generally flower continuously regardless of daylength. Although temperature also modifies the photoperiodic response of this second group of cultivars, they are less sensitive to high temperatures than SD cultivars. Some authorities recognize three groups, distinguishing between EB and DN cultivars on the basis of long day (LD) flower promotion in EB cultivars and no daylength effect in DN cultivars.[26,54] However, others consider the division between these photoperiodic types to be subjective and ill-defined.[6,7] In reviewing the work of Durner et al.,[54] Galletta and Bringhurst,[6] noted similar floral development patterns for all photoperiodic types in response to a given photoperiod × temperature regime. They concluded that the major difference among photoperiodic groups was a decreasing sensitivity to high temperatures from SD to EB to DN. Thus, flowering in strawberries ranges in a continuous gradient from obligate single-cropping ('Fairfax') to facultative short-day through the various degrees of everbearing or day neutral.[62,64]

Photoperiodic effects on flowering of strawberries were first reported by Sudds,[65] who observed increased flowering under an 8-hr photoperiod compared to a normal midsummer photoperiod of about 15 hrs in Pennsylvania. Darrow and Waldo[26] recognized that temperature was probably as important as photoperiod in regulating plant growth. At temperatures above 16°C, the critical photoperiod for floral initiation was 10 hrs or less, whereas at lower temperatures initiation occurred with photoperiods longer than 10 hrs. Dormancy or rest in some cultivars was considered to be caused by a combination of short photoperiods and low temperatures.[26] This rest period was broken by exposure to freezing temperatures, or, in some cultivars, simply by increasing the daily photoperiod (see Section III).

For plants maintained with LD (15-hr photoperiod) but for which varying percentages of leaf area were maintained with SD (10-hr photoperiod), floral initiation was directly proportional, and stolon development inversely proportional, to the percentage of total leaf area exposed to SD.[57] Entire plants maintained under LD failed to bloom, but flowers were produced in runner plants under LD with the attached mother plants under SD (inductive) conditions.[57] These experiments were considered to be indicative of a flowering stimulus that was produced in the leaves and translocated from the mother plant to the stolon under the appropriate photoperiodic conditions[57]. Other studies, however, failed to confirm the presence of a translocatable flowering stimulus.[7,66–69] The discrepancies among studies are difficult to explain, but may be due to cultivar differences, or differences in plant preconditioning.[66]

Flowering in strawberry was observed to occur under the following photoperiodic regimes: short light period (10 hr) with long dark period (14 hr), short light and dark periods (10 hr), and long light and dark periods (14 hr).[57] Floral inhibition only occurred when the daily light period was longer than the daily dark period (14 and 10 hrs, respectively). Based on this and other studies, it was hypothesized that a flower-inhibiting substance was produced in the leaves during the light period; flowering only occurred if the inhibitor was depleted during the subsequent dark period.[57] Thus, the duration of the dark period, rather than the light period, was considered to be the factor controlling floral initiation.[57] However, Vince-Prue and Guttridge[70] hypothesized that strawberry floral initiation was not promoted by SD conditions per se, but by a floral inhibitor produced under LD conditions. Thus, rather than SD stimulation of flowering, the absence of LD conditions releases strawberry from floral inhibition.[8]

Flowering has been observed to occur under continuous light,[8,71] in apparent contradiction to most studies of photoperiodic regulation of floral induction in strawberry. This phenomenon may have resulted from the type of light source used,[71] or from an unrecognized stress.[8]

Defoliation[72] and paired receptor-donor (plants joined by a stolon) experiments[66–69,73] demonstrated translocation of a floral inhibitor and the promotion of vegetative growth in donor plants maintained under LD or night interruption conditions to receptor plants maintained under SD (inductive) conditions.

Floral initiation in response to defoliation varies with photoperiodic type. Summer defoliation of a SD cultivar failed to promote floral initiation, whereas defoliation of an everbearing cultivar enhanced floral initiation.[74,75] Moore[76] reported yield reductions following defoliation, and suggested that leaf pruning eliminated the source of a floral-promoting substance. In another study, defoliation resulted in increased fruit production due to an increase in the number of crowns initiating bloom.[77] In this case, increased bloom was attributed to the removal of the source of a floral-inhibiting, growth-promoting substance located in the leaves.

Borthwick and Parker,[78] using daylength extension and night interruption, found that floral initiation was regulated by the duration of the dark period rather than by the length of the light period, and that flowering was, at least in part, phytochrome-mediated. This was evidenced by the fact that floral initiation occurred under 8- and 11-hr photoperiods, but was inhibited by a 3-hr night interruption, or by photoperiods of 14 hrs or greater.

Photoperiod has a marked effect on strawberry vegetative growth and morphology, in that stolon formation, petiole length and leaf area increase with increasing photoperiod.[5,13,14,79–82] Photoperiodic effects on stolon formation are cultivar-specific; 'Missionary' was more sensitive to LD or night interruption than 'Sparkle' or 'Tennessee Beauty'.[81] Maas and Cathey[82] reported that stolon formation in response to supplemental lighting varied with the photoperiodic type of the cultivar; with exposure to LD, stolon development was greatest for a SD cultivar, was intermediate for an EB cultivar, but did not occur for a DN cultivar, although an increase in crown branching was observed.

Regardless of plant preconditioning, the period of cell division in expanding strawberry leaves increases with increasing photoperiod. Cell numbers per leaf, leaf size, and petiole length all increase with LD, whereas with SD, cell division ceases shortly after leaf emergence, resulting in fewer cells per leaf, and smaller leaves with shorter petioles.[13,79,80] The effect of photoperiod is mainly due to an increase in cell number, rather than an increase in individual cell size.[13,79] Cell division in higher plants is promoted by gibberellic acid (GA),[83] and attempts have been made to explain photoperiodic regulation of strawberry growth and development on the basis of GA or other hormonal activity. Exposure to LD or to exogenous gibberellic acid (GA_3) elicited similar responses, i.e., floral inhibition, and increased stolon formation, petiole length and leaf size.[7,84–86] Furthermore, GA-like activity has been observed in strawberry tissue extracts,[60,87,88] and GA activity increased in strawberry plants exposed to LD or night interruption.[89] However, exogenous gibberellins applied in the absence of LD or chilling did not completely substitute for the endogenous factors that promoted petiole elongation.[86] A marked reduction in endogenous auxin at the time of floral induction in strawberry was determined to be the result of floral induction, rather than a causative factor.[90] Durner and Poling[55] suggested that floral initiation in strawberry is regulated by two substances, a reproductive growth promoter produced under SD conditions, and a vegetative growth promoter produced under LD or night interruption conditions. Production of these substances is proportional to the amount of leaf area exposed to SD or LD conditions, and translocation via the stolon occurs only from older to younger plants.[55]

Cold damage to strawberry plants maintained in the dark was greater than that of plants maintained under a 12-hr photoperiod, suggesting a light requirement for development of cold-hardiness in strawberry.[91] Other studies show that for some plants, including strawberry, translocatable, cold-hardiness promoters are produced in the leaves under SD conditions.[92–94] Thus, the observed lack of cold-hardiness in non-illuminated strawberry plants may have resulted from non-inductive photoperiodic conditions, rather than a lack of exposure to light per se.

C. PHOTOPERIOD × TEMPERATURE INTERACTIONS

Floral *induction* refers to the processes that occur in a leaf upon exposure to the environmental conditions that result in the eventual production of a flower bud; floral *initiation* refers to the physiological and anatomical changes that occur at a meristem in response to floral induction; and floral *differentiation* refers to the events that occur in the meristem from initiation through anthesis.[95]

There is a strong relationship between photoperiod and temperature in the regulation of strawberry reproductive and vegetative responses. Darrow[53] found that SD conditions favored floral initiation and inhibited stolon formation, regardless of temperature. However, the optimal temperature for floral initiation or stolon development was photoperiod dependent (Table 1). Under SD conditions, floral initiation from November through March was greatest at 21°C, intermediate at 15.6°C, and lowest at 12.8°C. There was no stolon initiation at any temperature. With a 14-hr photoperiod, the optimal temperature for floral initiation was 15.6°C, followed by 12.8 and 21°C. There was a five-fold increase

Table 1 **Influence of temperature and photoperiod on stolon and flower development in strawberry**

Treatments		Stolons (S) and flower panicles (F) produced per month											
Temp. (°C)	Photo-period	Nov		Dec		Jan		Feb		Mar		Total	
		S	F	S	F	S	F	S	F	S	F	S	F
21	16 hrs	91	2	95	0	27	0	40	0	59	12	312	14
	14 hrs	66	5	27	0	7	1	4	10	6	24	110	40
	Ambient	0	24	0	41	0	74	0	61	0	81	0	281
15.5	16 hrs	57	5	15	0	0	0	0	11	0	38	71	54
	14 hrs	19	13	4	7	0	0	0	24	0	71	23	115
	Ambient	0	26	0	16	0	14	0	53	0	70	0	179
12.8	16 hrs	40	6	8	1	1	0	0	0	0	61	49	68
	14 hrs	18	5	2	1	0	0	0	4	0	63	20	73
	Ambient	0	18	0	7	0	2	0	14	0	92	0	133

Adapted from Darrow, 1936.[53]

in the number of stolons produced at 21°C than at 12.8 or 15.6°C. With a 16-hr photoperiod, floral initiation during the five-month period was greatest at 12.8°C and lowest at 21°C, with stolon development inversely proportional to floral initiation. Cultivar differences were also noted, particularly in regard to floral initiation under LD conditions.[53] Under LD (14–16 hrs) and at 21°C, 'Klondike' produced only stolons, but 'Burrill' produced equal numbers of stolons and flower clusters, whereas at 15.6°C, 'Klondike' produced both flowers and stolons, and 'Burrill' was almost entirely sexually reproductive.

Hartman[56] studied floral initiation as a function of temperature and photoperiod in four cultivars. All cultivars bloomed at 15.6°C, regardless of photoperiod, and no cultivar flowered under a 15-hr photoperiod at 21°C. However, only three of the four cultivars flowered in a 10-hr photoperiod at 21°C, leading to the conclusion that temperature was as important as photoperiod in regulating floral induction in some cultivars. Similarly, for 'Marshall' plants maintained at 6°C, floral initiation occurred with exposure to 8-, 16-, or 24-hr photoperiods, but only occurred with 8- or 16-hr photoperiods, or an 8-hr photoperiod, for plants maintained at 10 and 14°C, respectively.[46] Floral initiation in 'Robinson' maintained at 9°C occurred with 8-, 12-, 16-, 20-, or 24-hr photoperiods, whereas at 17°C, initiation was limited to photoperiods of 12 hrs or less.[59] For Scandinavian cultivars adapted to long summer photoperiods, flower initiation occurred at 12 or 18°C with 10-, 12-, 14-, 16-, or 24-hr photoperiods, but at 24°C, floral initiation did not occur with photoperiods greater than 14 or 16 hrs.[5] The critical inductive temperature/photoperiod combination may vary among SD cultivars, but higher temperatures and longer photoperiods generally inhibit bloom and result in increased stolon development.

Although the optimal daylength for floral induction in SD cultivars appears to be between 8 and 11 hrs,[6,59,78,96,97] the role of photoperiod is more critical at temperatures above 15°C.[8] Thus, the critical photoperiod for floral induction is variously reported as between 12 and 16 hrs,[59,96] 12 and 14 hrs,[98] 11 and 14 hrs,[78] 11 and 15 hrs,[97] 13 and 15 hrs,[99] or 13 and 16 hrs,[5,8] depending on the cultivar, experimental temperature, and photoperiod. Fruiting seasons for SD cultivars tend to be longer in areas with mild winter climates (i.e., Florida and Southern California) due to a longer growing season in which photoperiods are less than 14 hrs, as well as to less winter chilling, because chilling (vernalization) can inhibit subsequent floral induction through promotion of vigorous vegetative growth[7] (see Temperature, Section III). In general, the critical photoperiod for inhibition of floral induction and that for promotion of vegetative growth responses such as increased petiole length, leaf size, and stolon formation are similar.[5,78]

Once floral induction and initiation have occurred, further development is enhanced by exposure to LD conditions.[7,90,100] For SD cultivars, floral cymes developing under SD conditions tend to be very large and basally branched, producing a maximum number of large fruit, but when photoperiods are lengthened, floral cymes tend to branch at a higher position and produce smaller fruit.[26,101] Differences in cyme morphology due to temperature or winter chilling have been observed due to vegetative growth promotion following low-temperature exposure[7] (see Section III).

The number of SD photoinductive cycles required for floral initiation in SD strawberry cultivars is highly variable, due to temperature × photoperiod interactions, as well as seasonal and pretreatment factors.[8] The minimum number of SD cycles required for floral induction is proportional to temperature, and usually ranges between 7 and 15,[46,57,59,70,90,102] but may be as many as 24[70,78] or more, and is greater at higher temperatures. For example, with a 16-hr photoperiod, more than 16 cycles were required for floral induction at 17°C, but only 10 cycles were needed at 9°C.[59] Similarly, with an 8-hr photoperiod, only 10 cycles were required for floral induction at 24°C, but more than 20 cycles were required at 30°C.[59]

D. SPECTRAL QUALITY

Variations in spectral quality can affect strawberry plant physiology. Under floral inductive conditions (8-hr days), flower initiation in a SD strawberry cultivar was delayed or inhibited, and leaf petiole length increased, by extending daylength with far-red light (FR, 700–800 nm) during the first half but not during the second half of the daily dark period.[70,103,104] Similar responses were obtained with exposure to red light (R, 600–700 nm) during the second half, but not the first half of the dark period. Combinations of R and FR were inhibitory at either time. Stolon formation, and leaf laminae and petiole length, all increased with exposure to FR during the first half of the daily dark period, or with exposure to R during the latter half of the dark period.[70,103,104] Flowering was delayed when filters were used to remove R from sunlight, but filtering out both R and FR resulted in earlier bloom than in control plants.[105] Daylength extension with incandescent light, which emits a high proportion of FR, inhibited flowering at irradiances of 0.1 J m^{-2} s^{-1}, whereas fluorescent lighting, which emits little FR, did not inhibit flowering, even at irradiances of 25 J m^{-2} s^{-1}.[59,102]

Sensitivity to FR, or to R plus FR before or after the daily light period, or to R during the latter half of the dark period are typical of LD, phytochrome-mediated plant responses to spectral quality, but atypical of SD plant responses.[70] Based on the work with spectral light responses, strawberry exhibits spectral sensitivity patterns and processes typical of LD plants, but exhibits an opposite floral response, in that LD conditions inhibit flowering in strawberry.[8,106]

III. TEMPERATURE

Strawberry growth and development are highly sensitive to variations in air and soil temperature. Simple effects of temperature as well as interactions often are cultivar and species dependent.

A. GROWING SEASON TEMPERATURES

1. Leaf Gas Exchange

A temperature optimum of 15°C for A has been reported for 'Bogota' strawberry plants, although mesophyll conductance was highest at 20°C.[107] Photorespiration and transpiration (E) increased as air temperature increased from 10 to 30°C, but stomatal conductance to water vapor (g_s) decreased, probably as a result of an increased vapor pressure deficit (VPD) between the leaf and air.[107] Van Elsacker et al.[37] reported a temperature optimum for A of 'Primella' strawberry to be between 15 and 25°C. Net CO_2 assimilation (on a leaf area basis) decreased as root temperature increased from 8 to 23°C[108]; however, on a whole plant basis, A increased due to increased leaf production at the higher temperatures. Transpiration and g_s also increased with increasing root temperature.[108]

Photosynthetic temperature optima may vary among *Fragaria* species and with preconditioning treatments. For plants maintained at either 20/10°C or 30/20°C (day/night), A_{max} for *F. chiloensis* was between 16 and 20°C.[109] However, for plants of *F. virginiana* and for two cultivars of *F. x ananassa* grown at 20/10°C or 30/20°C, A_{max} was between 18–24 and 24–30°C, respectively, indicating an increase in temperature optima for A with increased growth temperature regime. For plants of *F. vesca* grown at either 10/20 or 30/20°C, A_{max} was 15–20 and 25°C, respectively, again indicating photosynthetic acclimation to temperature.[47] The ability to acclimate to higher temperatures may account for the wider geographical adaptation of *F. x ananassa, F. virginiana* and *F. vesca* compared to that of *F. chiloensis.*

2. Growth and Development

a. Soil Temperatures

There was no significant difference in root growth (dry weight) in 'Robinson' strawberry plants maintained at soil temperatures of 7.2, 12.8, 18.3, or 23.9°C, although there was a trend toward greater root growth with decreasing soil temperature.[110] The soil temperature optima for vegetative growth of strawberry appears to vary with cultivar. For example, for the cultivars 'Shasta' and 'Lassen', root dry weights were greatest with soil temperatures of 12.8 and 7.2°C, respectively.[111] Crown dry weights were greatest with soil temperatures of 18.3 and 7.2°C, respectively, although leaf and total plant dry weights for both cultivars were greatest with soil temperatures of 12.8°C.[111] For 'Shasta' and 'Lassen', fruiting occurred only at 7.2 and 12.8°C, and was greatest at 7.2°C, while stolon development only occurred at or above soil temperatures of 18.3°C, and was greatest at 23.9°C. The use of black, polyethylene mulch increased diurnal soil temperatures at a 10-cm depth from 21 to 23°C at night, and from 27 to 30°C during the day, resulting in increased leaf elongation rates.[112] Voth et al.[113] observed a 4°C increase in soil temperature in January in southern California as a result of clear polyethylene bed mulch. For mulched plants, increased winter soil temperatures promoted plant growth, resulting in a 30% yield increase over plants on nonmulched beds.[113] Conversely, under high temperature conditions, opague polyethylene or organic fiber mulches are often used to reduce soil temperatures. The influence of various bed mulches on soil temperature, and on reproductive and vegetative growth of two day-neutral cultivars, was studied during the summer in Iowa.[114] Soil temperatures, plant dry weights, and production were reduced, but fruit yields increased with straw or white-on-black polyethylene mulch compared to clear polyethylene.[114]

b. Ambient Temperatures

Nine cultivars from diverse geographical areas had high growth rates with ambient temperatures between 20 and 26°C; maximum growth rates occurred at about 23°C.[115] Similarly, leaf and total plant dry weights for 'Robinson' were greatest at 23.9°C,[110] but leaf development in 'Marshall' was greatest between 14 and 17°C.[46] In a short-term growth chamber study, petiole and leaf elongation rates of 'Olympus' strawberry increased exponentially with increasing air temperatures from 10 to 28°C, although a subsequent study of longer duration suggested that optimal temperature for leaf elongation was 22°C.[116] Moderate leaf production rates in the cultivar 'Royal Sovereign' occurred at a minimum of 5°C,[13] and temperatures above 35°C inhibited stolon growth, and decreased fruit fresh weight and fruit soluble solids content.[6,117] During periods of high temperature, evaporative cooling from mist or sprinkler irrigation can result in increased flower and fruit production, and increased vegetative growth.[118–120]

In general, temperature effects on shoot growth of strawberry are similar to those of other temperate fruit crops, but optimal soil temperature for strawberry root growth appears to be lower than that for many temperate fruit crops.

Temperature effects on fruit set may be cultivar dependent.[121,122] For example, fruit set of 'Deutsch Evern' was good with temperatures between 10 and 26°C, whereas fruit set did not occur below 17°C for 'Jucunda',[122] possibly as a result of low temperature-induced stamen abortion.[6,121] Temperatures below 15.6°C may inhibit pollen germination and pollen tube growth, resulting in misshapen fruit.[123] Temperature can also affect fruit set and fruit quality due to indirect effects on pollinator (bee) activity. Although the cultivated strawberry has perfect flowers and is self-fruitful, pollination is enhanced by bees.[124–126] Bee activity decreases at temperatures below 10°C.[22] Thus cool temperatures, rain, or strong winds that inhibit bee flights, and result in decreased pollination and misshapen fruit.

The length of the fruit development period decreases with increasing temperatures.[2,15] In Maryland, the average period from anthesis to fruit maturity is about 31 days in early spring, but can be as short as 20–25 days in late spring or summer, or as long as 60 days in autumn, due to differences in prevailing temperature and photoperiod during these periods.[2] For 'Reiko' strawberry plants maintained in a growth chamber with a 14-hr photoperiod and night/day temperatures of 17/20°C, fruit matured 29 days after anthesis.[127] Fruit size at harvest was negatively correlated with the prevailing soil temperature 6 weeks prior to harvest.[128] During the summer, a 10°C decrease in soil temperature resulted in increases of 0.9–1.6 g/berry, suggesting that mulching, shading, or irrigating the soil to decrease soil temperatures can enhance fruit size and yield.[128] Strawberry fruit continue to increase in size until full maturity. Fruit size was reported to increase 14% between the stages of full red color and full maturity,[2] and fruit size and dry weight increased exponentially up to full maturity.[127]

Spring or autumn frosts often result in injury to strawberry reproductive organs. Cold injury to open strawberry flowers often occurs at temperatures of about −2°C,[2,129–131] although flowers of several cultivars have survived exposure to −4.5°C.[131] Susceptibility to cold injury may vary with cultivar,[131–133] and with the degree of ice nucleation that occurs in the tissue.[132] Supercooling occurs when the temperature of a tissue or liquid decreases below the freezing point without ice crystal formation[22] (see Chapter 8, Volume I for a discussion of supercooling). In the absence of supercooling (i.e., with ice crystal formation in the xylem), the T_{50} (temperature required to kill 50% of the samples) for open strawberry flowers was −2.1°C.[132] Immature berries were more cold-sensitive (T_{50} = −1.6°C), whereas unopened flower buds were more cold-hardy (T_{50} = −3.1°C) than opened flowers. Supercooling (i.e., no ice crystallization in the xylem) resulted in a T_{50} that was 1.3 to 2.8°C lower than when ice crystallization occurred, depending on the tissue. Style and receptacle tissues are reportedly more susceptible to cold injury than anthers,[2,133] although the opposite has also been observed[130,134]; this discrepancy may be due to cultivar differences.

B. COLD HARDINESS

Strawberry species and cultivars vary in their ability to withstand low temperatures.[2,130,131,135–139] Although certain cultivars reportedly withstand winter temperatures of −50°C,[2,6] strawberries are generally not as cold hardy as most temperate fruit crops,[140] and are frequently injured by freezing temperatures.[141,142] For a given cultivar, severity of injury depends on plant developmental stage, temperature preconditioning, climatic conditions, plant nutritional status, soil and plant moisture status, the minimum temperature and its duration, the rate of freezing and thawing, supercooling, and the presence of snow cover or mulch and the time of its application.[132,137–140,143–147] For 'Catskill' strawberry plants, low temperature injury to crown tissue increased as freezing rate increased from 1.7°C hr^{-1} to 5.6°C hr^{-1}.[144] Crown tissue injury also increased as the duration of low temperature increased from 1 hr to 3 days.[144]

Studies with several cultivars have indicated that the critical low temperature for survival of dormant strawberry plants is ca. −12°C. Harris[148] observed severe plant injury in four cultivars at −12°C, and Brierley and Landon[149] reported injury for several cultivars following exposure to temperatures of −6°C and mortality after exposure to −12°C. Angelo[143] reported severe injury or death of plants exposed to temperatures below −10°C. Seven cultivars were killed with temperatures of −12°C in October or November, although some cultivars survived exposure to −12°C in December.[138] For 'Blakemore' and 'Premier' strawberry plants, there was little or no plant mortality following exposure to −4.4 or −7.8°C, but considerable mortality occurred at −11°C.[146] Plant mortality of dormant 'Catskill' plants increased with decreasing temperature from 0 to −20°C.[140] All plants survived exposure to −4°C, 50% died at −12.5°C, while none survived exposure to −20°C. However, even mild freezing temperatures (−4°C) resulted in abnormal plant growth. For all subfreezing (below 0°C) temperature treatments, leaf emergence was delayed, although there was no difference among temperature treatments in leaf number after 12 weeks of growth. Exposure to subfreezing temperatures resulted in increased development of stolons and of misshapen leaves, and plants exposed to −12 or −16°C had smaller laminae. The percentage of plants that subsequently flowered and the total number of flowers per plant decreased with decreasing temperatures below −4°C. For 'Earliglow' plants exposed to various subfreezing temperatures, there was little or no root regrowth after exposure to −12°C, and no leaf regrowth after exposure to −14°C.[139] However, 'Honeoye' plants continued to produce new leaves and roots after exposure to −14°C.[139] For actively growing plants maintained for 2 weeks at fluctuating temperatures between about 1.7 and 26°C, there was no mortality for any of eight cultivars following exposure to −3°C for 24 hrs, but mortality after exposure to −6°C was cultivar dependent, and ranged from 0 to 100%.[135] All cultivars were killed with exposure to temperatures of −9°C or lower.

Water status of strawberry plants affected plant survival under low temperature conditions. Fruit yields and total plant biomass of two strawberry cultivars exposed to subfreezing temperatures increased with decreasing soil moisture.[138] Similarly, for strawberry plants exposed to −10°C, plants that were watered sparingly had greater survival than plants that received more frequent irrigation.[143] However, the influence of plant moisture status on cold tolerance may be cultivar dependent. Whereas hydrated (root water potential (Ψ_r) = −0.3 MPa) and dehydrated (Ψ_r = −1.4 MPa) 'Earliglow' plants survived temperatures of −8 and −12°C, respectively, there was no difference in survival temperature between hydrated and dehydrated 'Honeoye' plants, which survived temperatures of −14°C.[139]

Development of cold hardiness in strawberry plants occurs quickly. For 'Blakemore' and 'Premier' strawberry plants preconditioned at 0/13°C (day/night) for 0, 7, or 14 days, and then exposed to −7.8°C, survival was greater for plants preconditioned for 7 days than for plants preconditioned for 0 or 14 days.[146] The resumption of vegetative growth results in a rapid loss of cold hardiness.[6]

C. DORMANCY AND CHILLING

In autumn, exposure to cool temperatures and short photoperiods result in floral induction, branch crown development, reduced leaf size, and with prolonged exposure, the onset of dormancy in strawberry.[8,15,26,61,150,151] Unlike most temperate-zone fruit species, dormancy in strawberry is controlled solely by external conditions (ectodormancy),[6] such as temperature and photoperiod; dormant (resting) plants retain green leaves, and can resume growth when environmental conditions are favorable.[6,8,15,17] Thus, there is no true dormancy in strawberry.[152] However, in the absence of sufficient chilling, dormant strawberry plants have low vegetative vigor, produce short petioles and small leaflets, have a low, compact growth habit, and yield poorly.[8,15,17,26,61,153–155]

Exposure to low temperatures results in vegetative invigoration, thereby reversing the effects of rest.[8] Complete satisfaction of the chilling requirement promoted vigorous vegetative growth, and resulted in rapid leaf production, increased petiole length, larger laminae, increased stolon formation, and an inhibition of floral induction.[15,26,60,61,150,153,156,157] Partial satisfaction of the chilling requirement promoted leaf production and sustained flowering, but did not promote stolon development.[6,8,15] The growth and fruiting response of strawberry plants for which the chilling requirement is only partially satisfied is somewhat analagous to the phenomenon of "delayed foliation" that occurs in deciduous fruit trees that receive inadequate winter chilling.[6,7] In such trees, vegetative growth is reduced relative to fully chilled trees, and flowering often extends through the growing season.

In the absence of low temperatures, the rest period in some strawberry cultivars can be broken by extending the daily light period.[26] Also, cultivars from the southern United States or similar regions grow adequately at relatively low temperatures under SD and apparently have little or no rest period.[26] Growth during SD of October, November, and December is considered an indication of a cultivar's regional adaptation; most temperate-zone cultivars grow poorly at that time of year and enter a rest period, while cultivars adapted to the southern United States will grow adequately if temperatures are warm enough.[26] The duration of the inductive conditions required for the onset of rest has not been quantified, but may be about 4 to 6 weeks.[8] However, this period appears to be cultivar-dependent, and may be shorter for some cultivars.[158]

In Florida and certain other areas, annual strawberry production systems utilize nondormant ("green") plants that have been exposed to SD, but that have received little or no chilling in the nursery prior to transplanting.[158] Transplants are usually grown in high latitude (40–45° N) nurseries and are exposed to SD in the nursery in early autumn. Although exposure to SD results in increased root carbohydrate (starch) content,[159] these low-chill plants have adequate vigor only if a portion of the leaves remain intact (hence the term "green plants") during transplant and establishment.[158] Optimum nursery digging date for early fruit yield for such low-chilled or nonchilled plants was observed to be a photoperiodic response.[158] Chilling enhanced early yields for some cultivars, but only if plants were dug at or before a cultivar-specific optimum date. Digging after the optimum date resulted in decreased yields, even with a lengthy chilling period, suggesting a SD-induced dormancy.[61,151,158]

Annual production systems used in mild-winter areas such as California rely on cultivars that are bred and selected for growth and high yields with only partial chilling satisfaction.[15] In such systems, planting stock is typically produced in high-latitude (40–42° N), high elevation (>1,000 m elevation) nurseries.[6,160] In the nursery, exposure to high temperatures and LD during summer results in vigorous stolon production, but by late-September or mid-October, photoperiods decrease, and temperatures are sufficiently low to impart partial, but incomplete, chilling. Presumably, each cultivar has a particular photoperiodic and chilling requirement for optimal performance.[158,161] In the nursery, low temperatures and SD conditions reduced vegetative growth and increased stored carbohydrates in the primary roots, crown, and petioles.[6,159,162] Depending on the cultivar, nursery plants are dug and transplanted immediately, or given supplemental cold storage to impart added vigor.[163] Plants harvested at the proper time have adequate stored carbohydrates to handle the rigors of digging, transport, bare-root transplant, or supplemental cold storage, and to support rapid growth during autumn. While these plants have sufficient vigor to sustain fruit production during a 6- to 8-month fruiting season, they lack the excessive vigor that results in stolon development and inhibition of flowering.[61,157,161]

Bringhurst et al.[162] observed that the vigor response of plants exposed to a period of cold storage after digging from the nursery was equal to that of plants exposed to a similar period of chilling in the nursery. Leaves are removed when plants are dug from the nursery, thus harvested, cold-stored plants are physiologically distinct from plants left in the nursery. After a period of time, carbohydrate status of the two sets of plants should be different, since photosynthate will continue to accumulate in plants left in the nursery, whereas photoassimilate in stored plants will gradually decrease as a result of respirational losses during cold storage. The observations of Bringhurst et al.[162] suggest that at the time of harvesting plants for cold storage, carbohydrate levels were already adequate in both sets of plants, and that continued accumulation of photosynthate gave no advantage. They also suggest that after a certain period of chilling, any benefit of additional chilling is due mainly to an influence on the activity or concentration of endogenous growth regulators.

In addition to stimulating vegetative growth, chilling also stimulates the emergence of previously differentiated inflorescences,[8,150] and results in increased peduncle, pedicel, and inflorescence branch lengths.[8] For dormant 'Catskill' plants, petiole length, leaf number per plant, and number of flowers per plant all increased with increasing exposure to temperatures below 7.2°C.[150] For 'Sovereign' plants preconditioned in SD, stored at 1.7–4.4°C for 36 or 72 days, and then forced at about 15.5°C, petiole lengths were 1.4 and 1.6 times greater, respectively, than petioles of plants that were similarly treated but that received no cold storage.[61] For 'Climax' plants similarly maintained, petiole lengths were 1.5 and 2.2 times greater, respectively, than petioles of nonchilled plants.[61]

For strawberry, reported effective chilling temperatures (the temperature required for breaking rest for resumption of vigorous vegetative growth) range from −2 to 6.5°C.[60,63,153,158,164,165] However, chilling temperatures of 9.5 to 10°C apparently are effective for cultivars adapted either to regions with moderate winters (i.e., California or the southern United States), or to glasshouse forcing during the winter.[153,165]

High temperatures, or large diurnal temperature fluctuations, may counteract the cumulative chilling influence of low temperatures in deciduous fruit species.[166] Tanabe et al.[167] studied the effect of chilling temperature regime on vegetative growth of 'Hokowase' strawberry, and on levels of endogenous growth regulators. Plants were either left in the field under natural chilling conditions, or exposed to growth chamber temperatures of 4°C day/night, 15/4°C day/night, 25/4°C day/night, or constant glasshouse temperatures of 15–25°C. Vegetative growth, calculated as the product of petiole length x laminae width, was determined for all plants during a 60-day period following temperature treatment. Vegetative growth was greatest for plants left in the field. For plants in growth chambers or in the glasshouse, vegetative growth increased with increased exposure to low temperatures; plants maintained at 25/4°C (day/night) or in the glasshouse had the lowest vigor. Vigor appeared to be correlated with concentrations of kinetin and abscisic acid (ABA), in that kinetin concentrations were greatest for field grown plants, and decreased with increased temperature regime. Conversely, ABA concentrations were least for plants in the field, and increased with increased temperature regime. Lavee[168] depicted similar changes in ABA and cytokinins in association with the termination of rest.

The amount of chilling required to terminate rest in strawberry is cultivar dependent.[61,164] Temperate-zone cultivars require longer periods of chilling to overcome rest, while subtropical cultivars require little or no chilling.[8,26,61,154] The duration of the chilling period required for breaking rest ranges from as little as 2 to 4 weeks for cultivars such as 'Tioga' that have a low chilling requirement, to 8 weeks or more for cultivars such as 'Red Gauntlet'.[8]

IV. WATER

Due to an evergreen, semi-herbaceous growth habit, rapid growth rates, high fruit productivity, and a relatively shallow root system characterized by adventitious roots and rapid root turnover (see Section IV, Flooding),[169,170] strawberry plant growth and development are sensitive to variations in soil moisture. Growth and yield responses to soil moisture deficits and irrigation are well documented; however, relatively few studies have focused on the physiological responses of the strawberry to variations in soil moisture content or plant water status.

A. WATER RELATIONS

Plant water status is a function of the dynamic interaction between the rates of water uptake and water loss by the plant.[171] Decreases in leaf water potential (Ψ_l) in response to soil moisture deficits have been observed for strawberry[172]; Ψ_l may also decrease due solely to increased atmospheric VPD.[173] Maximum plant water status occurs pre-dawn, and determinations of Ψ_l at that time most closely reflect soil moisture status; mid-day determinations of Ψ_l are more negative, and reflect atmospheric conditions as well as plant interactions with soil moisture[172,174] (Figure 3).

Reductions in stomatal aperture in response to water deficits have been observed for many plant species, including strawberry[172,175] (Figure 3). However, leaf age and morphology, osmotic adjustment, VPD, irradiance level, and temperature may all influence g_s,[173,175] and as a result, g_s is not always well correlated with plant water status. Although studies of some species indicate a Ψ_l threshold for stomatal closure,[175,176] other studies indicate a gradual decrease in g_s with decreasing Ψ_l.[172,177–179] Although the minimum Ψ_l values reported in the literature for field-grown strawberries are less negative than the

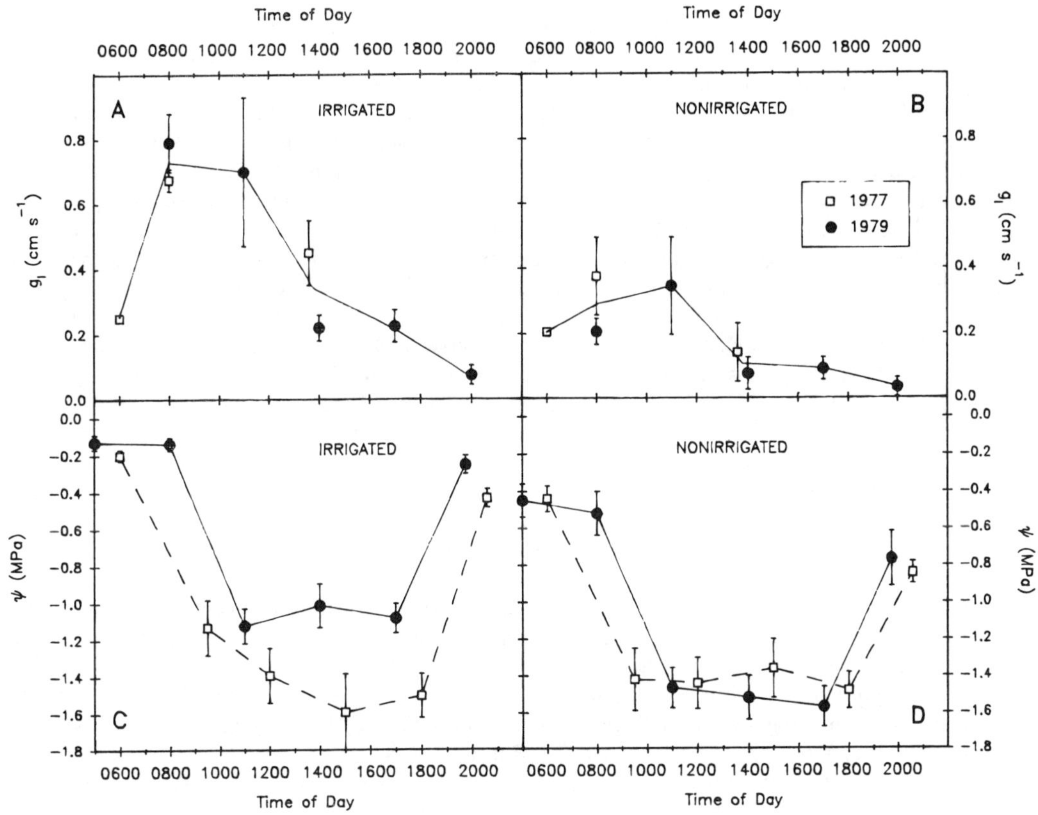

Figure 3 Diurnal leaf conductance (g_l) and leaf (xylem) water potential (Ψ)for irrigated and nonirrigated strawberry plants in first-year plantings in 1977 and 1979. Graphs A and B represent g_l of irrigated and nonirrigated plants, respectively, with g_l data averaged for the two years. Graphs C and D represent Ψ of irrigated and nonirrigated plants. The symbols represent means ± SE. Redrawn from Renquist, A. R., Breen, P. J., and Martin, L. W., *Scientia Hort.*, 18, 101, 1982.

minimum values reported for most woody fruit species, the diurnal patterns of Ψ_l and g_s for strawberry appear similar to the general diurnal patterns of other fruit species.

Renquist et al.[172] observed diurnal fluctuations in Ψ_l for irrigated and nonirrigated 'Olympus' strawberry plants during two years (1977, 1979) in the field in Oregon, and during one season in a glasshouse. In the field, regardless of irrigation treatment, there was a typical mid-morning decrease in Ψ_l, with a minimum Ψ_l value between late morning and late afternoon, and a gradual increase in Ψ_l after about 1700 hrs (Figure 3). Predawn Ψ_l was greater for irrigated than for nonirrigated plants (−0.18 and −0.4 MPa, respectively), reflecting the higher soil moisture content for that treatment. Also, for nonirrigated plants, Ψ_l decreased more rapidly in the morning, and increased more slowly in the evening than that for irrigated plants.[172] As a result of lower irradiance levels and a moderated environment in the glasshouse, Ψ_l of irrigated glasshouse plants was less negative throughout the day, and increased more rapidly in late afternoon compared to that of irrigated plants in the field.[172] After a 3-day drying cycle in the glasshouse, minimum Ψ_l was −2.3 MPa, which was more negative than minimum Ψ_l values for nonirrigated plants in the field, presumably due to the smaller rooting volume for nonirrigated glasshouse plants.

For irrigated and nonirrigated 'Olympus' strawberry plants, g_s ranged between about 45 and 360 mmol m^{-2} s^{-1}, and 45 and 180 mmol m^{-2} sec^{-1}, respectively (Figure 3).[172,176] Throughout most of the day, g_s of irrigated plants was 2 to 4 times greater than that of nonirrigated plants. For all plants, g_s was greatest between 0800 and 1100 hr, and then decreased steadily over the remainder of the day.[172,176] A close relationship between g_s and Ψ_l was not observed; maximum g_s often occurred when Ψ_l was at or near the minimum value.[172]

Darrow and Dewey[180] observed differences between irrigation regimes in regard to diurnal patterns of stomatal aperture in leaves of strawberry plants. For nonirrigated plants, stomata only opened slightly for a brief period in early morning, and were closed for the remainder of the day. For irrigated plants, 80% of the stomata were visibly open at 0900 hr, 25% were open at 1300 hr, 50% were open at 1500 hr, and all had closed at 1700 hr. For 'Bogota' strawberry plants maintained in growth chambers, A, g_s, and E remained relatively constant (16 μmol CO_2 m^{-2} s^{-1}, 225 mmol H_2O m^{-2} s^{-1}, and 4.0 mmol H_2O m^{-2} s^{-1}, respectively) for Ψ_l values between -0.4 to -1.0 MPa, but decreased sharply as Ψ_l decreased below -1.0 MPa.[181] Net CO_2 assimilation, g_s and E all approached zero as Ψ_l decreased below -2.4 MPa. In contrast, Renquist et al.,[172] working with plants in the field and in growth chambers, found that g_s decreased gradually over a wide range of Ψ_l, with no sharp decrease in g_s at a single Ψ_l value. Over the entire range of Ψ_l, g_s of field-grown plants at a particular Ψ_l value was consistently greater than that of growth-chamber plants.[172] For example, a Ψ_l of -1.5 MPa was associated with almost complete stomatal closure for growth-chamber plants, whereas g_s of field-grown plants at Ψ_l of -1.5 MPa was still about 50% of maximum value.[172,176] The differences in g_s at similar Ψ_l values for growth chamber and field-grown plants may have been due to larger root systems for field-grown plants.[172,176] Also, in the field, environmental stresses such as high irradiance levels and increased evaporative demand may have influenced osmotic potential, stomatal density, and cuticle thickness, thereby affecting Ψ_l and g_s.[172,176]

In a glasshouse study, 'Raritan' and 'Surecrop' strawberry plants were either drought-stressed (soil moisture tensions ranging from -0.028 to -0.074 MPa) or were non-drought stressed (soil moisture tensions ranging from -0.002 to -0.010 MPa).[182] For a given irrigation treatment, 'Surecrop' plants had less negative Ψ_l, and greater A and E, than 'Raritan' plants, possibly as a result of a larger root system and greater utilization of available soil moisture for 'Surecrop' than for 'Raritan' plants.[182] For 'Raritan', drought-stress reduced A and E by 24 and 38%, respectively, relative to that of nonstressed plants; for 'Surecrop', drought-stress reduced A and E by only 17 and 32%, respectively, again suggesting greater drought tolerance for 'Surecrop' than for 'Raritan'.

Stomatal morphology and density may influence g_s. In strawberry, stomata occur abaxially on petioles, leaves, and fruit, but not on leaf veins.[180] For several field-grown strawberry cultivars, stomatal density ranged between 200 and 500 stomata mm^{-2}, with the greatest density observed near the leaf mid-rib.[180] Stomatal density did not appear to be correlated with drought tolerance, but stomatal density for glasshouse-grown strawberry plants tended to be less than that for field-grown plants,[180] suggesting that light or other environmental factors may influence stomatal density. Cultivar and species differences were observed in regard to stomatal size and position, with stomata of drought-tolerant cultivars being more recessed than drought-sensitive cultivars.[180] However, in another study, no cultivar or species differences were observed in regard to stomatal position.[176]

Under similar glasshouse conditions, stomata of two putatively drought-tolerant *F. chiloensis* clones opened earlier, and remained open longer, than stomata of drought-sensitive cultivars or species.[180] Maintenance of stomatal aperture under drought conditions may be due to osmotic adjustment, which has been observed in *F. virginiana, F. chiloensis*, and some *F. x ananassa* cultivars.[116,183–186] With osmotic adjustment, active accumulation of cell solutes results in a more negative cell osmotic potential (Ψ_π), enabling continued water absorption and the maintenance of cell turgor and physiological and growth processes.[184] In Oregon, turgor potential (Ψ_p) (calculated as the difference between Ψ_l and Ψ_π) of 'Olympus' strawberry leaves was not affected by irrigation regime despite a more negative Ψ_l recorded at night and predawn, perhaps as a result of osmotic adjustment in nonirrigated plants.[116] Despite similar Ψ_p in irrigated and nonirrigated plants, leaf expansion rates were reduced for nonirrigated plants. Osmotic adjustment in *F. virginiana* was dependent on leaf age, with intermediate age leaves exhibiting the greatest capacity for adjustment.[184] Although a single drought cycle reduced Ψ_π by 0.1 to 0.2 MPa, and repeated stress cycles reduced Ψ_π by about 1.0 MPa, Ψ_π reductions were insufficient for maintaining growth rates similar to those of irrigated plants, or for preventing senscence of older leaves.

After 4 drying cycles to near the wilting point lasting 2.25 and 9.75 days for plants of *F. virginiana* and *F. chiloensis*, respectively, osmotic adjustment was observed to occur in *F. chiloensis*, but not in *F. virginiana*.[183] For *F. chiloensis* plants exposed to 3 wilting cycles, total soluble carbohydrate concentration (TSC) was 2.4-fold greater than that of nonstressed plants, while TSC of stressed *F. virginiana* plants was never more than 1.3-fold greater than that of nonstressed plants.[186] Glucose and fructose were the major soluble sugars contributing to the decrease in Ψ_π.[186] Despite the observed reductions in Ψ_π for water-stressed *F. chiloensis*, drought resulted in reduced leaf growth rates and total leaf area.[183]

Although water-stressed *F. virginiana* plants did not osmotically adjust, they exhibited no significant reduction in leaf area.

With rapidly-induced water stress, plants of *F. virginiana* did not osmotically adjust, whereas *F. chiloensis* plants exhibited osmotic adjustment of 0.42 MPa.[185] The observed lack of osmotic adjustment for *F. virginiana* may have been due to the rapid onset of water stress, since osmotic adjustment occurs as a response to a gradually increasing soil moisture stress.[175,187] In one comparative study, plants of *F. virginiana* had 70% greater leaf area and double the root dry weight of the *F. chiloensis* plants,[183] and thus may have had greater E (on a whole plant basis), contributing to a higher rate of soil desiccation. Gradual imposition of water stress resulted in osmotic adjustment in plants of both species, but the magnitude of adjustment was about 0.35 MPa greater in *F. chiloensis* than in *F. virginiana*.[185]

A slower rate of wilting for *F. chiloensis* also may have been related to Darrow and Sherwood's[188] observation that rates of water loss for plants of *F. chiloensis* were consistently lower than those for *F. virginiana*. Similarly, Darrow and Dewey[180] observed that, following a period of drought stress in the field, the number of stomata open at early morning was 4-fold greater for a drought-sensitive strawberry cultivar ('Marshall') than for a drought-tolerant cultivar ('Ettersburg 121'). Thus, drought tolerance in certain strawberry species or cultivars may be due to a greater degree of stomatal control. With such a conservative water-use "strategy," stomata close early in response to water deficits, thereby maintaining Ψ_l and Ψ_p, although g_s and A are reduced. While osmotic adjustment appears to be an important factor in strawberry drought tolerance, other factors, such as control of stomatal aperture in response to water stress may be of equal or greater importance.

Stomatal conductance and Ψ_l in strawberry appear to be sensitive to increased VPD. An unspecified strawberry cultivar maintained at an ambient temperature of 35°C and a VPD of either -0.5 or -3.5 kPa manifested reductions in g_s as VPD was increased from -0.5 to -5.0 kPa.[173] Presumably, as evaporative demand increased, Ψ_l decreased, resulting in reduced guard cell turgor and stomatal closure. Regardless of measurement VPD, g_s of plants preconditioned at a VPD of -0.5 kPa was consistently greater than that for plants maintained at -3.5 kPa.[173] For strawberry plants maintained at 20°C and an ambient VPD of -1.5 kPa, g_s was greatest at a measurement VPD of -1.5 kPa, and decreased as VPD decreased or increased.[173] In contrast, for 'Bogota' strawberry plants maintained at 20°C, g_s remained constant as VPD increased from -1.45 to -2.14 kPa before decreasing slightly at greater VPD.[189] Higgs and Jones[190] observed a strong correlation between seasonal values of Ψ_l and VPD, and a moderate correlation between seasonal values of g_s and Ψ_l, but g_s was not correlated with VPD. Although VPD was not determined, O'Neill[184] observed that pre-dawn Ψ_l of *F. virginiana* was most negative at high ambient temperatures, probably as a result of increased VPD. Discrepancies in the stomatal responses to VPD may be due to differences in cultivar, environmental preconditions or experimental conditions.

B. SOIL MOISTURE, VEGETATIVE AND REPRODUCTIVE GROWTH

Strawberry vegetative growth is sensitive to soil moisture deficits. In general, optimum yield and growth responses of strawberry have been obtained when soil moisture is maintained above 70–80% of field capacity.[176] For strawberry, decreased plastochron interval[191] and increased dry matter production[176,182] were associated with increased soil moisture level and increased water use. Leaf length of irrigated 'Olympus' strawberry plants was greater than that nonirrigated plants 12 days after leaf unfolding.[116] Leaf expansion rates were greatest between 1800 and 2300 hr, apparently being limited by low Ψ_l or Ψ_p in the afternoon, and by low temperatures later at night. For 'Olympus' strawberry plants, leaf area of irrigated plants was 205% greater than that of nonirrigated plants after 62 days, due to an effect of irrigation on leaf size and number.[116] Although soil moisture deficit had no effect on leaf growth of 'Tioga' during a drought period, reduced leaf growth rates were observed after drought stress was relieved.[192] For *F. virginiana* plants in containers, short-term water stress resulted in reduced leaf dry weight, leaf area, leaf cell length, and increased specific leaf weight and cell density.[184]

The influence of irrigation may be less noticeable in humid regions. For example, irrigation resulted in increased leaf number, leaf area, and biomass accumulation in a dry year, but there was little effect of irrigation in a wet year.[193] In a year characterized by cool, cloudy growth conditions, leaf area and leaf dry weight were unaffected by irrigation regime, although irrigation resulted in increased crown and leaf number.[194]

Stolon production and early stolon rooting has been reported to increase as a result of irrigation.[195] However, decreased stolon formation with increased soil moisture has also been observed, possibly as a result of cooler soil and plant temperatures or of excessive soil moisture with irrigation.[196] Rom and Dana[197] found that the initial number of primary roots of developing stolon plants was independent of soil moisture; however, root length and branching increased as soil water potential (Ψ_s) increased to field capacity ($\Psi_s \geq 0.03$ MPa). Petiole length, leaf dry weight, leaf growth rates, and top:root ratio were greater, and secondary root dry weight was reduced for developing stolons attached to mother plants maintained at low soil moisture tensions (0.05 or 0.2 MPa) compared to a high soil moisture tension (1.0 MPa).[198] For developing stolons, leaf area decreased with decreasing soil moisture, yet root development was greatest at intermediate soil moisture, possibly as a result of poor aeration in the wettest treatment.[198]

Irrigation has been shown to increase strawberry fruit yields.[176,182,196,199–206] However, irrigation in years of high rainfall had no effect on fruit yield.[207–209] Although flower and fruit number generally are dependent on adequate soil moisture,[176] irrigation promoted excessive vegetative growth in some studies, resulting in reduced flower bud formation,[210,211] or increased inter-crown competition.[203] Alternatively, soil moisture stress during the period of flower bud initiation in autumn has decreased fruit production the subsequent year.[176] However, soil moisture deficit prior to flower emergence resulted in increased flower numbers, but had no effect on fruit set or fruit weight.[192] Soil moisture deficits after flower emergence reduced flower number, fruit set, fruit weight, leaf number, and leaf size.[192] The observed differences among studies in yield response to irrigation may be due to differences in cultivars, soil moisture levels, timing of the moisture stress in relation to developmental stage, and environmental or preconditioning effects.

Irrigation has resulted in increased fruit size in some studies,[176,192,199,202,204,205] although decreased fruit size in response to irrigation has also been observed, possibly as a result of increased fruit number.[176] Fruit size of strawberry decreased diurnally as a consequence of fruit transpiration,[176] and such shrinkage is reportedly due to fruit transpirational water loss, rather than movement of water from the fruit to leaves or other plant organs.[176,212]

Soil moisture stress can result in an increase in percentage dry weight of strawberry fruit.[213] Similarly, fruit sugar, dry matter and acid content decrease with irrigation, probably due to dilution effects as a result of increased fruit size.[176] Increased fruit size in response to irrigation, and a concomitant dilution of soluble solids and thinning of cell walls, may reduce fruit firmness.[176] Okasha et al.[214] reported increased fruit firmness in light-or intermediate-textured soils compared to heavier-textured soils, probably due to decreased soil moisture availability in lighter soils. Soil moisture deficits have resulted in fruit dehydration and have had an adverse effect on fruit color.[209] Although irrigation promoted early fruit yield in one study,[202] it delayed ripening in another,[176] perhaps due to excessive vegetative growth or reduced temperatures.[176]

The occurrence of guttation from leaf hydathodes of strawberry has been correlated with high predawn Ψ_l (i.e., $\Psi_l > -0.08$ MPa).[215] Plants exhibiting guttation had greater g_s, and lower leaf temperatures, suggesting that the presence or absence of leaf guttation could be used as an indicator of plant moisture status.[215]

C. SALINITY

In areas with low rainfall and marginal water quality, excessive salt accumulation may occur in the soil.[216,217] Strawberry is one of the most sensitive fruit crops to salinity.[163,217–220] High salt concentrations inhibit new root initiation and root growth, resulting in poorly anchored plants, loss of feeder roots, leaf necrosis, and stunted growth or plant death,[163,219] but reductions in fruit yield often occur before visual symptoms of salt damage are apparent.[163,216] Strawberry yield decreases have been observed with a substrate electrical conductivity (EC_e) as low as 1.0 dS m^{-1},[220] and yield reductions of 50% were reported with an EC of 2.2 dS m^{-2}.[219] High concentrations of Cl are especially damaging; Na is less damaging since it is accumulated much less readily in strawberry plants.[217,219] Proper irrigation is important for minimizing salt damage to strawberry. Furrow irrigation resulted in greater salt accumulation at the soil surface than either sprinkler or drip irrigation.[216]

Differential responses to salinity have been observed among strawberry cultivars. 'Shasta' accumulated more Cl and Na, and exhibited leaf necrosis, yield reductions and plant mortality at lower salt concentrations than 'Lassen'.[217,219] As substrate EC_e increased from 1.2 dS m^{-2} to 2.2 dS m^{-2}, yields of 'Shasta' decreased by 50%, with a 29% plant mortality after 1 year, but no yield decrease or plant mortality was observed for 'Lassen'.[219] Increased salinity resulted in reduced vegetative growth for both cultivars, but shoot growth for 'Shasta' was reduced more than root growth. Salinity promoted early flowering for 'Tioga',[221] possibly as a result of a reduction in vegetative growth. Salinity has also decreased fruit size,[214,219] and increased titratable acids and sugars.[219]

Crop sensitivity to salinity can vary depending on environmental conditions. For example, detrimental effects of salinity are often more pronounced under hot, dry conditions than under cool conditions.[219,220] Increased evapotranspiration as a result of high temperature and high VPD increases the concentration of salts in the soil, resulting in a more negative soil Ψ_π and a decrease in available soil moisture.[220]

D. FLOODING

Strawberry is highly sensitive to excess soil moisture and poor soil aeration.[170,222,223] Moreover, strawberry roots are susceptible to a number of pathogenic fungi and bacteria, including fungi such as *Phytophthora* spp. and *Pythium* spp. that thrive in poorly aerated, waterlogged soils.[223] Although damage to strawberry from poor aeration and pathogen infection under waterlogged soil conditions is widely recognized,[6,163,170,222,223] relatively few studies have been conducted to determine the growth and physiological responses of strawberry to flooding.

The root system of the cultivated strawberry plant is comprised of perennial structural roots, and short-lived feeder roots. Feeder roots are comprised solely of primary tissues and have a maximum lifespan of about two weeks, but under optimal conditions are continually replaced by new roots.[170,222,223] Feeder roots are highly sensitive to poor soil aeration, and die rapidly when exposed to even short periods of flooding.[170,222,223] Strawberry plants may survive short durations of flooding, developing new feeder roots after waterlogging subsides, but long-term survival is dependent on environmental conditions. Environmental conditions conducive to high atmospheric VPD following a period of excessive soil moisture are especially deleterious.[170,223] For containerized strawberry plants, continuous waterlogging of a portion of the root system resulted in root death, reductions in leaf size and petiole length, and rapid leaf senescence.[224]

V. MISCELLANEOUS STRESSES

A. WIND

During a three-year period, growing strawberries outdoors in sheltered plots reduced mean wind velocity from 1.6 to 1.1 m s^{-1}, and increased yield by 56% compared to non-sheltered plots (Table 2).[225] Wind protection resulted in increased plant size and crown number, but had no influence on plant survival, number of flower trusses per crown, or fruit size (Table 1). Yield reductions for the non-sheltered plants were attributed to a reduction in vegetative growth due to leaf tissue damage[225]; reductions in yield may also have been due to adverse effects of wind on g_s and A.[6]

B. ACIDIC FOG

Acidic fogs can occur during the winter strawberry production season in southern California, but strawberry plants appear to be relatively tolerant of acid fog. Musselman et al.[226] reported little or no

Table 2 **Influence of wind shelter on mean yield, plant diameter, and number of flower trusses per plant during 1967–70 for 'Cambridge Favourite' strawberry plants**

	Date	Exposed	Sheltered	Increase due to shelter
Yield tonnes/ha	1968	9.8	16.6	69**
	1969	15.8	19.1	21
	1970	9.8	17.3	77**
Canopy width (cm)	1967	39.8	46.6	17**
	1968	42.1	50.8	21***
	1969	45.2	47.5	5
Trusses/plant	1968	9.0	15.9	77**
	1969	17.6	22.9	30*
	1970	13.5	19.6	45
Crowns/plant		9.6	12.9	34*

Note: *, **, *** indicate significant difference at $p = 0.05$, 0.01, and 0.001, respectively.
Adapted from Waister, 1972.

effect of short-term exposure to acid fog at pH of 2.4 or higher on leaf necrosis, vegetative growth, or yield of five strawberry cultivars. Acid fog at pH of 1.6 resulted in plant death after 8 weeks, and marketable yields were reduced due to calyx injury at pH of 2.8 or less.[226] Takemoto et al.[227] observed necrosis of leaves, petioles, flowers and fruit, and reductions in leaf area, and stem and leaf dry wt. for strawberries exposed to acid fog at pH 1.68 relative to control plants, but there was no effect on fruit number or g_s or A. The differential effects of acid fog between studies may have been due to differences in exposure periods, or to differences in experimental conditions or plant preconditioning.

C. ELEVATED AMBIENT CO_2

Exposure to elevated CO_2 concentrations typically enhance A, although concentrations above 1000 μmol mol^{-1} may result in stomatal closure and decreased A.[83] For 'Midway' and 'Raritan' strawberry plants, exposure to 700 μmol mol^{-1} CO_2 for 6 weeks reduced A, g_s, carboxylation efficiency, and apparent quantum yield relative to plants maintained at ambient (350 μmol mol^{-1}) CO_2 concentrations.[228] Cultivar differences were observed, in that the Rubisco activation state decreased for 'Midway' and increased for 'Raritan', and soluble Rubisco decreased 10–15% for 'Midway' and decreased 40–45% for 'Raritan'. Sruamsiri and Lenz[229] observed a reduction in g_s and E as ambient CO_2 concentration increased from 300 to 900 μmol mol^{-1}, although A increased 70%. The influence of elevated ambient CO_2 on physiology and growth of strawberry remains unresolved.

VI. CONCLUSIONS

Fragaria x ananassa is a highly variable and adaptive species that exhibits a greater sensitivity to environmental factors than most other fruit species. With the appropriate photoperiod, temperature, and plant preconditioning, many strawberry cultivars will flower and fruit almost continuously.

There are disparate reports of the influence of environmental factors such as atmospheric VPD, drought, elevated ambient CO_2, irradiance level, temperature, and chilling on the physiology, growth, and development of the strawberry, indicating the need for additional investigations. The physiological bases for photoperiodic and temperature regulation of strawberry growth and development have not been elucidated. There is little information regarding A, E, and water use efficiency among species and cultivars. Investigations of photoassimilate partitioning and of overall carbon balance of the strawberry plant are needed, particularly in regard to fruit yield and quality in DN cultivars.

REFERENCES

1. Childers, N. F., Foreword, *The Strawberry*, Childers, N. F., Ed., Horticultural Publications, Gainesville, FL, 1981.
2. Darrow, G. M., *The Strawberry*, Holt, Rinehart and Winston, New York, 1966, chap. 20.
3. Scott, D. H. and Lawrence, F. J., Strawberries, in *Advances in Fruit Breeding*, Janick, J. and Moore, J. N., Eds., Purdue Univ. Press, West Lafayette, IN, 1975.
4. Anonymous, *FAO Production Yearbook*, FAO Statistics Series no. 104, Rome, 1991.
5. Heide, O. M., Photoperiod and temperature interactions in growth and flowering of strawberry, *Physiol. Plant.*, 40, 21, 1977.
6. Galletta, G. J. and Bringhurst, R. S., Strawberry management, in *Small Fruit Crop Management*, Galletta, G. J. and Himelrick, D. G., Eds., Prentice Hall, Englewood Cliffs, N.J., 1990, chap. 3.
7. Guttridge, C. G., Fragaria, in *The Induction of Flowering*, Evans, L. T., Ed., MacMillan, Inc., Melbourne, 1969, chap. 10.
8. Gutridge, G. C., *Fragaria x ananassa*, in *CRC Handbook of Flowering*, Vol. 3, Halevy, A. H., Ed., CRC Press, Boca Raton, FL, 1985, chap. 1.
9. Angevine, M. W., Variations in the demography of natural populations of the wild strawberries *Fragaria vesca* and *Fragaria virginiana, J. Ecol.*, 71, 959, 1983.
10. Hancock, J. F., Jr. and Bringhurst, R. S., Ecological differentiation in perennial, octoploid species of *Fragaria, Amer. J. Bot.*, 66, 367, 1979.
11. Bringhurst, R. S. and Voth, V., Breeding strawberries for high productivity and large fruit size, in *The Strawberry*, Childers, N. F., Ed., Horticultural Publications, Gainesville, FL, 1981, 156.

12. Sjulin, T. M. and Dale, A., Genetic diversity of North American strawberry cultivars, *J. Amer. Soc. Hort. Sci.*, 112, 375, 1987.
13. Arney, S. E., Studies of growth and development of the genus *Fragaria*. II. The initiation, growth and emergence of leaf primordia in *Fragaria, Ann. Bot.*, 17, 477, 1953.
14. Arney, S. E., Studies of growth and development in the genus *Fragaria*. IV. Winter growth, *Ann. Bot.* (NS), 19, 265, 1955.
15. Dana, M. N., The strawberry plant and its environment, in *The Strawberry*, Childers, N. F., Ed., Horticultural Publications, Gainesville, FL, 1981, 33.
16. Guttridge, G. C., Observations on the shoot growth of the cultivated strawberry plant, *J. Hort. Sci.*, 30, 1, 1955.
17. Jahn, O. L. and Dana, M. N., Crown and inflorescence development in the strawberry, *Fragaria ananassa, Am. J. Bot.*, 57, 605, 1970.
18. Mann, C. E. T., Studies in root and shoot growth of the strawberry, *Ann. Bot.*, 44, 55, 1930.
19. Robertson, M. and Wood, C. A., Studies in the development of strawberry. I. Flower-bud initiation and runner development in early- and late-formed runners in 1951 and 1952, *J. Hort. Sci.*, 29, 104, 1954.
20. White, P. R., Studies of the physiological anatomy of the strawberry, *J. Agr. Res.*, 35, 481, 1927.
21. Jepson, W. H., *Manual of Flowering Plants in California*, Univ. of California Press, Berkeley, 1951.
22. Westwood, M. N., *Temperate-Zone Pomology*, W.H. Freeman and Co., San Francisco, 1993.
23. Gross, L. J. and Chabot, B. F., Time course of photosynthetic response to changes in incident light energy, *Plant Physiol.*, 63, 1033, 1979.
24. Jurik, T. W., Chabot J. F., and Chabot, B. F., Ontogeny of photosynthetic performance in *Fragaria virginiana* under changing light regimes, *Plant Physiol.*, 63, 542, 1979.
25. Jurik, T. W., Reproductive effort and CO_2 dynamics of wild strawberry populations, *Ecology*, 64, 1329, 1983.
26. Darrow, G. M. and Waldo, G. F., Responses of strawberry varieties and species to duration of the daily light period, *Tech. Bull. No. 453, U.S. Dept. Agric.*, 1934.
27. Reed, C. F., Wild strawberry species of the world, in *The Strawberry*, Darrow, G.M., Ed., Holt, Rinehart and Winston, New York, 1966, chap. 8.
28. Cameron, J. S., Hancock, J. F., and Flore, J. A., The influence of micropropagation on yield components, dry matter partitioning and gas exchange characteristics of strawberry, *Scientia Hort.*, 38, 61, 1989.
29. Cameron, J. S. and Hartley, C. A., Gas exchange characteristics of *Fragaria chiloensis* genotypes, *HortScience*, 25, 327, 1990.
30. Hancock, J. F., Flore, J. A., and Galletta, G. J., Gas exchange properties of strawberry species and their hybrids, *Scientia Hort.*, 40, 139, 1989.
31. Hancock, J. F., Flore, J. A., and Galletta, G. J., Variation in leaf photosynthetic rates and yield in strawberries, *J. Hort. Sci.*, 64, 449, 1989.
32. Cameron, J. S. and Hartley, C. A., Gas exchange and photosynthate partitioning characteristics of *Fragaria chiloensis* germplasm, *HortScience*, 23, 821 (abstract), 1988.
33. Hancock, J. F., Photosynthesis in strawberries and the possibility of genetic improvement, in *The Strawberry into the 21st Century*, Dale, A., and Luby, J. J., Eds., Timber Press, Inc., Portland, OR, 1991, chap. 27.
34. Campbell, D. E. and Young, R., Short-term CO_2 exchange response to temperature, irradiance and CO_2 concentration in strawberry, *Photosynthesis Res.*, 8, 31, 1986.
35. Ceulemans, R., Baets, W., Vanderbruggen, M., and Impens, I., Effects of supplemental irradiation with HID lamps, and NFT gutter size on gas exchange, plant morphology and yield of strawberry plants, *Scientia Hort.*, 28, 71, 1986.
36. Ferree, D. C. and Stang, E. J., Seasonal plant shading, growth, and fruiting in 'Earliglow' strawberry, *J. Amer. Soc. Hort. Sci.*, 113, 322, 1988.
37. Van Elsacker, P., Impens, I., and Liesse, H., Fotosyntheseresponsoppervlak van aardbei in functie van licht, CO_2 en temperatuur, *Revue de l'Agiculture*, 42, 649, 1989.
38. Sruamsiri, P. and Lenz, F., Photosynthesis and stomatal behavior of strawberries. I. Effect of light, *Gartenbauwissenschaft*, 50, 78, 1985.
39. Chandler, C. K., Miller, D. D., and Ferree, D. C., Shade during July and August reduces growth but not fruiting of strawberry plants, *HortScience*, 27, 1044, 1992.

40. Smeets, L., Effects of light intensity on stamen development in the strawberry cultivar 'Glasa', *Scientia Hort.*, 4, 255, 1976.
41. Smeets, L., Effect of light intensity on forcing of the strawberry cultivar 'Glasa', *Scientia Hort.*, 13, 33, 1980.
42. Garrison, S. E., Williams, J. M., Barden, J. A., and Kushad, M. M., Shade decreases fruit yield of strawberry, *Adv. Strawberry Prod.*, 10, 53, 1991.
43. Smeets, L., Effect of light intensity during flowering on stamen development in the strawberry cultivars 'Karina' and 'Sivetta', *Scientia Hort.*, 12, 343, 1980.
44. Dennis, F. G., Lipecki, J., and Kiang, C., Effects of photoperiod and other factors upon flowering and runner development of three strawberry cultivars, *J. Amer. Soc. Hort. Sci.*, 95, 750, 1970.
45. Smeets, L., Runner formation on strawberry plants in autumn and winter: II. Influence of light intensity on the photoperiodical behavior, *Euphytica*, 4, 240, 1955.
46. Went, F. W., *The Experimental Control of Plant Growth*, Vol. 17, Chronica Botanica, Waltham, MA, chap. 9, 1957.
47. Chabot, B. F., Environmental influences on photosynthesis and growth in *Fragaria vesca, New Phytol.*, 80, 87, 1978.
48. Chabot, B. F. and Chabot, J. F., Effects of light and temperature on leaf anatomy and photosynthesis in *Fragaria vesca, Oecologia*, 26, 363, 1977.
49. Chabot, B. F., Jurik, T. W., and Chabot, J. F., Influence of instantaneous and integrated light-flux density on leaf anatomy and photosynthesis, *Amer. J. Bot.*, 66, 940, 1979.
50. Jurik, T. W., Chabot, J. F., and Chabot, B. F., Effects of light and nutrients on leaf size, CO_2 exchange, and anatomy in wild strawberry (*Fragaria virginiana*), *Plant Physiol.*, 70, 1044, 1982.
51. Scott, D. H. and Draper, A. D., Light in relation to seed germination of blueberries, strawberries, and *Rubus, HortScience*, 2, 107, 1967.
52. Nakamura, S., Germination of strawberry seeds, *J. Jpn. Soc. Hort. Sci.*, 41, 367, 1972.
53. Darrow, G. M., Interrelation of temperature and photoperiodism in the production of fruit-buds and runners in the strawberry, *Proc. Am. Soc. Hort. Sci.*, 34, 360, 1936.
54. Durner, E. F., Barden, J. A., Himelrick, D. G., and Poling, E. B., Photoperiod and temperature effects on flower and runner development in day-neutral, June-bearing, and everbearing strawberries, *J. Amer. Soc. Hort. Sci.*, 109, 396, 1984.
55. Durner, E. F. and Poling, E. B., Strawberry developmental responses to photoperiod and temperature: A review, *Adv. Strawberry Prod.*, 7, 6, 1988.
56. Hartman, H. T., The influence of temperature on the photoperiodic response of several strawberry varieties grown under controlled environment conditions, *Proc. Amer. Soc. Hort. Sci.*, 50, 243, 1947.
57. Hartman, H. T., Some effects of temperature and photoperiod on flower formation and runner production in the strawberry, *Plant Physiol.*, 22, 407, 1947.
58. Smeets, L., Effect of temperature and day-length on flower initiation and runner formation in two everbearing strawberry cultivars, *Scientia Hort.*, 12, 19, 1979.
59. Ito, H. and Saito, T., Studies on the flower formation in the strawberry plants. I. Effects of temperature and photoperiod on the flower formation, *Tohoku J. Agr. Res.*, 13, 191, 1962.
60. Avigdori-Avidov, H., Goldschmidt, E. E., and Kedar, N., Involvement of endogenous gibberellin in the chilling requirements of strawberry (*Fragaria x ananassa* Duch.), *Ann. Bot.*, 41, 927, 1977.
61. Guttridge, C. G., The effects of winter chilling on the subsequent growth and development of the cultivated strawberry plant, *J. Hort. Sci.*, 33, 119, 1958.
62. Guttridge, C. G. and Anderson, H. M., Promoting second cropping in strawberry by avoiding chilling or advancing spring growth, *J. Hort. Sci.*, 51, 225, 1976.
63. Voth, V. and Bringhurst, R. S., Fruiting and vegetative response of Lassen strawberries in southern California as influenced by nursery source, time of planting, and plant chilling history, *Proc. Amer. Soc. Hort. Sci.*, 72, 186, 1958.
64. Nicoll, M. F. and Galletta, G. J., Variation in growth and flowering habits of Junebearing and everbearing strawberries, *J. Amer. Soc. Hort. Sci.*, 112, 872, 1987.
65. Sudds, R. H., Fruit-bud formation in the strawberry, *Penn. Ag. Expt. Sta. Bull.*, no. 230, 1928.
66. Jahn, O. L. and Dana, M. N., Dormancy and growth of the strawberry plant, *Proc. Am. Soc. Hort. Sci.*, 89, 322, 1966.

67. Guttridge, C. G., Photoperiodic promotion of vegetative growth in the cultivated strawberry plant, *Nature*, 178, 50, 1956.
68. Guttridge, C. G., Evidence for a flower inhibitor and vegetative growth promoter in the strawberry, *Ann. Bot.*, 23, 351, 1959.
69. Leshem, R. and Koller, D., The control of flowering in the strawberry *Fragaria x ananassa* Duch. I. Interaction of positional and environmental effects, *Ann. Bot.*, 28, 569, 1964.
70. Vince-Prue, D. and Guttridge, C. G., Floral initiation in strawberry: Spectral evidence for the regulation of flowering by long day inhibition, *Planta*, 110, 165, 1973.
71. Collins, W. B. and Barker, W. G., A flowering response of strawberry to continuous light, *Can. J. Bot.*, 42, 1309, 1964.
72. Thompson, P. A. and Guttridge, G. C., The role of leaves as inhibitors of flower induction in strawberry, *Ann. Bot.*, 24, 482, 1960.
73. Guttridge, C. G., Further evidence for a growth-promoting and flower-inhibiting hormone in strawberry, *Ann. Bot.*, 23, 612, 1959.
74. Mason, D. T., Inflorescence initiation in the strawberry. I. Initiation in the field and its modification by post-harvest defoliation, *Hort. Res.*, 6, 33, 1966.
75. Mason, D. T., Inflorescence initiation in the strawberry. II. Some effects of date and severity of post-harvest defoliation, *Hort. Res.*, 7, 97, 1967.
76. Moore, J. N., Effects of post-harvest defoliation on strawberry yields and fruit size, *HortScience*, 3, 45, 1968.
77. Guttridge, C. G., Anderson, M. M., Thompson, P. A., and Wood, C. A., Post harvest defoliation of strawberry plantations, *J. Hort. Sci.*, 36, 93, 1961.
78. Borthwick, H. A. and Parker, M. W., Light in relation to flowering and vegetative development, Rpt. 13th Intl. Hort. Cong. 1952, 2, 801.
79. Arney, S. E., Studies of growth and development in the genus *Fragaria*. III. The growth of leaves and shoot, *Ann. Bot.*, 18, 349, 1954.
80. Nishizawa, T., Effects of daylength on cell length and cell number in strawberry petioles, *J. Jpn. Soc. Hort. Sci.*, 59, 533, 1990 (Japanese, English summary).
81. Piringer, A. A. and Scott, D. H., Interrelation of photoperiod, chilling, and flower-cluster and runner production by strawberries, *Proc. Amer. Soc. Hort. Sci.*, 84, 295, 1964.
82. Maas, J. L. and Cathey, H. M., Photomorphogenic responses of strawberry to photoperiodic and photosynthetic radiation, *J. Amer. Soc. Hort. Sci.*, 112, 125, 1987.
83. Salisbury, F. B. and Ross, C. W., *Plant Physiology*, Wadsworth Publishing Co., Inc., Belmont, Calif., 1985, chapter 16.
84. Guttridge, C. G. and Thompson, P. A., The effect of gibberellins on growth and flowering of *Fragaria* and *Duchesnea*, *J. Exp. Bot.*, 15, 631, 1964.
85. Thompson, P. A. and Guttridge, C. G., Effect of gibberellic acid on the initiation of flowers and runners in the strawberry, *Nature*, 184, 72, 1959.
86. Guttridge, C. G., Interaction of photoperiod, chilling and exogenous gibberellic acid on growth of strawberry petioles, *Ann. Bot.*, 34, 349, 1970.
87. Goodwin, P. B. and Gordon, A., The gibberellin-like substances and growth inhibitors in developing strawberry leaves, *J. Exp. Bot.*, 23, 970, 1972.
88. Porlingis, I. C. and Boynton, D., Growth responses of the strawberry plant, *Fragaria chiloensis* var. *ananassa* to gibberellic acid and to environmental conditions, *Proc. Amer. Soc. Hort. Sci.*, 78, 256, 1961.
89. Uematsu, Y. and Katsura, N., Changes in endogenous gibberellin level in strawberry plants induced by light breaks, *J. Japan Soc. Hort. Sci.*, 51, 405, 1983.
90. Moore, J. N. and Hough, L. F., Relationships between auxin levels, time of floral induction and vegetative growth of the strawberry, *Proc. Amer. Soc. Hort. Sci.*, 81, 255, 1962.
91. Campbell, R. W. and Lingle, J. C., Some effects of low temperatures on the flower primordia of the strawberry, *Proc. Amer. Soc. Hort. Sci.*, 64, 259, 1954.
92. Boyce, B. R. and Marini, R. P., Cold acclimation of everbearing strawberry blossoms, *HortScience*, 13, 543, 1978.

93. Fuchigami, L. H., Weiser, C. J., and Evert, D. R., Induction of cold acclimation in *Cornus stolonifera* Michx., *Plant Physiol.*, 47, 98, 1971.

94. Steponkus, P. L. and Lanphear, F. O., Light stimulation of cold acclimation: Production of a translocatable promoter, *Plant Physiol.*, 42, 1673, 1967.

95. Durner, E. F. and Poling, E. B., Comparison of three methods for determining the floral or vegetative status of strawberry plants, *J. Amer. Soc. Hort. Sci.*, 110, 808, 1985.

96. Benoit, F., Further observations on the induction of a second flowering in the strawberry in the strawberry cultivar Redgauntlet, *Agricultura*, 23, 29, 1975.

97. Austin, M. E., Shutak, V. G., and Christopher, E. P., Response of Sparkle strawberry to inductive cycles, *Proc. Amer. Soc. Hort. Sci.*, 77, 372, 1961.

98. van den Muijzenberg, E. W. B., The influence of light and temperature on the photoperiodic development of the strawberry and its significance in cultivation, *Meded. Lab. TuinbPlTeelt Wageningen*, 37, 1942.

99. Downs, R. J. and Piringer, A. A., Differences in photoperiodic responses of everbearing and June-bearing strawberries, *Proc. Amer. Soc. Hort. Sci.*, 66, 234, 1955.

100. Durner, E. F. and Poling, E. B., Flower bud induction, initiation, differentiation and development in the 'Earliglow' strawberry, *Scientia Hort.*, 31, 61, 1987.

101. Foster, J. C. and Janick, J., Variable branching patterns in the strawberry infloresence, *J. Amer. Soc. Hort. Sci.*, 94, 440, 1969.

102. Jonkers, H., On the flower formation, the dormancy and the early forcing of strawberries, *Meded. Landbouwhogesch. Wageningen*, 65, 1, 1965.

103. Guttridge, C. G., Growth, flowering and runnering of strawberry after daylength extension with tungsten and fluorescent lighting, *Scientia Hort.*, 4, 345, 1976.

104. Vince-Prue, D., Guttridge, C. G., and Buck, M. W., Photocontrol of petiole elongation in light-grown strawberry plants, *Planta*, 131, 107, 1976.

105. Kadman-Zahavi, A. and Ephrat, E., Opposite response groups of short-day plants to spectral composition of the main light period and end-of-the-day red or far-red irradiations, *Plant Cell Physiol.*, 15, 693, 1974.

106. Vince-Prue, D., *Photoperiodism in Plants*, McGraw-Hill, Maidenhead, UK, 1975.

107. Sruamsiri, P. and Lenz, F., Photosynthesis and stomatal behaviour of strawberries. II. Effect of temperature, *Gartenbauwissenschaft*, 50, 84, 1985.

108. Udagawa, Y., Ito, T., and Gomi, K., Effects of root temperature on some physiological and ecological characteristics of strawberry plants 'Reiko' grown in nutrient solution, *J. Japan. Soc. Hort. Sci.*, 58, 627, 1989 (Japanese, English summary).

109. Caldwell, J. D., Hancock, J. F., and Flore, J. A., Strawberry leaf photosynthetic acclimation to temperature, *HortScience*, 25, 1166 (abstract), 1990.

110. Roberts, A. N., and Kenworthy, A. L., Growth and composition of the strawberry plant in relation to root temperature and intensity of nutrition, *Proc. Amer. Soc. Hort. Sci.*, 68, 157, 1956.

111. Proebsting, E. L., The effect of soil temperature on the mineral nutrition of the strawberry, *Proc. Amer. Soc. Hort. Sci.*, 69, 278, 1957.

112. Renquist, A. R., Breen, P. J., and Martin, L. W., Effects of black polyethylene mulch on strawberry leaf elongation and diurnal leaf water potential, *J. Amer. Soc. Hort. Sci.*, 107, 640, 1982.

113. Voth, V., Bringhurst, R. S., and Bowen, H. J., Effect of bed system, bed height and clear polyethylene mulch on yield, salt accumulation and soil temperature in California strawberries, *Proc. Amer. Soc. Hort. Sci.*, 91, 242, 1967.

114. Fear, C. D. and Nonnecke, G. R., Soil mulches influence reproductive and vegetative growth of 'Fern' and 'Tristar' day-neutral strawberries, *HortScience*, 24, 912, 1989.

115. Darrow, G. M., Experimental studies in the growth and development of strawberry plants, *J. Agr. Res.*, 41, 307, 1930.

116. Renquist, A. R., Breen, P. J., and Martin, L. W., Influences of water status and temperature on leaf elongation in strawberry, *Scientia Hort.*, 18, 77, 1983.

117. Hellman, E. W. and Travis, J. D., Growth inhibition of strawberry at high temperature, *Adv. Strawb. Prod.*, 7, 36, 1988.

118. Chesness, J. L. and Braud, H. J., Sprinkling to reduce heat stressing of strawberry plants, *Agr. Eng.*, 51, 140, 1970.
119. Goulart, B. L. and Gardner, J. F., Low volume irrigation: potential uses for strawberries, *Penn. Fruit News*, 69, 14, 1989.
120. Nonnecke, G. R. and Fear, C. D., Influence of mist irrigation on growth and development of day-neutral strawberry, *HortScience*, 22, 1044 (abstract), 1987.
121. Kronenberg, H. G., Poor fruit setting in strawberries. I. Causes of poor fruit set in strawberries in general, *Euphytica*, 8, 47, 1959.
122. Kronenberg, H. G., Braak, J. P., and Zeilinga, A. E., Poor fruit in strawberries. II. Malformed fruit in Jucunda, *Euphytica*, 8, 245, 1959.
123. Garren, R., Causes of misshaped strawberries, in *The Strawberry*, Childers, N.F., Ed., Horticultural Publications, Gainesville, FL, 1980, 326.
124. Bagnara, D. and Vincent, C., The role of insect pollination and plant genotype in strawberry fruit set and fertility, *J. Hort. Sci.*, 63, 69, 1988.
125. Blasse, W. and Haufe, M., The influence of bees on yield and fruit quality in strawberry (*Fragaria X ananassa* Duch.), *Archiv-fur-Gartenbau*, 37, 235, 1989.
126. Goodman, R. D. and Oldroyd, B. P., Honeybee pollination of strawberries (*Fragaria X ananassa* Duchesne), *Aust. J. Exp. Agric.*, 28, 435, 1988.
127. Miura, H., Imada, S., and Yabuuchi, S., Double sigmoid growth curve of strawberry fruit, *J. Japan. Soc. Hort. Sci.*, 59, 527, 1990.
128. Galletta, G. J., Draper, A. D., and Swartz, H. J., New everbearing strawberries, *HortScience*, 16, 726, 1981.
129. Boyce, B. R. and Marini, R. P., Cold acclimation of everbearing strawberry blossoms, *HortScience*, 13, 543, 1978.
130. Havis, L., Freezing injury to strawberry flower buds, flowers, and young fruits, *Ohio Agr. Expt. Sta. Bimonthly Bull.*, 23, 168, 1938.
131. Ourecky, D. K. and Reich, J. E., Frost tolerance in strawberry cultivars, *HortScience*, 11, 413, 1976.
132. Boyce, B. R. and Strater, J. B., Comparison of frost injury in strawberry buds, blossoms, and immature fruit, *Adv. Strawberry Prod.*, 3, 8, 1984.
133. Ki, W. K. and Warmund, M. R., Low-temperature injury to strawberry floral organs at several stages of development, *HortScience*, 27, 1302, 1992.
134. Peery, K. B. and Poling, E. B., Field observations of frost injury in strawberry buds and blossoms, *Adv. Strawberry Prod.*, 5, 31, 1986.
135. Brierley, W. G. and Landon, R. H., Cold resistance of strawberry plants in the early stages of growth, *Proc. Amer. Soc. Hort. Sci.*, 42, 432, 1943.
136. Daubeny, H. A., Norton, R. A., Schwartze, C. D., and Barritt, B. H., Winterhardiness in strawberries for the Pacific Northwest, *HortScience*, 5, 152, 1970.
137. Harris, R. E., Relative hardiness of strawberry cultivars at three times of the winter, *Can. J. Plant Sci.*, 53, 147, 1973.
138. Steele, T. A., Waldo, G. F., and Brown, W. S., Conditions affecting cold resistance in strawberries, *Proc. Amer. Soc. Hort. Sci.*, 32, 434, 1934.
139. Warmund, M. R. and Ki, W. K., Fluctuating temperatures and root moisture content affect survival and regrowth of cold-stressed strawberry crowns, *Adv. Strawberry Prod.*, 11, 40, 1992.
140. Marini, R. P. and Boyce, B. R., Influence of low temperatures during dormancy on growth and development of 'Catskill' strawberry plants, *J. Amer. Soc. Hort. Sci.*, 104, 159, 1979.
141. Marini, R. P. and Boyce, B. R., Susceptibility of crown tissues of 'Catskill' strawberry plants to low-temperature injury, *J. Amer. Soc. Hort. Sci.*, 102, 515, 1977.
142. Boyce, B. R. and Reed, R. A., Effects of bed height and mulch on strawberry crown temperatures and winter injury, *Adv. Strawberry Prod.*, 2, 12, 1983.
143. Angelo, E., Factors relating to hardiness in the strawberry. Part I. Development of cold resistance in strawberry varieties, *Minn. Agr. Expt. Sta. Tech. Bull.*, 135, 1939.
144. Boyce, B. R. and Smith, C. R., Low temperature crown injury of dormant 'Catskill' strawberries, *J. Amer. Soc. Hort. Sci.*, 91, 261, 1967.

145. Brierley, W. G. and Landon, R. H., The effect of time of mulching on the cold resistance of strawberry plants, *Proc. Amer. Soc. Hort. Sci.*, 38, 424, 1941.
146. Campbell, R. W. and Lingle, J. C., Some effects of low temperatures on flower primordia of the strawberry, *Proc. Amer. Soc. Hort. Sci.*, 64, 259, 1954.
147. Zurawicz, E. and Stushnoff, C., Influence of nutrition on cold tolerance of 'Redcoat' strawberries, *J. Amer. Soc. Hort. Sci.*, 102, 342, 1977.
148. Harris, R. E., Laboratory technique for assessing winter hardiness in strawberry (*Fragaria ananassa* Duch.), *Can. J. Plant Sci.*, 50, 249, 1970.
149. Brierley, W. G. and Landon, R. H., Winter behavior of strawberry plants, *Minn. Agr. Expt. Sta. Bull.*, 375, 1, 1944.
150. Bailey, J. S. and Rossi, A. W., Effect of fall chilling, forcing temperature and day length on the growth and flowering of Catskill strawberry plants, *Proc. Amer. Soc. Hort. Sci.*, 87, 245, 1965.
151. Darrow, G. M. and Waldo, G. F., Photoperiodism as a cause of the rest period in strawberries, *Science*, 77, 353, 1933.
152. Saure, M. C., Dormancy release in deciduous fruit trees, *Horticultural Reviews*, Vol. 7, AVI Publishing Co., 1985, chap. 6.
153. Kronenberg, H. G., Wassenaar, L. M., and Lindeloof, C. P. J. van de, Effect of temperature on dormancy in strawberry, *Scientia Hort.*, 4, 361, 1976.
154. Kronenberg, H. G. and Wassenaar, L. M., Dormancy and chilling for early forcing, *Euphytica*, 21, 454, 1972.
155. Guttridge, C. G. and Anderson, H. M., Promoting second cropping in strawberry by avoiding chilling or advancing spring growth, *J. Hort. Sci.*, 51, 225, 1976.
156. Lee, B. Y., Takahashi, K., and Sugiyama, T., Studies on dormancy in strawberry plants. II. Vegetative and flowering response in Donner variety transferred from the open to a greenhouse at different dates in autumn and winter, *J. Jpn. Soc. Hortic. Soc.*, 39, 232, 1970 (Japanese, English summary).
157. Smeets, L., Effect of chilling on runner formation and flower initiation in the everbearing strawberry, *Sci. Hortic.*, 17, 43, 1982.
158. Durner, E. F., Poling, E. B., and Albregts, E. A., Early season yield responses of selected strawberry cultivars to photoperiod and chilling in a Florida winter production system, *J. Amer. Soc. Hort. Sci.*, 112, 53, 1986.
159. Maas, J. L., Photoperiod and temperature effects on starch accumulation in strawberry roots, *Adv. Strawberry Prod.*, 6, 22, 1987.
160. Voth, V., The effect of nursery location latitude on California winter planted strawberries, *Acta Hortic.*, 265, 1989.
161. Voth, V. and Bringhurst, R. S., Influence of nursery harvest date, cold storage, and planting date on performance of winter planted California strawberries, *J. Amer. Soc. Hort. Sci.*, 95, 496, 1970.
162. Bringhurst, R. S., Voth, V., and Van Hook, D., Relationship of root starch content and chilling history to performance of California strawberries, *Proc. Amer. Soc. Hort. Sci.*, 75, 373, 1960.
163. Welch, N. C., Beutel, J. A., Bringhurst, R. S., Gubler, D., Otto, H., Pickel, C., Schrader, W., Shaw, D., and Voth, V., *Strawberry production in California, Leaflet 2959*, University of California Division of Agriculture and Natural Resources, Oakland, 1989.
164. Craig, D. L. and Brown, G. L., Influence of digging date, chilling, cultivars and culture on glasshouse strawberry production in Nova Scotia, *Can. J. Plant Sci.*, 57, 571, 1977.
165. Takai, T., Effective temperatures for chilling, and interaction of chilling and photoperiod on growth response of strawberry varieties, *Bull. Hortic. Res. Stn. Morioka, Jpn. Ser. C*, 6, 91, 1970 (Japanese, English summary).
166. Overcash, J. P. and Campbell, J. A., The effects of intermittent warm and cold periods on breaking the rest period of peach leaf buds, *Proc. Amer. Soc. Hort. Sci.*, 66, 87, 1955.
167. Tanabe, K., Hayashi, S., and Banno, K., Physiological studies on dormancy of strawberry (*Fragaria grandiflora* Ehrh.) III. Influence of temperature conditions in the period of chilling requirement on growth regulating substance in dormant strawberry crown and root, *Bull. Fac. Agric., Tottori Univ.*, 37, 1, 1985.
168. Lavee, S., Dormancy and bud break in warm climates: Considerations of growth-regulator involvement, *Acta Hort.*, 34, 225, 1973.

169. Nelson, P. E. and Wilhelm, S., Some anatomic aspects of the strawberry root, *Hilgardia*, 26, 631, 1957.
170. Wilhelm, S. and Nelson, R. D., Fungal diseases of strawberry, in *The Strawberry*, Childers, N.F., Ed., Horticultural Publications, Gainesville, FL, 1981
171. Smart, R. E. and Barrs, H. D., The effect of environment and irrigation interval on leaf water potential of four horticultural species, *Agric. Meteorol.*, 12, 337, 1973.
172. Renquist, A. R., Breen, P. J., and Martin, L. W., Stomatal behavior and leaf water status of strawberry in different growth environments, *Scientia Hort.*, 18, 101, 1983.
173. Johnson, J. D. and Ferrell, W. K., Stomatal response to vapour pressure deficit and the effect of plant water stress, *Plant Cell Environ.*, 6, 451, 1983.
174. Klepper, B., Diurnal patterns of water potential in woody plants, *Plant Physiol.*, 43, 1931, 1968.
175. Hsiao, T. C., Plant reponses to water stress, *Annu. Rev. Plant Physiol.*, 24, 519, 1973.
176. Davies, F. S. and Albrigo, L. G., Water relations of small fruit, *Water Deficits and Plant Growth*, Vol. 7, Kozlowski, T. T., Ed., Academic Press, Orlando, FL, 1983, chap. 3.
177. Castel, J. R. and Fereres, E., Responses of young almond trees to two drought periods in the field, *J. Hort. Sci.*, 54, 175, 1982.
178. Johns, G. G., Transpirational, leaf area, stomatal and photosynthetic responses to gradually induced water stress in four temperate herbage species, *Aust. J. Plant Physiol.*, 5, 113, 1978.
179. Jordan, W. R., Brown, K. W., and Thomas, J. C., Leaf age as a determinant in stomatal control of water loss from cotton during water stress, *Plant Physiol.*, 56, 595, 1975.
180. Darrow, G. M. and Dewey, G. W., Studies on the stomata of strawberry varieties and species, *Proc. Amer. Soc. Hort. Sci.*, 32, 440, 1934.
181. Sruamsiri, P. and Lenz, F., Photosynthesis and stomatal behavior of strawberries (*Fragaria X ananassa* Duch.). VI. Effect of water deficiency, *Gartenbauwissenschaft*, 51, 84, 1986.
182. Chandler, C. K. and Ferree, D. C., Response of 'Raritan' and 'Surecrop' strawberry plants to drought stress, *Fruit Var. J.*, 44, 183, 1990.
183. Archbold, D. D. and Zhang, B., Drought stress resistance in *Fragaria* species, *The Strawberry into the 21st Century*, Dale, A., and Luby, J., Eds., Timber Press, Portland, 1991, chap. 28.
184. O'Neill, S., Role of osmotic potential gradients during water stress and leaf senescence in *Fragaria virginiana, Plant Physiol.*, 72, 931, 1983.
185. Zhang, B. and Archbold, D. D., Water relations of a *Fragaria chiloensis* and a *F. virginiana* selection during and after water deficit stress, *J. Amer. Soc. Hort. Sci.*, 118, 274, 1993.
186. Zhang, B. and Archbold, D. D., Solute accumulation in leaves of a *Fragaria chiloensis* and a *F. virginiana* selection responds to water deficit stress, *J. Amer. Soc. Hort. Sci.*, 118, 280, 1993.
187. Jones, M. M. and Rawson, H. M., Influence of rate of development of leaf water deficits upon photosynthesis, leaf g, water use efficiency, and osmotic potential in sorghum, *Physiol. Plant.*, 45, 103, 1979.
188. Darrow, G. M. and Sherwood, H., Transpiration studies on strawberries, *Proc. Am. Soc. Hort. Sci.*, 28, 225, 1931.
189. Sruamsiri, P. and Lenz, F., Photosynthesis and stomatal behaviour of strawberries. III. Effect of humidity, *Gartenbauwissenschaft*, 50, 187, 1985 (German, English summary).
190. Higgs, K. H. and Jones, H. G., Water use by strawberry in south-east England, *J. Hort. Sci.*, 64, 167, 1989.
191. Arney, S. E., Studies in growth and development in the genus *Fragaria*. I. Factors affecting the rate of leaf production in Royal Sovereign strawberry, *J. Hort. Sci.*, 28, 73, 1953.
192. Martinez-Carrasco, R. and Sanchez de la Puente, L., Growth, yield and mineral nutrition of the strawberry under water deficits at various stages of development, *Anales de Edafologia y Agrobiologia*, 38, 7, 1979.
193. Rom, R. C. and Dana, M. N., Development and nutrition of strawberry plants prior to fruit bud differentiation, *Proc. Amer. Soc. Hort. Sci.*, 81, 165, 1962.
194. Renquist, A. R., Breen, P. J., and Martin, L. W., Vegetative growth responses of 'Olympus' strawberry to polyethylene mulch and drip irrigation regimes, *J. Amer. Soc. Hort. Sci.*, 107, 369, 1982.
195. Waldo, G. M., Effects of irrigation and plant spacing upon runner production and fruit yield of the Corvallis strawberry, *Proc. Amer. Soc. Hort. Sci.*, 44, 289, 1944.
196. Cannell, G. H., Voth, V., Bringhurst, R. S., and Proebsting, E. L., The influence of irrigation levels and application methods, polyethylene mulch, and nitrogen fertilization on strawberry production in Southern California, *Proc. Amer. Soc. Hort. Sci.*, 78, 281, 1961.

197. Rom, R. C. and Dana, M. N., Strawberry root growth studies in fine sandy soil, *Proc. Amer. Soc. Hort. Sci.*, 75, 367, 1960.
198. Collins, W. B. and Smith, C. R., Soil moisture effects on rooting and early development of strawberry runners, *J. Amer. Soc. Hort. Sci.*, 95, 417, 1970.
199. Blatt, C. R., Irrigation, mulch, and double row planting related to fruit size and yield of 'Bounty' strawberry, *HortScience*, 19, 826, 1984.
200. Dwyer, L. M., Stewart, D. W., Houwing, L., and Balchin, D., Response of strawberries to irrigation scheduling, *HortScience*, 22, 42, 1987.
201. Locascio, S. J. and Myers, J. M., Trickle irrigation and fertilization method for strawberries, *Proc. Fla. State Hort. Soc.*, 88, 185, 1975.
202. Locascio, S. J., Myers, J. M., and Martin, F. G., Frequency and rate of fertilization with trickle irrigation for strawberries, *J. Amer. Soc. Hort. Sci.*, 102, 456, 1977.
203. Renquist, A. R., Breen, P. J., and Martin, L. W., Effect of polyethylene mulch and summer irrigation regimes on subsequent flowering and fruiting of 'Olympus' strawberry, *J. Amer. Soc. Hort. Sci.*, 107, 373, 1982.
204. Rossi, P. P. and Rosati, P., Lysimetric measurements of water consumed by day-neutral strawberry cv. Fern, *Acta Hort.*, 265, 251, 1989.
205. Schrader, A. L. and Haut, I. C., Spacing studies on several strawberry varieties with and without irrigation, *Proc. Amer. Soc. Hort. Sci.*, 34, 355, 1936.
206. Staebner, F. E., The response of strawberries to irrigation in a dry harvest season, *Proc. Amer. Soc. Hort. Sci.*, 34, 349, 1936.
207. Greve, E. W., Some effects of nitrogen fertilizer and irrigation on the growth and blossoming of the Howard 17 strawberry, *Proc. Amer. Soc. Hort. Sci.*, 33, 397, 1935.
208. Myers, J. M. and Locascio, S. J., Efficiency of irrigation methods for strawberries, *Proc. Fla. State Hort. Soc.*, 85, 114, 1972.
209. Albregts, E. E. and Howard, C. M., Influence of fertilizer sources and drip irrigation on strawberries, *Proc. Fla. State Hort. Soc.*, 37, 159, 1978.
210. Judkins, W. P., The effect of training systems and irrigation on the yield of everbearing strawberries grown with sawdust mulch, *Proc. Amer. Soc. Hort. Sci.*, 55, 277, 1950.
211. Naumann, von W.-D., Timing of irrigation and fruit bud differentiation of strawberry, *Gartenbauwissenschaft*, 29, 21, 1964.
212. Goncharova, E. A., Saakov, V. S., Udovenko, G. V., and Yakovlev, A. F., Transport and exchange of T_2O in the leaf-fruit system of strawberry plants under normal and extreme conditions, *Fiziologiya Rastenii*, 23, 977, 1976 (English Translation).
213. Darrow, G. M. and Waldo, G. F., The moisture content of strawberries as influenced by growing conditions, *Proc. Amer. Soc. Hort. Sci.*, 33, 393, 1935.
214. Okasha, K. A., Helal, R. M., Khalaf, S., and El-Hifny, I. M., Effect of some soil types on fruit quality of Tioga strawberry cultivar, *Egypt. J. Hort.*, 13, 71, 1986.
215. Glenn, D. M. and Takeda, F., Guttation as a technique to evaluate the water status of strawberry, *HortScience*, 24, 599, 1989.
216. Brown, J. G., and Voth, V., Salt damage to strawberries, *Calif. Agr.*, 9, 11, 1955.
217. Ehlig, C. F., Salt tolerance of strawberries under sprinkler irrigation, *Proc. Amer. Soc. Hort. Sci.*, 77, 376, 1961.
218. Bernstein, L., Salt tolerance of fruit crops, *Agric. Information Bull. 292*, U.S. Dept. Agric., 1964.
219. Ehlig, C. F. and Bernstein, L., Salt tolerance of strawberries, *Proc. Amer. Soc. Hort. Sci.*, 72, 198, 1958.
220. Maas, E. V. and Hoffman, M., Crop salt tolerance—current assessment, *J. Irrig. Drainage Div., ASCE*, 103, 115, 1977.
221. Okasha, K. A., Helal, R. M., Khalaf, S., and El-Hifny, I. M., Effect of some soil types on flowering, fruit setting and yield of Tioga strawberry cultivar, *Egypt. J. Hort.*, 13, 61, 1986.
222. Wilhelm, S. and Nelson, P. E., A concept of rootlet health of strawberries in pathogen-free field soil achieved by fumigation, in *Root Diseases and Soil-Borne Pathogens*, Toussoun, T. A., Bega, R. V., and Nelson, P. E., Eds., Univ. of Calif Press, Berkeley, 1981, p. 208.
223. Wilhelm, S., The healthy strawberry root system, in *Compendium of Strawberry Diseases*, Maas, J. L., Ed., APS Press, St. Paul, Minn., 1984, p.78.
224. Ball, E. and Mann, C. E. T., Root and shoot development of the strawberry. IV. The influence of some cultural practices on the normal development of the strawberry plant, *J. Pom. Hort. Sci.*, 6, 104, 1927.

225. Waister, P. D., Wind as a limitation on the growth and yield of strawberries, *J. Hort. Sci.*, 47, 411, 1985.
226. Musselman, R. C., Sterrett, J. L., and Voth, V., Effects of simulated acidic fog on strawberry productivity, *HortScience*, 23, 128, 1988.
227. Takemoto, B. K., Bytnerowicz, A., and Olszyk, D. M., Physiological responses of field-grown strawberry (*Fragaria x ananassa* Duch.) exposed to acidic fog and ambient ozone, *Env. Exp. Bot.*, 29, 379, 1989.
228. Moon, J. W., Jr., Photosynthetic inhibition in strawberry grown under twice ambient CO_2, *HortScience*, 25, 1166 (abstract), 1990.
229. Sruamsiri, P. and Lenz, F., Photosynthesis and stomatal behaviour of strawberries. IV. Effect of carbon dioxide and oxygen, *Gartenbauwissenschaft*, 50, 221, 1985.

Chapter 11

Temperate Nut Species

Peter C. Andersen

CONTENTS

0-8493-175-0/94/$0.00+$.50

I. ALMOND

A. DISTRIBUTION AND BOTANY

The almond [*Prunus amygdalus* Batsch syn. *P. dulcis* (Mill.) D. A. Webb] (family Rosacea) is closely related to other stone fruits such as peaches (*P. persica* L. Batsch), plums (*P. salicina* Lindl, or *P. domestica* L.), and cherries (*P. avium* L. and *P. cesasus* L.).[1] Genetically, almond is more similar to roses (*Rosa* spp.), and apples (*Malus domestica* Borkh.) than other nut species. However, within *P. amygdalus* there is much more genetic diversity than in closely related species such as peach (*P. persica* (L.) Batsch). Consequently, there is great conspecific variability in phenology, growth, and bearing habit.[2]

Almond originated in southwest Asia. The cultivation of almond has spread to northern Africa, southern Europe, the Mediterranean region, California, and other locations around the world.[3] Climatic and edaphic requirements are exacting and successful culture is limited to select locations.[4]

The almond is a vigorous, medium-sized tree and is generally long-lived compared to other *Prunus* spp. such as peach.[3] Flowers are borne laterally on spurs or shoots. Flowering typically occurs prior to the onset of vegetative growth. The flowers are perfect (i.e., possessing both pistils and stamins); however, the flowers of most cultivars are self-incompatible and cross pollination is required to obtain good yields.[4] The almond fruit consists of the edible seed (kernel), a shell (endocarp), and the hull (mesocarp). Trees typically bear fruit during the third or fourth year.[3,4]

B. LIGHT

Leaf gas exchange characteristics of container-grown 'Nonpareil' almond trees were unique among five *Prunus* spp. tested.[5] Net CO_2 assimilation (A) of almond leaves did not approach an asymptote at a photosynthetic photon flux (PPF) up to 1600 μmol m^{-2} s^{-1}, whereas maximum A (A_{max}) occurred at a PPF of 400 to 700 μmol m^{-2} s^{-1} for leaves of peach [*Prunus persica* (L.) Batsch], plum (*P. cerasifera* Ehrh.), and cherry (*P. avium* L.). Significantly higher A_{max} values were recorded for almond (up to 18 μmol m^{-2} s^{-1}) compared to apricot (*P. armeniaca* L.) (7 to 8 μmol m^{-2} s^{-1}) and other *Prunus* spp. (13 to 14 μmol m^{-2} s^{-1}). Leaf nitrogen concentration (on a leaf area basis) was also greater for almond than the other species tested. A positive correlation between A and leaf nitrogen content was expected since a high percentage of leaf nitrogen is in the form of ribulose biphosphate carboxylase/oxygenase

(RUBISCO), the enzyme responsible for the synthesis of CO_2 and ribulose biphosphate to form hexoses. Net CO_2 assimilation, when calculated on a leaf nitrogen basis, was similar for almond, plum, peach, and cherry.[5]

C. WATER

1. Plant Water Deficits

Stomatal conductance to water vapor (g_s) and leaf water potential (Ψ_L) of almond varied diurnally with maximum g_s and minimum Ψ_L recorded during midday.[6,7] Maximum g_s (up to 375 μmol m^{-2} s^{-1}) and minimum Ψ_L (as low as -2.1 MPa) have been recorded during midday in Spain.[6] Leaf gas exchange and Ψ_L of almond was reduced considerably with increases in the leaf to air vapor pressure deficit (VPD).[8,9] Turner et al.[9] established that the A and g_s of almond and other woody plant species was reduced to a greater extent than that of herbaceous plants in response to increasing VPD. They invoked a lower hydraulic resistance in herbaceous compared to woody species as the factor responsible for the greater $\Delta g_s/\Delta\Psi_L$. Torrecillas et al.[8] found a progressive decrease in Ψ_L in almond as transpiration (E) increased which was also attributed to a substantial internal plant resistance to water flow.

Genotypic variations in plant water relations have been reported within *P. amygdalus.* For example, Sanchez-Blanco et al.[6] observed that the g_s of cv. Garriques was lower and more dependent on VPD than cv. Ramillete.

The concept of a non-hydraulic signal that is responsible for stomatal closure[10,11] in the absence of a reduction in Ψ_L has been investigated in almond. Wartinger et al.[7] evaluated the concentration of abscisic acid (ABA) in the xylem fluid and plant water relation of young 'Ne Plus Ultra' almond trees growing in lysimeters in Israel. Diurnally, g_s was maximum (200 to 400 mmol m^{-2} s^{-1}) at mid-morning and Ψ_L was minimum at midday (-3.0 MPa), although neither g_s or Ψ_L were correlated with the concentration of ABA in xylem fluid. By contrast, maximum g_s and ABA were inversely correlated when plants experienced gradual soil drying (i.e., predawn Ψ_L declined from ca. -0.3 MPa during the spring to as low as -2.0 MPa during mid-summer). The authors suggested that a very narrow range of ABA concentrations in xylem fluid may regulate stomatal behavior and proposed that fine roots exposed to soil drying are the source of ABA production and responsible for ABA in xylem fluid. In an adjunct study on 1- to 4-year-old almond trees grown in a lysimeter, FuBeder et al.[12] documented that diurnal or seasonal changes in plant water status (i.e., g_s, Ψ_L) were not related to changes in cytokinin concentration in the xylem fluid. Rather the influence of cytokinins appeared to be dependent on ABA concentration, which in turn, was dependent on soil-water status. They concluded that while some cytokinins in xylem fluid may influence stomatal behavior in the short term, the modulation of stomatal aperture is mainly via ABA concentrations.

2. Flooding

Prunus spp. are less tolerant to waterlogged and/or anaerobic soil conditions than many other genera in the family Rosacea (e.g., *Pyrus* or *Malus*).[13] Interspecific comparisons within *Prunus* have shown that almond and apricot are among the least tolerant to waterlogged soil conditions of all fruit tree species.[13–15] After 2 days of flooding, root growth of 'Chellaston' almonds and 'Nemaguard' peach was severely and moderately inhibited, respectively.[16] The susceptibility of almond rootstocks is exacerbated by a low resistance to fungal infection by *Phytophthora* spp.[15,16] Almonds, regardless of rootstock, should be grown in well-drained soils that are not prone to flooding for even a short period of time.[17] Although Marianna plum rootstock is tolerant to flooded soil conditions and to soil-borne diseases compared to peach, almond, or almond-peach hybrids rootstocks, it has compatibility limitations with almond scions.[18]

The mechanisms responsible for flooding sensitivity among fruit trees is poorly understood.[13] No physiological data are available for almond; however, peach (a closely related species) has been shown to undergo cyanogenesis (i.e., the release of toxic hydrogen cyanide from cyanogenic glycosides) under conditions of anaerobiosis.[17]

D. TEMPERATURE

The relationship between temperature and flowering of almond has been studied in Australia. Rattigan and Hill[19] used the models developed for peach by Richardson et al.[20] and Ashcroft et al.[21] to predict the date of flowering of 12 almond cvs. over a 7-year period. The model described flowering as a two-stage process. In the first stage, dormant flower buds perceive the cumulative exposure to low temperature (chilling) until a genotype-dependent level is reached. A chilling requirement of 220 to 320 chilling units was calculated for twelve almond cultivars. In the second stage, flowers began to develop at a rate proportional to temperature. The heat summation requirement ranged from 5300 to 8900 growing-degree-hours above 4.5°C.[20]

High leaf temperatures may have been a contributing factor for a low percentage of successful bud take of almond. Weinbaum et al.,[22] working with young almond trees prone to bud failure during propagation, documented an increased leaf temperature and a reduced g_s compared to plants not prone to bud failure. Although leaf temperatures of young-grafted trees prone to bud failure were as much as 4°C higher than ambient temperatures (37 versus 33°C) there was no evidence that higher leaf temperatures were lethal.

E. SALINITY

In many semi-arid regions where almonds are cultivated, quality or quantity of irrigation water are major limitations to production.[23] Olivar[24] reported that one third of the irrigated almond acreage worldwide is adversely affected by salinity. The distribution of salt as a result of irrigation with saline water in a given soil is highly dependent upon the method and rate of irrigation. For example, at low-drip irrigation rates [50% of evapotranspiration (ET)] in an almond orchard with saline irrigation water [electrical conductivity of the saturation soil extract (ECe) = 1.5 dS m^{-1}], the greatest ECe (i.e., 5.7 dS m^{-1}) was directly below the drip line.[25] Increased rates of irrigation moved the zone of salt accumulation farther from the drip zone.[25]

Symptoms of salt stress (yield reductions, dwarfing, foliar burn, and premature senescence) in *Prunus* spp. may occur at an ECe of 1.5 to 1.7 dS m^{-1},[26] despite the fact that soils are not classified as saline until the ECe reaches 4 dS m^{-1}.[27] Maas[28] reported that tree growth of almond was reduced by 19% for every 1 dS m^{-1} increase in soil salinity. Salinity damage may be due to an osmotic effect or to the deleterious effect of specific ions (e.g., NaCl, SO_4^-). In *Prunus* spp. the effects of Cl and Na were greater than the osmoticum effect.[29,30] Sulfate salts were more inhibitory to growth of *Prunus* spp. than isosmotic concentrations of Cl salts.[29,30]

There is sufficient variation in salt tolerance among *Prunus* rootstocks to justify the consideration of salt tolerance as a selection criterion in regions with saline soils. Almond, peach, plum, and almond-peach hybrids have been used as rootstocks for almonds. 'Elberta' peach on 'Lovell' rootstocks accumulated less Cl and were more salt tolerant than trees on 'Shalil' rootstocks.[30] Bernstein et al.[31] also reported that 'Lovell' conferred a greater salinity tolerance to the scion than 'Shalil' rootstocks. Evaluation of the performance of *Prunus* spp. on 'Lovell' rootstocks confirmed that salinity tolerance was determined by both the rootstock and the scion.[31] Genotype-specific rootstock/scion interactions in Cl accumulation and growth also occurred.

Application of a 50:50 mixture of 4000 ppm NaCl and $CaCl_2$ over a 3-year period to field-grown almond trees near the Rio Grande river resulted in minor changes in cationic composition of leaves.[31] Sodium concentrations were highest in wood, roots, and bark and lowest in shoots and leaves. Leaf Na levels were lower in Texas almond than in peach, plum, prune, apricot, or 'Nonpareil' almond. Based on growth and Cl contents, it was concluded that approximately one half the salinity-induced reduction in growth was due to Cl toxicity and one half was due to an osmotic effect.

Maas and Hoffman[26] ranked salt-tolerance of *Prunus* spp. from high to low as plum, almond, peach, apricot. The salinity tolerance of seedlings of *Prunus* spp. was evaluated in sand-culture experiments in a glasshouse.[23] Trees were irrigated with half-strength Hoagland's solution supplemented with chloride and sulfate salts of Na^+, Ca^{++} and Mg^{++} to yield an ECe of 1.5, 4.5, or 6.0 dS m^{-1}. Based upon growth (height, dry weight, etc.) and foliar appearance, salinity tolerance from high to low was as follows: 'Titan' almond × 'Nemaguard' and *P. mexicana* Wats., 'Nemaguard' and 'Nemared', 'Myrobolan' plum (*P. cerasifera* J. F. Ehrh.), and bitter almond (*P. amygdalus*).

F. MISCELLANEOUS STRESSES

Very little information is available concerning effects of other environmental stresses on almond. DeJong[5] found that A of almond leaves increased linearly with increases in intercellular CO_2 concentration (C_i) when ambient CO_2 concentrations were increased. The slope of the A/C_i curve was greater for almond than the other four *Prunus* species tested. Unfortunately, there are no data concerning growth and productivity of almond as a function of ambient CO_2 concentrations.

Two additional reports worthy of mention concern frost injury and ozone. Lindow and Connell[32] showed that ice nucleation active (INA) strains of the bacteria *Pseudomonas syringae* accounted for over 99% of the ice nucleation present on almond leaves. Application of antagonistic non-INA strains of bacteria or bactericides reduced both the density of INA bacteria and frost injury at $-3°C$.

Ozone sensitivity is another characteristic of almonds. When seedlings were subjected to 0.25 μl l^{-1} ozone for 4 hrs each day for 4 months per year over a 2-year period, growth was reduced more for almond than for peach or apricot.[33]

REFERENCES: ALMOND

1. Bailey, L. H. and Bailey, E. Z., *Hortus Third,* A concise dictionary of plants cultivated in the United States and Canada, Macmillan Publishing Co., New York, N. Y., 1976.
2. Socias i Company, R. and Felipe, A. J., Almond: A diverse germplasm, *HortScience,* 27, 718, 1992.
3. Kester, D. E. and Asay, R., Almonds, Advances in fruit breeding, in: J. Janick and J. N. Moore (Eds.). Purdue Univ. Press, West Lafayette, Ind., 387, 1975.
4. Griggs, W. H., Pollination requirements of fruits and nuts, *Calif. Agr. Exp. Sta. Circ.,* 424, 1953.
5. DeJong, T. M., CO_2 assimilation characteristics of five *Prunus* tree fruit species, *J. Amer. Soc. Hort. Sci.,* 108, 303, 1983.
6. Sanchez-Blanco, M. J., Sanchez, M. C., Planes, J., and Torrecillas, A., Water relations of two almond cultivars under anomalous rainfall in non-irrigated culture, *J. Hort. Sci.,* 66, 403, 1991.
7. Wartinger, A., Heilmeier, H., Hartung, W., and Schulze, E. D., Daily and seasonal courses of leaf conductance and abscisic acid in the xylem sap of almond trees [*Prunus dulcis* (Miller) D. A. Webb] under desert conditions, *New Phytol.,* 116, 581, 1990.
8. Torrecillas, A., Ruiz-Sanchez, M. C., and Hernansaez, A., Response of leaf water potential to estimated transpiration in almond trees, J. *Hort. Sci.,* 64, 667, 1989.
9. Turner, N. C., Schulze, E. D., and Tollan, T., The responses of stomata and leaf gas exchange to vapour pressure deficits and soil water content, *Oecologia,* 63, 338, 1984.
10. Zhang, J. and Davies, W. J., Abscisic acid produced in dehydrating roots may enable the plant to measure the water status of the soil, *Plant Cell Environ.,* 12, 78, 1989.
11. Zhang, J. and Davies, W. J., Sequential response of whole plant water relations to prolonged soil drying and the involvement of xylem sap ABA in the regulation of stomatal behaviour of sunflower plants, *New Phytol.,* 113, 167, 1989.
12. FuBeder, A., Wartinger, A., Hartung, W., and Schulze, E. D., Cytokinins in the xylem sap of desert-grown almond (*Prunus dulcis*) trees: daily courses and their possible interactions with abscisic acid and leaf conductance, *New Phytol.,* 122, 45, 1992.
13. Schaffer, B., Andersen, P. C., and Ploetz, R. C., Responses of fruit crops to flooding, in *Horticultural Reviews,* Vol. 13, Janick, J., Ed., John Wiley & Sons, Inc., NY, 1992, chap. 7.
14. Norton, R. A., Hansen, C. J., O'Reilley, H. J., and Hart, W. H., Rootstocks for peaches and nectarines in California, Calif. *Agr. Exp. Sta. Ext. Leaf.,* 157, 1963.
15. Wicks, T. and Lee, T. C., Effects of flooding, rootstocks and fungicides on *Phytophthora* crown rot of almonds, *Aust. J. Exp. Agric.,* 25, 705, 1985.
16. Wicks, T. J., Susceptibility of almond and cherry rootstocks and scions to *Phytophthora* species, *Aust. J. Exp. Agric.,* 29, 103, 1989.
17. Rowe, R. N. and Catlin, P. B., Differential sensitivity to waterlogging and cyangenesis by peach, apricot, and plum roots, *J. Amer. Soc. Hort. Sci.,* 96, 305, 1971.
18. Micke, W. and Kester, D., *Almond Orchard Management,* Agric. Sci. Pub., Div. Agr. Sci., Univ. California, Berkeley, CA, 1981.
19. Rattigan, K. and Hill, S. J., Relationship between temperature and flowering in almond, *Aust. J. Exp. Agric.,* 26, 399, 1986.
20. Richardson, E. A., Seeley, S. D., and Walker, D. R., A model for estimating the completion of rest for 'Redhaven' and 'Elberta' peach trees, *HortScience,* 9, 331, 1974.
21. Ashcroft, G. L., Richardson, E. A., and Seeley, S. D., A statistical method of determining chill unit and growing degree hour requirements for deciduous fruit trees, *HortScience,* 12, 347, 1977.
22. Weinbaum, S. A., Even-Chen, Z., and Kester, D. E., Increased stomatal resistance in two cultivars of almond sensitive to bud failure, *HortScience,* 15, 583, 1980.

23. Ottman, Y. and Byrne, D. H., Screening rootstocks of *Prunus* for relative salt tolerance, *HortScience,* 23, 375, 1988.
24. Olivar, M., Salinity: a problematic by-product of irrigation, *The Grower* (March), 8, 1985.
25. Nightingale, H. I., Hoffman, G. J., Rolston, D. E., and Biggar, J. W., Trickle irrigation rates and soil salinity distribution in an almond (*Prunus amygdalus*) orchard, *Agric. Water Management,* 19, 271, 1991.
26. Maas, E. V. and Hoffman, G. J., Crop salt tolerance-current assessment, J. Irr. Drain. Div., *Proc. Amer. Soc. Civil Eng.,* 103, 115, 1977.
27. Anonyomus., U. S. Salinity Laboratory Staff, Diagnosis and improvement of saline and alkali soils, U. S. Dept. Agr. Handbook, 60, 1954.
28. Maas, E. V., Salt tolerance of plants, *Appl. Agric. Res.,* 1, 12, 1986.
29. Brown, J. W., Wadleigh, C. H., and Hayward, H. E., Foliar analysis of stone fruit and almond trees on saline substrates, *Proc. Amer. Soc. Hort. Sci.,* 61, 49, 1953.
30. Hayward, H. E., Long, E. M., and Ulvits, R., The effect of chloride and sulfate salts on the growth and development of the Elberta peach on Shalil and Lovell rootstocks, *U. S. Dept. Agr. Tech. Bull.,* 922, 1946.
31. Bernstein, L., Brown, J. W., and Hayward, H. E., The influence of rootstock on growth and salt accumulation in stone-fruit trees and almonds, *Proc. Amer. Soc. Hort. Sci.,* 86, 1956.
32. Lindow, S. E. and Connell, J. H., Reduction of frost injury to almond by control of ice nucleation active bacteria, *J. Amer. Soc. Hort. Sci.,* 109, 48, 1984.
33. McCool, P. M. and Musselman, R. C., Impact of ozone on growth of peach, apricot, and almond, *HortScience,* 25, 1384, 1990.

II. CHESTNUT

A. DISTRIBUTION AND BOTANY

The genus *Castanea* (family Fagaceae) consists of 12 species of deciduous trees and shrubs native to northern temperate regions; *Castena* species have separate staminate and pistillate flowers and one to seven nuts enclosed in a prickly, dehiscent involucre.[1] All *Castanea* species have a somatic chromosome number 2n = 24. Chestnuts are monoecious and usually protandrous.[1,2] Flowers are borne on current year's wood. Chestnuts rarely self-pollinate and isolated trees produce few nuts. Pollination is by wind. *Castanea* spp. generally grow fast and have rot-resistant wood.[3] Unlike most other tree nuts, chestnuts are relatively low in protein and fat, yet high in carbohydrates (Table 1, Section IV, A). Essential amino acids represent a fairly high proportion of the amino acid profile.[4] Unlike many temperate zone nut species *Castena* do not exhibit a strong tendency to bear heavy and light crops in alternate years,[3] which may be a function of the lower lipid/carbohydrate ratio in nuts of *Castena* species compared to other nut crops. In many regions chestnut trees are still seedling propagated; however, the development of a commercial industry requires the utilization of successful techniques for mass clonal propagation.

The native range of *Castanea dentata* (Marsh) Borkh. (The American chestnut) is characterized by ca. 75 and 120 cm rainfall.[4] The American chestnut was once the dominant hardwood species in eastern North America until the introduction of Chestnut blight, caused by the fungal pathogen *Cryphonectria* (*Endothia*) *parasitica,* which ultimately destroyed ca. 3.5 billion trees.[5] Symptoms of infection by *C. parasitica* include bark cankers, leaf wilting, senescence and eventual tree death. Reductions in hydraulic conductance of stems and g_s, and transpiration (E) of leaves result from localized canker infection.[6,7] American chestnut trees can be extremely long lived (i.e., over 500 years).[5] There is no cure for Chestnut blight, although American chestnut × Chinese chestnut hybrids have been bred with disease resistance.[5]

Chinese chestnut (*C. mollissima* Bl.) is currently the major species of commerce worldwide. The Chinese chestnut has a high degree of blight resistance, sufficient cold hardiness, and acceptable nut quality, but the wood quality is inferior to that of *C. dentata.*[2] Other chestnut species of commercial importance include the Japanese chestnut (*C. crenata* Sieb. and Zucc.) and the European chestnut (*C. sativa* Mill.).[2]

Physiological data on *Castanea* spp. is fragmentary, as is the case for many temperate nut species.

B. LIGHT

Photosynthetic characteristics of *C. sativa* have been measured via quantifying the assimilation of $^{14}CO_2$.[8] The A_{max} was ca. 16 to 17 μmol m^{-2} s^{-1} when leaves were exposed to 300 ppm CO_2. A specific point of light saturation was not achieved at a PPF up to 1800 μmol m^{-2} s^{-1}, although a relatively high A was recorded at a PPF as low as 700 μmol m^{-2} $s^{-1.}$ Apparent quantum yield generally ranged from 0.03 to 0.06 mol CO_2 mol $quanta^{-1}$.[8] Stomatal conductance to CO_2 vapor was ca. 10 times that of total residual conductance to CO_2 when PPF exceeded 200 μmol m^{-2} s^{-1}, (i.e., residual conductance is the reciprocal of the sum of the resistances for CO_2 diffusion from intercellular spaces to fixation in the chloroplast).[8] These data underscore the importance of nonstomatal limitations for CO_2 diffusion and fixation in *Castena* as reported for citrus.[9]

C. WATER

Approximately two thirds of the total daily quantity of water used by 6- or 7-year-old *C. sativa* in Germany was transpired between 1000 and 1800 hr.[10] Parenchyma tissue ostensibly contributed water to the transpiration stream between 600 and 1000 hrs since water uptake from the soil was minimal during this period. In China, A_{max} of *C. mollissima* occurred during the spring after leaves were fully expanded then declined as leaves aged.[11]

Leaf gas exchange and plant water relations of xeric (*Quercus prinus* and *Q. ilicifolia*) and mesic (*Q. vibra* and *C. dentata*) species were compared seasonal where soil moisture varied from 7 to 13% v/v in a sandy loam soil in the northeastern United States.[12] Net CO_2 assimilation, g_s, and Ψ_L were reduced for all species during peak drought periods (i.e., soil moisture 7% v/v, VPD = 3.5 kPa). Leaf water potential was generally highest for *C. dentata,* yet similar for the other species. Predawn Ψ_L was typically greater than −0.4 MPa, but fell to −0.7 MPa during a drought period. The A_{max} and g_s were ca. 6 μmol m^{-2} s^{-1} and 140 mmol m^{-2} s^{-1} during early morning, while during drought periods A and g_s were near zero. One characteristic contributing to the drought sensitivity of *C. dentata* is an extremely thin leaf (i.e., leaf thickness 160 μ*M*);[13] only *Acer plantanoides* L. and *Alnus incano* (L.) Moench had thinner leaves of 52 forest trees tested.[13] Abrams et al.[12] concluded that stomatal closure was responsible for maintaining a higher midday Ψ_L in *C. dentata* (i.e., −1.5 MPa) than in *Quercus* spp.

Osmotic adjustment and changes in tissue elasticity have often been invoked as mechanisms associated with drought tolerance.[12] *C. dentata* manifested a 0.7 MPa reduction in leaf osmotic potential (Ψ_π) (i.e., −1.6 to −2.3 MPa) from spring to late summer; during this period relative leaf water content (RWC) declined only 3%.[12] Therefore an active accumulation of solutes and not simply dehydration was responsible for osmotic adjustment. *C. dentata* maintained a lower bulk modulus of elasticity (ϵ), (i.e., greater tissue elasticity) and contributed to a higher leaf turgor pressure (Ψ_P) at a lower RWC than would occur during non-drought conditions.[12] Alternatively, it can be argued that tissue with a high ϵ will develop low Ψ_L during episodes of water stress and thereby retain a large gradient which will facilitate continued water uptake with soil drying. Despite osmotic adjustment and increased tissue elasticity, *C. dentata* had lower values of A and g_s exchange than the *Quercus* species in dry soil, and as such, should be considered less tolerant to periods of drought, and less competitive under these conditions.

The hydraulic conductance of water flow (Lp) through *C. dentata* may be a limiting factor to maintaining high rates of leaf gas exchange. Water flow in xylem vessels of woody spp. is much less than that predicted by Poiseuille's law due to: (1) nonconstant vessel diameter and a finite vessel length; (2) vessel walls are neither smooth nor rigid; and (3) water flow in xylem vessels is often through relatively small pits in axial walls or through perforations in end walls.[14,15] Further difficulty in the assessment of stem Lp is that only a small percentage of xylem vessels may actually function in water conduction (especially in old trees), and the hydraulic characteristics of the remainder may vary widely. In ring porous species such as *C. sativa* or *Juglans nigra* essentially only the outermost growth ring functions in water transport.

Zimmerman[16] reported that hydraulic constrictions (i.e., more narrow vessels) often exist between the junction of stem/branch, branch/twig, and twig/leaf which cause a reduced pressure drop just apical to the restriction. Salleo[15] determined that a small proportion of vessels ended in nodes for *C. sativa* and *J. nigra* which would minimize the deleterious impact of a gas embolism. The relatively short and narrow vessels formed later in the growing season (i.e., summer) for *C. sativa* and *J. nigra* may substitute as a mechanism to resist embolism formation.[15] Valancogne and Nasr[17,18] estimated the rate of sap flow by a method of heat transference to be within 30% of the actual value of E (determined by differences in container weight). Estimates of E by heat transferences tended to be overestimated when sap flow rates were low.[17,18]

D. TEMPERATURE

Castena spp. usually adapt to a wide range of temperatures. *Castena* spp. are generally cold-hardy and generally avoid frost injury to vegetative and reproductive buds because they bloom relatively late in the spring.[2] Sawano[19] reported that cold hardiness of shoots of *C. mollissima* increased from ca. −7 to −12°C from 5 Nov. to 25 Dec.; defoliation of leaves in the fall reduced cold hardiness. A decline in cold hardiness with early defoliation was associated with a high relative water content and a low carbohydrate content in shoots of *C. mollissima.*[19] *C. dentata* and *C. mollissima* have been reported to survive to at least −30°C when properly acclimated to the cold.[2]

E. RESPONSE TO ELEVATED CO_2

The response of *C. sativa* seedlings to increases in atmospheric CO_2 was studied in ventilated chambers over a 2-year-period.[20] CO_2 enrichment to 700 μmol m^{-2} s^{-1} CO_2 resulted in the termination of shoot elongation by mid season compared to those grown in 350 μmol m^{-2} s^{-1} ambient CO_2. Early leaf senescence and a reduction in leaf chlorophyll content was associated with the dilution of nutrients due to the increased growth of seedlings in high CO_2. An increase in root dry weight accounted for nearly all of the carbon gain in CO_2-enriched seedlings; however, shoot weight and leaf area were reduced in response to elevated CO_2. These unexpected results indicated that the influence of the increased carbon dioxide concentration occurring globally may be extremely species specific.

REFERENCES: CHESTNUT

1. Bailey, L. H. and Bailey, E. Z., *Hortus Third,* A concise dictionary of plants cultivated in the United States and Canada, Macmillan Publishing Co., New York, N. Y., 1976.
2. Jaynes, R. A., Chestnuts, in *Nut Tree Culture in North America,* 2nd Ed., Jaynes, R. A. (Ed.), Northern Nut Growers Assoc., Inc., Camden, Conn., 1981, chap. 9.
3. Rutter, P. A., Miller, G., and Payne, J. A., Chestnuts (*Castanea*), *Acta Hort.,* 290, 759, 1990.
4. McCarthy, M. A. and Meredith, F. I., Nutrient data on chestnuts consumed in the United States, *Econ. Bot.,* 42, 29, 1988.
5. Rutter, P. A., Chestnut ecology and the developing orcharding industry, in *Proc. Second Pacific Northwest Chestnut Cong., Chestnuts and Creating a Commercial Chestnut Industry:* Burnett, M. S. and Wallace, R. D. Eds., Chestnut Growers Exchange, Portland, OR, pp. 33–51, 1987.
6. Ewers, F. W., McManus, P. S. and Goldman, A., The effect of virulent and hypovirulent strains of *Endothia parasitica* on hydraulic conductance in American chestnut, *Canad. J. Bot.,* 67, 1402, 1989.
7. McManus, P. S. and Ewers, F. W., The effect of *Cryphonectria parasitica* on water relations of American chestnut, *Physiol. Molec. Plant Pathol.,* 36, 461, 1990.
8. Deweirdt, C. and Carlier, G., Photosynthesis of chestnut (*Castanea sativa*) leaves measured in situ in coppice stands by incorporation of $^{14}CO_2$, *Acta Oecologia,* 9, 145, 1988.
9. Lloyd, J., Syvertsen, J. P., Kriedemann, P. E., and Farquhar, G. D., Low conductance for CO_2 diffusion from stomata to the sites of carboxylation in leaves of woody species, *Plant Cell Environ.,* 15, 873, 1992.
10. Braun, H. J. and Marte, I., Concerning the water absorption of trees after the end of daily transpiration and thoughts on the transpiration-principle, *Allgemeine-Forst-und.,* Jagdzeitung. 158, 67, 1987.
11. Mu, Y. and Li, X., Studies on the photosynthetic characteristics of some deciduous fruit trees, *Acta. Hort. Sinica,* 13, 157, 1986.
12. Abrams, M. D., Schultz, J. C., and Kleiner, K. W., Ecophysiological responses in mesic versus xeric hardwood species to an early-season drought in central Pennsylvania, *For. Sci.,* 36, 970, 1990.
13. Carpenter, S. B. and Smith, N. D., Variation in shade leaf thickness among urban trees growing in metropolitan Lexington, Kentucky, *Castanea,* 44, 94, 1979.
14. Giordano, R., Salleo, A., Salleo, S., and Wanderlingh, F., Flow in xylem vessels and Poiseuille's law, *Canad. J. Bot.,* 56, 333, 1978.
15. Salleo, S., Functional aspects of water conduction pathways in vascular plants, *Giorn. Bot. Ital.,* 118, 53, 1984.
16. Zimmerman, M. H., Hydraulic architecture of some diffuse-porous trees, *Can. J. Bot.,* 56, 2286, 1978.
17. Valancogne, C. and Nasr, Z., A method of measuring the natural sap flow in small trees, *Agronomie,* 9, 609, 1989.

18. Valancogne, C. and Nasr, Z., Measuring sap flow in the stem of small trees by a heat balance method, *HortScience,* 24, 383, 1989.
19. Sawano, M., Studies on the frost hardiness of Chestnut trees. VI. Hardiness of shoots as affected by defoliation, Hyogo-Naka-Daigaku-Kenkyu-Hokoku, *Sci. Rep.,* 9, 15, 1971.
20. Mosseau, M. and Enoch, H. Z., Carbon dioxide enrichment reduces shoot growth in sweet chestnut seedlings (*Castanea sativa* Mill.), *Plant Cell Environ.,* 12, 927, 1989.

III. FILBERT

A. DISTRIBUTION AND BOTANY

The European filbert (*Corylus avellana* L.), family Betulacea (birch family) was the dominant tree species in much of northern Europe from 8,000 to 5,500 B.C.[1] Currently, the distribution of filbert is limited to climates moderated by large bodies of water such as the southern coast of the Black Sea and the northern coast of the Mediterranean Sea. Turkey accounts for two thirds of the production worldwide; Spain, Italy, and the west coast of the United States are responsible for most of the remaining production.

All but two *Corylus* spp. are native to southern Europe and/or Asia minor; the remaining two species (*C. americana* and *C. cornuta*) are native to north America.[1,2] Other species of lesser economic importance include the Turkish filbert, *C. colurna* and the Chinese hazel, *C. chinensis* which may grow to over 20 and 35 m in height, respectively.

All nine *Corylus* spp. are deciduous shrubs or trees with a somatic chromosome number of 22, with the possible exception of *C. coburna* (n = 18 or 22).[1,2] All species are monoecious with imperfect (separate male and female) flowers and most species are self sterile. Pollination is by wind and most *C. avellana* cultivars require cross pollination. Thus, two or more cultivars should be planted together. There is conspicuously little physiological data available for filbert.

B. ANATOMICAL/MORPHOLOGICAL CHARACTERISTICS OF LEAVES

Valuable baseline data for *C. avellana* have been presented concerning leaf chlorophyll concentration (432 mg m^{-2}), chlorophyll a/chlorophyll b (3.2), chlorophyll diameter (2.8 μM), leaf nitrogen content (1.52 g m^{-2}), abaxial stomatal density (155 mm^{-2}) and the ratio of mesophyll cell surface area to leaf area (24.4).[3] Mehlenbacher and Thompson[4] reported that chlorophyll deficiency in filbert was due to a simple recessive gene conferring chlorophyll deficiency. They also found that a gene controlling anthocyanin development was inherited independently. Anthocyanin-deficient plants survived in the field, but chlorophyll deficient seedlings lacking anthocyanin died, presumably in part due to photoinhibition.

C. LEAF GAS EXCHANGE AND LEAF WATER RELATIONS

Stomatal aperture increases in response to conditions that enhance rapid A and decreases in response to conditions that enhance rapid transpiration.[5] Cowan and Farquhar[6] suggested that stomatal behavior is "optimal" if a minimum quantity of water is lost per unit of carbon gained. One of the first species for which this theory was tested was *C. avellana.* Farquhar et al.[7] showed that stomata of *C. avellana* respond to humidity, or more specifically, the difference in the water concentration difference between the inside of the leaf and the atmosphere, VPD. When leaves of *C. avellana* were exposed to increasing VPD (at 28°C), g_s was adjusted (i.e., decreased) to maintain proportionality between E and A. This constant is known as the gain ratio ($\partial E/\partial A$). Maximum E occurred at 2.0 kPa and declined with larger VPD. Thus, the direct response of stomata to humidity was the mechanism responsible for the optimization of carbon gain with respect to water loss. Under severe water stress $\partial E/\partial A$ appeared to increase implying a greater reduction in A relative to E or g_s (i.e., there was an increase in the calculated value of C_i).[7] However, these data must be viewed with caution without confirmation of stomatal homogeneity.[8,9] (A brief discussion on this topic can be found in the Walnut section (IV) of this chapter; a more detailed discussion appears in the Citrus chapter of Volume II.)

The photosynthetic responses of filbert leaves to light and temperature have not been adequately tested. Harbinson and Woodward[3] evaluated the leaf gas exchange of *C. avellana* and seven other forest tree species growing in the field to simulated sunflecks of low intensity (up to a PPF of 230 $\mu mol\ m^{-2}\ s^{-1}$). Of the eight tree species, only *C. avellana* and *Ulmus glabra* were not light saturated at a PPF of 230 $\mu mol\ m^{-2}\ s^{-1}$.[3] Maximum A, g_s, and E for *C. avellana* under the conditions described were 7.6 $\mu mol\ m^{-2}\ s^{-1}$, 300 $m^{-2}\ s^{-1}$, and 3.4 mmol mol^{-1}, respectively. Under the experimental conditions of

Farquhar et al.[7] (i.e., PPF = 400 to 500 μmol m^{-2} s^{-1}, temperature = 28°C) maximum values of A and g_s for *C. avellana* were 9 μmol m^{-2} s^{-1} and 140 mmol m^{-2} s^{-1}, respectively.

There is a paucity of data concerning the leaf water relations of *C. avellana.* The response of A, g_s, and E to changes in VPD under well-watered conditions was examined for several herbaceous (i.e., *Helianthus annuus, Pisum sativum,* etc.) and woody (i.e., *C. avellana* and *Pistacia vera*) species.[10] A VPD-induced decrease in g_s and Ψ_L occurred for all species, but was greatest for woody species. Turner et al.[10] concluded that the VPD around the leaf decreased gas exchange through a direct effect on the leaf epidermis (i.e., guard cells). Turner et al.[11] in a related study found that for the woody species *in situ* psychrometer-derived values of Ψ_L did not correlate with those measured with a pressure chamber apparatus which they attributed to very low epidermal conductance of water vapor.

D. TEMPERATURE

Westwood[12] stated that *C. avellana* is cold hardy when dormant. Plants are not usually injured by temperatures that may occur during bloom, although pistillate flowers may be killed at temperatures of −10°C.[12] Westwood[12] indicated that vegetative buds require ca. 800–1200 hrs of chilling below 45°, although reproductive buds require considerably less chilling.

REFERENCES: FILBERT

1. Lagerstedt, H. B., Filbert, in *Nut Tree Culture in North America,* 2nd ed., Jaynes, R. A., Ed., Northern Nut Growers Assn., Inc., Camden, Conn., 1981, chap. 10.
2. Bailey, L. H. and Bailey, E. Z., *Hortus Third,* A concise dictionary of plants cultivated in the United States and Canada, Macmillan Publishing Co., New York, N. Y., 1976.
3. Harbinson, J. and Woodward, F. I., Field measurements of the gas exchange of woody plant species in simulated sunflecks, *Ann. Bot.,* 53, 841, 1984.
4. Mehlenbacher, S. A. and Thompson, M. M., Inheritance of a chlorophyll deficiency in hazelnut, *HortScience,* 26, 1414, 1991.
5. Cowan, I. R., Stomatal behaviour and environment, *Adv. Bot. Res.,* 4, 117, 1977.
6. Cowan, I. R. and Farquhar, G. D., Stomatal function in relation to leaf metabolism and environment, *Symp. Soc. Exp. Biol.,* 31, 471, 1977.
7. Farquhar, G. D., Schulze, E. D., and Kuppers, M., Responses to humidity by stomata of *Nicotiana glauca* L. and *Corylus avellana* L. are consistent with the optimization of carbon dioxide uptake with respect to water loss, *Aust. J. Plant Physiol.,* 7, 315, 1980.
8. Downton, W. J. S., Loveys, B. R., and Grant, W. J. R., Non-uniform stomatal closure induced by water stress causes putative non-stomatal inhibition of photosynthesis, *New Phytol.,* 110, 503, 1988.
9. Mansfield, T. A., Hetherington, A. M., and Atkinson, C. J., Some current aspects of stomatal physiology, *Annu. Rev. Plant Physiol. Plant Mol. Biol.,* 41, 55, 1991.
10. Turner, N. C., Schulze, E. D., and Gollan, T., The responses of stomata and leaf gas exchange to vapour pressure deficits and soil water content, *Oecologia,* 63, 338, 1984.
11. Turner, N. C., Spurway, R. A., and Schulze, E. D., Comparison of water potentials measured by in situ psychrometry and pressure chamber in morphologically different species, *Plant Physiol.,* 74, 316, 1984.
12. Westwood, M. N., *Temperate-Zone Pomology,* W. H. Freeman and Co., New York, 1978, chap. 15.

IV. PECAN

A. INTRODUCTION

1. Distribution

The pecan [*Carya illinoensis* (Wagenh.) K. Koch] (family Juglandaceae) is a large (i.e., >55 m height and >2 m diameter) deciduous, monoecious tree native to North America.[1] The pecan tree evolved in deep, fertile, well-drained soils of the Mississippi floodplain. The natural habitat of the pecan in the United States is primarily along rivers from latitude 26° to 42° and longitude 84° to 103° where it may exist as a pure stand.[2] It is not uncommon for pecan trees to be flooded for short periods of time during the dormant season, although flooded soil conditions are less common during the growing season.

The climate of the native range of pecan is characterized by long hot humid summers and moderately cool winters.[2] Regions having an arid/semi-arid climate and with access to abundant irrigation water are optimum for commercial pecan cultivation. Currently, the cultivated range of pecan extends considerably west and southeast of it's native range. The major production areas in the United States are Georgia, Texas, Alabama, Louisiana, Oklahoma, and Arizona. Mexico is also increasing production at a rapid rate. Limited quantities of pecans are produced in Australia, Brazil, Israel, and South Africa.

The acreage of pecans in the southwestern United States has been increasing at a very rapid rate. In the United States the potential for successful pecan production is greater in the Southwest than in the Southeast for three reasons: (1) cloudy/hazy atmospheric conditions prevalent in the Southeast result in much lower light intensities than in the Southwest; (2) high humidity and abundant precipitation during the growing season in the southeast greatly enhances insects and disease pressures; and (3) new cultivars released over the last 30 years are better adapted to the Southwest than the Southeast.

2. Botany

Pecan leaves are alternate and odd-pinnate. Flowers are unisexual; male flowers are a drooping catkin; female flowers are a star-shaped terminal raceme.[1] Pistillate flowers are borne terminally in clusters on current years wood and staminate flowers are produced on one-year-old wood. Pistillate flowers arise from the most apical one (or two) primary bud(s) on each shoot, while staminate flowers develop from most primary and secondary buds except for the terminal bud.[3,4] Wood and Payne[5] demonstrated that primary, secondary, and tertiary buds in a given node have the potential to produce staminate and pistillate flowers (and nuts) along the length of the one-year shoot. Staminate flower opening and pistillate flower receptivity is typically asynchronous for a given seedling or cultivar, thus cross pollination is normally essential for maximum productivity. The fruit is a drupe with a stone or nut enclosed in a thick green husk that splits into four parts at maturity.[1]

The period of juvenility (i.e., the phase in the life cycle of a plant that is limited to vegetative growth), is especially prolonged for pecan trees and may last 10 to 12 years. Juvenility in the native habitat of pecan is an adaptative characteristic facilitating rapid vegetative growth and a competitive position in the forest canopy.[6] The desirability for a rapid return on investments after orchard establishment has prompted pecan breeders to choose precocity as an important selection criterion. A high degree of precocity however, has been correlated with a low or variable kernel percentage (i.e., the fraction of nut weight accounted for by the kernel), particularly with older trees.[7]

3. Cultivation

Pecans are propagated by budding or grafting the desired cultivar on seedling rootstocks. In the United States 'Curtis', 'Moore' (Southeast), 'Riverside' (West), and 'Giles' (North) are often used as seed for rootstocks, although any "available" seed source has often been used. The reliance on seedling rootstocks (i.e., genotypic variability) and the uncertainty as to the identity of seed source have contributed to the variability in tree performance of a given cultivar.

Pecans have been cultivated for only a very short period of time and are considered 'wild' since plant growth and development resembles a forest tree species rather than a "domesticated" crop.[6] The first pecan cultivar, 'Centennial' was grafted in 1846, and not until the early to mid-1900's were a significant proportion of the pecans named cultivars.[8] Today ca. 500 pecan cultivars have been named and propagated. The pecan industry in many states is still largely based on seedlings and pecan seedlings account for a substantial proportion of total yield in North America.[8]

Lack of tree size control is a major impediment to pecan production.[9] Dwarfing rootstocks are not available and cultivars with a dwarfing growth habit such as 'Cheyenne' have had cultural problems. A low density planting scheme (i.e., plant spacing >15.2 m) is not economically feasible since a prolonged period of time (i.e., ca. 15 years or more) is required for a positive return on investments. In a high density configuration (i.e., plant spacing <12.2 m) trees under intensive management may become overcrowded with a concomitant reduction in productivity after only 10 to 12 years. Since pecan trees do not respond well to pruning, alternate tree removal is often required before a positive return of investments is realized.[9]

Heavy and light crop production in alternate years is a serious limitation to successful pecan production, especially for certain cultivars.[10–16] Native stands of pecans typically bear a good crop once every several years due to competition for limiting resources (i.e., light, water, nutrients) and the extremely high amount of energy contained in oil-rich nuts (Table 1).[6] Only native trees located in an open-environment are capable of producing nuts on a somewhat regular basis, but owing to a large tree

Table 1 **Protein, fat, and total carbohydrate content of chestnuts and selected tree nuts**

Tree nuts	Protein (%)	Fat (%)	Total carbohydrate (%)
Almonds, dried, unblanched	19.95	52.21	20.40
Brazilnuts, dried, unblanched	14.34	66.22	12.80
Cashew nuts, dry roasted	15.31	46.35	32.69
Chestnuts, raw			
American	4.83	1.32	48.58
Chinese	4.20	1.11	49.07
European	1.98	1.63	40.28
Coconut meat, raw	3.33	33.49	15.23
Filberts (hazelnuts), dried, unblanched	13.04	62.64	15.30
Macadamia nuts, dried	8.30	73.72	13.73
Pecans, dried	7.75	67.64	18.24
Pistachio nuts, dried	20.58	48.39	24.81
Walnuts, English or Persian, dried	14.29	61.87	18.34

Note: American, Chinese, and European chestnut data.[104] Data for other selected tree nuts.[105]

size, yields per hectare are low. Even under optimum cultural conditions yield of pecans seldom exceeds 2250 kg/ha.[17,18]

Although environmental factors have a profound influence on alternate bearing, little quantitative data are available due to the difficulty in isolating the contribution of each environmental variable. Sparks[19] concluded that no single variable can alleviate alternate bearing in pecan, although a total management program can minimize its' intensity. Factors that tend to enhance the tendency for alternate bearing include: a frost or freeze-damage induced crop loss; insufficient irrigation, fertilization, or irradiance levels; inadequate pest control; and premature defoliation. Alternate bearing has often been correlated with the level of assimilate reserves, particularly the carbohydrate concentration in the root.[12–14,16] Fruiting shoots typically produce fewer flowers the following year compared to vegetative shoots[20] despite containing a similar concentration of carbohydrates.[12] The involvement of plant hormones is suspected, but no quantitative data are available.

The environmental constraints to successful pecan cultivation are location dependent. In the southeastern United States reduced levels of irradiance and high relative humidity during the growing season are important limitations. In the western United States, or in arid regions, the availability of irrigation water can be a major concern. In the northern extremes of the cultivated range the greatest limitation is a growing season of adequate duration to ripen nuts, and to a lesser extent minimum winter temperatures. By contrast, in extreme southern areas a lack of winter chilling can limit productivity.

B. LIGHT

1. Plant Growth and Development

Light is the resource/input that has been least successfully manipulated in pecan production. With the implementation of culture and management practices such as irrigation, fertilization, pest control, etc., the lack of available sunlight may often limit yield and quality in pecan orchards. The total amount of quanta intercepted per hectare of land cannot be manipulated. Unfortunately, for pecan there has been virtually no progress in enhancing the efficiency of light utilization for several reasons.

The inefficiency of light utilization is largely a function of biological constraints associated with growth habit and physiology. Pecan orchards over 60 years old may be overcrowded at a density of 35 trees per hectare. Orchards established at high densities (i.e., >75 trees per ha) and subjected to intensive management may become overcrowded in as little as 15 to 20 years. In overcrowded orchards, the light levels are sufficient for nut production only in the tops of trees since flower initiation and nut development are light dependent. Shading may reduce yield as a result of the light environment of a given year, or the effect may be cumulative.[21] Excessive shading is also associated with a gradual loss of lower limbs, enhanced insect and disease pressures and a reduction in nut quality. For example, Higdon[22] noted that nut size decreased and the number of pops (unfilled nuts) increased with increased tree density. Andersen[23] also found that trunk cross-sectional area was reduced and limb breakage was increased for 14-year-old pecan trees of 6 cultivars at a high density (112 trees per ha) compared to a lower density (i.e., 43 trees per ha).

A dwarfing rootstock for pecan is not currently available. The growth retardant paclobutrazol has been shown to be an effective growth retardant of pecan,[24,25] but it is not registered for use in many countries including the United States. Thus, improved light distribution can only be accomplished by pruning or by tree removal. However, pruning has either reduced or has had no effect on yield of pecan.[26–31] Crane et al.[27] showed that cumulative yield was essentially zero for the 3 years following severe pruning where scaffold limbs were cut back to stubs. Pruning by the removal of shoot terminals increased kernel quality and shoot length, yet reduced yield.[29,32] Worley[9] found that yield was less than or equal to control trees when 20-year-old 'Elliot', 'Desirable', and 'Farley' (spacing = 14 × 14 m) pruned for 8 years by: (1) removing competing wood from alternate temporary trees as needed; (2) top and side hedging; and (3) selective limb removal. Yields of 4-year-old 'Western Schley' have been increased by 50% after relatively light pruning; however, the significance of these data is unclear as 4-year-old trees do not ordinarily have a high percentage of the canopy heavily shaded.[33]

2. Leaf Gas Exchange

The influence of light levels on A and the associated plant carbohydrate status, has received much attention due to the tendency of pecan to bear heavy and light crops in alternate years.[10–15]

Factors that may influence the effect of light (and other environmental variables) on leaf gas exchange should be noted prior to a detailed discussion on irradiance. First, A_{max}, g_s, and E have been recorded for recently expanded leaves; leaf gas exchange was greatly reduced for leaves that were 4-months old compared to recently expanded leaves.[34] A decline in A and g_s of pecan was reported during late season compared to mid season; and the presence of a crop resulted in a 10 to 40% increase in A compared to nonfruiting trees.[35] Evidence that leaf carbohydrate status influences A of pecan was obtained by Marquard.[36] Girdling fruiting and non-fruiting shoots reduced A to 30 and 3%, respectively, of the ungirdled control shoots. Leaf gas exchange measured on excised branches or leaves should be avoided since xylem plugging may induce partial stomatal closure in as little as 3 hrs.[34] The above factors plus environmental preconditioning must be considered when interpreting gas exchange data of pecan leaves.

Loustalot and Hamilton[37] first provided evidence that A_{max} of pecan leaves was reached at relatively high light levels of (i.e., under clear conditions), although light response curves were not reported. Crews et al.,[38] working with excised shoots of 'Brooks', 'Stuart', and 'Mobile' pecan reported that light saturated-A (i.e., 7 $\mu mol\ m^{-2}\ s^{-1}$) occurred at a PPF of ca. 700 $\mu mol\ m^{-2}\ s^{-1}$. However, 2- to 3-fold higher A_{max} have been subsequently reported for leaves from field-grown trees.[34,39,40]

Recent data have confirmed that light saturation is achieved at relatively high PPF. Diurnal measurements showed that A_{max} for leaves of 'Choctaw' and 'Cape Fear' occurred in the morning when PPF approached 1500 $\mu mol\ m^{-2}\ s^{-1}$; after late morning, A remained high (Figure 1) or declined slightly (Figure 2), until early evening. 'Cape Fear' trees that were preconditioned for 15 days to 30% full sunlight by using a neutral filter polyethylene shadecloth, a neutral filter[41] manifested much lower A during daylight hours (Figure 2).

Light response curves for sun- and shade-preconditioned trees have been established after exposing trees to full sunlight for 1 hr (Figure 3). The light response curves for pecan leaves preconditioned to 100 and 30% full sunlight was similar at low PPF, but diverged at ca. 200 $\mu mol\ m^{-2}\ s^{-1}$. Light saturation (i.e., 95% A_{max}) occurred at a PPF of ca. 1500 and 1300 $\mu mol\ m^{-2}\ s^{-1}$ for leaves preconditioned in 100 and 30% full sunlight, respectively (Figure 3). Apparent quantum yield determined by the slope of A/PPF at PPF $< 150\ \mu mol\ m^{-2}\ s^{-1}$ was not influenced by light preconditioning, although light compensation points and dark respiration rates were ca. 18 and 35% lower, respectively, for shade- compared to sun-preconditioned leaves.

Net CO_2 assimilation is greatly reduced in the shaded interior of pecan. For leaves of 'Elliot' naturally shaded in the canopy interior exposed to ca. 15% full sunlight PPF (i.e., PPF = 312 $\mu mol\ m^{-2}\ s^{-1}$), A was ca. 30% that recorded for leaves in full sunlight (Figure 4). However, g_s and E of leaves in the canopy interior were ca. 70% of full sun values, hence water use efficiency (WUE) declined. Steinberg et al.[42] also noted that g_s was similar for shaded and sun-exposed leaves of 'Wichita'. Evidence for a high light requirement of pecan leaves was obtained by Wood.[39] A moderate level of shading associated with a layer of sooty mold on adaxial leaf surfaces (i.e., PPF = 1320 $\mu mol\ m^{-2}\ s^{-1}$ compared to 1738

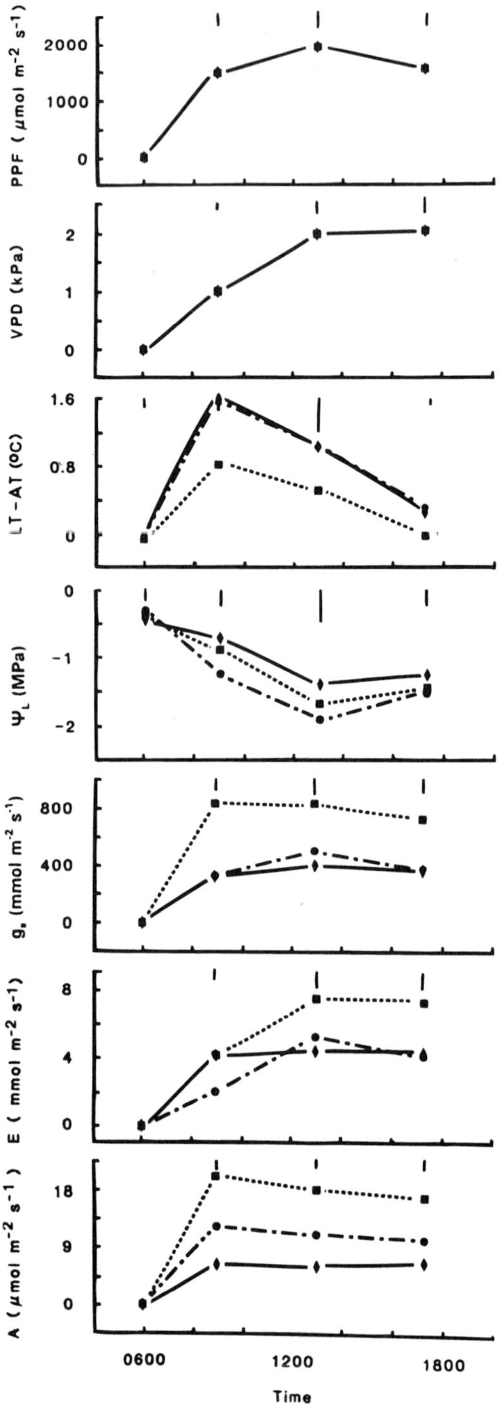

Figure 1 Diurnal cycle of photosynthetic photon flux (PPF), vapor pressure deficit (VPD), leaf temperature-air temperature (LT-AT), leaf water potential (Ψ_L), stomatal conductance to water vapor (g_S), transpiration rate (E), and net CO_2 assimilation rate (A) of pecan leaves (Andersen and Brodbeck[34]). Spring-flush expanded leaves are represented by ●; summer-flush expanded leaves by ■; and summer-flush expanding leaves by ♦. Vertical bars at the top of the PPF and VPD graphs represent ±1 SE; vertical bars represent LSD, 0.05 level.

μmol m^{-2} s^{-1} in full sunlight) reduced A by 20%; full sun values of A occurred after the film of sooty mold was peeled off the leaf.[39]

The above experiments have characterized the response of pecan leaves to relatively constant PPF. However, under natural conditions plants are typically exposed to frequent and sometimes drastic fluctuations in light level.[43–45] Intermittent cloud cover (which can reduce PPF to 10 to 30% of full sunlight), wind-generated leaf movements and mutual leaf shading may result in changes in PPF lasting less than a second to hours.

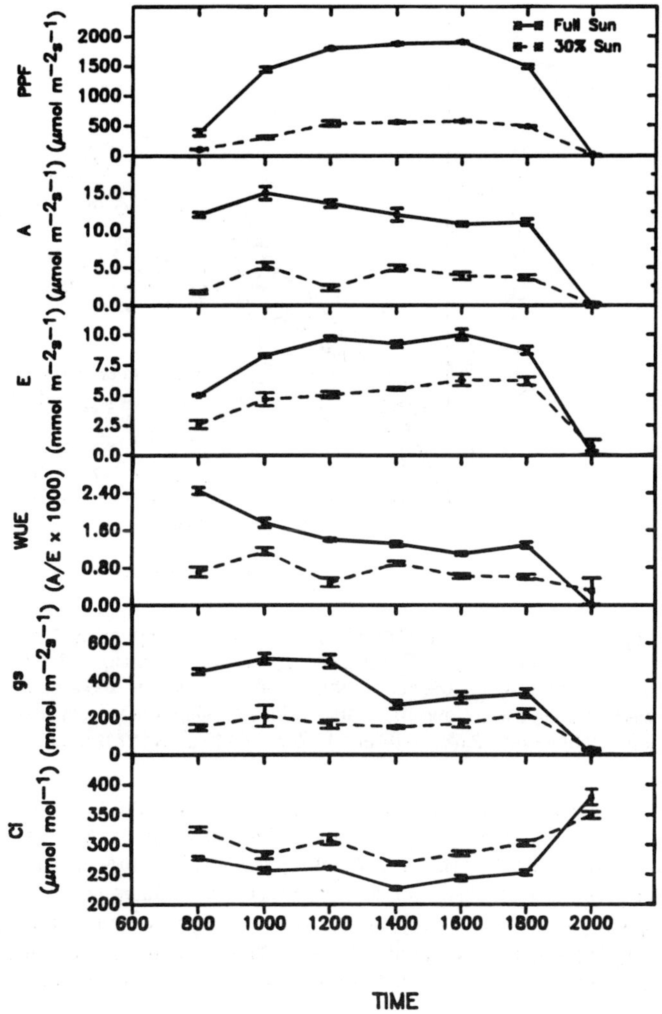

Figure 2 Diurnal characteristics of leaf gas exchange of 'Cape Fear' pecan leaves preconditioned to 100 or 30% sun for 14 days. Photosynthetic photon flux (PPF), net CO_2 assimilation (A) transpiration (E), water use efficiency (WUE), stomatal conductance to water vapor (g_S) and intercellular CO_2 concentration (C_i) from 800 to 2000 hr (Andersen, 1990 unpublished data).[106] Circle and error bars represent mean $\pm$ SE, n = 4. Leaf temperatures for sun and shade leaves, respectively were as follows: 800 hr, 21.1 $\pm$ 0.1 and 22.0 $\pm$ 0.3; 1000 hr, 29.2 $\pm$ 0.4 and 28.3 $\pm$ 0.3; 1200 hr, 31.6 $\pm$ 0.2 and 31.4 $\pm$ 0.5; 1400 hr, 36.3 $\pm$ 0.4 and 34.0 $\pm$ 0.2; 1600 hr, 36.6 $\pm$ 0.3 and 35.2 $\pm$ 0.4; 1800 hr, 33.1 $\pm$ 0.1 and 31.4 $\pm$ 0.2; 2000 hr, 25.8 $\pm$ 0.1 and 26.0 $\pm$ 0.2 (mean $\pm$ SE).

Net CO_2 assimilation of higher plants declines quickly if the PPF declines below light-saturated PPF (with the exception of crassulacean acid metabolism plants) and increases rapidly with a return to full sunlight. The stomatal response to fluctuating irradiance, however, varies widely among species.[43–46] Plants having a stomatal aperture strongly dependent on light level have been classified as "sun tracking" and those having a stomatal aperture essentially independent of short-term changes in irradiance have been designated "non-sun-tracking".[45]

Pecan is a non-sun-tracking species when supplied with non-limiting soil moisture.[47] Pecan leaves were shaded with polyethylene shadecloth in the following sequence to produce 100, 66, 33, 10 and 100% the PPF in full sunlight with each sequence lasting 80 seconds (Figure 4). Net CO_2 assimilation was highest in full sunlight consistent with above data documenting a light saturation of ca. 1500 μmol m^{-2} s^{-1}. With increasing shade A declined; however, g_s remained unaffected. Consequently, WUE decreased and C_i increased with increasing shade. When leaves were reexposed to full sunlight A, g_s, E, and C_i were nearly identical to initial values recorded in full sunlight.

When gas exchange of pecan leaves was measured in full sunlight for 18 min then in 30% full sunlight (i.e., PPF = 583 μmol m^{-2} s^{-1}) for 54 min, an abrupt 50% reduction in A occurred just after shading; however, g_s declined very slowly such that g_s was reduced by only 25% after a 54 min period (Figure 5). Consequently, C_i increased from a steady state 225 to 230 μmol m^{-2} s^{-1} in full sun to 255 to 265 μmol m^{-2} s^{-1} in 30% full sunlight. Similarly, Woods and Turner[48] found that g_s of beech, maple, oak, and yellow poplar reached stable values 12, 18, 20, and 36 min, respectively, after shade was imposed.

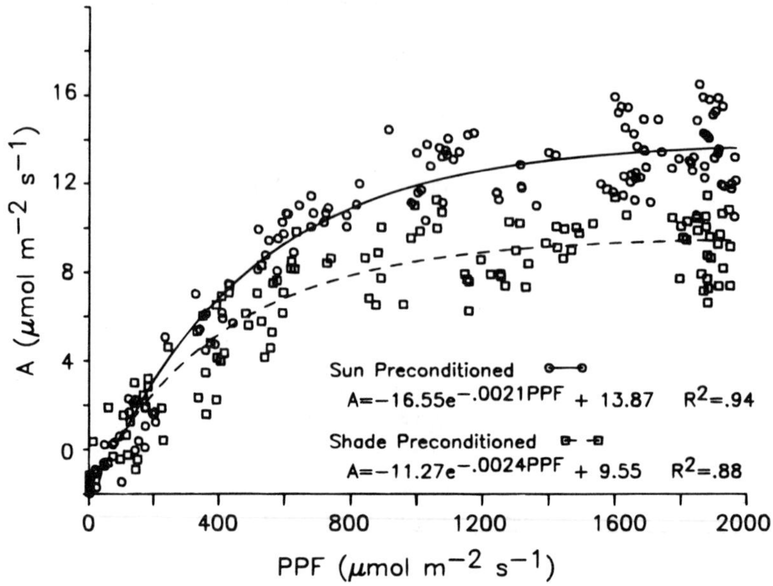

Figure 3 Net CO_2 assimilation (A) at different levels of photosynthetic photon flux (PPF) for 'Cape Fear' pecan leaves preconditioned to 100 or 30% (shade) of full sunlight PPF for 15 days (Andersen, 1990 unpublished data).[106] Trees of the 30% sun treatment were exposed to 100% sun for 1 to 3 hr prior to measurements. Data were collected from four 100% sun- and 30% shade-preconditioned trees. Vapor pressure deficits and leaf temperatures were 2.52 ± 0.04 and 2.46 ± 0.04 kPa, and 27.0 ± 0.2 and 26.7 ± 0.4°C, respectively (mean ± SE).

For pecan, A and stomatal aperture were not tightly coupled resulting in great variations in WUE and C_i with changes in light intensity (Figures 4 and 5). A good correlation between A and g_s has often been reported in leaves of herbaceous species,[49–51] although nonconstant C_i has been reported in leaves of peach,[52] grapevine,[53] olive,[54] and blueberry.[55] Knapp and Smith[43–45] have classified most herbaceous species tested as sun-tracking and most woody species tested as non-sun-tracking. In variable sunlight, it appears that for pecan and many woody species carbon gain is maximized at the expense of water loss, whereas for many herbaceous species water loss is minimized at the expense of carbon gain.[45]

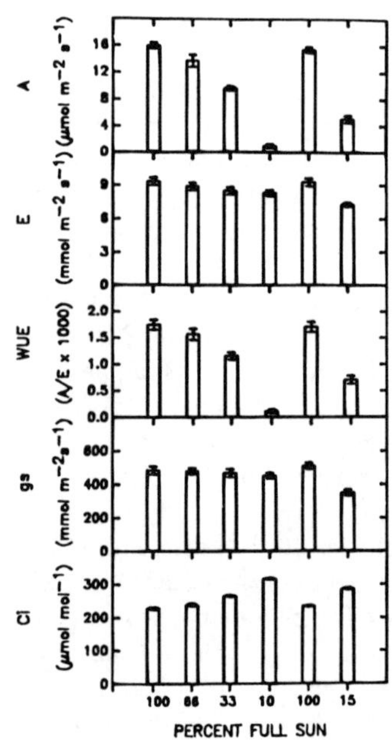

Figure 4 Net CO_2 assimilation (A), transpiration (E), water use efficiency (WUE), stomatal conductance to water vapor (g_S), and intercellular CO_2 concentration (C_i) of 'Elliot' pecan leaves exposed to 100, 66, 33, 10, and 100% full sun with each sequence lasting ca. 80 sec. (i.e., full sun PPF = 2083 ± 43 μmol m^{-2} s^{-1}) (Andersen 1991).[47] The last column of 15% full sun represents values for naturally shaded leaves located in the canopy interior. Values are means ± SE, n = 9.

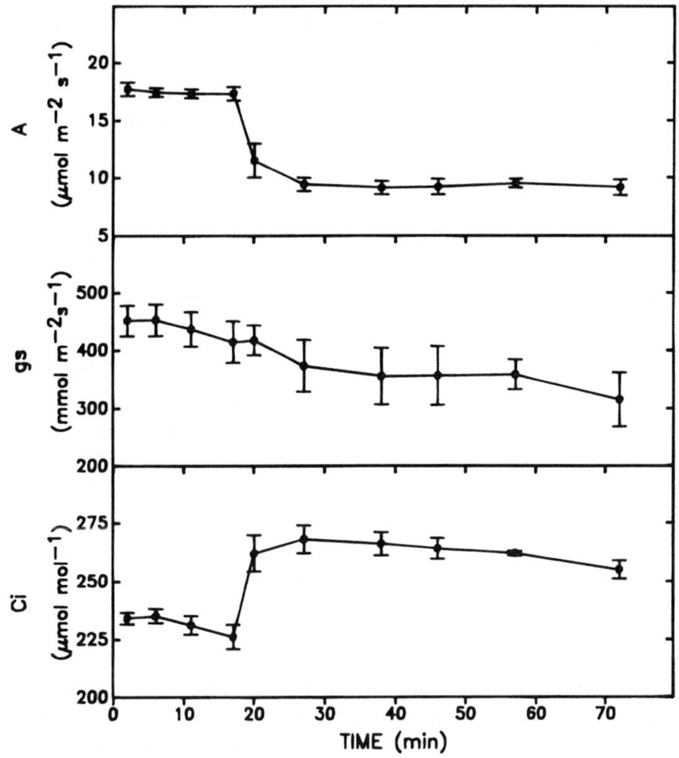

Figure 5 Effect of variable light intensity on net CO_2 assimilation (A), stomatal conductance to water vapor (g_S) and intercellular CO_2 concentration (C_i) of pecan leaves. Leaves were exposed to 20 minutes of full sunlight, (i.e., PPF = 1924 ± 24 μmol m^{-2} s^{-1}), then 50 min of 33% sunlight (i.e., PPF = 583 ± 10 μmol m^{-2} s^{-1} (Andersen 1991).[47] Circles ± error bars represent means ± SE, n = 4.

The physiologic response to fluctuating irradiance levels for a given species may be influenced by the level of water stress whether caused by low soil moisture conditions or high VPD.[44] When water conservation becomes more critical than maintaining high levels of A, presumably a plant would tend to behave in a more sun-tracking mode.[43–45] Although whole tree carbon gain has not been adequately quantified for pecan given that light saturation of individual leaves occurred at a PPF of 1500 μmol m^{-2} s^{-1} (Figure 3), Andersen[56] proposed that whole canopies of mature pecan trees are never light saturated under non-limiting conditions of soil moisture and VPD.

C. WATER

1. Water Requirements

Native stands of pecan trees are generally confined to regions with deep alluvial soils (i.e., bottomlands). Upon germination, pecan seedlings repartition the rich energy reserves in the nut to a deep, well-developed taproot. The taproot may extend considerable distances to the water table and allow the tree to survive extended periods of drought; however, a more steady supply of soil moisture in the upper soil horizons is required for optimum tree performance. Irrigation has been shown to be cost-effective in the southeastern United States, the major pecan producing region in the world, which is characterized by a high humidity and abundant but often erratic rainfall.[57]

Irrigation in numerous locations has increased tree growth, yield, nut size, and percentage kernel.[57–61] Drought occurring early in the season have reduced nut size, while deficits during mid- to late-season have reduced percentage kernel, delayed shuck split, and harvest date. Late season water stress has also induced premature defoliation,[58] fruit abortion,[62] and greatly enhanced the percentage of nuts showing vivipary (preharvest germination) or a condition known as sticktight (delayed shuck opening).[60,61]

The water requirements of pecan trees are dependent on tree size/density and climatic and edaphic conditions. Once optimal tree size is achieved for a given tree density per hectare, irrigation requirements should be similar regardless of orchard density. Most (i.e., 95%) of the soil moisture depletion in several high density orchards (60 to 120 trees per ha) in west Texas occurred in the upper 1 m of soil.[63] Soil moisture depletion was independent of distance from the tree.[63] However, Miyamoto[63] cautioned that these data may not apply to pecans grown in deep, well-drained soils. Roots of pecan trees extended

Table 2 **Osmotic potential (Ψ_π) at turgor point loss and turgor pressure (Ψ_p) and Ψ_π at full hydration of spring- and summer-flush leaves**

Leaf age	Turgor point loss Ψ_π (MPa)	Full hydration Ψ_p(MPa)	Full hydration Ψ_π(MPa)
Spring-Flush (Expanded)	−2.7	1.8	−2.3
Summer-Flush (Expanded)	−2.4	1.7	−2.1
Summer-Flush (Expanding)	−1.6	0.9	−1.3
LSD 0.05	0.2	0.2	0.2

Note: Measurements were performed 25–29 August.

far beyond the drip line, and when optimum tree size is achieved for a given density, roots may occupy the upper soil horizons of the entire orchard.

Estimates of water consumption in Texas ranged from 18 cm water per month[64] to 100 cm water per season.[65] (Note: 1 cm per ha is equivalent to 100,000 l ha^{-1}). Calculated values of water consumption by Miyamoto[63] were 100 to 130 cm per season; daily water consumption was dependent on tree number × tree diameter (and not cross-sectional area). During mid to late summer daily water consumption was up to 1 cm per day.[63] Irrigation is terminated in many regions in September or October to prepare for the orchard floor for harvest; however, Stein et al.[60] advised against terminating irrigation until shuck split under dry conditions.

2. Plant Water Deficits

Although relative humidity during the growing season is typically very high in much of the native range of pecan (or in the southeastern United States), pecans have been cultivated successfully in arid regions (i.e., western United States and Israel) when provided with irrigation. Rieger and Daniell[66] reported that three-month-old, container-grown 'Curtis' seedlings were very sensitive to increases in VPD; an increase in VPD from 0 to 2 kPa was associated with a 50% reduction in g_s. The g_s of field-grown 'Stuart' trees was reduced slightly less than 50% when VPD was increased from 1 to 3 kPa.[34] By contrast, expanding, recently expanded, and 4-month-old leaves of 'Choctaw' trees in the field manifested consistently high A and g_s when exposed to VPD up to 2 kPa (Figure 1). The variable response of g_s to VPD for pecan is likely related to differences in the extensiveness of rooting, the ratio of leaf area to root absorbing area and tree L_p.

Pecan seedlings grown in containers maintained an absorptive condition despite a prolonged period of drought. Container-grown pecan seedlings subjected to soil water potentials (Ψ_s) of −0.3, −0.6 or −1.1 MPa manifested progressive reductions in g_s, E, and Ψ_L.[66] At a Ψ_s of −1.1 MPa Ψ_L declined to −3.0 MPa. Upon rewaterings Ψ_L recovered within 15 min to pre-treatment values, although leaf gas exchange remained somewhat low for at least a day perhaps due to the residual effects of ABA.[66] Similarly, Taylor and Fenn[67] noted that pecan seedlings regained turgor 1 hr after terminating a 30-day drought period. In a split root experiment on pecan seedlings water was absorbed in wet soil and transferred across the stem or crown tissue and eventually exuded by roots growing in a dry soil.[67] Presumably, the deep taproot of field-grown pecan trees in well-drained soils ameliorates the effects of high VPD or of moisture variations in the upper soil horizons. For example, minimum Ψ_L reported for field-grown 'Choctaw' occurred during midday and ranged from −1.4 MPa for expanding leaves to −1.9 MPa for leaves that were ca. 4-months-old (Figure 1).

Many physiological changes are associated with the aging of pecan leaves. Leaf gas exchange rates increased for expanding leaves until full expansion then decreased until leaf abscission.[34] The slope of g_s/Ψ_L also increased and minimum Ψ_L decreased with increasing leaf age. Stomata of expanded and expanding leaves closed in response to reductions in Ψ_L; however, basal leaflets on leaves that were 4-months old did not respond to reductions in Ψ_L to −4.1 MPa.[34] These data do not provide evidence for a critical threshold for stomatal closure.

The reduction in the Ψ_L required for stomatal closure with increasing leaf age was mediated, in part, by osmotic adjustment (Table 2). Leaf Ψ_π at turgor point loss or at full hydration decreased and Ψ_p increased with leaf aging.

Andersen and Brodbeck[34] measured the rate of elongation of pecan leaves as a function of Ψ_p and found that growth rate declined exponentially with a decrease in Ψ_p; leaf elongation ceased at a Ψ_p =

0.2 MPa (Figure 6). The biophysical basis for leaf expansion was investigated by relating ϵ of different-aged pecan leaves to Ψ_p.[34] The ϵ of all leaf-age categories increased with increases in Ψ_p (Figure 7). A lower ϵ (i.e., higher elasticity) of expanding leaves was recorded only at a $\Psi_p = 0.1$ MPa, (i.e., well below the minimum Ψ_p required for leaf elongation). Thus, leaf-averaged values of ϵ did not provide a biophysical explanation for expansion growth.

Up to one third of the fruit on some pecan trees may be lost when nut walls split during the water stage of nut filling.[68] Prussia et al.[69] determined the ϵ of developing nut wall material of four pecan cultivars. The ϵ increased (i.e., tissues became more rigid) with age. 'Wichita' was the cultivar with the highest ϵ, yet it was the cultivar most likely to rupture.[69]

Andersen and Brodbeck[34] stated that the hydraulic system of pecan trees was efficient in maintaining high rates of leaf gas exchange and high Ψ_p and Ψ_L under high levels of VPD and leaf temperatures. Steinberg et al.[42] found that the Lp of pecan trees, as measured by a heat pulse method, was higher than most woody species but similar to values recorded for some herbaceous species. Transpiration and sap flow were closely coupled and Lp did not vary during the day indicative of a very low level of tree capacitance. Steinberg et al.[42] concluded that the efficient water transport system of pecan facilitates a depression of Ψ_L down to -2.0 MPa without a reduction in E.

3. Flooding

Pecan trees are often exposed to waterlogged soil conditions in their native and cultivated range, particularly during the dormant season. The survival of pecan seedlings after 4 weeks of flooding during the growing season was greater than black walnut, siberian elm, boxelder, and cottonwood but less than bald cypress, green ash, or silver maple.[70] Bitter pecan [*Carya aquatica* (Michaux) Nuttall] has been used as a rootstock for *C. illinoensis* and has had good growth in wet sites.[71] However, there are no quantitative data concerning the intraspecific flood tolerance within *C. illinoensis.* Flood tolerance is determined primarily by the rootstock and not the scion.[72] Since the source of seedlings used for rootstocks is often variable (and sometimes unknown) differential flood tolerance is extremely difficult to quantify on mature trees in the field.

Plant tolerance to waterlogged soil conditions is seasonally dependent.[73] Flooding 'Dodd' pecan seedlings for 28 days while dormant did not affect growth or leaf elemental concentration.[74] Flooding pecan seedlings for 28 days during budbreak reduced root and shoot growth and leaf N and Fe concentrations, while flooding during active growth reduced root and shoot growth and concentrations of most leaf elements. Several months after the termination of flooding, leaf elemental concentrations were similar to the nonflooded controls. Loustalot[75] reported that 35 days of flooding killed as much as 30% of the root system of pecan seedlings. Leaf chlorosis, leaf necrosis, and leaf abscission occurred in response to summer and fall flooding for a 'Stuart' orchard in Louisiana.[76]

Leaf gas exchange declines in response to soil flooding. Net CO_2 assimilation decreased after 1 day[77] and 5 days[75] of soil flooding. Stomatal conductance (or E) and A were reduced concomitantly in flooded

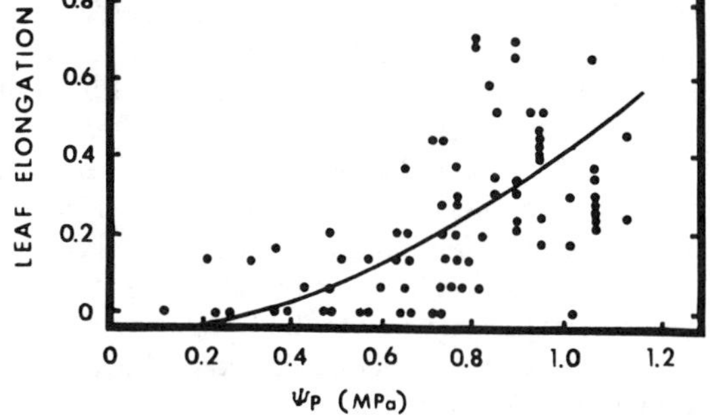

Figure 6 The effect of turgor pressure (Ψ_p) on leaf elongation rate over a 12-hr period (Andersen and Brodbeck, 1988).[34] Leaves were excised at 1900 hr, transferred to the laboratory and slowly dehydrated to variable degrees under fluorescent lights (PPF = 256 ± 87 μmol m^{-2} s^{-1}). After 12 hr, leaf length and leaf water potential was determined. The relationship between Ψ_L, Ψ_π, and Ψ_p were determined on seven expanding leaflets by pressure-volume curves, and calibration curve was constructed relating Ψ_p to Ψ_L.

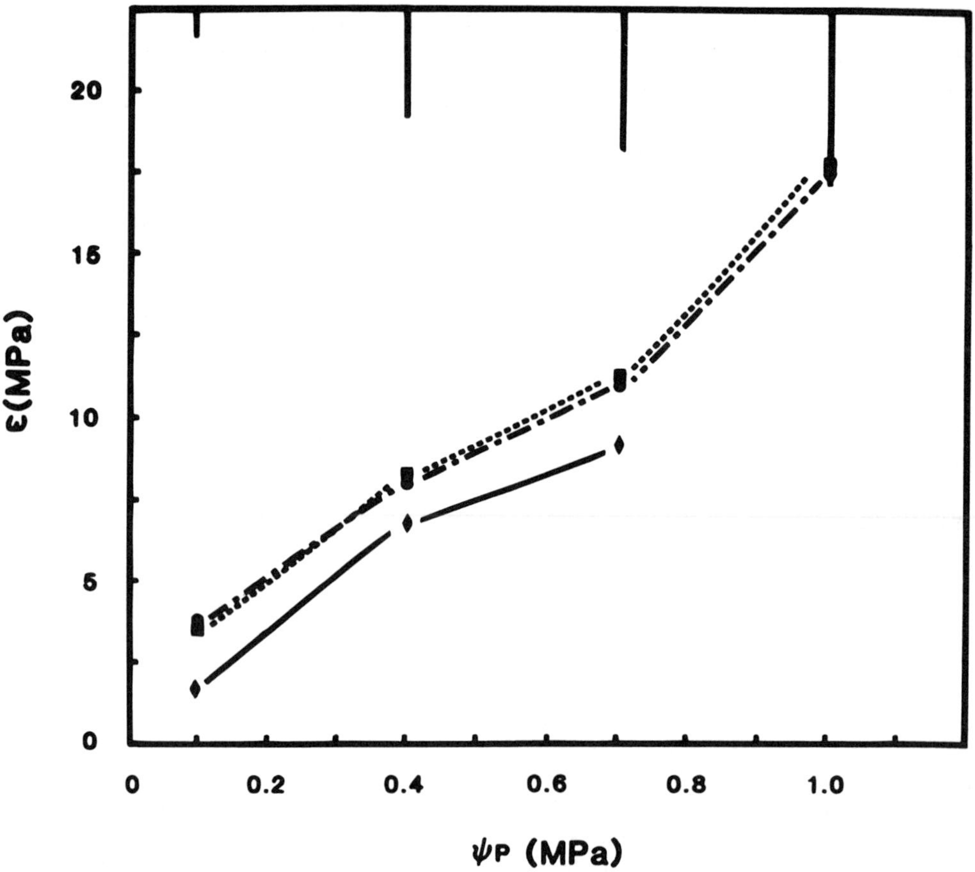

Figure 7 Bulk modulus of elasticity (ε) of pecan leaves at leaf turgor pressure (Ψ_p) = 0.1 to 1.0 MPa (Andersen and Brodbeck, 1988).[34] Spring-flush expanded leaves are represented by ●; summer-flush expanded leaves by ■; and summer-flush expanding leaves by ♦. Vertical bars at the top of the graph represent LSD, 0.05 level.

plants,[75,77] although Ψ_L was similar or increased compared to nonflooded controls.[77] Thus, as in many species,[73] flood-induced stomatal closure in pecan was not mediated by a reduction in Ψ_L.

D. TEMPERATURE

1. Cold Hardiness

a. Symptoms of Cold Damage

Freeze damage in pecan is typically most severe in regions of the trunk just above (1 m) the soil line; plant tissue below the soil is usually not directly damaged by low temperatures.[78] Trunks of pecan trees are most susceptible to cold injury since trunks are the last part of the tree to become dormant in the fall, and with very low wind velocities, cold (dense) air tends to settle near the ground. Freeze damage can be identified by discoloration of the region surrounding the vascular cambium. The bark may also split longitudinally or may become separated from the trunk.[78]

When frost or freeze damage occurs on shoots of pecans, buds or leaf scars first become discolored followed by a general discoloration of the inner bark. Pecan buds sustaining damage as a result of an early fall freeze tend to break earlier the following spring than non-damaged buds.[79] However, bud injury sustained during mid-winter or late spring tends to delay bud break.[80]

Symptoms of cold injury or plant mortality can occur at various time intervals after the date of cold injury. Cold injury may result in bud mortality prior to budbreak, shortly after budbreak or by midsummer. Alternatively, tree mortality may be gradual over a several year period or the tree may simply be stunted for a prolonged period of time.[78]

b. *Influence of Environmental Factors*

Lang et al.[81] described 3 stages of dormancy as: (1) paradormancy—the initiation of dormancy controlled by environmental factors; (2) endodormancy—the period of deep rest controlled by internal plant factors; and (3) ectodormancy—late winter or early spring dormancy which is controlled by environmental conditions unfavorable for growth.

The extent to which a tree is in endodormancy will have a profound effect on cold tolerance. Pecan trees are particularly vulnerable to cold damage early (paradormancy) or late (ectodormancy) during the dormant period compared to when the tree is in endodormancy.[81] Fluctuating late fall to early spring temperatures will accentuate the damaging effects of a given temperature since fluctuating temperatures do not promote the physiological state of endodormancy. For example, on 19 Nov. 1969 pecan trees in south Georgia sustained cold damage when temperatures dropped to −8°C because prior fall temperatures were unseasonably mild.[78] By contrast, cold injury does not usually occur in the native range (or most of the cultivated range) of pecan if temperatures are consistently low. Again, during the spring cultivars that tend to come out of dormancy and break bud earlier (e.g., 'Moore', 'Schley') are more likely to incur cold damage than cultivars that come out of dormancy late (e.g., 'Curtis', 'Stuart').[82]

Grauke and Pratt[83] reported that tree damage due to a spring freeze was proportional to the degree of bud development which was influenced by both rootstock and scion. For example, growth was more advanced for 'Elliot' and 'Curtis' seedling rootstocks compared to 'Apache', 'Sioux', 'Riverside' and 'Burkett', and for 'Candy' or 'Cape Fear' scions compared to 'Stuart'. Payne and Sparks[84] noted that cold injury was much more severe for pecan trees on non-juvenile (grafted) compared to juvenile (budded seedling) trunks. They suggested that the greater resistance of juvenile tissue to cold damage may be an adaptation while the tree is young to enhance competitiveness in its native range.

Any stress that weakens a tree going into dormancy such as drought, poor nutrition, premature defoliation, and inadequate insect and disease control will tend to exacerbate the extent of cold sensitivity. Sparks and Payne[78] ascribed the reduced cold tolerance of stressed pecan trees to a failure to enter deep dormancy (i.e., endodormancy) and a tendency toward earlier than normal bud break. A balanced nutritional program minimizes the likelihood of cold injury. Low levels of K or high levels of N fertilization dramatically increased cold injury to young 'Moore' trees.[85] Freeze injury is less likely in bare soil compared to soil with a cover crop.[78] Extreme temperature fluctuations and winter injury (i.e., sunscald damage) to tree trunks have been reduced by painting trunks with white latex paint.[86]

2. Chilling and Heat Units in Relation to Dormancy and Harvest Date

Waite[87] first suggested that pecan buds require an exposure to chilling temperatures in order to break dormancy. Lack of sufficient chilling may delay foliation, cause irregular foliation, result in shorter shoots and smaller leaves, and may result in reduced yields.[88–90] For convenience, chilling hours are often calculated as the cumulative number of hours <7.2°C, although it is known that temperatures vary in effectiveness in determining the onset and termination of winter dormancy.[90,91] Arnold[92] concluded from a survey of pecan production in locations in Florida, USA with only 100 chilling hours of chilling temperatures below 7.2°C that pecan may have a very low chilling requirement. He suggested that budbreak was mainly regulated by the accumulation of heat units. By contrast, the chilling requirement of stem cuttings of 'Desirable', 'Mahan', and 'Stuart' was 500, 500, and 600 hrs, respectively,[93] while Amling and Amling[94] determined that the chilling requirement of pecan depended on the climatic conditions during the fall (i.e., the phase of paradormancy). It was proposed that the "rest intensity" (i.e., deep dormancy) increases if trees are exposed to fall temperatures ≥1.2°C.[94] Amling and Amling[95] later concluded that exposure to low temperature was required for pistillate flower formation. Smith et al.[96] found that 6°C resulted in ≥50% budbreak of Dodd seedlings occurred after 900 hrs at 6°C, although 1000 to 2500 hrs of chilling increased percentage budbreak and reduced the time to budbreak. The duration of chilling required to break dormancy was dependent on bud location and temperature (between 1 to 9°C). Lateral buds had a higher chilling requirement than terminal buds. Lateral budbreak of 'Dodd' seedlings was enhanced more by 1000 hours of chilling at 5°C rather than 1 or 9°C; however, with 2500 hours of chilling at a temperature of 1°C was more effective in promoting budbreak.

The temperature requirements for budbreak of pecan were examined in a comprehensive study incorporating both experimental data and data derived from records obtained from a wide geographic range.[97] Sparks[97] found that budbreak was determined by the interaction of low followed by high temperatures; a meaningful specific chilling or heating requirement could not be established. The heat requirement decreased with increasing chill hour accumulation. Budbreak occurred in the absence of chilling with sufficient heating; however, this is accompanied by a variability in the timing of budbreak (i.e., consistent budbreak on a tree does have a critical chilling requirement). Heating degree days with minimum temperatures <2.2°C were inefficient in promoting budbreak; 3.9°C was the most efficient temperature on which to base chill and heat use accumulation.[97]

The regulation of budbreak by the interaction of chilling and heating may be an adaptation that facilitates a broad native and cultivated range of pecan.[97] In the north central United States a high degree of chilling received during the winter is followed by a low heat requirement for budbreak, thus allowing for nut maturation with a minimum of growing degree days. Conversely, pecan trees grown in the southern United States or in regions with little or no chilling (i.e., Brazil or Mexico) manifest delayed budbreak, thus minimizing the likelihood of frost damage during the winter or early spring. In these cases the dormant period is prolonged in the absence of chilling temperature; however, this is not necessarily concomitant with a great degree of cold hardiness.

Sparks[97] found that the fruit development period for pecan cultivars was dependent on geographic origin and varied from 137 to 198 days. Nut weight, and to a lesser degree percentage kernel was proportional to the time required for fruit development. Pecan nut and kernel characteristics were dependent on genotype and environment and genotype/environment interactions.

3. High Temperature

Maximum air temperatures in the native range of pecan seldom exceed 38°C, although in arid areas of cultivation air or leaf temperatures can approach 44°C. Leaves appear to tolerate these high temperatures, although 44°C is above the temperature optimum for many processes of growth and development of many temperate zone plant species. For a related species, shagbark hickory [*Carya ovata* (Mill.) K. Koch] damage to the chloroplast membrane was detected by nuclear magnetic resonance at temperature thresholds between 53 and 57°C,[98] ca. 10 degrees higher than is likely to occur for pecan leaves in nature.

Temperature optimums for growth, development, and maximum yield of pecan have not been determined. Lipid reserves in pecan seeds are converted to starch prior to germination.[99] Lipid hydrolysis in germinating seeds is temperature dependent and raising the incubation temperature to 30°C for 3 days substituted for 60 days of stratification at 4°C.

Yates and Sparks[100] quantified the germination of pecan pollen as a function of temperature (i.e., 25, 35, or 45°C) and relative humidity (5, 50, and 97%). Maximum germination percentage increased as temperature decreased and relative humidity increased. Similarly, Marquard[101] found that germination rate and rate of pollen tube growth was maximum at 27°C and was ca. 0 at 34°C. Pollen can be stored for a prolonged period of time provided that it is oven-dried at 35°C and stored in a moisture-proof bag.[102] Pollen stored at −12°C for 2 years under these conditions was as viable as fresh pollen.

Crews et al.,[38] working with excised shoots of 'Mobile', 'Stuart', and 'Brooks' pecan, found that A_{max} occurred at a leaf temperature of 27°C (8 μmol m^{-2} s^{-1}); at 40°C A declined to 3 μmol m^{-2} s^{-1}. In that study A_{max} was low for a C_3 mesophyte. Subsequent studies[34,47,103] have shown that A_{max} for pecan leaves is ca. 16 to 20 μmol m^{-2} s^{-1}. The low temperature optimum for A and the low values of A_{max} reported on excised branches[38] may have been a result of xylem plugging.[34] Andersen and Brodbeck[34] found that there was no evidence of a decline in A, g_s, and E at leaf temperature up to 41°C when trees were supplied with adequate soil moisture; in fact, A was linearly related to leaf temperatures in the range of 30.5 to 41°C leading to the suggestion that maximum temperatures encountered in the field are seldom limiting A on a leaf area basis. Whole-tree A, however, may be reduced with increasing temperature above a given threshold since the effect of temperature on respiration of non-photosynthetic tissues is not considered when A is measured on a single leaf. There is ample justification for more experimentation concerning the effects of both high and low temperature on physiological processes on the whole tree level.

E. MISCELLANEOUS ENVIRONMENTAL STRESSES

There are little or no quantitative data concerning physiological response of pecan to salt, wind, or atmospheric pollutants (i.e., sulfur dioxide, nitrous oxide, ozone elevated CO_2 concentrations, etc.). While the limited physiological data has focused on the major environmental stresses existing in the current range of production, there is clearly a need for research on other environmental stresses that may occur presently or in the future.

REFERENCES: PECANS

1. Bailey, L. H. and Bailey, E. Z., A concise dictionary of plants cultivated in the United States and Canada, *Hortus Third,* Macmillan Publishing Co., New York, N. Y., 1976.
2. Little, E. L., Jr., *Atlas of United States trees,* Vol 1. Conifers and important hardwoods, Misc. Publ., U. S. Dept. Agric., Forest Serv., 1146, 1971.
3. Shuhart, D. V., Morphological differentiation of pistillate flowers of the pecan, *J. Agr. Res.,* 34, 687, 1927.
4. Woodruff, J. G. and Woodruff, N. C., Abnormalities in pecans: I. Abnormalities in pecan flowers, *J. Hered.,* 21, 39, 1930.
5. Wood, B. W. and Payne, J. A., Flowering potential of pecan, *HortScience,* 18, 326, 1983.
6. Wolstenholme, B. N., The ecology of pecan trees I & II, *Pecan Q.,* 13, 14, 32, 1979.
7. Sparks, D., Inter-relationship of precocity, prolificacy, and percentage kernel in pecan, *HortScience,* 25, 297, 1990.
8. Sparks, D., *Pecan cultivars—the orchard's foundation,* Pecan Production Innovations, Watkinsville, GA, 1992.
9. Worley, R. E., Effects of hedging and selective limb pruning of Elliott, Desirable, and Farley pecan trees under three irrigation regimes, *J. Amer. Soc. Hort. Sci.,* 110, 12, 1985.
10. Davis, J. T. and Sparks, D., Assimilation and translocation patterns of carbon-14 in the shoot of fruiting pecan trees, *J. Amer. Soc. Hort. Sci.,* 99, 468, 1974.
11. Lockwood, D. W. and Sparks, D., Translocation of ^{14}C in 'Stuart' pecan in the spring following assimilation of $^{14}CO_2$ during the previous growing season, *J. Amer. Soc. Hort. Sci.,* 103, 38, 1978.
12. Smith, M. W., McNew, R. W., Ager, P. L., and Cotten, B. C., Seasonal changes in the carbohydrate concentration in pecan shoots and their relationship to flowering, *J. Amer. Soc. Hort. Sci.,* 111, 558, 1986.
13. Sparks, D. and Brack, C. E., Return bloom and fruit set of pecan from leaf and fruit removal, *HortScience,* 7, 131, 1972.
14. Wood, B. W., Pecan production responds to root carbohydrates and rootstock, *J. Amer. Soc. Hort. Sci.,* 114, 223, 1989.
15. Wood, B. W. and McMeans, J. L., Carbohydrate changes in various organs of bearing and nonbearing pecan trees, *J. Amer. Soc. Hort. Sci.,* 106, 758, 1981.
16. Worley, R. E., Fall defoliation date and seasonal carbohydrate concentration of pecan wood tissue, *J. Amer. Soc. Hort. Sci.,* 104, 195, 1979.
17. Crocker, T. F., *Commercial pecan production in Georgia,* The Cooperative Extension Service, The Univ. of GA, Bulletin 609, 1988.
18. Westberry, G., Crocker, T., Harrison, K., and Givan, W., *Estimated costs of producing pecans,* The University of Georgia College of Agric. and Env. Sciences, Cooperative Ext. Work, AG Econ 92-022, 1992.
19. Sparks, D., Alternate fruit bearing—a review, *Pecan South,* 2, 44, 1975.
20. Malstrom, H. L. and McMeans, J. L., Shoot length and previous fruiting affect subsequent growth and nut production of 'Money-maker' pecan, *HortScience,* 17, 970, 1982.
21. Sparks, D., Physiology-site, growth, flowering, fruiting, and nutrition, in *Nut Tree Culture in North America,* Jaynes, E., Ed., Northern Nut Growers Assoc., W. F. Humphrey Press, Inc., Geneva, N. Y., 1981, chap. 16.
22. Higdon, R. J., Too close spacing of pecan trees reduce size and quality of the nuts, *Proc. S. E. Pecan Grow. Assn.,* 44, 32, 1951.
23. Andersen, P. C., Tree growth and yield of seven pecan cultivars at three orchard spacings, *Proc. S. E. Pecan Growers Assoc. Mtg.,* 80, 111, 1987.
24. Andersen, P. C. and Aldrich, J. H., Effect of soil-applied Paclobutrazol on 'Cheyenne' pecans, *HortScience,* 22, 79, 1987.
25. Wood, B. W., Influence of paclobutrazol on selected growth and chemical characteristics of young pecan seedlings, *HortScience,* 19, 837, 1984.
26. Crane, H. L., Results of pecan pruning experiments, *Proc. Georgia-Florida Pecan Grow. Assn.,* 27, 11, 1933.
27. Crane, H. L., Hardy, M. B., Dodge, F. N., and Loomis, N. H., The effects of thinning and stand of trees and other orchard practices on the growth and yield of pecans, *Proc. S. E. Pecan Grow. Assn.,* 29, 27, 1935.

28. Hardy, M. B., Progress report on attempts to control biennial bearing in pecans, *Proc. S. E. Pecan Grow. Assn.*, 40, 54, 1947.
29. Storey, J. B., Madden, G., and Garza-Falcon, G., Influence of pruning and growth regulators on pecans, Texas Agr. Expt. Stat., PR-2709, in *1970 Pecan Research 1965–1969,* Texas Agr. Expt. Sta. Consolidates, PR-2709-2722, 1970.
30. Worley, R. E., Effects of hedging and selective limb pruning of Elliott, Desirable, and Farley pecan trees under three irrigation regimes, *J. Am. Soc. Hort. Sci.*, 110, 12, 1985.
31. Worley, R. E., Effect of pruning old non-crowded 'Stuart' pecan trees to three heights, *Proc. S. E. Pecan Grow. Assoc.*, 80, 123, 1987.
32. Overcash, J. P. and Kilby, W. W., Growth responses to pruning young pecan trees, *Proc. S. E. Pecan Grow. Assn.*, 65, 95, 1972.
33. Kuykendall, J. R., Tate, H. F., and Clark, L., Pecan pruning experiments in Arizona—1969, *Proc. S. E. Pecan Grow. Assn.*, 63, 98, 1970.
34. Andersen, P. C. and Brodbeck, B. V., Net CO_2 assimilation and plant water relations characteristics of pecan growth flushes, *J. Amer. Soc. Hort. Sci.*, 113, 444, 1988.
35. Wood, B. W., Tedders, W. L., and Reilly, C. C., Sooty mold fungus on pecan foliage suppresses light penetration and net photosynthesis, *HortScience,* 23, 851, 1988.
36. Marquard, R. D., Influence of leaf to fruit ratio on nut quality, shoot carbohydrates, and photosynthesis of pecan, *HortScience,* 22, 256, 1987.
37. Loustalot, A. J. and Hamilton, J., Effects of downy spot on photosynthesis and transpiration of pecan leaves in the fall, *Proc. Amer. Soc. Hort. Sci.*, 39, 80, 1941.
38. Crews, C. E., Worley, R. E., Syvertsen, J. P., and Bausher, M. G., Carboxylase activity and seasonal changes in CO_2 assimilation rates in three cultivars of pecan, *J. Amer. Soc. Hort. Sci.*, 105, 798, 1980.
39. Wood, B. W., Fruiting affects photosynthesis and senescence of pecan leaves, *J. Amer. Soc. Hort. Sci.*, 113, 432, 1988.
40. Wood, B. and Payne, J., Influence of single applications of insecticides on net photosynthesis of pecan, *HortScience,* 19, 265, 1984.
41. Yates, D. J., Shade factors of a range of shade materials, *Ag. Eng. Aust.*, 15, 22, 1986.
42. Steinberg, S. L., McFarland, M. J., and Worthington, J. W., Comparison of trunk and branch sap flow with canopy transpiration in pecan, *J. Exp. Bot.*, 41, 653, 1990.
43. Knapp, A. K. and Smith, W. K., Effect on water stress on stomatal and photosynthetic responses in subalpine plants to cloud patterns, *Amer. J. Bot.*, 75, 851, 1988.
44. Knapp, A. K. and Smith, W. K., Influence of growth form and water relations on stomatal and photosynthetic responses to variable sunlight in subalpine plants, *Ecology,* 70, 1069, 1989.
45. Knapp, A. K. and Smith, W. K., Stomatal and photosynthetic responses to variable sunlight, *Physiol. Plant.*, 78, 160, 1990.
46. Chazdon, R. L. and Pearcy, R. W., Photosynthetic responses to light variation in rainforest species. I. Induction under constant and fluctuating light conditions, *Oecologia,* 69, 517, 1986.
47. Andersen, P. C., Leaf gas exchange of 11 species of fruit crops with reference to sun-tracking/non-sun-tracking responses, *Can. J. Plant Sci.*, 71, 1183, 1991.
48. Woods, D. B. and Turner, N. C., Four tree species of varying shade tolerance, *New Phytol.*, 70, 77, 1971.
49. Farquhar, G. D., Dubbe, D. R., and Raschke, K., Gain of the feedback loop involving carbon dioxide and stomata: theory and measurement, *Plant Physiol.*, 62, 406, 1978.
50. Wong, S., Cowan, I. R., and Farquhar, G. D., Stomatal conductance correlates with photosynthetic capacity, *Nature,* 282, 424, 1979.
51. Wong, S., Cowan, I. R., and Farquhar, G. D., Leaf conductance in relation to rate of CO_2 assimilation. II. Effects of short-term exposures to different photon flux densities, *Plant Physiol.*, 78, 826, 1985.
52. DeJong, T. M., Leaf nitrogen content and CO_2 assimilation capacity in peach, *J. Amer. Soc. Hort. Sci.*, 107, 955, 1982.
53. Downton, W. J. S., Loveys, B. R., and Grant, W. J. R., Stomatal fully accounts for the inhibition of photosynthesis by abscisic acid, *New Phytol.*, 108, 263, 1988.
54. Bongi, G., Mencuccini, M., and Fontanazza, G., Photosynthesis of olive leaves: effects of light flux density, leaf age, temperature, peltates and H_2O vapor pressure deficit on gas exchange, *J. Amer. Soc. Hort. Sci.*, 112, 143, 1987.
55. Moon, J. W., Jr., Flore, J. A., and Hancock, J. F., Jr., A comparison of carbon and water vapor gas exchange characteristics between a diploid and highbush blueberry, *J. Amer. Soc. Hort. Sci.*, 112, 134, 1987.

56. Andersen, P. C., Photosynthetic characteristics of pecan and ten species of fruit crops with emphasis on sun tracking/non-sun tracking responses, *First National Pecan Workshop Proc.*, pp. 168–172, 1991.
57. Daniell, J. W., Moss, R. B., and Dean, B., The use of irrigation in pecan orchards, *Pecan South,* 6, 84, 1979.
58. Alben, A. O., Results of an irrigation experiment on Stuart pecan trees in East Texas in 1956, *Proc. Texas Pecan Grow. Assn.,* 36, 16, 1957.
59. Magness, J. R., Moisture supply as a factor in pecan production, *Proc. Natl. Pecan Grow. Assn.,* 30, 18, 1931.
60. Stein, L. A., McEachern, G. R., and Storey, J. B., Summer and fall moisture stress and irrigation scheduling influence pecan growth and production, *HortScience,* 24, 607, 1989.
61. Worley, R. E., Tree yield and nut characteristics of pecan with drip irrigation under humid conditions, *J. Amer. Soc. Hort. Sci.,* 107, 30, 1982.
62. Sparks, D., Predicting nut maturity of the pecan from heat units, *HortScience,* 24, 454, 1989.
63. Miyamoto, S., Consumptive water use of irrigated pecan, *J. Amer. Soc. Hort. Sci.,* 108, 676, 1983.
64. Romberg, L. D., Irrigation of pecan orchards, *Proc. S. E. Pecan Grow. Assoc.,* 53, 20, 1960.
65. Thomson, W. C., Scheduling pecan irrigations by computer, *Proc. Western Pecan Conf.,* in M. D. Bryant, ed., New Mexico State Univ., Las Cruces, 1974, 7.
66. Rieger, M. and Daniell, J. W., Leaf water relations, soil-to-leaf resistance, and drought stress in pecan seedlings, *J. Amer. Soc. Hort. Sci.,* 113, 789, 1988.
67. Taylor, R. M. and Fenn, L. B., Translocation of water within root systems of pecan, grape, and tomato, *HortScience,* 20, 104, 1985.
68. Worley, R. E. and Taylor, G. G., An abnormal nut splitting problem of pecan (*Carya illinoensis* Koch), *HortScience,* 7, 70, 1972.
69. Prussia, S. E., Campbell, D. T., Tollner, E. W., and Daniell, J. W., Apparent modulus of elasticity of maturing pecans, *Transac. ASAE,* 28, 1290, 1985.
70. Loucks, W. L. and Keen, R. A., Submersion tolerance of selected seedling trees, *J. For.,* 71, 496, 1973.
71. Dyer, J. M. and Cantrell, B. G., Field grafting of sweet pecan to bitter pecan rootstock in seasonally flooded bottomlands, *Tree Planters Notes,* 40, 25, 1989.
72. Andersen, P. C., Leaf gas exchange characteristics of eleven species of fruit crops in north Florida, *Proc. Fla. State Hort. Soc.,* 102, 229, 1989.
73. Schaffer, B., Andersen, P. C., and Ploetz, R. C., Responses of fruit crops to flooding, in *Horticultural Reviews,* Vol. 13, Janick, J., Ed., John Wiley & Sons, Inc., NY, 1992, chap. 7.
74. Smith, M. W. and Bourne, R. D., Seasonal effects of flooding on greenhouse-grown seedling pecan trees, *HortScience,* 24, 81, 1989.
75. Loustalot, A. J., Influence of soil moisture conditions on apparent photosynthesis and transpiration of pecan leaves, *J. Agr. Res.,* 71, 519, 1945.
76. Alben, A. O., Waterlogging of subsoil associated with scorching and defoliation of Stuart pecan trees, *Proc. Am. Soc. Hort. Soc.,* 72, 219, 1958.
77. Smith, M. W. and Ager, P. L., Effects of soil flooding on leaf gas exchange of seedling pecan trees, *HortScience,* 23, 370, 1988.
78. Sparks, D. and Payne, J. A., Winter injury in pecans, a review, *Pecan South,* 5, 57, 1978.
79. Sparks, D., Payne, J. A., and Horton, B. D., Effect of sub-freezing temperatures on bud break of pecan, *HortScience,* 11, 415, 1976.
80. Sparks, D. and Payne, J. A., Freeze injury susceptibility of non-juvenile trunks in pecan, *HortScience,* 12, 497, 1977.
81. Lang, G. A., Early, J. D., Martin, G. C., and Darnell, R. L., Endo-, para- and ectodormancy: physiological terminology and classification for dormancy research, *HortScience,* 22, 371, 1987.
82. Anonymous, United States Department of Agriculture, Bureau of Plant Industry, Division of Fruit and Vegetable Crops and Diseases, Albany, Ga, and Shreveport, La, 1959.
83. Grauke, L. J. and Pratt, J. W., Pecan bud growth and freeze damage are influenced by rootstock, *J. Amer. Soc. Hort. Sci.,* 117, 404, 1992.
84. Payne, J. A. and Sparks, D., unpublished, 1977.
85. Sharpe, R. H., Blackmon, G. H., and Gammon, N., Jr., Relation of potash and phosphate fertilization to cold injury of Moore pecans, *Proc. S. E. Pecan Grow. Assoc.,* 45, 81, 1952.
86. Jensen, R. E., Savage, E. F., and Hayden, R. A., The effect of certain environmental factors on cambium temperatures of peach trees, *J. Amer. Soc. Hort. Sci.,* 95, 286, 1970.

87. Waite, M. D., Factors influencing the setting of nuts and fruits, *Proc. Natl. Pecan Growers' Assoc.*, 24, 122, 1925.
88. Finch, A. H. and Van Horn, C. W., Notes on the relation of warm winter temperatures to blossoming and nut setting of the pecan, *Proc. Amer. Soc. Hort. Sci.*, 37, 493, 1939.
89. Gammon, N., Jr. and Sherman, W. B., Bud development evaluation, *Pecan Quart.*, 6, 11, 1972.
90. Couvillon, G. A. and Erez, A., Effect of level and duration of high temperatures on rest in the peach, *J. Amer. Soc. Hort. Sci.*, 110, 579, 1985.
91. Richardson, E. A., Seeley, S. D., and Walker, D. A., A model for estimating the completion of rest for 'Redhaven' and 'Elberta' peach trees, *HortScience*, 9, 331, 1974.
92. Arnold, C. E., Pecans in central and south Florida, *Proc. Fla. State Hort. Soc.*, 84, 345, 1971.
93. McEachern, G. R., Wolstenholme, B. N., and Storey, J. B., Chilling requirements of three pecan cultivars, *HortScience*, 13, 694, 1978.
94. Amling, H. J. and Amling, K. A., Onset, intensity, and dissipation of rest in several pecan cultivars, *J. Amer. Soc. Hort. Sci.*, 105, 536, 1980.
95. Amling, H. J. and Amling, K. A., Physiological differentiation of pistillate flowers of pecan and cold requirements for their initiation, *J. Amer. Soc. Hort. Sci.*, 108, 195, 1983.
96. Smith, M. W., Carroll, B. L., and Cheary, B. S., Chilling requirement of pecan, *J. Amer. Soc. Hort. Sci.*, 117, 745, 1992.
97. Sparks, D., Chilling and heating model for pecan budbreak, *J. Amer. Soc. Hort. Sci.*, 118, 29, 1993.
98. McCain, D. C., Croxdale, J., and Markley, J. L., Thermal damage to chloroplast envelope membranes, *Plant Physiol.*, 90, 606, 1989.
99. Van Staden, J., Gilliland, M. G., and Dimalla, G. G., The effect of temperature on the mobilisation of food reserves of pecan nuts, *Z. Pflanzenphysiol. Bd.*, 91, 415, 1979.
100. Yates, I. E. and Sparks, D., Hydration and temperature influence in vitro germination of pecan pollen, *J. Amer. Soc. Hort. Sci.*, 114, 599, 1989.
101. Marquard, R. D., Pollen tube growth in *Carya* and temporal influence of pollen deposition on fertilization success in pecan, *J. Amer. Soc. Hort. Sci.*, 117, 328, 1992.
102. Yates, I. E., Sparks, D., Conner, K., and Towill., L., Reducing pollen moisture simplifies long-term storage of pecan pollen, *J. Amer. Soc. Hort. Sci.*, 116, 430, 1991.
103. Wood, B. W. and Payne, J. A., Net photosynthesis of orchard grown pecan leaves reduced by insecticide sprays, *HortScience*, 21, 112, 1986.
104. McCarthy, M. A. and Meredith, F. I., Nutrient data on chestnuts consumed in the United States, *Econ. Bot.*, 42, 29, 1988.
105. Anonymous, United States Department of Agriculture, Composition of foods: nut and seed products; raw, processed, prepared. *U.S.D.A. Agric. Handb.*, 8–12, 1984.
106. Andersen, P. C., unpublished data, 1990.

V. PISTACHIO

A. DISTRIBUTION AND BOTANY

The common pistachio, *Pistacia vera* L., is one of perhaps ten species with edible nuts in the genus *Pistacia.* The pistachio belongs to the family (Anacardiaceae), which also includes the cashew nut, mango, and poison ivy. Most *Pistacia* species are indigenous to the Old World; only two are indigenous to the United States.[1]

Pistachio probably originated in Central Asia, although it has often been referred to throughout recorded history in Western Asia, Asia Minor, and in the Mediterranean countries.[2] Wild pistachio trees remain an important component of the ecosystems in these regions.[2] The native range of pistachio is characterized by long, hot, dry summers and moderately cool or cold winters. Although pistachio will grow in rocky shallow soil, productivity is enhanced in a deep well-drained soil and by proper cultural practices. Turkey, Iran, Afghanistan, Italy, Syria, and the United States (Sacramento and San Joaquin Valleys of California) account for much of the production; lesser quantities are produced in Greece, India, Lebanon, Pakistan, and Tunisia.

The pistachio is a moderate-sized deciduous tree (ca. 7 to 10 m) characterized by imparipinnate leaves.[1] Trees are dioecious and the inflorescences (male and female) are panicles that may consist of several hundred individual flowers. The fruit is borne laterally on the previous season's growth. Pistillate and staminate flowers must bloom concurrently to allow for adequate cross-pollination.[2] Pistachio inflorescences exhibit strong apical dominance. Terminal flowers of the inflorescence are responsible for most of the crop, although they represent only 8% of the flowers.[3] This, and the tendency to bear heavy and light crops in alternate years[4–6] are two characteristics that limit productivity. The reader is referred to Crane and Maranto[4] or Joley[2] for further general information on pistachio.

Quantitative data concerning the environmental physiology of pistachio has been essentially limited to the most obvious variables that would tend to limit productivity (i.e., drought and salinity) in the cultivated range. Several physiological adaptations have been demonstrated that facilitate growth, yield, and survival of pistachio in regions where few other fruit or nut crops can be successfully grown.

B. LIGHT

Net CO_2 assimilation characteristics of *P. vera* have not been vigorously evaluated under controlled environment conditions, although the available data indicate that light-saturated values of A occur at relatively high levels of irradiance. Behboudian et al.[7] reported that light-saturated A was ca. 17 μmol m^{-2} s^{-1} and occurred at a PPF of ca. 1000 μmol m^{-2} s^{-1}, although detailed data were not presented. Net CO_2 assimilation of *P. vera* and *P. atlantica* Desf. under field conditions in California, USA was maximum (ca. 20 μmol m^{-2} s^{-1}) during midday when PPF exceeded 1700 μmol m^{-2} s^{-1}.[8] Both species are indigenous to arid regions of Asia. Rates of leaf gas exchange of a *Pistacia* species indigenous to regions of high summer rainfall (*P. integerrima* Stewart) were much lower, although A_{max} also occurred at a PPF > 1700 μmol m^{-2} s^{-1}.[8] High rates of leaf gas exchange of *P. vera* and *P. atlantica* may be based on a tendency to benefit from relatively short periods of favorable soil moisture.[9]

C. WATER

1. Water Requirements

P. vera is drought tolerant, although productivity may be enhanced by maintaining moderate levels of soil moisture. The influence of soil moisture tension on the growth and morphological characteristics of *P. vera* cvs. Fandoghi and Badami has been evaluated in a glasshouse.[10] Leaf, stem and root dry weight, stem length, the number of leaves, and root dry weight were generally not reduced when seedlings were irrigated at a soil moisture tension of -100 kPa compared to -35 kPa, although irrigation at -500 kPa tended to reduce the growth variables.

The performance of *P. vera* cv. Kerman on *P. atlantica* was evaluated under five irrigation regimes [0, 25, 50, 75, and 100% of ET] over a three-year period in a commercial orchard.[11] Although nut weight, nut size, number of nut clusters per tree, nuts per cluster, shoot length, and rachis size were reduced at 0, 25, and 50% ET; tree performance was similar when irrigation was scheduled at 75 or 100% ET. Phene et al.[12] applied drip irrigation to *P. vera* cv. Kerman on *P. atlantica* in the San Joaquin Valley and recorded a reduction in yield at an ET of 25% compared to 50, 70, or 100% ET; however nut quality (as quantified by the number of blank, aborted, non-split and split nuts) was reduced at 25 and 50% ET compared to 75 or 100% ET. Goldhamer et al.[11] concluded that nut abortion and the formation of blank nuts were the physiological processes most sensitive to mild drought stress (75% ET scheduling), while trunk growth, shoot growth, and yield were affected by more severe drought stress.

In the Sacramento Valley of California mature pistachio trees with canopies covering 60% of the orchard floor use about 100 cm of water during an average season.[4] Spiegel-Roy et al.[13] evaluated the growth and yield of *P. vera* Israel 502, ‘Kerman’ and ‘Larsen’ on two seedling rootstocks *P. atlantica* and *P. vera* over an 8-year period in the Negev Desert (Israel). During extreme drought (1977), when soil moisture content was below 5% v/v (i.e., Ψ_s -1.5 MPa) for a 19-month period, shoot growth was curtailed by May 1, while trunk cross-sectional area (TCA) increased through August. Despite a low yield during a year with extreme drought, flower buds differentiated to allow for a high yield the following year.

Yield of *P. vera* cvs. on ‘Kerman’ and ‘Larsen’ rootstocks in the Negev desert were not greatly influenced by the rootstock (*P. vera* vs. *P. atlantica*), although all scion/rootstock combinations exhibited a clear pattern of alternate bearing.[13]

There may be sufficient genetic variation to warrant selection of superior seedling rootstocks for a particular environment. Crane and Iwakiri[3] identified several trees of ‘Kerman’ on *P. atlantica* seedling rootstocks with superior growth and yield characteristics. Johnson and Weinbaum[15] evaluated 113 of ‘Kerman’/*P. atlantica* seedlings and found that TCA, yield, and yield efficiency (yield/TCA) varied among individual trees 4-, 8-, and 3-fold, respectively. Great variations also occurred in the tendency to bear heavy and light crops in alternate years.

2. Water Deficits

The evolution of *P. vera* in a semi-arid environment is consistent with the acquisition of various physiological and morphological adaptations to drought. *Pistacia* spp. possess pinnately compound leaves comprised of 1 to 20 pairs of individual leaflets and may be oriented horizontally, vertically or randomly to sunlight.[7] Leaves of *P. vera* are isolateral and randomly oriented (which facilitate maximum light penetration), in contrast to the dorsoventral morphology of most *Pistacia* spp.[8] A xerophytic characteristic of leaves of *Pistacia* spp. is a well-developed palisades tissue.[16] The ratio of abaxial to adaxial stomata varied from 1.3 in (*P. vera*) to 13.3 (*P. integersima*).[8] (A 3.1 ratio for *P. vera* has also been reported.[7]) In some species (i.e., *P. atlantica, P. chinensis*) adaxial stomata are restricted to areas adjacent to leaf veins; some *Pistacia* spp. only possess abaxial stomata (i.e., *P. lentiscus, P. mexicana, P. weinmannifolia*).

Stomatal conductance to water vapor of most *Pistacia* spp. was generally several times greater for abaxial compared to adaxial leaf surfaces; however, consistent with isolateral leaf morphology abaxial and adaxial g_s were similar for *P. vera.*[8] Maximum reported values of g_s of *P. vera* (400 mmol m^{-2} s^{-1})[8,17] are rather high compared to many woody C_3 plants;[18] yet similar to maximum g_s of many species of temperate fruit or nut trees.[19–23] The maximum g_s of *P. vera* was greater than or equal to the g_s of other *Pistacia* spp., in part due to a high adaxial conductance.[8]

Lin et al.[8] noted that *Pistacia* species in California did not manifest a midday depression in g_s or E, although leaf to air VPD were not given. The relationship of A and g_s was curvilinear for *P. vera, P. atlantica,* and *P. integerrima*. Turner et al.[17] working with container-grown *P. vera* showed that g_s and Ψ_L was reduced linearly with increasing VPD from 1.0 to 3.0 kPa, typical of many woody species; A declined from 14 to 11 μmol m^{-2} s^{-1} under these conditions.

Turner et al.[17] showed that reductions in g_s and Ψ_L with increases in VPD were greater for woody species (including *P. vera*) compared to herbaceous species tested; this is indicative of a relatively low hydraulic conductance of woody species. They also found that the reduction in g_s with a decline in Ψ_L ($\Delta g_s/\Delta\Psi_L$) of *P. vera* was greater than other woody species tested which would result in a greater degree of water conservation.[17] Consistent with a high degree of drought tolerance, Behboudian et al.[7] reported a positive A was recorded at a Ψ_L of -5.0 MPa.

Behboudian et al.[7] monitored changes in the ratio of total resistance to diffusion of H_2O vapor compared to CO_2 vapor as an index of WUE and found that a reduction in WUE occurred at a $\Psi_L <$ -3.0 MPa. Also, during later stages of drying cycle they reported the occurrence of nonstomatal inhibition of photosynthesis; however, this interpretation should be viewed with caution in the absence of data confirming stomatal homogeneity.[24,25]

Turgor maintenance at low Ψ_L is a xerophytic characteristic of *P. vera.* Behboudian et al.[7] calculated a Ψ_P of 0.3 MPa from thermocouple psychrometer-derived values of Ψ_L (-6.0 MPa) and Ψ_π (-5.7 MPa). In another study, *P. vera* had the lowest Ψ_π (i.e., -4.5 MPa) of any of over 30 native Asian species tested.[26] Positive turgor at such negative Ψ_L support the suggestion that *P. vera* is able to tolerate drought conditions better than other fruit or nut crops.[13]

D. TEMPERATURE

A lack of sufficient chill unit accumulation during the winter, or heat units accumulation during the summer, and the occurrence of spring frosts and minimum winter temperatures are environmental factors that may limit the range of pistachio cultivation. However, few controlled studies concerning the influence of temperature on growth and physiological processes have been conducted.

The amount of winter chilling required for proper growth and development is cultivar dependent. Pistachio trees cv. Kerman require ca. 1000 hrs of chilling below 5.8°C for satisfactory vegetative and reproductive growth.[27] Thus, pistachio cultivation should be restricted to regions averaging at least 1500 hrs of winter temperatures 5.8°C or less. Symptoms of insufficient winter chilling on cv. Kerman included delayed and irregular bloom and foliation, the production of leaves with a reduced number of leaflets, stem dieback, reduced yield in the current year and perhaps in subsequent years.[27]

Pistachio trees require long hot summers; however, the critical number of heat units required for nut ripening has not been determined. Based on locations where production has been successful a minimum of 6 or 7 months averaging over 15.6°C are probably required to complete the fruit development period.[2] With insufficient heat-hour accumulation pistachio will set fruit that will not ripen.[2]

Spring frosts or minimum winter temperatures may not often be a limiting factor for *P. vera* in its native or cultivated range. For example, pistachio trees were not injured by −8°C in California, −18°C in Tehran, and as low as −30°C in Maryland, USA.[2]

The effects of high temperatures on growth and physiological processes of *Pistacia* species has received little attention. In Australia, Behboudian et al.[7] reported that A had an optimum temperature range of 25 to 28°C, although at higher temperatures data concerning VPD were not given. In California, A of three *Pistacia* spp. was maximum during midday at a temperature of 30°C.[8]

E. SALINITY

1. Growth Responses

Pistachio is often cultivated in locations (i.e., Central Asia, Asia Minor, Australia) with high soil salinity. Soil salinity may arise from native soils with a high soluble salt content, irrigation with water possessing a high salt content (i.e., high EC), long-term use of fertilizers with a high salt content, or by salt water intrusion.

Sepaskhah and Maftoun[10] studied the growth of *P. vera* 'Fandoghi' and 'Badami' at three irrigation intervals, i.e., water at a Ψ_s of (−35, −100, or −500 kPa) and five salinity levels (EC = 0.5, 1.5, 2.5, 3.5, or 4.5 dS m^{-1}) in a glasshouse. Salinity reduced leaf dry weight, leaf area, stem length, and plant ET; however there was an interaction between salinity and irrigation. Shoot growth for plants in the two higher irrigation regimes was reduced at an EC_e of 12 dS m^{-1}; when plants were irrigated at a $\Psi_s$of −500 kPa shoot growth was reduced at an $EC_e > 15$ dS m^{-1}.[10] In another glasshouse study, a 50% reduction in shoot and root growth was achieved at an EC_e of 7.9 dS m^{-1} for 'Fandoghi' and an EC_e of 9.3 to 10 dS m^{-1} for 'Badami' and 'Kale-ghoochi'; shoot growth ceased at an EC_e of 15.5 to 15.9 dS m^{-1} for 'Fandoghi' and 18.7 to 20.6 dS m^{-1} for 'Badami' and 'Kale-ghoochi'.[28] Sepaskhah et al.[29] suggested that the lower salinity tolerance of 'Fandoghi' compared to 'Badami' may be due to a reduced ability to osmoregulate, and/or the differential uptake, transport, or partitioning of Na and Cl in different tissues.

The remarkably high level of salt tolerance of *Pistacia* spp. was underscored by a study that showed no significant change in dry matter production of *P. atlantica, P. vera,* and *P. terebinthus* seedlings when treated with 150 m*M* Cl and 90 m*M* Na for a three-week period.[30] *P. atlantica* had higher rates of Na but not Cl uptake and transport compared to *P. vera*; however, in all three species the accumulation of Na and Cl was diluted by growth.[30] Under certain circumstances a critical concentration of Cl may eventually be reached as Ashworth et al.[31] found leaf scorch of pistachio was associated with a Cl concentration of 1.9%.

The accumulation of Na and Cl in *P. vera* was proportional to the level of salinity,[32] but was inversely related to irrigation frequency. Sodium accumulated predominantly in the roots, whereas Cl accumulated predominantly in stems and leaves. Salinity did not influence the tissue concentration of macroelements with the exception of a decline in root K concentration.[32] In another study, Ca but not K or Mg increased in leaves of several *Pistacia* spp. with increased salinity.[33]

Picchioni and Miyamoto[33] evaluated the intraspecific salt tolerance of *Pistacia* spp. in an outdoor lysimeter and found that in all species an EC_e of 8.7 dS m^{-1} was associated with a reduction of root and stem growth; and leaf growth was more sensitive to salinity than root growth. Although increasing salinity resulted in a higher root to stem ratio, particularly for *P. terebinthus*; clear differences in salinity tolerance among species were not demonstrated. Differences in leaf Na and Cl concentrations occurred between *Pistachia* spp.; however, all partitioned most of the Na to the roots and most of Cl to the leaves.[30,33] Unlike many glycophytes, *P. vera* does not exhibit a positive growth response with increasing salinity,[34] and does not accumulate Cl or Na in leaves to the extent of mangroves *Avicennia marina*,[35] but instead regulates ion distribution more similar to spinach.[36]

2. Physiological Responses

Salinity tolerance of *Pistacia* spp. has been quantified based on the quantity of ultraviolet (UV)-absorbing solutes leaking from root apices.[37] In lysimeter experiments solute leakage occurred at a comparable EC_e that resulted in an increased leaf Na (18.1 dS m^{-1}).[37] Cell injury was proportional to salinity and was greatest in a *P. terebinthus* selection with the least Na exclusion ability.[37] No differences in the leakage of UV-absorbing solutes occurred among the other *Pistachia* spp. or selections at a given salinity level. Most importantly, increases in the Na:Ca ratio greatly enhanced the leakage of UV-absorbing solutes at a given Ψ_s. Intensification of root damage with low Ca may be related to the stabilizing influence of Ca on cell membranes.[38]

The influence of higher levels of salt applications on the physiology of *P. vera* cv. Kerman have been tested under glasshouse conditions.[7,39] Walker et al.[39] raised the concentration of Cl by 25 m*M* daily up to 100 or 175 m*M* Cl (accompanied by Na:Ca:Mg in a 6:1:1 ratio). Although growth was reduced at 100 m*M* Cl and had ceased at 175 m*M* Cl; a positive Ψ_p in leaves was maintained due largely to an increase in the concentration of sucrose and monosaccharides in the short-term (i.e., 2 days). In the long-term (i.e., 40 days), sugars, Na, Cl and to a lesser extent proline and shikimic acid contributed to osmoregulation.[39] Leaf emergence eventually was terminated after treatment with 175 m*M* Cl and leaf Na and Cl increased to 255 and 142 m*M*, respectively; however, A and E were unaffected and total plant carbohydrate content increased.[39] They concluded that growth cessation was not due to limitations of photosynthesis, assimilate content, plant water relations, and Na or Cl concentration in expanding leaves.

Net CO_2 assimilation, leaf chlorophyll content and plant water relations (i.e., WUE, Ψ_l, Ψ_s, Ψ_p) of *P. vera* cv. Kerman were not influenced by Cl treatments of 75 m for 84 days and then 225 m*M* for 23 days.[7] When the concentration of Cl was increased to 400 m*M* in daily increments of 25 m*M*, both A and g_s were reduced substantially.[7] Behboudian et al.[7] suggested that reduced photosynthesis was not due to effects on the photosynthetic apparatus (i.e., similar reductions in A and g_s, and slight increase in WUE and no change in chlorophyll content), but rather a result of partial stomatal closure and a lowered Ψ_L (determined during the morning). In all cases Ψ_p was greater than 1.0 MPa suggesting that leaf dehydration (or loss of guard cell turgor) may not have been directly responsible for partial stomatal closure, although data concerning midday plant water status were not presented.

All available data indicates that *P. vera* (as well as many of the other *Pistacia* cultivars tested) are most tolerant of all fruit or nut crops to both drought or salinity, with the possible exception of date palm (*Phoenix dactylifera*).[40]

REFERENCES: PISTACHIO

1. Bailey, L. H. and Bailey, E. Z., *Hortus Third,* A concise dictionary of plants cultivated in the United States and Canada, Macmillan Publishing Co., New York, N. Y., 1976.
2. Joley, L. E., Pistachios, *Nut Tree Culture in North America,* 2nd ed., Jaynes, R. A., Ed., Northern Nut Growers Assoc., Inc., 1973, chap. 12.
3. Crane, J. C. and Iwakiri, B. T., Vegetative and reproductive apical dominance in Pistachio, *HortScience,* 20, 1092, 1985.
4. Crane, J. C. and Moranto, J., *Pistachio Production,* California pistachio industry annual report—Crop Year 1986–87, 1988, 1–13.
5. Crane, J. C., Catlin, P. B. and Al-Shalan, I., Carbohydrate levels in the Pistachio as related to alternate bearing, *J. Amer. Soc. Hort. Sci.,* 101, 371, 1976.
6. Takeda, F., Ryugo, K. and Crane, J. C., Translocation and distribution of C-photosynthates in bearing and nonbearing pistachio branches, *J. Amer. Soc. Hort. Sci.,* 105, 642, 1980.
7. Behboudian, M. H., Walker, R. R., and Torokfalvy, E., Effects of water stress and salinity on photosynthesis, *Scientia Hortic.,* 29, 251, 1986.
8. Lin, T. S., Crane, J. C., Ryugo, K., Polito, V. S., and DeJong, T. M., Comparative study of leaf morphology, photosynthesis, and leaf conductance in selected *Pistacia* species, *J. Amer. Soc. Hort. Sci.,* 109, 325, 1984.
9. Maximov, N. A., The physiological significance of the xeromorphic structure of plants, *J. Ecol.,* 19, 272, 1931.
10. Sepaskhah, A. R. and Maftoun, M., Growth and chemical composition of Pistachio cultivars as influenced by irrigation regimes and salinity levels of irrigation water. I. Growth, *J. Hortic. Sci.,* 56, 277, 1981.
11. Goldhamer, D. A., Phene, B. C., Beede, R., Scherlin, L., Mahan, S., and Rose, D., Effects of sustained deficit irrigation on Pistachio tree performance, *Proc. Fourth International Micro-Irrigation Congress,* Albury-Wodonga, Australia, Oct. 23–28, pp. 4B(1–8), 1988.

12. Phene, R. C., Menezes, J., Jr., Goldhamer, D. A., Aitkens, G. J., Beede, R., and Kjelgren, R., Irrigation scheduling of drip irrigated Pistachios, *Proc. of the Third International Drip/Trickle Irrigation Congress,* Nov. 18–21, 1985, pp. 805–810.
13. Spiegel-Roy, P., Mazigh, D., and Evenari, M., Response of Pistachio to low soil moisture conditions, *J. Amer. Soc. Hort. Sci.,* 102, 470, 1977.
14. Crane, J. C. and Iwakiri, B. T., Pistachio yield and quality as affected by rootstock, *HortScience,* 21, 1139, 1986.
15. Johnson, R. S. and Weinbaum, S. A., Variation in tree size, yield, cropping efficiency, and alternate bearing among 'Kerman' Pistachio trees, *J. Amer. Soc. Hort. Sci.,* 112, 942, 1987.
16. Shields, L. M., Leaf xeromorphy as related to physiological and structural influences, *Bot. Rev.,* 16, 399, 1950.
17. Turner, N. C., Schulze, E. D., and Gollan, T., The responses of stomata and leaf gas exchange to vapor pressure deficits and soil water content. I. Species comparisons at high soil water contents, *Oecologia,* 63, 338, 1984.
18. Korner, C. J., Scheel, A., and Bauer, H., Maximum leaf diffusive conductance in vascular plants, *Photosynthetica,* 13, 45, 1979.
19. Andersen, P. C., Leaf gas exchange of 11 species of fruit crops with reference to sun-tracking/non-sun-tracking responses, *Can. J. Plant Sci.,* 71, 1183, 1991.
20. Avery, D. J., Maximum photosynthetic rate—A case study in apple, *New Phytol.,* 78, 55, 1977.
21. DeJong, T. M., CO_2 assimilation characteristics of five *Prunus* tree species, *J. Amer. Soc. Hort. Sci.,* 108, 303, 1983.
22. Jones, H. G. and Cumming, I. G., Variation of leaf conductance and leaf water potential in apple orchards, *J. Hort. Sci.,* 59, 329, 1984.
23. Lakso, A. N., Precautions on the use of excised shoots for photosynthesis and water relations measurements of apple and grape leaves, *HortScience,* 17, 368, 1982.
24. Downton, W. J. S., Loveys, B. R., and Grant, W. J. R., Non-uniform stomatal closure induced by water stress causes putative non-stomatal inhibition of photosynthesis, *New Phytol.,* 110, 503, 1988.
25. Mansfield, T. A., Hetherington, A. M., and Atkinson, C. J., Some current aspects of stomatal physiology, Annu. Rev. Plant Physiol., *Plant Mol. Biol.,* 41, 55, 1990.
26. Sveshnikova, V. M., Osmotic pressure in plants of the Badkhyz Hill country, *Problems of Desert Development,* 2, 68, 1988.
27. Crane, J. C. and Takeda, F., The unique response of the Pistachio tree to inadequate winter chilling, *HortScience,* 14, 135, 1979.
28. Sepaskhah, A. R. and Maftoun, M., Relative salt tolerance of Pistachio cultivars, *J. Hortic. Sci.,* 63, 157, 1988.
29. Sepaskhah, A. R., Maftoun, M., and Yasrebi., J., Seedling growth and chemical composition of three Pistachio cultivars as affected by soil applied boron, *J. Hortic. Sci.,* 63, 743, 1988.
30. Walker, R. R., Torokfalvy, E., and Behboudian, M. H., Uptake and distribution of chloride, sodium and potassium ions and growth of salt-treated Pistachio plants, *Aust. J. Agric. Res.,* 38, 383, 1987.
31. Ashworth, L. J., Jr., Gaona, S. A., and Surber, E., Nutritional diseases of Pistachio trees: potassium and phosphorus deficiencies and chloride and boron toxicities, *Phytopathology,* 75, 1084, 1985.
32. Sepaskhah, A. R. and Maftoun, M., Growth and chemical composition of Pistachio seedlings as influenced by irrigation regimes and salinity levels of irrigation water. II. Chemical composition, *J. Hortic. Sci.,* 57, 469, 1982.
33. Picchioni, G. A. and Miyamoto, S., Salt Effects on growth and ion uptake of Pistachio rootstock seedlings, *J. Amer. Soc. Hort. Sci.,* 115, 647, 1990.
34. Flowers, T. J., Troke, P. F., and Yeo, A. R., The mechanisms of salt tolerance in halophytes, *Annu. Rev. Plant Physiol.,* 28, 89, 1977.
35. Downton, W. J. S., Growth and osmotic relations of the mangrove *Avicennia marina,* as influenced by salinity, *Aust. J. Plant Physiol.,* 9, 519, 1982.
36. Downton, W. J. S., Grant, W. J., and Robinson, S. P., Photosynthetic and stomatal responses of spinach leaves to salt stress, *Plant Physiol.,* 77, 85, 1985.
37. Picchioni, G. A. and Miyamoto, S., Rapid testing of salinity effects on Pistachio seedling rootstocks, *J. Amer. Soc. Hort. Sci.,* 116, 555, 1991.
38. Levitt, J., Salt and ion stresses, *Responses of plants to environmental stresses,* 2nd ed., Vol. 2., Academic, New York, 1980, 365.

39. Walker, R. R., Torokfalvy, E., and Behboudian, M. H., Photosynthetic rates and solute partitioning in relation to growth of salt-treated Pistachio plants, *Aust. J. Plant Physiol.*, 15, 787, 1988.
40. Maas, E. V. and Hoffman, G. J., Crop salt tolerance-current assessment, Irri. Drainage Div., *Amer. Soc. Civil Eng. J.*, 103, 115, 1977.

VI. WALNUTS

A. DISTRIBUTION AND BOTANY

The genus *Juglans* (family Juglandaceae) consists of ca. 20 species of deciduous, monoecious trees that are native to North and South America, southeastern Europe, and eastern Asia.[1] *Juglans* spp. possess large aromatic pinnate leaves. Male flowers (catkins) are borne on previous season's wood, while female flowers (racemes) are borne on current year's wood. The fruit is a drupe enclosed within a thick indehiscent husk. The edible seed is 2 to 4 lobed, and is retained within the shell until germination. The somatic chromosome number ($2n = 32$) is similar to that of pecan. *Juglans* spp. contain jugalone, (a naphthoquinone), mainly in leaves, fruit, and bark.[2,3]

The Persian walnut (*Juglans regia* L.) is believed to have originated near the Caspian Sea in Iran and was transported to many regions of Europe and Asia, and more recently to the Americas.[4] Persian walnut trees are medium-sized but tend to have a very spreading growth habit and grow best on fertile, well-drained soils with a pH 6 to 7. Clones of *J. regia* are self-fertile, although asynchronous production of pollen and receptivity of the stigma (dichogamy) can reduce pollination efficiency and yield.[5] Grafted Persian walnut trees are fairly precocious for a nut tree and may bear the first crop in 2 to 5 years. A unique and desirable characteristic of *J. regia* compared to other *Juglans* spp. is that the husk separates from the nut as the nut ripens.

The black walnut (*J. nigra* L.) is a large tree (height > 30 m) native to the eastern United States and southern Canada and is important for its valuable timber and nut crop. *J. nigra* typically occurs on mesic sites with deep well-drained soils.[7] Native stands of *J. nigra* account for a substantial proportion of lumber and nuts harvested. Often plantings are established at relatively close spacings and eventually alternate trees are cut for lumber, while remaining trees are retained for nut production and cut at a later date. Seedling trees are often planted, although many cultivars have been clonally propagated.[6] Like *J. regia, J. nigra* is self-fruitful but dichogamy can reduce self-pollination efficiency.

Other *Juglans* spp. of some commercial importance include the Japanese walnut (*J. sieboldiana*), the Manchurian walnut (*J. mandschurica*), the southern California black walnut (*J. californica*), the northern California black walnut (*J. hindsii*), and the butternut (*J. cinerea*); however, there is a paucity of data concerning the environmental physiology of these species. The majority of studies on environmental physiology of *J. regia* and *J. nigra* concern the influence of light exposure and drought; the impact of other environmental stresses have not been adequately investigated.

B. LIGHT

1. Vegetative and Reproductive Growth

The influence of irradiance level on fruiting, specific leaf weight (SLW), subsequent spur productivity and nitrogen and carbon allocation has been studied for *J. regia* in California. Light distribution in the interior canopy, although not studied, is likely a growth limitation of *J. nigra* given that the amount of light incident on the ground below the canopy is low relative to other species.[9]

Mature Persian walnut orchards have been reported to be among the most shaded of any fruit or nut crop.[8] Ryugo et al.[10] reported that a single walnut leaf transmitted only ca. 50 of 2000 $\mu mol\ m^{-2}\ s^{-1}$ incident PPF, and the second leaf transmitted only 10 of 50 $\mu mol\ m^{-2}\ s^{-1}$. Although no data were given on differences in light quality after passage through a leaf, the ratio of far red to red light is generally enhanced after passage through a leaf. Heavily shaded (PPF = 10 to 30 $\mu mol\ m^{-2}\ s^{-1}$) spurs of *J. regia* cv. Hartley had a lower SLW, leaf chlorophyll concentration, number of leaves per spur, leaf area per spur, spur length, spur dry weight, dry weight per unit length of spur, dry weight per nut, number of nuts per spur, and dry weight of nuts per dry weight of spur than sun-exposed (PPF = 600 to 1240 $\mu mol\ m^{-2}\ s^{-1}$) spurs.[10] However, dry weight per kernel, oil weight per kernel, percentage oil content in nuts, percentage starch, and soluble carbohydrate concentration in spurs were not greatly influenced by light exposure.[10] In shaded, but not in sun-exposed spurs, the accumulation of starch ceased when nuts began accumulating dry matter; in sun-exposed spurs starch increased up until nut harvest in the fall.

Klein et al.,[11] found that spurs of *J. regia* cvs. Serr and Hartley could be segregated into two categories: (1) sun-exposed for a large portion of the day (i.e., PPF > 700 μmol m^{-2} s^{-1}), or (2) shaded for most of the day (i.e., total PPF between 100 and 700 μmol m^{-2} s^{-1} occurring for only 1 hr per day and the remainder of the day PPF < 100 μmol m^{-2} s^{-1}). Both leaf nitrogen and SLW were allocated on the basis of light exposure. For example, leaf nitrogen was about 2 and 1 μg mm^{-2} for sun and shade leaves, respectively; SLW was ca. 6 to 9 and 2 to 5 mg cm^{-2} for sun and shade leaves, respectively.

Spurs of *J. regia* cv. Hartley that were sun-exposed the previous season had a higher return bloom and growth rate.[10] Similarly, spurs of *J. regia* cvs. Serr and Hartley that were shaded the previous season (inferred by a SLW < 4 mg cm^{-2}) sustained greater mortality during the winter, which was especially great on shaded spurs that had borne fruit the previous season.[12] Fruit set was enhanced on spurs with the highest SLW and N per unit leaf area (i.e., those that were sun-exposed the previous season), although kernel weight, kernel N, and kernel oil concentration did not vary with SLW in either cultivar. Thus, after fruit set, the level of light exposure the previous season did not generally have an overriding effect on kernel size or quality.[9,11]

The importance of light distribution on bearing of *J. regia* has also been supported by studies on pruning.[13] Flower bud initiation occurred about 4 weeks after they were formed on *J. regia* cv. Chico. Pruning accelerated flower-bud differentiation, and flower-bud differentiation was not reduced by defoliation and/or etiolation. Pruning also increased light penetration and nut distribution through the canopy, nut size and percent edible kernel of *J. regia* cv. Ashley.[13] Yield of biennially pruned trees was similar to that of annually pruned or non-pruned trees in the year following pruning, but was often greater in the alternate year.

2. Leaf Gas Exchange

Photosynthetic characteristics of *J. regia* leaves are very similar to those reported for many fruit species. Maximum rates of A of *J. regia* cv. Serr measured on leaves of excised shoots in the laboratory or on mature trees in the field were approximately 13 μmol m^{-2} s^{-1}.[14] Light-saturated A occurred at a PPF of ca. 800 μmol m^{-2} s^{-1}. Net CO_2 assimilation and g_s were reported to be linearly related, although the relationship appeared to be curvilinear at high values of A.

C. WATER

1. Vegetative and Reproductive Growth

Based upon native tree distribution *Juglan* spp. are mesophytes. Irrigation is required in many areas of cultivation to achieve maximum productivity.

J. regia grown in the arid climate of the central valleys of California are often exposed to atmospheric conditions conducive to high levels of moisture stress (i.e., high light and temperature and low relative humidity), which when coupled with inadequate irrigation may reduce tree growth, yield, and quality.[15] Martin et al.[15] compared Ψ_L, tree growth, and kernel weight and quality of *J. regia* cv. Ashley after one growing season without irrigation. Reductions in Ψ_L at sunrise (ca. 0.2 MPa), trunk growth (ca. 50%), kernel weight (ca. 20%), and no difference in kernel quality occurred by the end of the growing season without irrigation, although levels of soil moisture tension were not recorded. Martin et al.[15] reported that the wilting point of container-grown seedlings occurred at a Ψ_L of −1.9 MPa. The following year, when all trees received irrigation, kernel weights were actually greater for trees not receiving irrigation the previous year; thus the initiation of flower buds was not inhibited by drought. Kernel quality, as determined by the percentage of lightly colored and edible kernels was not affected by drought. However, in a similar trial with cv. Serr, both kernel weight and kernel quality were reduced without irrigation and poor quality nuts occurred in sun-exposed positions of the tree.[16]

In France, yield of *J. regia* cv. Franquette irrigated at 100% ET by sprinkler irrigation or 100, 75, or 50% ET by drip irrigation was increased from 53 to 78% compared to the non-irrigated treatment.[17] Irrigation improved kernel color but had no effect on leaf mineral composition.

Return bloom of field-grown trees,[15,16] and container-grown trees,[18] was unaffected by drought occurring during the previous season. However, phenology and bud dormancy can be influenced by periods of drought. For example, a period of spring drought promoted the growth of normally quiescent buds during the summer.[18]

2. Plant Moisture Deficits

The available database for *J. nigra* has been expanded over other *Juglan* spp. largely due to the fact that it is one of the most valuable timber species in North America. In its native range it is often codominant on moist (not wet) sites with species such as sugar maple (*Acer saccharum* Marsh.).

The sensitivity of leaf stomata of *J. nigra* to drought has typically been assessed on: (1) excised leaves or shoots exposed to extremely rapid drying; (2) container-grown plants exposed to fairly rapid drying; and (3) field-grown trees experiencing gradual drying. The more rapid the desiccation, the less likely that physiological and biochemical changes will occur that will influence stomatal behavior. Thus, plant responses to water stress will depend on the rate of imposition as well as the severity of the water deficit.

Davies and Kozlowski[19] evaluated the response of six forest tree species grown in containers to a single drying cycle in a growth chamber, (i.e., 25°C/20°C day/night temperatures and 70–90% relative humidity). Prior to the imposition of drought, g_s of *J. nigra* (ca. 160 mmol m^{-2} s^{-1}) was at least twice that of *Fraxinus americana, Acer rubrum, Cornus amomum,* and *Ulmus americana* and 75% greater than *Acer saccharium.* Stomata of *Acer* spp. were most sensitive to water stress and closed at relative high Ψ_L, while stomata of *J. nigra* and remained relatively open until a Ψ_L of <-1.0 MPa. *J. nigra* exhibited the least control over stomatal aperture, and upon rewatering was the only species tested that did not regain turgor. Net CO_2 assimilation for *J. nigra* has been reported to fall to zero at an Ψ_L = -1.9 to -2.2 MPa.[19] Similarly, nearly complete stomatal closure for container-grown, three-month-old seedlings of *J. nigra* occurred at a Ψ_s of -0.5 MPa; leaf abscission ensued at a Ψ_L of -2.2 MPa.[20] Based on the tendency to undergo leaf abscission and the above data the authors concluded that *J. nigra* could only be competitive in sites where water was readily available throughout the growing season.[19]

In contrast, field-grown *J. nigra* in Missouri were unique among 8 forest tree species in that low Ψ_L (predawn or mid day) was never achieved and g_s remained relatively high as drought increased.[21] Drought avoidance by deep rooting and by a gradual abscission of older leaves (starting in late July) was invoked as drought avoidance mechanisms for *J. nigra.*[21,22] An acropetal progression of leaf abscission has been considered to be an adaptive response to drought since basal leaves with a lower photosynthetic potential are shed first while the root to shoot ratio is enhanced.[23,24]

The anatomy of xylem vessels is largely based on a compromise between a high efficiency of water conduction (favoring long and wide vessels) and resistance to the formation and passage of gaseous emboli (favoring short and narrow vessels often with bordered pits conferring high resistance).[25] The greater resistance to water flow between soil and leaves (R) and water potential gradients between the root and crown of mature *J. nigra* compared to a gymnosperm, *Juniperus virginiana* has been attributed to a high resistance to water flow through the bordered pits of tracheids of *J. virginiana.*[22] In diffuse porous wood such as in *Vitis vinifera* L. with long wide vessels, zones of protection against cavitation reside largely in the nodes, whereas in ring porous trees (i.e., *J. nigra*) short narrow vessels accompany the larger vessels along the whole shoot.

Ni and Pallardy[26] compared the R of glasshouse-grown seedlings of J. nigra to R of *Quercus alba, Q. stellata* Wangenh. and *Acer saccharum* under high ($\Psi_s \geq -0.3$ MPa) and low ($\Psi_s < -1.5$ MPa) soil moisture conditions. *J. nigra* had greater R at high soil moisture conditions and a greater rate of increase in R with soil drying than the other spp., despite drastic reductions in E. There was no evidence of the formation of a substantial number of emboli in xylem vessels of these species; a great portion of the increased resistance to water flow was attributed to increasingly prevalent air gaps formed across the soil/root interface.[26] The contradictory results discussed for container-grown seedlings[26] compared to mature trees[22] was likely due to: (1) the rate of soil drying; (2) the probability that the young seedlings had not yet adequately explored the soil; and (3) differences in the root to shoot ratio and carbon partitioning between seedlings and mature trees.

The potential to resist damage during dehydration is thought to be associated with maintenance of membrane integrity during dehydration.[27] A stress-induced loss of membrane integrity results in an efflux of solute from cells.[28] Dehydration tolerance of *J. nigra* and other forest tree species have been evaluated by quantifying electrolyte leakage from leaf discs.[23] Drought-tolerant *Quercus* spp. manifested less electrolyte leakage than drought-susceptible *J. nigra* or *A. saccharum.*[23] As drought intensified, seedlings of *J. nigra* and *A. saccharum* showed no and little tendency for drought hardening, respectively.

J. nigra seedlings obtained from seven geographic origins were subjected to six different levels of drought stress.[29] The survival of stem cambial tissue of *J. nigra* to a predawn Ψ_L (Ψ_{LPD}) as low as -4.0 MPa was facilitated by leaf abscission in an acropetal direction. More than 80% of leaves abscised from nearly all plants from each geographic region at a Ψ_{LPD} of −3.5 MPa or less. No differences in mortality (near 0%) or leaf abscission patterns occurred between genotypes. The capacity for both osmotic adjustment and increased tissue elasticity in eight seedling sources of *J. nigra* varied from absent to substantial, depending upon seed source. Osmotic potential at turgor point loss was reduced by 0.7 to 0.8 MPa which would facilitate growth by cell expansion and permit physiological processes such as continued leaf gas exchange at a lower Ψ_L than would be possible in the absence of changes in solute concentration. The relative contribution of increased solute accumulation and tissue elasticity to dehydration tolerance was also genotype-specific.

3. Water Use Efficiency

Ni and Pallardy[20] tested the hypothesis that higher rates of A, g_s, and WUE (defined as A/g_s) under conditions of low soil moisture occur for xeric (i.e., *Quercus alba* and *Q. stellata*) compared to mesic (*J. nigra* and *A. saccharum*) species. They found that seedlings of *Quercus* spp. had higher gas exchange rates at both high and low Ψ_s; however, there was no trend for increasing WUE in the more xeric species. Rates of leaf gas exchange were relatively low and WUE relatively high for *J. nigra* seedlings. They suggested that a high WUE under dry conditions may actually serve as a disadvantage since water may ultimately be used by competing plant species.[20]

Parker and Pallardy[30] measured a decrease in WUE with drought progression for four seedling sources of *J. nigra,* which they attributed to nonstomatal inhibition of A. This conclusion must be viewed with caution without data confirming stomatal homogeneity and data characterizing chlorophyll fluorescence.[31-35] Stomatal conductance, E, or A measured by infrared gas analyzers is a volume-averaged value. If leaves are not homogeneous (or relatively homogeneous) in respect to stomatal aperture calculated values of WUE and intercellular CO_2 concentrations are erroneous as a result of the curvilinear relationship between A and g_s. The consequence of "stomatal patchyness" is to underestimate WUE and overestimate C_i if measurements/calculations are made with conventional gas exchange instrumentation. Since A and g_s were more closely related to Ψ_s than to Ψ_L Parker and Pallardy[30] invoked the possibility of a nonhydraulic sensing mechanism or signal occurring from the root to the shoot as has been previously reported for root synthesized abscisic acid in corn.[36]

4. Root Growth

J. nigra grows best in deep, well-drained soils with a high moisture holding capacity. Although deep rooted, the majority of the root system of *J. nigra* is located in the upper 60 cm of the soil.[37] The growth of roots of deciduous trees may occur all year which may be particularly important for *J. nigra* given the high minimum Ψ_L, the tendency for leaves to abscise under conditions of drought and the importance of depth of rooting for this species.[22] Indeed, Frossard et al.[38] found that the taproot of *J. regia* cv. Franquette seedlings was a major storage organ (particularly the central parenchyma) for both water and carbon.

Increases in root length, number of growing roots, and suberization were evaluated in a rhizotron in relation to Ψ_s and soil temperature of mature *J. nigra.*[39] Root growth, the number of growing roots, and the rate of suberization decreased markedly as Ψ_s decreased from 0 to −0.5 MPa, and reached zero at a Ψ_s of −0.5 to −1.0 MPa. Root growth began at a soil temperature of 4°C and maximum root growth and the number of growing roots were recorded at 17 to 19 and 21°C, respectively.[39]

For *J. nigra* the Ψ_π at full hydration of leaf tissue was much greater than that of root tissue, indicating that the root system is capable of releasing a larger quantity of water than leaves in response to decreasing Ψ_L.[24] This "capacitance" of root tissue, would serve as a substantial reservoir of water available to shoot tissue, and should be viewed as both a seasonal and a diurnal adaptation to plant moisture stress. In addition, the reduction in root Ψ_π reported in genotypes of *J. nigra* would favor turgor maintenance and permit the absorption of soil moisture at a lower Ψ_s.[29] The authors concluded that there was sufficient variation in tissue water relations of genotypes of *J. nigra* to justify selection of specific genotypes to specific regions.

Soil amelioration via soil mixing, subsoiling, or the addition of soil amendments have not resulted in a cost effective increase in tree growth.[37] Dey et al.[41] found that decreases in plant growth of *J. nigra* were associated with reduced Ψ_s and Ψ_L when multicropped with either tall fescue (*Festuca arundinaceae* Schreb.), orchard grass (*Dactyles glomerata* L.), or lespedeza (*Lespedeza striata* Thunb. cv. Kobe).

Several studies have documented the importance of mycorrhizae on growth of *J. nigra.*[42,43] The inoculation of *J. nigra* with vesicular-arbuscular mycorrhizal fungi (VAM) improved growth and increased leaf retention at low levels of available soil phosphorous (i.e., ≤ 50 ppm), but not at high levels of available phosphorous (i.e., 75 ppm).[42] Ponder[44] suggested that the frequency of irrigation altered the ratio of introduced and indigenous mycorrhizal fungi on *J. nigra.*

5. Flooding

The tolerance of *Juglans* spp. to flooded soil conditions is dependent on various biotic (e.g., tree age, nutritional status, and phenological stage of developments) and abiotic factors (e.g., temperature, VPD, soil porosity, and chemistry).[45] Flooding tolerance is often assessed on a relative basis. For example, plant survival after 4 weeks of soil flooding was greater for pecan [*Carya illinoensis* (Wagenh.) K. Koch] than for *J. nigra.*[46] Survival and growth of *J. siebaldiana, J. cinerea,* and *J. mandshurica* has been reported to be good in a floodplain habitat with 20 days or less of flooding annually.[47] Observational data collected in France has indicated that *J. nigra* may be more tolerant to waterlogging than *J. regia.*[48] By contrast, survival of container-grown *J. nigra* and *J. hindsii* was about equal with 50% of the trees surviving after 12 to 16 days of flooding at a root-zone temperature of 23°C.[49] Catlin et al.[50] reported that seedlings of *J. hindsii* and *J. regia* were much more sensitive to flooded soil conditions than *Pterocarya stenoptera* DC; Paradox seedlings (hybrids between *J. hindsii* and *J. regia*) were more flood tolerant than either parent but more flood sensitive than *P. stenoptera.* After 4 to 5 days of flooding at a root-zone temperature of 33°C, one half of the *J. regia* or *J. hindsii* survived while all *P. stenoptera* survived 26 days of flooding. *P. stenoptera* was also the most resistant of the walnut rootstocks to infection by six *Phytophthora* spp.[51–53]

Abscisic acid has often been invoked as a possible signal emanating from the root that may account for plant symptoms associated with soil flooding (i.e., reduced growth stomatal closure, leaf chlorosis, leaf abscission).[45] Although ABA may be a component of symptom expression, it increased only in leaves of *J. hindsii* but not in *J. regia* with soil flooding.[50] In *P. stenoptera,* ABA concentration was the same in leaves from flooded or control treatments.[50] It was concluded that ABA could not be used to screen genotypes for sensitivity to flooding. A leakage of phenolic substances from the flooded root systems of *Juglans* spp. has been demonstrated; however, the leakage of vacuolar phenolic compounds was viewed as a phenomenon secondary to an anaerobiosis induced decline in energy content (i.e., ATP) and membrane integrity.[50]

D. TEMPERATURE

1. Carbon Assimilation

Tombesi et al.[14] reported that the optimum temperature for A_{max} of leaves of *J. regia* cv. Serr was about 20 to 24°C (at PPF = 900 μmol m^{-2} s^{-1}, VPD = 1.5 kPa); a sharp decline in A was noted above 30°C. Sun-exposed leaves of *J. regia* in California manifested reductions in A as ambient temperatures (and presumably VPD) increased above ambient levels.[14] Leaf gas exchange of *J. nigra* has not been rigorously evaluated as a function of temperature.

2. Cold Hardiness

Minimum winter temperature is an important environmental limitation to the successful culture of *Juglans* spp. For example, the successful culture of *J. regia* in north and mid-Western United States, Canada, northern Europe, and northern Asia is precluded by minimum winter temperatures. In Russia, *J. nigra* and *J. mandshuria* were reported to have a lower shoot water content during the dormant season than *J. regia.*[54] *J. nigra* and *J. mandshuria* withstood −30 to −36°C and *J. regia* withstood −18 to −30°C without appreciable damage. Similarly, a minimum temperature of −24°C resulted in ca. 25 to 50% survival of pistillate and staminate flowers of cold-hardy clones of *J. regia.*[55] In Pennsylvania, *J. regia* sustained cold injury (i.e., stem dieback) during 1983–1984 (minimum temperatures of −26 to 29°C).[56] Among the 12 most cold hardy selections identified in Iowa there was no relationship between the timing of fall defoliation and cold hardiness.[57]

3. Dormancy

There is a paucity of quantitative data concerning the requirement of chilling temperatures for releasing dormancy of *Juglans* spp. From September to January, Mauget[58] transferred 3- to 4-year-old *J. regia* cv. Franquette from the orchard during dormancy to a greenhouse maintained at 15 to 20°C. All trees transferred from the orchard before January manifested delayed foliation indicating that the total chilling units accumulated before January were insufficient to permit normal budbreak and foliation. In any event, in most of the cultivated range the inadequate winter chilling is not as great a limitation to production as minimum winter temperatures.

E. WIND

Wind is a selection criterion that may direct leaf morphology one way when winds velocity is predominantly low, and another way when wind velocity is high.[59] Instability of leaves at low wind velocities may result in the loss of an optimal skyward orientation for A_{max}; at high wind velocities leaves may tear, or if they have a high drag coefficient may result in limb breakage or tree uprooting. The compound leaves of *J. nigra* formed an elongated cylinder which decreased in diameter with increasing velocity.[54] Vogel[59] studied the drag coefficient of 8 forest tree species and found that species with pinnately compound leaves (i.e., *J. nigra* and *Robinia pseudoacacia* L.) had the lowest drag coefficient; however, the threshold wind velocities for physical damage to leaves were not generally greatly different among species.

F. ELEVATED CARBON DIOXIDE CONCENTRATIONS

The influence of elevated CO_2 concentrations on plant performance is a relevant topic considering the global increases in CO_2 occurring over the last 50 years. Tombesi et al.[14] reported that A of *J. regia* cv. Serr increased linearly with increasing C_i (i.e., in response to elevated ambient CO_2 concentrations), typical for a C_3 plant. Wood and Hanover[60] found that 700 to 2100 μmol mol^{-1} CO_2 increased dry weight and plant height of *J. nigra;* 1400 to 2100 μmol mol^{-1} increased leaf area and leaf number; 2100 μmol mol^{-1} CO_2 increased the root/shoot ratio more than the other treatments.

REFERENCES: WALNUTS

1. Bailey, L. H. and Bailey, E. Z., A concise dictionary of plants cultivated in the United States and Canada, *Hortus Third,* Macmillan Publishing Co., New York, N. Y., 1976.
2. Binder, R. G., Benson, M. E., and Flath, R. A., Eight 1,4-naphthoquinones from *Juglans, Photochemistry,* 28, 2799, 1989.
3. Borazani, A., Graves, C. H., Jr., and Hedin, P. A., A survey of juglone levels among walnuts and hickories, *The Pecan Quarterly,* 17, 9, 1983.
4. Grino, E., Carpathian (Persian) walnuts, *in Nut Tree Culture in North America,* 2nd ed., Jaynes, E., Ed., Northern Nut Growers Assoc., Inc., Hamden, Conn., 1981, chap. 6.
5. Serr, E. F., Jr., Persian walnuts in the western states, *Handbook of North American Nut Trees,* Jaynes, R. A., Ed., Northern Nut Growers Assoc., New Haven, Conn., 1973, chap. 18.
6. Zarger, T. G., Black walnuts—as nut trees, *Handbook of North American Nut Trees,* Jaynes, R. A., Ed., Northern Nut Growers Assoc., Inc., New Haven, Conn., 1973, chap. 14.
7. Fowells, H. A., Silvics of the forest trees of the United States, *U.S. Dept. Agric. Hdbk.,* 271, 1965.
8. Erez, A. and Weinbaum, S. A., Field estimation of leaf nitrogen by light transmittance, *J. Plant Nutr.,* 8, 103, 1985.
9. Brown, R. D. and Gillespie, T. J., Estimating radiation received by a person under different species of shade trees, *J. of Arboriculture,* 16, 158, 1990.
10. Ryugo, K., Marangoni, B., and Ramos, D. E., Light intensity and fruiting effects on carbohydrate contents, spur development, and return bloom of 'Hartley' walnut, *J. Amer. Soc. Hort. Sci.,* 105, 223, 1980.
11. Klein, I., Weinbaum, S. A., DeJong, T. M., and Muraoka, T. T., Relationship between fruiting, specific leaf weight, and subsequent spur productivity in walnut, *J. Amer. Soc. Hort. Sci.,* 116, 426, 1991.
12. Klein, I., DeJong, T. M., Weinbaum, S. A., and Muraoka, T. T., Specific leaf weight and nitrogen allocation responses to light exposure within walnut trees, *HortScience,* 26, 183, 1991.
13. Olson, W. H., Ramos, D. E., Ryugo, K., and Snyder, R. G., Annual and biennial pruning of mature lateral-bearing english walnuts, *HortScience,* 25, 756, 1990.
14. Tombesi, A., DeJong, T. M., and Ryugo, K., Net CO_2 assimiliation and characteristics of walnut leaves under field and laboratory conditions, *J. Amer. Soc. Hort. Sci.,* 108, 558, 1983.
15. Martin, G. C., Uriu, K., and Nishijima, C., The effect of drastic reduction of water input on mature walnut trees, *HortScience,* 15, 157, 1980.

16. Ramos, D. E., Brown, L. C., Uriu, K., and Marangoni, B., Water stress affects size and quality of walnuts, *Calif. Agr.,* 32, 5, 1978.
17. Charlot, G., Irrigation of walnuts, *Infos-Paris,* 62, 6, 1990.
18. Dreyer, E. and Mauget, J. C., Immediate and delayed effects of summer drought on the development of young walnut trees (*Juglans regia* L., cv. Pedro): growth dynamics and bud winter dormancy, *Agronomie,* 6, 639, 1986.
19. Davies, W. J. and Kozlowski, T. T., Variations among woody plants in stomatal conductance and photosynthesis during and after drought, *Plant Soil,* 46, 435, 1977.
20. Ni, B. and Pallardy, S. G., Response of gas exchange to water stress in seedlings of woody angiosperms, *Tree Physiol.,* 8, 1, 1991.
21. Pallardy, S. G., Parker, W. C., Whitehouse, D. L., Hinckley, T. M., and Teskey, R. O., Physiological responses to drought and drought adaptation in woody species, *Proc. Second Symposium,* Randall, D. D., Blevins, D. G., Larson, R. L. and Rapp, B. J., Eds., Univ. of Missouri, Columbia, Missouri, 1983, 185–199.
22. Ginter-Whitehouse, D. L., Hinckley, T. M., and Pallardy, S. G., Spatial and temporal aspects of water relations of three tree species with different vascular anatomy, *Forest Sci.,* 29, 317, 1983.
23. Martin, U., Pallardy, S. G., and Bahari, Z. A., Dehydration tolerance of leaf tissues of six woody angiosperm species, *Physiol. Plant.,* 69, 182, 1987.
24. Parker, W. C. and Pallardy, S. G., Genotypic variation in tissue water relations of leaves and roots of black walnut (*Juglans nigra*) seedlings, *Physiol. Plant.,* 64, 105, 1985.
25. Salleo, S., Functional aspects of water conduction pathways in vascular plants, *Giornale-Botanico-Italiano,* 118, 53, 1984.
26. Ni, B. and Pallardy, S. G., Response of liquid flow resistance to soil drying in seedlings of four deciduous angiosperms, *Oecologia,* 84, 260, 1990.
27. Bewley, J. D., Physiological aspects of desiccation tolerance, *Annu. Rev. Plant Physiol.,* 30, 195, 1979.
28. Levitt, J., Responses of plants to environmental stresses. II. *Water, Radiation, Salt and Other Stresses,* 2nd ed., Academic Press, New York, 1980.
29. Parker, W. C. and Pallardy, S. G., Drought-induced leaf abscission and whole-plant drought tolerance of seedlings of seven black walnut families, *Can. J. For. Res.,* 15, 818, 1985.
30. Parker, W. C. and Pallardy, S. G., Gas exchange during a soil drying cycle in seedlings of four black walnut (*Juglans nigra* L.) families, *Tree Physiol.,* 8, 339, 1991.
31. Chaves, M. M., Effects of water deficits on carbon assimilation, *J. Exp. Bot.,* 42, 1, 1991.
32. Downton, W. J. S., Loveys, B. R., and Grant, W. J. R., Non-uniform stomatal closure induced by water stress causes putative non-stomatal inhibition of photosynthesis, *New Phytol.,* 110, 503, 1988.
33. Lloyd, J., Syvertsen, J. P., Kriedemann, P. E., and Farquhar, G. D., Low conductance for CO_2 diffusion from stomata to the sites of carboxylation in leaves of woody species, *Plant, Cell Environ.,* 15, 873, 1992.
34. Mansfield, T. A., Hetherington, A. M., and Atkinson, C. J., Some current aspects of stomatal physiology, *Annu. Rev. Plant Physiol. Plant Mol. Biol.,* 41, 55, 1991.
35. Kaiser, W. M., Effects of water deficit on photosynthetic capacity, *Physiol. Plant.,* 71, 142, 1987.
36. Zhang, J. and Davies, W. J., Abscisic acid produced in dehydrating roots may enable the plant to measure the water status of the soil, *Plant, Cell Environ.,* 12, 73, 1989.
37. Pham, C. H., Yen, C. P., Cox, G. S., and Garrett, H. E., Slope position, soil water storage capacity and black walnut root development, in *Proc. Soil Moist-Site Prod. Symp.,* Balmer, W. E., Ed., Nov. 1–3, 1977, Myrtle Beach, S. C., 1978, 326–335.
38. Frossard, J. S., Cruiziat, P., Lacointe, A., Chat, J., Charron, A., Chenevard, D., and Mourton, C., Root biology of the young walnut: germination, growth, water relations, role in the management of reserves, Huitieme colloque sur les recherches fruitieres, *'La racine—le porte-greffe',* 15, 1989.
39. Kuhns, M. R., Garrett, H. E., Teskey, R. O., and Hinckley, T. M., Root growth of black walnut trees related to soil temperature, soil water potential, and leaf water potential, *For. Sci.,* 31, 617, 1985.
40. Ponder, F., Jr., Soil-water variations and black walnut growth, *Annual Report Northern Nut Growers Assoc.,* 76, 149, 1985.
41. Dey, D., Conway, M. R., Garrett, H. E., Hinckley, T. S., and Cox, G. S., Plant-water relationships and growth of black walnut in a walnut-forage multicropping regime, *Society of American Foresters,* Bethesda, MD., 33, 70, 1987.

42. Kormanik, P. P., Effects of phosphorus and vesicular-arbuscular mycorrhizae on growth and leaf retention of black walnut seedlings, *Can. J. For. Res.*, 15, 688, 1985.
43. Melichar, M. W., Garrett, H. E., and Cox, G. S., Mycorrhizae benefit growth and development of eastern black walnut seedlings, *Northern J. Appl. For.*, 3, 151, 1986.
44. Ponder, F., Jr., Soil moisture levels and mycorrhizal infection in black walnut seedlings, *Commun. Soil Sci. Plant Anal.*, 14, 507, 1983.
45. Schaffer, B., Andersen, P. C., and Ploetz, R. C., Responses of fruit crops to flooding, *Horticultural Reviews*, Vol. 13, Janick, J., Ed., John Wiley & Sons, Inc., NY, 1992, chap. 7.
46. Loucks, W. L. and Keen, R. A., Submersion tolerance of selected seedling trees, *J. For.*, 71, 496, 1973.
47. Kuz'min, E. N., Floodplain plantations of several *Juglans* species in the Novgorod-Severskii Poles'e, *Lesovodstvo-i-Agrolesomelio-ratsiya*, 70, 36, 1985.
48. Solignat, G., Note sur le compartement a l'lasphyzie radiculaire de porte-greffes du noyer, *J. regia. Pomol. Franc.*, 16, 75, 1974.
49. Catlin, P. B. and Olsson, E. A., Response of eastern black walnut and northern California black walnut seedlings to waterlogging, *HortScience*, 21, 1379, 1986.
50. Catlin, P. B., Martin, G. C., and Olsson, E. A., Differential sensitivity of *Juglans hindsii, J. regia*, Paradox Hybrid, and *Pterocarya stenoptera* to waterlogging, *J. Amer. Soc. Hort. Sci.*, 102, 101, 1977.
51. Matheron, M. E. and Miretich, S. M., Pathogenicity and relative virulence of *Phytophthora* spp. from walnut and other plants to rootstocks of English walnut trees, *Phytopathology*, 75, 977, 1985.
52. Matheron, M. E. and Mircetich, S. M., Influence of flooding duration on development of Phytophthora root and crown rot of *Juglans hindsii* and paradox walnut rootstocks, *Phytopathology*, 75, 973, 1985.
53. Matheron, M. E. and Mircetich, S. M., Relative resistance of different rootstocks of english walnut to six *Phytophthora* spp. that cause root and crown rot in orchard trees, *Plant Disease*, 69, 1039, 1985.
54. Kanivets, V. I., Frost resistance of walnut and methods of studying it, *Sadovodstvo*, 2, 25, 1987.
55. Khallaeva, S. N., Determining the winter hardiness of walnut by means of direct freezing. Methods of determining the winter hardiness of walnut, *Tsvetkovye rasteniya*, 3, 83, 1983.
56. Davie, B. and Davie, L., Winter hardiness of about 50 Persian walnut cultivars, *Annual Report Northern Nut Growers Assoc.*, Hamden, Conn., 75, 97, 1984.
57. Domoto, P. A., Selecting carpathian walnuts (*Juglans regia* L.) for cold hardiness, *Annual Report Northern Nut Growers Assoc.*, Hamden, Conn., 77, 21, 1986.
58. Mauget, J. C., Dormancy break and bud burst in walnut trees (*Juglans regia* L., cv. Franquette) subjected to temperatures above 15°C during their apparent rest period, *Agronomie*, 3, 745, 1983.
59. Vogel, S., Drag and reconfiguration of broad leaves in high winds, *J. Exp. Bot.*, 40, 941, 1989.
60. Wood, B. W. and Hanover, J. W., Accelerating the growth of black walnut seedlings, *Tree-Planters Notes*, 32, 35, 1981.

VII. CONCLUSIONS: TEMPERATE NUT SPECIES

The database concerning the environmental physiology of temperate nut crops is limited compared to many fruit crops, due to a number of factors. First, temperate nut crops are of limited economic importance on a worldwide scale, and current production is usually limited to specific geographic locations. Second, the cultivation of many nut crops is a very recent phenomenon and many species have not been genetically manipulated or domesticated to the extent of other food crops. Third, the very large size and long life span of many nut species increases the difficulty of experimentation (i.e., due to physical constraints, and the influence of environmental preconditions on plant physiology and plant assimilate/hormonal status). Fourth, most species are subjected to non-intensive cultivation, and often traditional/sustainable cultural practices are often employed.

Many nut crop species evolved as a climax forest tree species. To establish dominance in a forest canopy assimilates/nutrients are preferentially diverted into vegetative growth processes. Reproductive growth occurs when the assimilate/hormonal status of the tree is appropriate and when resources such as light, water, and nutrients are above a given threshold. Temperate nut species have often not been successfully manipulated from a horticultural standpoint. One example is the lack of progress concerning the efficiency of light utilization by pecan. Moreover, most nut species that are clonally propagated are still grown on seedling rootstocks. Quite often the seed source of seedling trees is not known and very few studies have been directed at quantifying the growth characteristics of seedlings from different seed sources.

The temperate nut crop species are very diverse genetically. The one commonality is the production of an energy-rich organ, typically high in protein and oil. The energy costs required to produce an organ so high in calories is much greater than most other reproductive (or vegetative) structures in the plant kingdom. Consequently, yields (i.e., kg/ha) are extremely low even under intensive management conditions compared to other fruit crops. For many nut species, compounding yield limitations are a relatively prolonged period of juvenility and a tendency to bear heavy and light crops in alternate years. These characteristics are also indicative of the amount of energy reserves that must be invested in reproductive structures. Several temperate nut species [i.e., almond (*Prunus amygdalus*), chestnut (*Castena* spp.), pecan (*C. illinoensis*), and pistachio (*Pistacia vera*)] have high rates of A compared to other C_3 forest tree or fruit crop species which would help meet the energy demands of a developing nut.

Agricultural sustainability has been increasing in importance. Although nut species produce a limited yield per hectare, in optimum habitats they require a minimum of inputs compared to most other fruit or vegetable crops. In the United States for example, seedling trees or native stands of pecans or black walnut (*J. nigra*) account for a substantial proportion of total nut production. In parts of Asia nuts of pistachio, filbert (*Corylus avellana*), and chestnut spp. (*Castena* spp.) are often collected from native stands and/or plantings that are not under intensive management. Irrigation, fertilization, or regimented pest control programs are often not regularly employed.

Few generalities can be applied to the environmental physiology of temperate nut species. Nut crops have colonized a wide range of habitats. For example, *Pistacia* spp. grow in arid regions and are among the most drought tolerant food crop species. By contrast, black walnut is a strict mesophyte, requiring a steady supply of moisture. It utilizes leaf abscission as a drought avoidance mechanism at a Ψ_L of ca. −2.0 to −2.5 MPa.

For many species (i.e., almond, chestnut, pecan, pistachio) net CO_2 assimilation is light saturated at relatively high PPF ($\geq$1,000 μmol m^{-2} s^{-1}). In addition, a relatively thick tree canopy may result in excessive shading for many species [almond, pecan, black walnut, Persian walnut (*J. regia*)] and limit productivity.

As has been emphasized throughout this volume that it is extremely difficult to interpret the influence of a given environmental stress under field conditions independent of other environmental factors. For example, high temperatures have often been cited as deleterious to physiological processes; however, the influence of increased VPD may be an overriding factor. There is no evidence to suggest that high temperatures occurring in the field approach lethal levels for species of nut crops. The extent to which physiological and growth processes are limited by high temperatures have not been adequately delineated.

Temperate nut species require a period of cool/cold winter temperatures followed by warm temperatures for normal budbreak, growth and development. The chilling requirement may vary from 200 to 300 hours for almond to more than 1000 hours of cumulative chilling for cultivars of pistachio. Pecan trees show elasticity in terms of a chilling requirement and may be grown in many geographic locations. Pecan trees will eventually bloom without a chilling period when sufficient heat units accumulate, although bloom period may be abnormally prolonged.

Salinity tolerance has been studied for those species frequently exposed to high salinity as a result of cultivation in inherently saline soil, irrigation with saline water, long-term use of fertilizers with a high salt content, wind blown salt spray, or salt water intrusion. Almond and pistachio are moderately and extremely tolerant of saline soils, respectively; an adequate database does not exist for the other temperate nut crops. Similarly, environmental stresses such as wind, atmospheric pollutants or elevated ambient CO_2 concentrations have received very little attention.

In summary, with the exception of pecan, the current database concerning the environmental physiology of temperate nut crops is not very comprehensive. Even for pecan the physiological database is an order of magnitude less than for other fruit crops such as apple, grape, or peach. Physiological data are critically needed to refine culture and management practices, to fully identify geographic locations of adaptability and to model tree growth and productivity.

INDEX

C

D

E

F

G

H

I

K

L

M

N

O

P

Q

R

S

T

U

V

W

X

Y

Z